SILVICULTURE AND ECOLOGY OF WESTERN U.S. FORESTS

Silviculture and Ecology *of Western U.S. Forests*

SECOND EDITION

John C. Tappeiner II, Douglas A. Maguire,
Timothy B. Harrington, and John D. Bailey

Oregon State University Press   Corvallis

The paper in this book meets the guidelines for permanence and durability of the Committee on Production Guidelines for Book Longevity of the Council on Library Resources and the minimum requirements of the American National Standard for Permanence of Paper for Printed Library Materials Z39.48-1984.

Library of Congress Cataloging-in-Publication Data

Tappeiner, J. C. (John C.)
Silviculture and ecology of western U.S. forests / John C. Tappeiner II, Timothy B. Harrington, Douglas A. Maguire, and John D. Bailey. — Second edition.
pages cm
Includes bibliographical references and index.
ISBN 978-0-87071-803-8 (pbk. : alk. paper)
1. Forests and forestry—West (U.S.) 2. Forest management—West (U.S.) 3. Forest ecology--West (U.S.) I. Maguire, Douglas A. (Douglas Alan) II. Harrington, Timothy B. (Timothy Brian) III. Bailey, John D., 1961- IV. Title.
SD391.T37 2015
634.9—dc23
2015008128

 First edition 2007.
Second edition 2015
Printed in the United States of America

Oregon State University Press
121 The Valley Library
Corvallis OR 97331-4501
541-737-3166 • fax 541-737-3170
www.osupress.oregonstate.edu

Contents

Conversion Factors
(USDA Miscellaneous Publication No. 225, 1949)

Length:

1.0 in = 2.540 cm

1.0 m = 0.394 ft

1.0 ft = 3.05 m

1.0 m = 3.281 ft

Area:

1.0 ft^2 = 0.093 m^2

1.0 m^2 = 10.764 ft^2

1.0 ac = 0.405 ha

1.0 ha = 2.47 ac

Volume

1.0 ft^3 = 0.028 m^3

1.0 m^3 = 35.315 ft^3

Basal area/area

1.0 ft^2/ac = 0.230 m^2/ha

1.0 m^2/ha = 4.356 ft^2/ac

Weight

1 lb = 0.454 kg

1 kg = 2.205 lb

Volume/area

1.0 ft^3/ac = 0.070 m^3/ha

1.0 m^3/ha = 14.291 ft^3/ac

Weight/area

1.0 lb/ac = 1.1208 kg/ha

1.0 kg/ha = 0.892 lb/ac

1.0 Mg/ha = 1000 kg/ha (metric ton)

= 892 lb/ac

Preface

This book focuses on the silviculture of western U.S. forests for two reasons. First, the authors' careers were spent mainly in the forests of California, Oregon, Washington, and Arizona, with some brief forays into southeastern Alaska, Montana, the Lake states, New England, and the Southeast. Most of our research and practical forest management work was done in Douglas-fir, ponderosa pine, lodgepole pine, true fir, and Sierra Nevada and southern Oregon mixed-conifer forests. Therefore, many of the examples we use are derived from either firsthand experience or the experience of colleagues in these forests. However, many of the species and forest types that we discuss occur throughout the western U.S., so the principles that we develop are valuable to the silviculture of western forests. With the help of excellent reviews and by consulting the literature, we have included considerable information from the interior west. Therefore our work is of value throughout the West.

Secondly, it is increasingly difficult to write a single book that can adequately cover the practice and science of silviculture for the different forest regions of North America. We believe that the science and art of silviculture, the extensive, recent silviculture and forest ecology literature, and the great variety of silvicultural practices that vary by forest type require region-specific treatment. The practice of silviculture is necessarily becoming more and more region specific because of each region's unique forests and ecosystems, and the ways in which local societies view and use the forest. Our intent is not to promote provincialism. Clearly silviculturists can learn from the concepts and practices developed across a range of forests (Nyland 2002; Smith et al. 1997). Nonetheless, although we use examples from other regions to illustrate concepts and creative practices, our focus remains on western forests.

This book contains a wide range of information based on research, much of it developed within the last 30 yr, as well as our observations of silvicultural practices and discussions with foresters, wildlife ecologists, and other forest resource specialists. Our purpose is to make this array of information available so that the practice of silviculture can grow to accommodate societies' continually changing demands for resources from their forests. It should provide a basis for informed discussion and debate. It is not intended to be prescriptive. Local practices are best developed and implemented by forest resource professionals who are knowledgeable about prevailing management policies and opportunities, working in their own forest ecosystems.

This second edition was prompted by new information and experience that has become available on the effects and uses of prescribed fire, on silviculture practices to promote and maintain characteristics of older forests, and new information on the nutrition of forest trees and stands and their responses to forest fertilization.

The photographs on the cover of the book illustrate some of the opportunities and challenges for silviculture in western forests. Mixed species often form stands with complex structures that produce commercial wood, provide habitat for a variety of species, and are aesthetically pleasing. However, because of their complex structures as well as their susceptibility to a variety of insects and pathogens, these forests can be extremely susceptible to severe fire. The productive Douglas-fir forests are just entering an era of balancing the growth of young stands for wood production and as habitat for early seral species with growing and maintaining older stands to provide habitat for old forest-associated species and for aesthetic reasons, as well as for wood production. We believe that the summary of information and experience presented in this book will help forest managers and society at large take advantage of these opportunities to provide many options for forest resources for the future.

Acknowledgments

During our careers there have been many professionals who, by their example and the encouragement they offered, stimulated us to be the best silviculturists we could be. Among those we especially thank for this gift are: Dietrich Muelder, John Helms, John Fiske, Phil Aune, Thaddeus Harrington, Bob Curtis, Carrol Powell, and David W. Smith.

The senior author thanks Mike Collopy, who provided the initial spark to begin the project, and Jack Walstad, who encouraged its completion. We also acknowledge the many foresters and silviculturists from public and private organizations who inspired us through their interests in forest ecology and silviculture, their passion for sound forest management, and their insightful questions,.

Gail Wells, who has authored several books on forestry, provided cheerful help and encouragement, thoroughly edited the manuscript, and improved the first edition immensely with her insightful questions. We especially thank Sandie Arbogast for her skill and patience in providing much improved figures for this edition. We thank Jim Boyle, who generously provided valuable consultation and review on forest soils; Robin Rose, Mike Newton, Kevin O'Hara, and several anonymous reviewers, whose thorough reviews and many comments improved the text throughout; and Kathy Maas-Hebner and John Moore, who provided extensive literature reviews on ecological principles related to silviculture. Doug Mainwaring, Sean Garber, Andrew Moores, Leslie Brodie, Peter Gould, Dave Peter, and Warren Devine provided a key review of the final draft of the book. Finally, thank you to several graduate and undergraduate classes in silviculture with whom we pilot tested early versions of the book.

CHAPTER 1
Silviculture

Silviculture and Its Evolving Objectives

Silviculture is the art and science of growing stands of forest vegetation. The practice of silviculture is bounded by ecological, social, and economic principles, state and federal forest-practice regulations, and the objectives and costs of silvicultural practices. It begins with a clear understanding of the objectives of the landowner—public or private, industrial or small forest landowner.

The practice and science of silviculture have evolved considerably during the past century. Graves (1911) defined silviculture as "the art of establishing, developing, and reproducing forests." In earlier times, the primary objective of forest management was timber production, although forage for wildlife and domestic animals and fuel wood were also important in some regions (Fernow 1914). Today silviculturists regard the forest as a biological community or, even more comprehensively, as an ecosystem, including soils, fire, microclimate (wind, temperature, relative humidity, light), animals, fungi, etc. The emphasis of silviculture has changed from a primary concern with watershed protection and timber at the beginning of the 20th century to a broader focus that currently includes non-game wildlife and forest biodiversity (table 1-1). This shift is attributable to local social, political, and economic conditions. Fernow (1914) stated that the ultimate goal of silviculture was to secure on a given area a high production of valuable material, so the owner might secure the largest possible returns in the long run. The emphasis was clearly on timber, as indicated by his five objectives of silviculture: (1) to secure quick reproduction after the removal of timber; (2) to produce valuable species instead of those having little or no market value; (3) to secure a full stock, in contrast to stands of small yield; (4) to produce trees of good form and quality; and (5) to accomplish the most rapid possible growth compatible with a full stand and good quality. These remain the goal of many of today's forest owners. Although few silviculturists would take issue with these objectives from strictly a timber management point of view, an increasing proportion of forestland (including commercial forestland) is managed in accordance with alternative or additional, non-timber, objectives and constraints. In many situations, particularly on public lands, timber production may be a secondary objective or a by-product of silvicultural practices performed to meet objectives such as wildlife-habitat improvement, maintenance of aesthetic quality, reduction in susceptibility to wildfire, and many others (Kohm and Franklin 1997; Curtis et al. 1998; O'Hara et al. 1994).

Table 1-1. Different views of silviculture as an art and science, in chronological order.

Graves (1911)

"... the art of establishing, developing, and reproducing forests. The ultimate goal of all silvicultural work is to secure on a given area a high production of valuable material, in order that the owner may secure the largest possible returns in the long run."

Broun (1912)

"Sylviculture is the art of applying the knowledge of the requirements of different trees, in tending and regenerating existing woods or in rearing fresh woodland crops, and in working them to the best advantage of the forest owner"

Schenck (1912)

"Sylviculture comprises all human activities by which trees, wood, bark, and any other forest product imaginable are raised and tended. Sylviculture is the art of the second growth, essentially."

Hawley (1921)

"... the art of producing and tending a forest; the application of the knowledge of silvics in the treatment of a forest. The purpose of silviculture is the production and maintenance of such a forest as shall best fulfill the objects of the owner."

Toumey and Korstian (1937)

"... various methods of raising and caring for forest crops; applied silvics; the essentials of modern silviculture can be characterized by four words: soil, reproduction, pruning, and thinning. By combining the influence of all four in rational silvicultural practice, the highest quality and quantity of timber may be produced, thus attaining maximum financial returns while maintaining the soil at its highest productive capacity. Rational silvicultural practice, then, is not attaining maximum tree development, but permitting trees to live only long enough to become large enough to yield the highest value per unit area."

Champion and Trevor (1938)

"theory and practice of raising forest crops"; "methods of raising tree crops and their growth and care up to the time of their final harvesting"

Baker (1950)

"... the methods of handling the forest in view of its silvics. Silviculture is aimed primarily at the production of timber crops and is concerned with the upbuilding of forests in the biological sense."

Kostler (1956)

"Silviculture is a biologically dependent technique by means of which treatments are so regulated that definite objectives are attained; within the framework of organized forestry these are for the most part economic objectives. Silviculture is therefore not an end in itself but a means to an end."

Daniel, Helms, and Baker (1979)

"Silviculture is concerned with controlling the establishment, growth, composition, and quality of forest vegetation. This can only be done in any given forest cover or locality if there is a clearly defined management objective that describes what is to be achieved."

Parkash and Khanna (1991)

". . . the art and science of cultivating forest crops."

Smith, Larson, Kelty, and Ashton (1997)

"the art of producing and tending a forest; the application of knowledge of silvics in the treatment of a forest; or the theory and practice of controlling forest establishment, composition, structure, and growth. . . . applied forest ecology is part of the biological technology that carries ecosystem management into action."

Helms (1998)

"the art and science of controlling the establishment, growth, composition, health, and quality of forests and woodlands to meet the diverse needs of land owners and society on a sustainable basis.

Nyland (2002)

". . . the methods for establishing and maintaining communities of trees and other vegetation that have value to people. Silviculture also ensures the long-term continuity of essential ecologic functions and the health and productivity of forested ecosystems . . . practical silviculture aims to create and maintain tree communities that serve specific objectives."

The theory and practice of silviculture in North America has its roots in central Europe in the 19th century, possibly earlier. The context was an increasingly urgent need to ensure a continued supply of wood for fuel and housing and to provide cover for game in forests that had been heavily depleted of merchantable wood, at a time when there was little in the way of planned systems for extraction and regeneration of future crops of wood. The concepts developed under these conditions were transported in the late 1800s to North America, where untouched native forests contained a relative abundance of wood. In fact, the surplus of forests was being cleared for agriculture and mining. Major forest fires had also impacted much of the forestland in North America. The European experience provided a sound basis upon which to begin the practice of silviculture, but the development of silviculture as a science has continually responded to contemporary issues in forestland management. Working definitions of silviculture have changed as the science has evolved (table 1-1), but silviculture as the art of producing and tending a forest (Hawley 1921) is as relevant and important today as it ever has been.

There are those who would argue that "nature knows best" and it can "manage" a forest better than humans can. However, we believe that intelligent human management based on a strong awareness of natural processes—and that includes forests with no human intervention—can generally meet society's goals better than total reliance on random, natural forest processes.

Taking a similar point of view, Harris (1984) concluded that forests managed by design will be superior to those managed by default. In other words, both forests and humans will benefit if stand and landscape structure—and their dynamics over time—are shaped by conscious design rather than by accident. The unfavorable accumulation of hazardous levels of fuels in many western forests is an example of the need for wise management by design. A policy of preventing fire was implemented in the early 1900s because large, severe fires occurred frequently at that time, and because there was little knowledge—and few resources—regarding the application of fire to forests for beneficial purposes. This fire-suppression policy worked well in the short run and is still appropriate to certain forest types. In many forests, however, the long-term suppression of wildfire has led to an accumulation of small trees and dead wood that can fuel very severe fires. Thus, by default, society has unusually flammable forests as a result of the widespread policy of controlling fire. A policy that encouraged thoughtful control of fire in some forests, and the use of fire and thinning to reduce fuels in others, would have left western U.S. forests in a better condition than we find them today.

The scope of silviculture in this book is consistent with the definitions of Helms (1998), Hawley (1921), and Graves (1911). The art and science of tending and regenerating forests remain a primary silvicultural activity. Because of the complexity of contemporary forest management issues, forests are viewed in this book as ecosystems operating at various spatial scales (Cissel and Swanson 1999).

Uncertainty and the Long-Term Nature of Silviculture

Silvicultural systems are by their nature long-term. It is sobering to remember the length of time required for a forest stand to develop and mature. This process may span the careers of several forest managers and, in contemporary society, several changes in forest policy as well. A 50-yr-old stand is comparatively young by western forest standards. Thus foresters—and society at large—have only modest "control" over silvicultural systems and stand and forest development. It is wise to view silvicultural systems as "working hypotheses" of stand development, as Smith et al. (1997) suggest. Silvicultural systems will often have to be modified for several reasons. Natural events such as windthrow, insect and pathogen outbreaks, fire, and unexpected regeneration of trees or shrubs occur frequently and may alter stand density and species composition. Shifts in policy, markets for forest products, landowners' need for income, and public attitudes often require reevaluation of, and changes to, systems. Furthermore, new information on silvicultural practices, plant biology, or forest ecology may provide new insights and reasons for modifying systems.

Given the inherent uncertainty in the enterprise of tending forests, we believe that one important principle is that a silvicultural system should preserve future options. An example is the shift in silvicultural theory and practice in response to an outbreak of Swiss needle cast disease (*Phaeocryptopus gaeumannii*) on thousands of acres of Douglas-fir in northwestern Oregon. Stands with mixtures of western hemlock and other species may come through the episode better than pure stands of Douglas-fir; these other species probably do not retard the spread of the disease, but if they are present they can replace Douglas-fir killed or damaged by it. Silviculturists who worked to produce pure Douglas-fir stands for economic efficiency and high yield are consequently reevaluating that practice on sites where needle cast is common. As a result, in disease-prone stands where western redcedar, red alder, and western hemlock can grow, managers may either plant these species along with Douglas-fir or favor them where they regenerate naturally. The outcome in terms of species composition, wildlife habitat, and forest yield is not yet known, but a mixture of species will minimize the risk of losing an entire stand. Even though some species in the mix are of less commercial value, this silvicultural strategy will maintain future options better than continuing to plant only commercially valuable trees susceptible to the disease.

Given the present rate of change in public sentiment toward forestry, and resulting changes in regulations constraining silvicultural practices, one may start to question whether the concept of a silvicultural system is even viable (Shindler et al. 2002). The context of the discussion often changes too rapidly for systems to be fully implemented, tested, and understood. Environmental issues change rapidly, and new regulations proliferate much faster than the typical cycle of thinning or harvest of an even-aged stand. From an even broader perspective, potential climate change and human population growth almost guarantee that the objectives guiding the design of current silvicultural systems will be modified.

Numerous creative solutions have been proposed for meeting concurrent commodity and amenity objectives, but many of these are silvicultural treatments designed to produce certain stand structures over a relatively short term. These innovative treatments are often not well integrated into comprehensive silvicultural systems. A given treatment may be designed to produce a certain type of stand or vegetation structure, but the longevity of that structure and the future dynamics of the stand may be only superficially considered and understood. For example, it has been proposed that groups of trees or single trees be left after harvesting of even-aged stands. These trees are intended to provide structural diversity in the next stand and help certain organisms survive from one stand to the next. However, it is not clear how the trees around these groups and individuals should be managed. For example, should they be thinned or underburned? The practice raises other questions. Is it necessary to carry over young trees in stands being managed on short rotations? If so, what species and how many trees should be left? Finally, it is an open question whether those trees will actually function as intended for conserving biodiversity.

Established silvicultural practices can often meet contemporary objectives with very little modification. The distinction between "traditional" and "new" approaches can be fuzzy. For example, as we will discuss in chapter 8, slight modifications in regular thinning practices can benefit a variety of organisms ranging from bryophytes to birds (Muir et al. 2002). Traditional shelterwood methods, such as irregular shelterwood and shelterwood with reserves (Troup 1928; Matthews 1991; Smith et al. 1997), contain features that seem identical to the more recently proposed "green-tree or variable retention treatments" (Franklin et al. 1997), although numerous distinctions can be made (Mitchell and Beese 2002). It seems unwise to discard conventional systems and the knowledge of stand dynamics they have generated; however, it is equally unwise to deny that conventional systems may require considerable modification to meet today's complex objectives.

Similarly, the creation and implementation of a viable silvicultural treatment or system should not be held hostage by endless debates about terminology. An effective prescription accomplishes its objectives without the need to agonize over the appropriate label. Smith et al. (1997) aptly state that: "Silvicultural systems should be descriptive, not prescriptive"; that is, silvicultural systems should outline a general approach to stand management and not prescribe the exact timing and details of future treatments.

Applying New Information from Forest Ecosystem Science in Silviculture

The effective practice of silviculture requires an understanding of major processes in forest ecosystems. Silviculturists pioneered research into many aspects of ecosystem dynamics, such as regeneration and growth of stands. For their part, forest ecologists uncovered many aspects of ecosystems, such as natural vegetation dynamics, the role of disturbance in structuring ecosystems, and the dynamics of nutrient pools and nutrient cycling (Swanson and Franklin 1992). Insights into natural vegetation dynamics and their role in wildlife habitat help predict and explain the efficacy of various silvicultural treatments. Our knowledge about the role of dead wood in forests and diverse forest structures expanded rapidly in the last few decades of the 20th century. Using this new information is not always straightforward, and regulations and guidelines for its use may be established before their likely consequences are well understood. For example, what is the trade-off between the benefit of leaving snags and wood on the forest floor and multiple layers of trees for wildlife relative to the risk of a stand-replacing fire? How does this concern vary by forest type? For most applications of ecosystem science to silviculture, an adaptive management approach is the most rational. Certain processes and species are apparently important for maintaining ecosystem integrity, but often there is little information about their precise role, and the long-term effects of silvicultural practices on these species and processes are uncertain.

Focus on Western Forests—The Purpose of This Book

This book was written as a reference and textbook for forest managers, teachers, graduate and undergraduate students, researchers, and others who want a contemporary, thorough account of the science and practice of silviculture in the western United States. Starting in about 1960, shifts in public attitudes toward western forests have prompted rapid growth in the level of silvicultural activity, the types of treatments applied, and research about various ecosystem responses in the western United States. Also, work in forest ecology directly related to silviculture continues to provide a fundamental basis for silviculture practices. Both research and implementation of new practices in all aspects of silviculture have produced a tremendous amount of valuable information. To date there has been no attempt to synthesize this information for western U.S. forests. Current silviculture texts are based mainly on experience gained in eastern forest ecosystems that contain species quite different from those in the West.

CHAPTER 2
Silvicultural Systems

In this book we define silvicultural systems as planned sets of treatments by which forest vegetation is regenerated, manipulated, and often harvested—a set of treatments that could be repeated in perpetuity. Silvicultural systems are developed after a stand evaluation. They are a series of silvicultural treatments or practices that are employed, if needed, to ensure that a forest stand will produce the desired outcome in terms of species composition, growth rate, density, and stand structure (Helms 1998). In most parts of the world, continued production of wood (for paper, fuel, and construction) remains the major objective of silviculture, but that objective is balanced to varying degrees by concerns for water, wildlife habitat, soil productivity, fire potential, and so on. Forestry and silviculture have a long tradition of contributing to watershed protection and conservation (see literature reviewed by Satterlund and Adams 1992).

In the past few decades, environmental concerns have prompted considerable innovation in individual silvicultural treatments, such as thinning or reforestation, but considerably less innovation in silvicultural systems. Silvicultural systems are designed to provide for regeneration and sustainability; a silvicultural treatment may or may not be part of a consistent strategy making up a silvicultural system. An exclusive focus on silvicultural treatments may overlook the dynamic nature of a forest and its potential for growth and stand development. Troup (1928) considered a silvicultural system to be the process by which the cohorts of trees constituting a forest are tended, removed, and replaced by new trees, resulting in the production of wood and a distinctive stand structure. He listed three distinctive features of a silvicultural system: (1) the method of regeneration; (2) the form or structure of the stand produced; and (3) the orderly arrangement of stands over the forest as a whole. These three elements are to be developed in the context that best suits the forest in question and with a commitment to the continued economic use of wood. The use of the term "system" is perhaps too restrictive, because it implies that once stand development is underway there is little possibility of change. This is certainly not the case, since unforeseen shifts in the owners' objectives or in public policy, or the effects of variables such as drought or insect epidemics, may force change and adaptation. Development can often be modified by applying treatments such as retaining live trees at harvest, thinning, prescribed fire, underplanting, etc. When these treatments are applied, however, their long-term effects should be evaluated and the need for future treatments (or a new "system") considered.

Traditional Silvicultural Systems

Many authors have provided descriptions and classifications of silvicultural systems (see table 2-1). At one level these systems aim to produce stands with specific size/age/species structures. The three basic options are even-aged stands, two-aged stands, and multi-aged or uneven-aged stands. Champion and Trevor (1938), emphasize that none of these systems is a "watertight compartment", because each stand has sufficiently unique features, or owners' objectives vary somewhat, For two very similar stands some modification, however slight, is often necessary to achieve the owner's objectives."

Silvicultural treatments, and the system under which a stand is managed, are therefore unique to some degree. One does not have to delve very far into the silvicultural literature to appreciate the extremely rich variety of techniques that have been applied (Troup 1928; Smith et al. 1997). Condensing the diversity of systems into a simplified classification, as we have necessarily done in table 2-1, masks this variety. This simplification nevertheless has merit, because the terminology of silviculture can otherwise hinder communication rather than efficiently convey similarities in approach. However, it is also important to realize that traditional silvicultural systems were never intended to obstruct innovation, circumscribe the range of possibilities, or limit the variety of silvicultural treatments. Silvicultural systems should be regarded as descriptive rather than prescriptive (Smith et al. 1997). In our opinion, there has been far too great a tendency to promote one particular system at the expense of others. Fashion and partisanship often play a great role in our society—even within forestry. A forester should make use of fundamental silvicultural knowledge and adapt systems to meet the special requirements of the case at hand, refusing to be led astray by any current fad. No text or reference book can adequately present the range of possibilities for managing forests.

As one gains a historical perspective on the development of silvicultural systems throughout the world, the line between traditional silvicultural systems and apparently modern innovations becomes increasingly fuzzy. One cannot help but believe that in many cases we are "reinventing the wheel." Most of the references to silvicultural systems throughout this book will rely on a simplified classification published by the Society of American Foresters (Helms 1998).

We will depart from the traditional approach, which is to provide systematic coverage of the different systems, and instead look at specific types of responses to silvicultural treatments. Our rationale is that an organized summary of this knowledge

Table 2-1. A sample of silvicultural systems described in the forestry literature. Also see Helms (1998).

System/Regeneration method	*References*
Even-aged	
Clearcutting (clearfelling): A stand is cut to reproduce a new stand of single or mixed species. Either natural or artificial regeneration is used.	Graves (1911); Hawley (1921); Troup (1928); Champion and Trevor (1938); Kostler (1956) Broun (1912); Matthews (1991); Parkash and Khanna (1991) Nyland (2002); Smith et al. (1997)
Strip: Trees are harvested in strips to facilitate natural regeneration. Strips may be cut in the direction of potentially damaging winds to reduce damage to trees along newly exposed edges.	Hawley (1921); Troup (1928); Parkash and Khanna (1991) **Progressive clear-strip:** Troup (1928) **Alternate clear-strip:** Troup (1928)
Seed tree: Trees are left scattered or in groups in otherwise clearcut stands to provide seed for natural regeneration. Trees are usually removed after regeneration is established	Graves (1911); Hawley (1921); Smith et al. (1997); Hawley (1921)
Shelterwood: Trees are left for protection from temperature extremes, sunscald, and often to provide seed. Trees may be scattered across the stand for shade and seed may come from the edges of strips or small openings. Trees are usually removed after regeneration is established	Kostler (1956); Hawley (1921); Toumey and Korstian (1937); Matthews (1991); Nyland (2002); Smith et al. (1997) **Uniform system:** Broun (1912); Graves (1911); Troup (1928); Champion and Trevor (1938); Matthews (1991); Parkash and Khanna (1991) **Compartment:** Broun (1912) **Group:** Broun (1912); Parkash and Khanna (1991); Smith et al. (1997) **Strip:** Graves (1911); Hawley (1921); Troup (1928); Champion and Trevor (1938); Matthews (1991); Parkash and Khanna (1991); Nyland (2002), Smith et al. (1997)

Table 2-1 cont.

System/Regeneration method	References
Coppice: Trees are cut and most of the regeneration occurs from sprouts on roots or at the base of the stem.	Broun (1912); Graves (1911); Hawley (1921); Troup (1928); Champion and Trevor (1938); Matthews (1991); Parkash and Khanna (1991); Nyland (2002); Smith et al. (1997)
Two-aged (two story)	
Seed tree, shelterwood with reserves, irregular shelterwood, coppice with standards, clearcutting with reserves (two-tiered); shelterwood coppice: Trees from the previous stand are left indefinitely (scattered or in groups) for aesthetics; wildlife habitat; production of large, valuable logs; dissemination of lichens and bryophytes, snags, etc.	Hawley (1921); Kostler (1956); Nyland (2002); Smith et al. (1997) Graves (1911); Hawley (1921); Troup (1928); Champion and Trevor (1938); Matthews (1991); Parkash and Khanna (1991); Franklin et al. 1997
Uneven-aged	
Single tree selection: An uneven-aged mixed or single species stand is maintained by cutting individual trees to form small canopy gaps to release seedlings, saplings and small trees. Usually regeneration is natural, but some shade-intolerant species may be planted.	Smith et al. (1997); Nyland (2002); Troup (1928)
Group selection: An uneven-aged mixed or single species stand is maintained by cutting small groups of trees to form canopy gaps. Natural regeneration, planting, and /or release of advanced regeneration follows cutting. Groups tend to merge over time.	Smith et al. (1997); Nyland (2002); Troup (1928); Schenck (1912); Graves (1911); Hawley (1921)

will provide the fundamental information required to design silvicultural treatments (and systems) to attain both conventional and new objectives.

Silvicultural systems were developed in Europe after forests had already been exploited for centuries (Kirby and Watkins 1998; Watkins 1998; Spurr 1956). Wood and forest cover were in demand for fuel, construction materials, forage and bedding for livestock, and as habitat for game. These systems were developed at a time when technology and understanding of forest and tree growth were minimal, while labor was plentiful and inexpensive.

Obviously, society in the western United States has changed dramatically since forest management began in the early 1900s. Population has increased, we are dependent upon technology, and the cost of labor has increased. Society has become increasingly urban. Many of today's communities value the forest not primarily for wood production or hunting, but as a source of aesthetic enjoyment, a place for recreation, as habitat for many species of wildlife, and—in the urban/forest interface—as a place to live. Silvicultural systems will most effectively serve contemporary society to the extent that they contribute to meeting this broad range of often-conflicting demands.

The practice of silviculture over the past several centuries has provided numerous examples of silvicultural systems (table 2-1). This variety of experience shows that forest stands with various compositions of species can be grown within a wide range of structures, and it suggests that, given the adaptability of trees and shrubs, new systems can readily evolve through the application of past experience to new situations. For example, the traditional even-age systems have been adapted to agroforestry systems directed at growing both young trees and food, in the earlier stages of stand development, and producing forage under small saw-timber in the later stages. Following is a brief account of the evolution of silvicultural systems in forests of the western United States. During initial settlement, wood was harvested with little thought of continued production or orderly harvesting and stand replacement. Forests and their ability to provide wood products seemed limitless. In general, the most accessible, highest-value trees were cut. In addition, many acres of forest were burned to clear the land for mining or grazing, or by accidental, severe wildfires.

The first attempt at silviculture in western forests took place early in the late 1800s and early 1900s, with selective cutting in stands of old trees. Generally, selective cutting meant harvesting valuable trees that were likely to die from insect attacks (Keen 1936, 1943) or other causes, or simply removing the less vigorous so that the more vigorous ones could grow (Dunning 1923, 1928; Kirkland and Brandstrom 1936; Pearson 1923). This practice probably should not be considered part of a silvicultural system, because there was no plan for replacing the large trees that were removed.

Selective or partial cutting worked reasonably well in ponderosa pine (Dolph et al. 1995) and mixed-conifer forests (Dunning 1923), including Sitka spruce-western hemlock forests (Deal and Tappeiner 2002; Deal et al. 2002). It was most successful on sites with vigorous, smaller trees and advanced conifer regeneration in the understory, where the terrain was not too steep, and where logging would not damage the overstory and understory trees. However, with the exception of some sites in southwestern Oregon, selective cutting in Douglas-fir forests (Isaac 1956; Smith 1970; Munger 1911, 1950; Curtis 1998) resulted in severe logging damage to remaining trees, windthrow, conversion of stands to shade-tolerant species, or a general lack of conifer regeneration because of logging damage, competition with shrubs, and lack of seed. Also, if the high-quality trees were cut, the result was an overall reduction in stand quality because of poor selection of species, poor growth potential of leave trees, and lack of regeneration. Partial cutting in older Douglas-fir stands was judged a failure (Curtis 1998; Munger 1950; Isaac 1956), and even-aged management with clearcutting as the regeneration method became the common system for managing these forests. In mixed conifer and ponderosa pine forests, both even- and uneven-aged systems proved to be effective approaches for stand regeneration.

Reforestation practices, including the ability to plant conifers after timber harvest and wildfires (Hobbs et al. 1992), aided the development of even-aged systems. Advances in growth-and-yield simulation (Hann 2005), logging technology (Kellogg 1980; Jarmer et al. 1992; Kellogg et al. 1996) and thinning methods aided the development of both systems. Training of foresters and landowners in these technologies was also important in the development of silviculture.

Today, there is considerable innovation and adaptation of systems in all forest types to special forest management objectives, particularly on public land and small ownerships. One prominent example in western North America is the proposed "variable-retention" system, which is designed primarily to accommodate timber production while minimizing impacts on forest biodiversity (Franklin et al. 1997, Beese and Bryant 1996). Silvicultural systems and practices continue to evolve and adapt in response to concerns for wildlife habitat, reduction of wildfire potential, and other forest values. An ecosystem analysis (chapter 3) that identifies the major environmental variables likely to affect the system, consideration of the owner's objective and costs, worker safety, forest practice rules, etc., are all part of implementing a silvicultural system.

Distinction Between Silvicultural Systems and Regeneration Methods

Silvicultural systems are often identified with their methods of regeneration, rather than with the stand structures they develop or maintain (Tesch and Helms 1992); however, a silvicultural system also includes treatments needed to fulfill management objectives. Regeneration in two-aged systems may use a combination of artificial and natural regeneration as described below. No particular regeneration method has been associated with this system in western forests.

Even-aged silvicultural systems grow stands of relatively uniform tree sizes and ages (figure 2-1). Clearcutting, the various shelterwood methods, the seed-tree method, and coppice are all regeneration methods for even-aged stands. Clearcutting provides a suitable environment for regeneration of many species, including Douglas-fir, ponderosa pine, lodgepole pine, Engelmann spruce, western hemlock, and seeding or sprouting of hardwoods. Douglas-fir and ponderosa pine are usually planted. Regeneration of lodgepole pine and western hemlock occurs mainly from natural seedlings already present in the stand before clearcutting and established thereafter. The shelterwood method provides a microclimate, seed, or both to make it easier for seedlings to become established. Leaving seed or shade trees scattered across the stand, or cutting strips or openings through a stand so that seed and shade are provided from the stand's edges, can also facilitate seedling establishment. The seed-tree method differs from the shelterwood method in that fewer trees are left, because the intent is for them to provide seed but not shade. In even-aged-systems, regeneration of the whole stand occurs at about the same time after a major disturbance, such as a harvest or a fire. These regeneration methods are not generally species-dependent, so the method used is often strictly a management decision.

Single-tree selection and group selection methods (figure 2-1) are ways of regenerating uneven-aged and other multi-aged stands. Small size classes (seedlings and saplings) are present throughout the stand and are released as larger trees are removed individually or in groups. Natural regeneration and release of advance regeneration are common in these systems. Planting shade-intolerant species, like ponderosa, may be used to maintain them, especially in mixed-conifer stands. Regardless of the regeneration method, in an uneven-aged stand the establishment of seedlings occurs among older trees and is a more or less ongoing process that maintains a stand with a wide range of tree sizes and usually ages.

Stand Growth in Even-Aged, Two-Aged, and Uneven-Aged Silvicultural Systems

The different silvicultural systems grow and maintain quite different stand structures. Traditionally, age has been the index used to describe these systems, and we will use this convention, even though size is a better descriptor, being more easily measured and observed than age, and in fact, not all trees in an even-aged stand are of exactly the same age. Even-aged stands are composed of trees of a relatively uniform size. Uneven-aged and two-aged stands are composed of trees of a broad range of sizes.

Even-aged stands can result from recruitment of seedlings or saplings that span an age range (Oliver and Larson 1996). As the younger trees grow, a relatively uniform stand structure occurs. The important point is that the stand will be managed for a single-canopy layer of trees of relatively uniform size. Even-aged stands generally

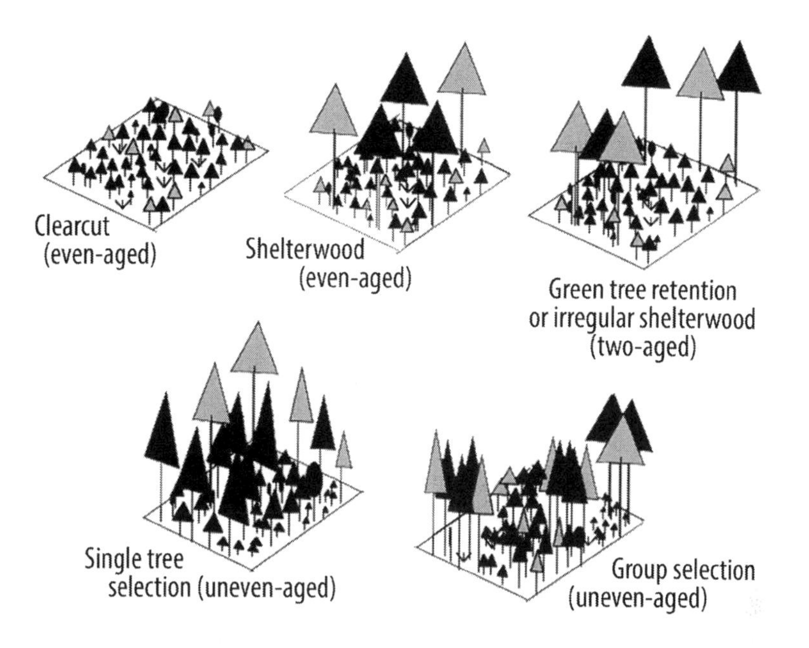

Figure 2-1. Regeneration methods commonly used in western forests. Even-aged methods, clearcut, shelterwood and irregular shelterwood or green tree retention methods are shown in early stages of regeneration. In the shelterwood method the overstory trees are removed after the regeneration is established. In irregular shelterwood or green retention they are left and a two-aged stands results. The uneven-aged methods show the stand structure developed and maintained by those methods. In the single tree and group selection methods the uneven-aged structure is maintained by regenerating trees throughout the stand or in groups, respectively.

start after a disturbance such as fire or clearcutting. Regeneration sources include new natural seedlings, planted seedlings, advance regeneration (seedlings already present), or coppice (sprouts from roots or stems). As these trees grow, the stages of stand development and dynamics described by Oliver and Larson (1996) occur: stand initiation, stem exclusion, and understory reinitiation. The number of trees decreases and their average size increases as the stand grows, but a relatively uniform diameter or tree size distribution is maintained (figure 2-2). Tree density and stand development can be regulated by controlling planting density and by thinning. At some point in its development, the stand reaches rotation age, based on landowners' management objectives, or sometimes the effects of other disturbance. Then the entire stand is harvested, and a new stand is started.

In two-aged or two-story stands, size classes or structure are more important than age. Two-age stands (figure 2-2) can occur in several ways:

(1) Large trees survive fire or wind or are left after harvest (shelterwood with reserves or irregular shelterwood; Smith et al. 1997). Regeneration occurs beneath them by natural seeding (figure 2-2b), planting, or coppice.

Figure 2-2. Examples of even-aged stands (A), two-aged (B), and uneven-aged stands (C and D). C is being managed by the group selection method, D by the single tree selection method.

(2) All large trees are gone after a disturbance. Regeneration of a species mixture occurs, with some species becoming established several years earlier than the rest. This head start in establishment plus their rapid height growth enables them to become overstory trees as other trees become established beneath them.

(3) A stand starts as an even-aged stand, and following a disturbance like windthrow, or thinning, trees are established in the understory and form a second story (figure 2-1).

How the overstory trees are managed, how long they are retained, and what treatment occurs in the second story all depend on the landowner's objectives. For example, overstory trees may be left indefinitely, partially removed, or harvested along with the understory during thinning or clearcutting. Ultimately, in a two-age system new cohorts of overstory trees are recruited periodically from the understory, and trees in both the understory and overstory are harvested to maintain the two-aged structure indefinitely.

In the northern Rocky Mountains, for example, the most vigorous trees of lodgepole pine and larch (30-35 ft) overtopped Douglas-fir (20 ft), subalpine fir (14ft), and Engelmann spruce to form a two-story stand, a result of their natural height growth potential (Cole and Schmidt 1986). Douglas-fir and western redcedar stands in western Oregon and Washington also form two-story stands in the same way. Delayed establishment of the shade-tolerant species may also promote two-story stands in both cases. Many other combinations of shade-intolerant and shade-tolerant species, such as ponderosa pine and white fir respectively, also develop two-story stands. However, without controlling the density of the more rapidly growing shade-intolerant species, the development of the second story of tolerant species will be limited (Garber and Maguire 2004).

Sometimes trees from the previous stand are left in the new stand indefinitely to meet some combination of timber, wildlife, and/or aesthetic objectives. The systems are then referred to as clearcut with reserves, or seed tree or shelterwood with reserves (figure 2-3). This might be considered a modification of classic even-age systems with reserves or variable-retention harvesting. It has been proposed as a way to conserve biodiversity while extracting timber in a regeneration harvest (Franklin et al. 1997). Details of silvicultural systems based on variable-retention harvesting are currently being worked out in several forest types and may evolve to two-aged systems (Maguire et al. 2006). Their efficacy in conserving biodiversity is also being assessed (Beese and Bryant 1999, Aubry et al. 2004).

In uneven-aged management, a range of tree sizes occurs throughout the stand (figures 2-1, 2). Cutting trees in all size classes maintains this uneven size structure so that there is a relatively continuous regeneration of new trees and growth of trees from smaller size classes into larger classes. There is no rotation age or distinct regeneration period as in even-aged stands. As wood is harvested, stand structure is maintained by the removal of trees from several size classes in a cutting cycle, often of 10 or more years. Remaining trees grow from one size class into the next. Regeneration is nearly always present, ensuring that enough trees are maintained in the smallest size class to continually "fuel" the system. The actual structures or diameter distributions of trees in uneven-aged and other multi-aged stands vary through time, especially before and after harvest. However, stand structure generally focuses on an ideal distribution or goal to ensure that there are enough trees in each size class to maintain a yield of wood and a cover of trees (figure 2-4; see chapter 10 for details of determining diameter distributions). When these ideal distributions are impractical or become too constraining, it may be more effective simply to maintain and manage three or more cohorts regardless of the diameter distribution (e.g., O'Hara 1996, 2014).

Diameter distributions are quite different in even- and uneven-aged stands. In even-aged systems, as stands grow they are often thinned (or self-thinning occurs); tree numbers decrease as tree size increases (figure 2-4). At rotation age, the stand is harvested and regenerated. In uneven-aged systems, a tree cover is maintained using stand structure defined with diameter distribution as a guide. Regeneration occurs through time, as does harvest of trees. The actual stand structure, shown as the dark line in figure 2-4, is compared to the distribution at the beginning of each cutting cycle to determine in what size classes there is a surplus or deficiency of trees. Thinning and

Figure 2-3. Comparison of methods to regenerate even-aged and two-aged stands. (A) clearcut (left) and a shelterwood (right center); and (B) green tree retention, in Douglas-fir forests. The shelterwood trees reduce the temperature extremes compared to the clearcut. This can aid seedling establishment on sites prone to frost in the early growing season and where there are extreme high summer temperatures. In these forests, shelterwoods are not needed for regeneration on most sites; like green tree retention they are left to grow large trees or to produce two-storied stands for wildlife habitat and aesthetics, especially on public lands.

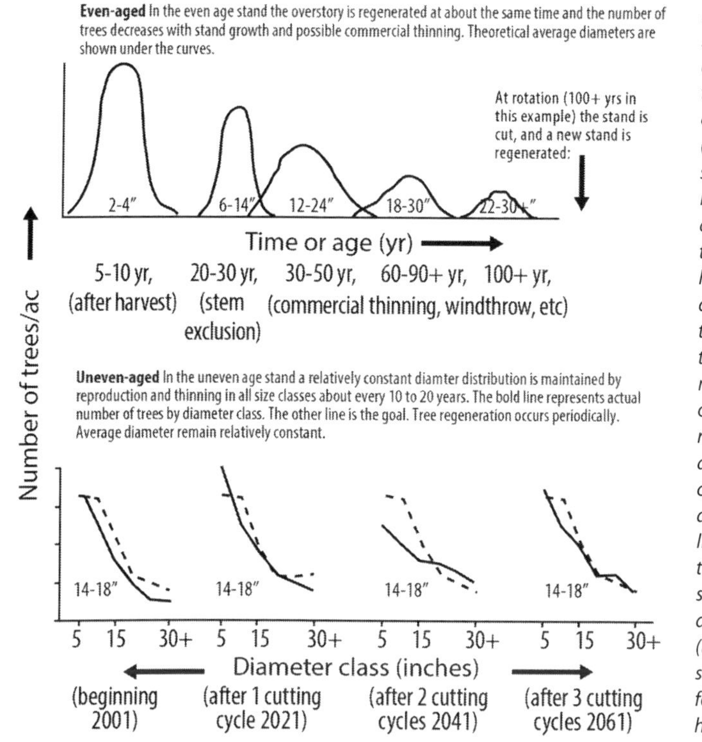

Figure 2-4. Theoretical structure and ranges of diameters (shown under the curves) through time of an even-aged stand (A) and an uneven-aged stand (B). In even-aged management, a single cohort of trees is grown through time and harvested. Diameter distributions change as trees grow and stands are thinned. In uneven-aged management, a cover of trees is maintained by relatively frequent harvest and regeneration of new cohorts of trees. Diameter distribution goal (dashed line) is used to guide the management of the stand. Comparing actual diameter distribution (black line) to the goal shows where there are too few trees or excess trees to harvest or thin.

regeneration are done on a cycle to harvest trees and to enable smaller trees to grow into larger size classes (figure 2-4). For example, at the first cutting cycle (figure 2-4). trees are deficient in all size classes. At the second cycle, trees are surplus in the smaller and larger size classes, and deficient in the middle classes. The structure is adjusted over time, using the diameter distribution as a guide, to provide the resources that meet the owner's objectives. This silvicultural system uses two regeneration methods: single-tree selection and group selection.

These three general types of silvicultural systems and stand structures represent a wide range of systems. Obviously many variations are possible, and all three systems might be used within a single forest property. Systems are being developed to convert even-aged stands to two-aged and uneven-aged stands (Miller and Emmingham 2001).

Many stands have been managed by removing trees because of their value, or low vigor, infestations of mistletoe or pathogenic fungi, or susceptibility to insects. "Sanitation/salvage" or "partial cutting" was a common practice until about 1960 in ponderosa pine and mixed-conifer forests (Keen 1936, 1943; Dunning 1923) and in western hemlock-Sitka spruce forests in southeast Alaska (Deal and Tappeiner 2002), as was salvage of trees damaged and blown down by windstorms. These practices were suitable for the forest management policies at that time, but they were not implemented with a silvicultural system in mind and often resulted in a complex uneven-aged structure (figure 2-5). Features included areas with no regeneration, patches of vigorous poles and saplings at various densities, and a range of sizes, densities, and species of larger trees. Options for bringing these stands into a silvicultural system include: a) starting a new even-aged stand by harvesting all merchantable trees and b) working with the current structure to provide future options. This might entail commercial thinning of larger trees, precommercial thinning of saplings, and clearing of shrubs and planting among larger trees and groups of saplings (figure 2-5). In the future, the landowner might decide to convert to either an even-aged or uneven-aged stand or to maintain the complex, irregular structure. All options are appropriate given the biology of the species, forest practice regulations, and the owner's objectives.

What Are the Major Differences Among Silvicultural Systems?

Silvicultural systems differ in their objectives—maintaining and managing stands of either uniform or variable structure—and in the methods of regeneration, thinning, prescribed fire, and other treatments used to accomplish those objectives. Following is a brief comparison of some important points to consider in the use of these three systems.

REGENERATION

Regeneration to ensure the continuation of the stand is a major objective of all silvicultural systems (see chapters 7 and 10 for detail of regeneration and its use in even and uneven-aged stands). Two examples of regeneration methods in even-aged systems are the clearcutting and shelterwood methods (figure 2-4). The regeneration phase may include site preparation, for planting or natural regeneration and control of competition to ensure growth and survival of seedlings (see chapter 10). Site preparation and planting of Douglas-fir and ponderosa pine are common. Natural regeneration commonly occurs in western hemlock, true fir, and lodgepole pine forests; because these species consistently produce large seed crops, the shrub communities are not as vigorous as those in other forest types, and advance regeneration and natural seeding occur readily on many sites. Natural regeneration often occurs in the understory of 50 yr+ even-aged stands, especially after thinning (Bailey and Tappeiner 1998). On many sites, this natural regeneration could become the next even-aged stand or the lower story of a two-aged stand.

Shelterwood methods of regeneration may be used for a number of reasons: a) shade for natural or planted seedlings to keep soils from reaching extreme temperatures lethal to seedlings; for example in true fir forests and on cold, frost-prone sites (Minore 1978, Stein 1984); b) seed for natural regeneration, and c) to produce a two-aged stand

Figure 2-5. Variation in structure in an approximately 20-ac stand of Douglas-fir, grand fir, and hardwoods. Damaged and wind-thrown trees were removed after a severe windstorm about 40 yr earlier. Two years before the photo, the stand was planted after shrubs were cleared from around (A) bigleaf maple, (B) red alder, and (C) a group of Douglas-fir saplings. Douglas-fir was planted in dry areas; western redcedar was planted near streams. Seedlings of both species were protected from browsing by plastic mesh tubes. In (D) Douglas-fir 6-8 in dbh were thinned and conifers were planted beneath them. Douglas-fir 18-20 in dbh and 10-14 in dbh were thinned in (E) and (F), respectively.

for aesthetics, wildlife habitat such as nest sites, or sites for lichens and bryophytes (Franklin et al 1997). If trees are left only for seed or shade, they are usually removed as soon as seedlings are established and no longer susceptible to microclimate extremes. It is best to remove these large trees while seedlings are small, flexible, and not so easily broken. After removal of overstory trees, the site appears the same as it would after clearcutting. Both scattered and clumped trees provide shade and seed (Troup 1928). Leaving trees in groups may be best if the goal is wildlife habitat or aesthetics (Franklin et al. 1997), as well as minimizing logging costs and disturbance that might injure overstory trees or new regeneration. Dispersed trees might be best for habitat and dispersal of canopy dwelling organisms.

If trees are retained for habitat and aesthetics, they will be left at least until the regenerated seedlings have formed a canopy. They may be left on the site indefinitely and become part of the next stand, emerging above the main canopy for many decades. Leaving these trees in groups may be best if the goal is wildlife habitat or aesthetics (Franklin et al. 1997), as well as minimizing logging costs and disturbance that might injure overstory trees or new regeneration. Dispersed trees might be best for habitat and dispersal of canopy dwelling organisms. Such residual trees may be considered a shift from an even-aged to two-aged system or an irregular shelterwood (Smith et al. 1997) (table 2-1). Shelterwood trees continue to grow (Latham and Tappeiner 2002), and even at low densities reduce the growth of the smaller trees beneath them (Zenner et al. 1998; Acker et al. 1998; Harrington 2006, Maguire 2006).

In uneven-aged and two-aged systems, regeneration is established among larger trees or in small openings in the forest. "Banks" of seedlings are established, and they may grow slowly—compared to seedlings in even-aged stands—until they are released when the trees around them are removed or their stem density is reduced through natural or prescribed thinning. Natural and planted regeneration methods are used in uneven-aged stands. Planting may be done in groups, especially in mixed-conifer stands, to maintain shade-intolerant species such as ponderosa pine along with the shade-tolerant true firs which are well adapted to uneven-aged stands.

It is often advantageous to be flexible and opportunistic in regenerating stands. From time to time it will probably be necessary to clear areas of tree and shrubs and establish small "plantations" in some places in an uneven-aged stand. Treatments typical of even-aged management may be needed to establish trees in areas of windthrow and insect mortality, for example, or in areas where a dense understory of shrubs occurs and no tree regeneration is present

FOREST TYPES AND SPECIES COMPOSITION

The plant community is a major factor in determining the appropriate silvicultural system. On productive sites with shade-intolerant overstory species such as Douglas-fir and red alder, shade and shrubs are common in the understory and even-aged management is likely to be most easily applied. Also, on sites being reclaimed from shrub fields following severe fires, for example, planting will produce even-aged systems for a number of decades. Uneven-aged management is favored in forests where the major tree species regenerate naturally. The conifer trees need to be relatively shade tolerant, or the canopy needs to be open enough for seedling establishment and growth. Generally shade-tolerant species such as white fir, incense-cedar, western hemlock, and drier-site

Douglas-fir are favored in uneven-aged or two-aged systems. In some dense mixed-conifer forests it may be necessary to maintain an open canopy and plant ponderosa pine or Douglas-fir, if the objective is to maintain these species in the stand. Understory shrubs and grasses are also important in all systems. In even-aged stands they are the major component during regeneration, after which their density may decrease markedly (Schoonmaker and McKee 1988). In uneven-aged stands, shrubs are kept at a reasonably low level by a continuous tree cover. However, on sites where shrubs readily invade small canopy gaps, it can be difficult to establish tree regeneration, especially natural regeneration, and uneven-aged management may be impractical. For example, on moist sites in the Coast Range of Oregon, the persistence of dense covers of shrubs such as vine maple and salal in openings, coupled with the intolerance of Douglas-fir to shade, make uneven-aged management difficult and impractical on many sites. However, on drier sites, especially in southern Oregon and California, shrub cover within stands is generally low, overstory canopies are less dense, and uneven-aged management is more easily accomplished.

CONTROL OF STAND DENSITY

In all systems, it is important to control tree density and prevent development of over-dense stands. Precommercial thinning is often done in young, even-aged stands of saplings and poles. Proper thinning ensures a future stand of vigorous trees with large crowns and stems resistant to stresses such as snow and ice damage, windthrow, and bark beetle attacks. Tending small trees in uneven-aged stands is also important for the same reasons and for the additional purpose of maintaining enough vigorous trees in the smaller size classes to replace trees harvested from the larger size classes.

ELEMENTS OF WILDLIFE HABITAT

Wildlife habitat varies among systems (Hunter 1999), and important elements of habitat can be provided in these systems. Probably the biggest difference among the systems is the open habitats that occur during the regeneration phase of even-aged stands, when shrubs and forbs are plentiful and there is little tree cover even if trees were retained after timber harvest. Patches of this type of habitat can be provided in uneven-aged stands by cutting groups of trees, but the habitats provided by the forest at the dense self-thinning stage, where the overstory trees are dense and the understory is sparse, would not be common in uneven-aged stands. The later stages of development of even-aged and two-aged stands are similar to the continuous cover found in uneven-aged systems.

Snags and dead wood can be provided in all systems, depending on habitat needs. Similarly, large, old-growth conifers or hardwoods such as bigleaf maple or California black oak, which provide nests and foraging sites, can be maintained in all systems. Thinning around bigleaf maple, Pacific madrone, or California black oak may be needed to maintain or develop them for habitat, or conifers will eventually overtop them.

RISK AND USE OF FIRE—DAMAGING AGENTS

Fire can severely damage both types of stands. Uneven-aged stands are probably more susceptible to fire and it may be more difficult to carry out prescribed burning in them. Maintaining a range of tree sizes produces "fuel ladders" that can carry fires from the smaller trees into larger ones. If uneven-aged and two-aged systems are used in fire-prone areas, maintaining areas with large trees and minimal dead wood and understory plants to act as fuel breaks in strategic locations might lower the chances of a stand replacing fire. Even-aged stands are most susceptible to fire from the regeneration stage through the stem exclusion stage. Stands of large, older trees with little fuel and flammable vegetation in the understory are more resistant to fire. Slash from thinning and natural tree mortality can add to the fuels and flammability in both systems.

Stem damage from ice and heavy wet snow, windthrow, pathogens, and insects occur in all three systems; and it seems that none of these systems is inherently more resistant or vulnerable to these ecosystem variables. Root, stem, and foliage diseases and defoliating insects can also occur in all three systems. Because uneven-aged and two-aged stands have a range of size classes, they may be somewhat more resistant to insects such as bark beetles that invade trees in certain size classes (>4" dbh). On the other hand, uneven-aged and two-aged stands could be more susceptible to dwarf mistletoe because it spreads easily from infected overstory trees to young understory trees. Because many pathogens and insects of western forests are host-specific (an exception is spruce budworm), mixed-species stands, whether even or uneven-aged, or two-aged, are more likely to be resilient to pathogens or insects than are pure stands of mixed sizes or ages.

FOREST OPERATIONS

Forest operations are different in each system. In even-aged systems, site preparation and planting, if required, are focused on a particular area. This enables the use of large ground-based or aerial equipment and broadcast prescribed fire for control of slash and competing vegetation. In uneven-aged systems, site preparation, planting, and shrub control occur on small areas scattered throughout the stand. However, if natural regeneration works well, then the cost of careful logging to save naturally established seedlings and saplings could offset the cost of site preparation and planting nursery-grown seedlings. Uneven-aged and two-aged management require that the growth of both seedlings and larger trees be considered during logging. Smaller trees have to be protected as the larger ones are felled; thus, careful tree selection and logging are required. Careful tree selection and logging are also required for thinning in even-aged stands. Minimizing soil compaction is also important in both systems, because often access every 10–20 yr is needed for thinning (in even- and two-aged stands) and harvest/releasing regeneration (in uneven-aged stands). Therefore careful layout and logging planning is required in all systems.

The choice of systems often depends on the scale of the operation and the resources available, as well as the personal preferences of the owners. In situations where careful professional attention and skilled labor can be used in planning and executing all phases of management, uneven-aged systems are likely to be successful. Very intensive management, including frequent thinning and management of even- and uneven-aged stands of mixed species, may be more realistic on small ownerships. On large ownerships or in situations in which only a few people are available to plan and carry out operations, or where most work is done by contract, managing even-aged stands is often the best choice.

Development of New Systems

In order to better understand the opportunities for managing forests for a range of resources, large-scale experiments testing new silvicultural systems have been established (Monserud 2002). Previous studies generally used small plots (<1.0 ac); while suitable for providing information on the regeneration and growth of trees and shrubs, studies at this scale cannot yield reliable information on the effects of treatments at an operational scale on variables such as plant-species diversity, wildlife populations, riparian systems, biodiversity, and social issues (Monserud 2002). Consequently, studies have been initiated that evaluate treatments such as thinning, retention of green trees in otherwise even-aged systems, retention of uncut areas, and cutting gaps within stands. These studies are new, generally <15 yr old, and have not yet yielded comprehensive information. Evaluation of 5-yr results in western Washington and Oregon show that leaving 15, 40, 75, and 100% of the basal area of the original stand in dispersed or aggregated patterns affects growth and mortality of the trees retained and the newly established seedlings (Maguire et al. 2006). For example, seedling mortality after 5 yr varied by species; for ponderosa pine it was greater at 40% retention than at 15% retention, while there was no difference in Douglas-fir survival. The combination of differences in growth of seedlings and older trees with the effects of mortality of older trees from wind and other factors will surely cause considerable variability in stand development and structure (Maguire et al. 2006). As the use of green tree retention and these studies progress, the results can be used to develop new systems or to modify existing ones. Even with their relatively large scale, these studies may not address ecosystem variables like fire, water quality and yield, the effects of insects and pathogens, and populations of wildlife with large home ranges. Along with these studies, forestry organizations are trying different methods of thinning and leaving trees following harvest and regeneration. The need to have variable stand structures for wildlife habitat and aesthetics, to maintain habitat of fish-bearing streams, and to reduce fire potential are some of the issues that motivate these new practices.

Silviculture Prescriptions

Silvicultural prescriptions are developed when applying one or more treatments to a stand. For example, if a thinning is to be done, silviculturists will consider the current species composition and density of the stand; the density after thinning; possible effects of wind, insects, and fungi; log markets for the species and sizes of trees to be thinned; and so on. Using these variables and the owner's objectives (table 2-2), a given treatment, several treatments, or no treatment at all will be recommended. Prescriptions for thinning might involve removal of certain tree species and size classes and retention of others, and possibly the removal of slash, or prescribed fire, following thinning and evaluation of stand density and growth in 5 to 10 yr. Along with treatment effectiveness, the cost, cash flow, and return on investment are major considerations when developing silvicultural prescriptions.

Table 2-2. The elements of a silviculture prescription. Evaluating these questions before beginning site preparation for regeneration, thinning, prescribed burning, or other treatments will help insure that major variables that control the outcome of the treatment are not overlooked.

1. What are the current stand structure and species composition, site class, density, growth-rate? Be sure to consider within stand variability.
2. What are the sources of seeds and sprouts that might occur after treatment and what will be their effects on stand structure, habitat, growth rates, etc.?
3. What are the environmental variables that could affect the outcome of the prescription? Use the "ecosystem check list (climate/microclimate, soil, subsurface water, mass movement, fire, fungi, insects, vertebrates. See Chapter 3 and Chapter 10 cases 1 and 2).
4. What will be the stand structure and species composition at some future date (in 10 to 25+yr) with no treatment and with proposed treatment (s)? Consider growth rates, habitat for wildlife, the appearance of the stand in the short- and long-term. Stand simulations and knowledge of growth and reproduction of shrubs and forbs will help with this projection. Sketches may help too.
5. Will the prescription meet owner's objectives for aesthetics, economics, and habitat? What volume of wood will the stand produce in the short and long term, and what is its value? What species of wildlife will it favor? What will the prescription cost to implement?
6. Does the prescription meet forest regulations and policies?
7. How will the prescription be implemented, and what future evaluations or treatments will be needed? Clarify any uncertainties or assumptions—what could go wrong?

Silvicultural prescriptions may depend upon the conditions in nearby or adjoining stands (Puttemann and Tappeiner 2013). For example, in figure 2-5 the patches of forbs, shrubs, and saplings (condition D), may be an important component for diversity in this landscape. This suggests not regenerating condition D and possibly expanding shrub patches in condition C. Thus patches of shrubs would not be needed elsewhere in the stand, and the cost of regeneration reduced. Similarly, the presence of well-developed hardwoods in D makes it unnecessary to maintain or manage for them in conditions B, C, and F. In the same way, prescriptions do not have to be executed precisely as stated. For example if the prescription calls for reforesting 100 acres but follow-up examinations show that there are several areas of about 1-3 acres that are not being reforested, it might best to not reforest them. Replanting for example, could be expensive, because of the cost of treating well-developed shrub cover, wildlife damage, lack of labor, and other factors. The costs might be offset by the habitat provided by the openings in the new stand. Costs, landowner's objectives, forest regulations—these all influence the decision to reforest or not.

The objectives of conventional silvicultural systems, developed when wood production and habitat for big game were the primary goals of forest management, are still important in today's society, but so are other objectives, including the reduction of fire risk and the maintenance of riparian and old-forest habitats. Forest stands are generally quite adaptable and can grow with a variety of species and structures. Disturbances (wind, fire, insects and pathogens, etc.) are common in forests. All this means that considerable flexibility is needed so that stand structures and silvicultural systems can be modified to rectify disturbances and meet new objectives of landowners, and for new silvicultural systems to emerge. As we discuss in the next several chapters, it is important that silviculturists know the ecology of the tree and shrub species they are growing and the ecosystem context of those stands, if both conventional systems and new ones are to be successful.

REVIEW QUESTIONS

1. Explain why silviculture is a science and an art.
2. Explain why silviculture cannot be practiced without an objective. What role do silviculturists play in setting objectives for a stand or forest?
3. What is a silviculture system?
4. What is the difference between a silviculture system and a regeneration method? What regeneration methods are used to regenerate even-aged and uneven-aged stands?
5. How do the structures of even-aged and uneven-aged stands change through time?
6. Explain several ways that a two-aged stand can occur.
7. How do volume or basal area, numbers of trees, and growth change through time in even- and uneven-aged systems? Discuss the possibilities for changes in tree sizes and numbers in a two-aged system.
8. Discuss some of the pros and cons of using each silvicultural system.
9. Are these the "only" systems that can be used to manage forests? Explain.
10. When and where were the first silvicultural systems developed? What were the principal forest management issues at that time in that place? Contrast the early development discussed above with the development of silvicultural systems in the West.
11. What are the elements of a silviculture prescription? Discuss how they might vary by forest type or forest management organization.

CHAPTER 3
Ecological Principles Basic to Silviculture

Introduction

Silviculture encompasses a number of disciplines, including plant, animal, and ecosystem ecology (Kimmins 2004; Perry 1994; Seymour and Hunter 1999; Waring and Schlesinger 1985); soils (Richter and Markewitz 2001); plant biology and silvics (Burns and Honkala 1990); forest entomology and pathology; and genetics. The practice of silviculture is often constrained by costs and driven by profit or conservation goals that stem from landowner objectives or agency policies. In one stand, certain aspects of the ecosystem—say, insects—may be highly important, whereas in other stands, the overriding concern may be soils, or the potential effects of fire, or the need to produce economic yields of wood in the presence of an endemic pathogen. In still other stands there may be no ecological impediments. Consequently, silviculturists must be generalists, with a strong background in a range of topics and practical experience, though it is difficult to define the proper scope and depth of knowledge appropriate for the successful practice of silviculture. That will vary with the forest ecosystems and social context within which silviculture is practiced. Clearly, silviculturists must be expert in areas such as forest regeneration, stand growth, thinning, and density management of forest stands; however, the details of these practices will vary considerably among forest types and ownerships. The purpose of this chapter is to present the basic information about forest ecology and biology that is necessary to practice silviculture effectively. The references should enable readers to find additional information they may need.

Forest Stands—Structure and Ecosystems

Effective forest management requires partitioning the forest into separate management units or stands. Stands are delineated according to their similarity in vegetation structure and species composition, topographic position, site history, soil type, geologic substrate, and other factors that will influence silvicultural treatments (figure 3-1). At the lower end of the size range, a stand is typically several acres. At the upper end, a stand can be as large an area as can feasibly be treated in a relatively uniform manner. Simply stated, differences in vegetation structure and site conditions create a mosaic of more or less distinct units with relatively uniform conditions, usually distinct from those in adjoining stands. An important step in designing silvicultural strategies for the forest and its constituent stands is to characterize the site (see chapter 5), the vegetation structure, and the current developmental stage of the stand in the context of short- and long-term dynamics (See O'Hara and Nagel 2013).

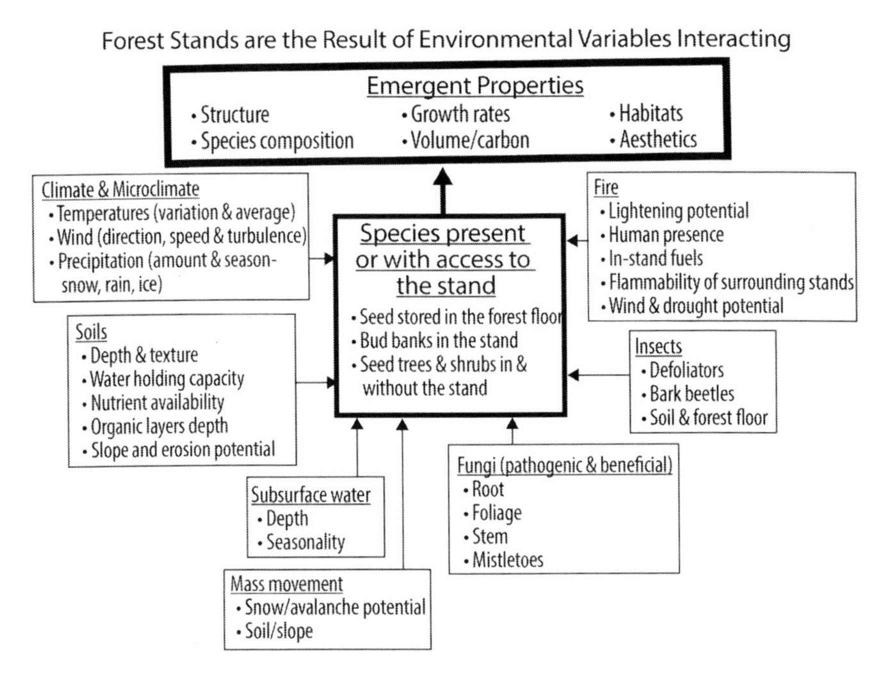

Figure 3-1. Forest stands are ecosystems; a diagram including the major ecological variables and sources of vegetation that affect forest stands.

Uniform species composition and stand structure are not criteria for determining stand boundaries. Uneven-aged stands are by their nature quite heterogeneous (see figures 2-4, 2-5). Currently, diverse, variable stand structures are often intentionally encouraged for wildlife habitat, and features like groups of hardwoods and shrubs, snags, uncommon tree species, etc. may all add to variation within both even- and uneven-aged stands.

In managed forests, there is nearly always a practical aspect to defining forest stands. They have to be discrete areas that can be easily located on maps and in the field. Features such as streams, ridgetops, roads, and property boundaries as well as changes in species composition, tree size and age, structure, density, etc. often determine stand boundaries. They are operational units of forest management for all forest values (O'Hara and Nagel 2013).

Forest stand boundaries may be determined with regard to silvicultural practices, and the potential effects of treatments may require the shifting of stand boundaries or modification of treatments. For example, if a heavy thinning in stand A might cause trees to blow down in stand B, the thinning may have to be modified in A, or the boundary between the two stands shifted, or perhaps B should be thinned first to develop its resistance to wind. Similar examples could be given for prescribed burning, use of logging systems, visual or scenic objectives, etc. Ideally, a prescription might call for thinning a young stand at the bottom of a slope and reforesting an unstocked stand immediately uphill. However, harvesting the stand at the bottom, when the time came, might entail logging through—and likely damaging—the young uphill stand. Therefore it might be best to defer either the regeneration of the uphill stand or the thinning of the downhill stand until the entire hillside can be managed as one stand. In another example, when carrying out prescribed burning, stand boundaries should be located so that fire can be controlled within a stand and not escape and damage adjacent stands. Thus, a criterion for establishing stand boundaries is to achieve relative independence among stands, so that activities in one stand will not have major negative effects on adjacent stands. Stand boundaries are therefore often determined from a practical viewpoint and not according to areas of homogenous vegetation. Consequently, there may be considerable variation in structure and species composition within forest stands.

Stand Structure

Stand structure or vegetation structure is a general term that refers to the vertical and horizontal arrangement of plant species and individuals and the important abiotic elements of a forest stand. Stand structure is commonly described as the distribution of numbers of trees across diameter classes (diameter distribution) in a stand table or a graph (figure 3-2; see also figures 2-1 and 2-2). Ideally the ages of the trees in each diameter class would be part of the description, but it is often too time-consuming or expensive to get an accurate age for many trees, especially for stands with trees of variable sizes and species. The density of tree seedlings and the cover of shrubs in the understory may also be part of a stand's description. There is a virtually infinite array of possible stand structures because of subtle variations in size class and distribution

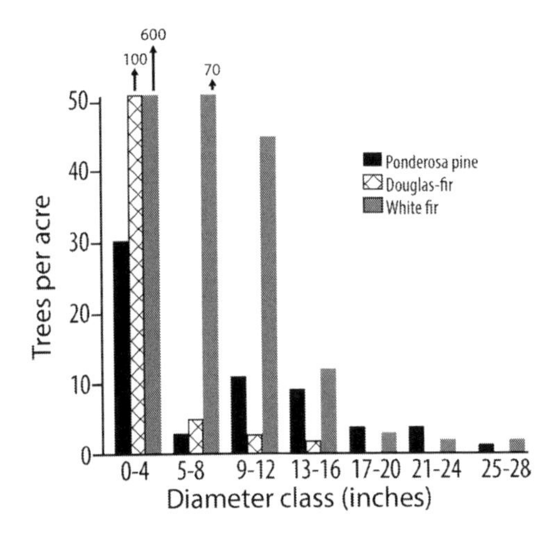

Figure 3-2. Stand structure—number of trees per acre and species by diameter class in a mixed-conifer stand in eastern Oregon. The numerous white fir and Douglas-fir in the smaller diameter classes are the result of fire exclusion. Adapted from Hodgson (1990).

(horizontal and vertical arrangement) of individual vascular plants, lichens, fungi, and plant parts such as branches and roots. Ideally stand structure might also include propagules, such as seed banks stored in the forest floor or on trees and bud banks at the bases of trees or shrubs and in roots and rhizomes below ground (Grime 1981). These are the source of new individuals that may have an important role in silvicultural prescriptions and stand structure goals. In addition, features such as large trees, dead wood, and wet areas might be included in a complete definition of stand structure because they may be important habitat for wildlife, or may affect fire behavior and the reproduction of certain plant species. Stand density—including trees and basal area per size class and area, stand density index, and cubic volume per area (see chapter 6)—is also an important part of stand structure. In addition, there are methods for quantifying the vertical arrangement of trees. The main point is that stand structure is multivariate, and the variables of interest are dictated by management objectives.

Stand structure is central to silvicultural practice because a diverse array of silvicultural objectives can be stated in terms of stand structure. The objective of reducing fire potential is related to the density of the overstory as well as the density and height of dead wood and vegetation in the understory (see chapter 8). The structure of a productive stand for timber management might be defined by the number of well-spaced trees of a certain diameter at certain ages—say 20 and 60 yr. Habitat requirements for the spotted owl, for example, include structural features of old-growth Douglas-fir such as a multi-layered canopy with a large amount of within-canopy open space. Aesthetically appealing recreation sites may require open stands of tall trees that allow long viewing distances, but campgrounds may require dense understory vegetation that serves as a visual screen. Elk hiding cover in lodgepole pine stands depends on the size and density of individual stems, whereas thermal cover depends on total canopy cover (Smith and Long 1987).

Forest Stands Are Ecosystems

Forest stands are ecosystems. Stand structure is the result of the interaction through time of the plants in the stand with the abiotic and biotic parts of the ecosystem and with natural disturbances or silvicultural treatments such as site preparation, thinning, prescribed burning, and fertilization. Thus, silviculture is practiced in an ecosystem context (figures 3-1, 3-3). Changing stand density alters the abiotic parts of the system (soil water, light, temperature, wind speed, arrangement of fuels and fire potential, etc.) and also may alter biotic parts (fungi, invertebrates, vertebrates).

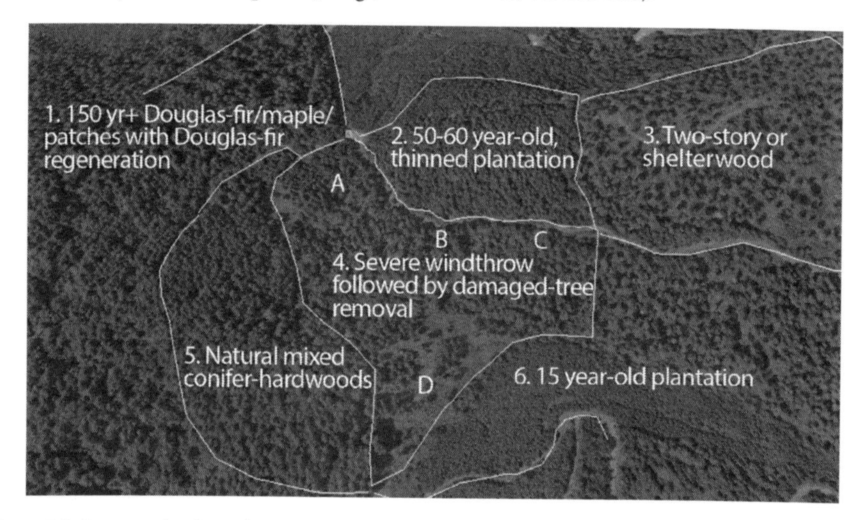

Figure 3-3. An example of stands in a managed forest. Stand boundaries are determined by a combination of tree sizes and ages, species composition and structure, and physical features—roads in this example. The stand impacted by wind is the most variable, including: A. dense conifers and some hardwoods, B. scattered young conifers, C. scattered large conifers and hardwoods, D. scattered conifer saplings, seedlings and grasses/shrubs. This stand can be managed to maintain a diverse structure or make it more uniform. The older stands to the left are less variable than the stand affected by wind; however, they include trees of different sizes and species. The plantations and two-story stand are the least variable. In this example structure and variability are determined by what can be seen from the air and does not include understory species, sizes and distributions.

The biotic and abiotic ecosystem components occur at scales ranging from a few yards or less within a stand to considerable distances beyond a stand's boundaries. The composition and density of understory species may vary considerably within a few yards. Differences in depth and rock content of the soil may occur from riparian zones to ridgetops. The potential effects of fire and potential ignition points within a stand are determined by the arrangement and type of fuels from shrubs, trees, and flammable woody material on the ground, as well as by topography. The habitat for small seed-eating mammals may include within-stand patches of shrubs, cavities in trees, and pieces of large logs over several acres, whereas habitat for ungulate populations might be evaluated over several thousand acres.

Viewing forest stands as ecosystems enables a rigorous evaluation when deciding whether or not a proposed treatment will meet forest management objectives. Because the goal of silviculture is to manage vegetation to achieve a desired stand structure and species composition, this evaluation focuses on the variables that affect vegetation. The following grouping of major environmental variables provides a basis for evaluating forest stands as ecosystems (figures 3-1, 3-3). Others may choose to group the variables differently or use other terminology. The main point is that a thorough, systematic evaluation is needed to determine the factors that might affect the forest's response to proposed silvicultural practices.

We consider the forest ecosystem variables as falling into three major categories:

1. Stand structure/vegetation variables include tree diameter and height distributions, understory tree-seedling density, and shrub and herb cover for important species; vegetative propagules at the base of stems, on rhizomes, burls, etc.; seed availability on shrubs or trees, in seed banks in the forest floor, or in serotinous cones; seed from adjoining stands; and snags, logs on the forest floor, fuel type, and arrangement.

2. Abiotic variables include climate/microclimate, soils, subsurface water, mass-movement, and fire potential.

3. Biotic variables include fungi (pathogenic and beneficial), insects (pests and beneficial), and vertebrates.

Development of vegetation and stand structure is the result of the interaction and dynamics of these variables, as well as those entering the stand as propagules (wind-blown seed, seed banks, and bud banks), acting through time (figure 3-3). The first three abiotic variables (climate/microclimate, soils, subsurface water) are fundamental to site productivity—the potential for a site to produce plant biomass. This potential is frequently estimated by site index—the height of the "best" dominant and codominant trees in a stand at a base age of 50 or 100 yr (Curtis et al. 1974; Dunning 1942; Hann and Scrivani 1987; King 1966; McArdle et al. 1961; Schumacher 1928; see chapter 5). Some variables (the potential for fire, effects of herbivores, wind, etc.) are evaluated both within and without the stand.

Use of Environmental Variables

These environmental variables provide the basis for evaluating silvicultural prescriptions (figure 3-3). Information about these variables and how they are likely to affect stands is obtained from a variety of sources, including observations within stands, the literature, short courses, personal experience, and consultation. At the current stage of development of ecological and silvicultural science it is often unfeasible to predict with certainty whether or to what extent a particular variable will affect stand development, especially given the long-term nature of forest stand development. But careful evaluation can reveal the variables most likely to affect a particular stand in light of the management being considered for that stand and help decide whether a planned course of action needs to be modified or abandoned. For example, if a root disease is prevalent in a large part of a stand, it might be best to abandon plans for thinning and instead regenerate the stand with a mixture of species with low susceptibility to the disease. No one can predict accurately how the disease might affect the regeneration as it grows, but using a species mix will provide options. The regeneration method used would take into account the distribution of the disease, the species to be regenerated, and the owner's objectives.

Frequently, variables interact. For example, a combination of root disease and mistletoe may make dense stands of pine susceptible to a severe outbreak of bark beetles during years of drought. Growing low-density stands on droughty south slopes might reduce the effect of such an outbreak but would not necessarily prevent it, because the extent or severity of drought cannot be predicted. Silviculturists should assess the likelihood of high rates of insect-related mortality in dense stands on dry sites, recognizing that the presence of pathogens could increase their severity.

The focus of an environmental evaluation depends on the management objectives for the stand and the matching of potential treatments with objectives. For example, an evaluation of the potential for reforestation after a fire will consider variables affecting the establishment and early growth of the seedlings: microenvironment, potential effect of shrubs and grass in reducing soil water and providing habitat for seed predators and gophers, seed sources for natural regeneration, and the consequent need for planting. In deciding whether to thin a stand, it would be appropriate to identify the variables that might cause failure, increase expense and loss of revenue, a major impact on soils or other resources, or have a negative impact on adjacent landowners. These might include wind; the possibility of heavy, wet snow or ice that might break treetops; and the risk of fire or bark beetles in built-up slash. Evaluations will not result in treatments that prevent all windthrow, browsing of seedlings, or root diseases, but will help a manager decide whether to treat or not treat, or to defer or modify treatments.

Climate and Microclimate

The regional climate has a strong effect on forest ecology and on many silvicultural practices. Dry summers and falls and wet winters and springs are characteristic of western forests (figure 3-4). Precipitation usually occurs as rain in the coastal mountains; occasionally there are heavy snow and ice storms. Snow is common at higher elevations in the Sierras and Cascades, with depths exceeding 12 ft. Total precipitation is plentiful in westside forests but much lower east of the Cascades and Sierras (figure 3-4), often less than half that of westside forests. Most precipitation comes in the winter; only about 5–10% from June through September (Baker 1944). The timing of precipitation has a profound effect on forest growth and silvicultural practices. Because of summer

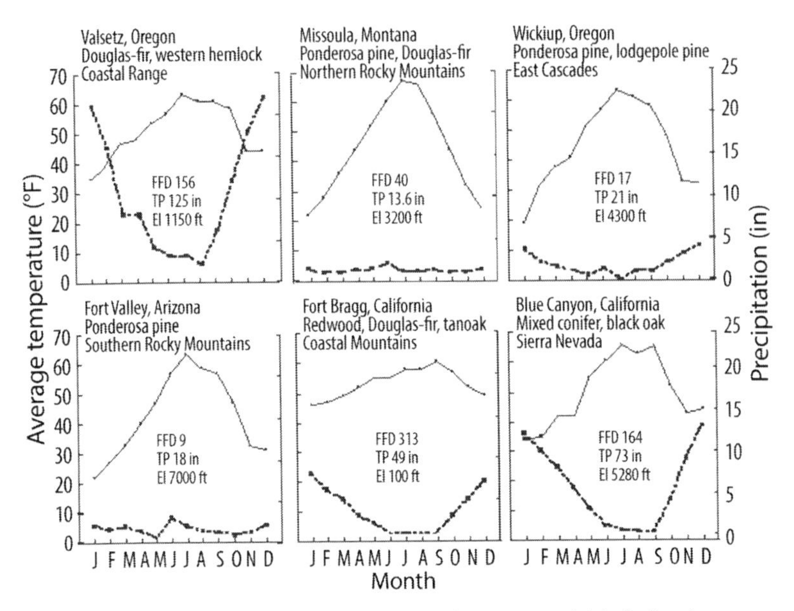

Figure 3-4. Average monthly precipitation (dashed line), and temperature (plain line) at six representative sites in western forests. FFD=frost free days, TP=total annual precipitation, El=elevation (ft).

drought, competition for water is a major consideration in many silvicultural treatments, especially reforestation. Also, dense stands of trees often undergo moisture stress, especially during prolonged drought. Soil properties and competition largely determine the amount of water available to plants. Because plant growth is limited by cold weather during much of the wet period, much growth relies on water stored in the soil toward the end of the rainy season.

Exceptions to this general pattern of winter precipitation occur in some interior forests. For example, precipitation from summer thundershowers occurs in the Rocky Mountains. In northern Arizona (figure 3-4), summer precipitation comes from weather systems in the Gulf of California. In these forests, precipitation is more uniformly distributed throughout the year, with about 20% occurring from June through September. Summer precipitation is critical for natural regeneration of ponderosa pine in Arizona (Pearson 1923).

The precipitation values in figure 3-4 do not take into account fog that condenses on the forest canopy and falls to the forest floor. On some sites near the coast (figure 3-4), fog may add appreciably to total precipitation. Even on sites where it does not add significantly to precipitation, it favors plant growth and forest productivity by reducing evapotranspiration.

Temperatures in western forests are generally mild. Summer averages seldom exceed 65°F. Lows are about 30°F in the coastal mountains. At higher elevations or at interior sites, winter, spring, and fall temperatures are cooler than at coastal sites,

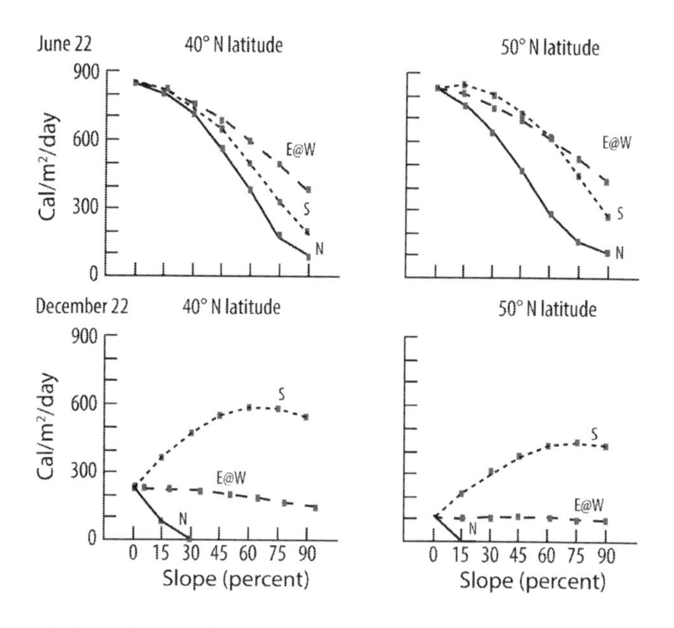

Figure 3-5. Potential solar radiation on four aspects on June 22 and December 22 at 40 and 50 degrees north latitude. Adapted from Kaufmann and Weatherred (1982).

and summer temperatures are often hotter. Frost-free days can number 150 or more in forests near the coast. On the east side, spring and fall temperatures are lower and frost-free days are fewer (9–20 per yr, figure 3-4). Mild temperatures in the fall, winter, and early spring at lower elevations in the westside forests enable evergreen plants to photosynthesize during this period (Waring and Franklin 1979a, b). The growing season may include the months of October through February on some sites, where seeds may germinate and photosynthesis occurs (Emmingham and Waring 1977; Harrington et al. 1994; Helms 1972), although cell division and elongation are minimal.

The climate is strongly affected by elevation, topography (slope, aspect, landform), and vegetation, resulting in a mosaic of microclimates (figure 3-5). Baker (1944) estimated that mean July temperatures in the Cascades decrease 7–10°F with each 2000-ft increase in elevation. He estimated the growing season (number of days with a minimum temperature of 42°F) to be 140 days at 6000 ft and 240 days at 2000 ft. Total precipitation also increases with elevation. However, high amounts of precipitation occur in the Coast Ranges at elevations of less than 2000 ft.

Microclimates vary considerably with slope and aspect, resulting in significant differences in plant moisture stress, the amount of thermal cover for wildlife, fuel drying and snowmelt, and other important processes on different slopes and aspects. Microclimate differs according to topography mainly because of the way slope and aspect affect incoming short-wave solar radiation.

Radiation indices (Frank and Lee 1966; Kaufmann and Weatherred 1982) provide a method for assessing the potential microclimate on a range of slopes and aspects at different latitudes (figure 3-5). Potential radiation on a site changes greatly with variations in latitude, aspect, and slope. For example, on June 22 at latitude 50°N, potential radiation on a south aspect with a slope of 45% is about 1.5 times greater than on a similar slope with a north aspect. At 40°N, it is 1.3 times greater on 45% south slopes than on 45% north slopes (figure 3-5). On both aspects, potential direct beam radiation decreases as slope increases. At the spring and fall equinoxes, there is more radiation on slopes of over 50% than at other times of the year. At the winter solstice, potential radiation is greatest on south aspects. North slopes of over 30% at 40°N receive no direct solar radiation at all on the winter solstice because the topography blocks the incoming radiation. East and west aspects may receive the most radiation on steep slopes in the summer.

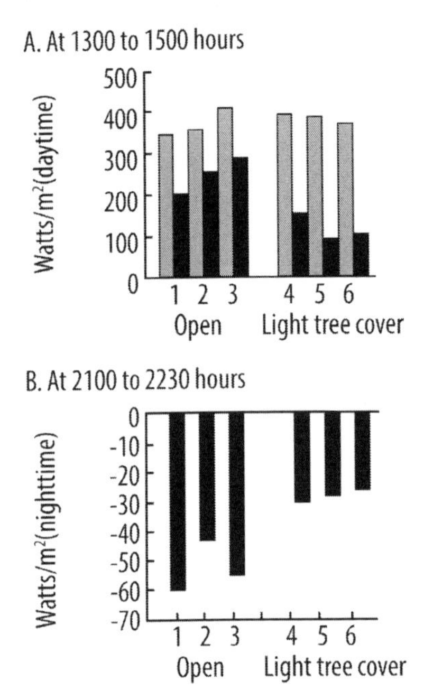

Figure 3-6. (A) Incoming and (B) outgoing radiation in August on open sites (1, 2, 3) and sites with a tree cover (4, 5, 6) in southwestern Oregon. The differences in incoming radiation at the canopy (gray bars) and at the ground (black bars) in (A) are caused by sparse cover of shrubs (1, 2, 3) and a tree cover (4, 5, 6) reflecting the incoming radiation back to the atmosphere. In (B) there is more outgoing radiation (black bars) at night in the open than under the tree cover. Adapted from Holbo and Childs (1987).

Topography has a major effect on minimum temperatures because of the phenomenon of cold air drainage and cooling from outgoing radiation. On calm nights, when there is little mixing of the air mass, cold air settles into small basins or onto level topography, producing a temperature inversion that may bring fog or severe frost. Frost may result in winterkill or top kill of vegetation that is not protected by a cover of snow or other vegetation. Damaging or lethal temperatures usually occur within 6–12 ft of the ground. In ponderosa and lodgepole pine forests, it has been suggested that lodgepole pine is often found on flat terrain and invades meadows because its seedlings and saplings tolerate frost (and high water tables) better than those of ponderosa pine.

Vegetation has a major effect on microclimate, affecting light, temperature, precipitation, and wind (figures 3-6 and 3-7). The effects depend on the density, height, and arrangement of the vegetation. The amount of radiation received beneath a forest canopy will vary with the phenology (time) of leaf fall and of shoot and leaf expansion in both conifer and deciduous forests. Vegetation affects the net radiation, which is defined as the balance between

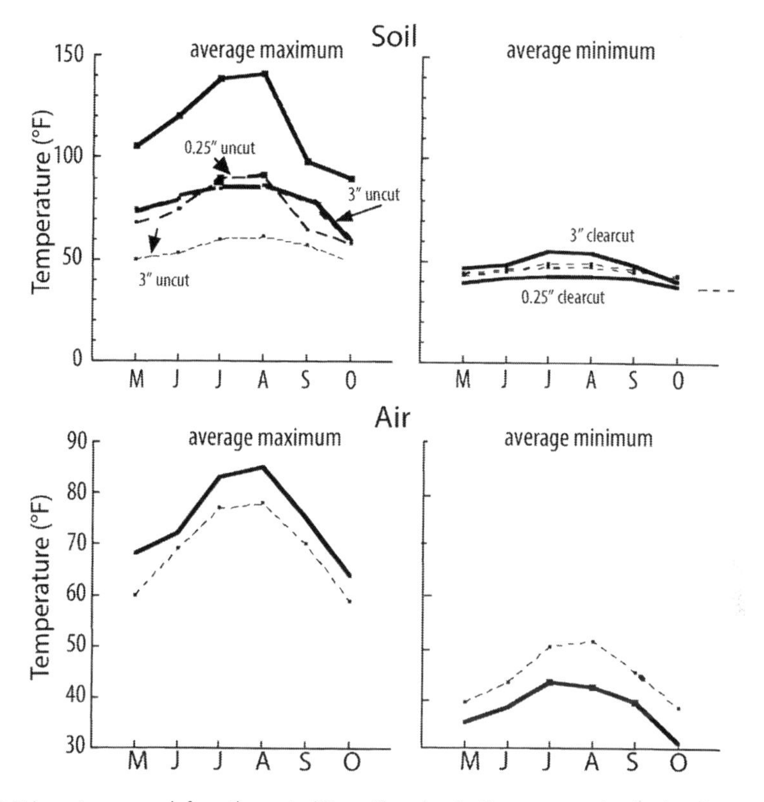

Figure 3-7. Long-term records from the central Sierra Nevada of soil temperature (at depths of 0.25 in and 3.0 in) and air temperatures (at 4 ft) in a forest stand (dashed line) and in a clearcut (solid line). The tree cover reduces the maximum and raises the minimum air and soil temperatures at 0.25 in. Adapted from Stark (1963).

incoming and outgoing radiation (Holbo and Childs 1987). Net radiation is the energy available for processes such as evapotranspiration: warming of the air, plant and soil surfaces; melting of snow; and so on (figure 3-6). Even a light forest canopy can prevent a high proportion of the incoming radiation from reaching the forest floor during the day (figure 3-6a), reducing the energy available for soil and air warming or other processes (figures 3-6b and 3-7). The situation is reversed at night, when there is a net loss of radiation; the forest cover reduces the rate of energy transfer to the atmosphere, and thus it is generally warmer at night in a forest than in the open (figure 3-6b). This effect of cover on outgoing radiation is the reason shelterwoods are used on frost-prone sites. However, wind and cold air drainage also affect the temperature. A forest cover also reduces air and soil temperature extremes, especially at ground level and just below the soil surface (figure 3-7).

Forest cover also affects light quality, or the wavelength of radiation. Forest canopies absorb visible or generally photosynthetically active radiation (from 0.4 to 0.7

microns) and reflect longer wavelengths. As a rule, hardwood forests reflect more far-red radiation than conifer forests. However, light intensity alone may have more of an effect than light quality on the growth of plants in the understory (Lieffers et al. 1999).

It is important to recognize that it is difficult to isolate environmental variables in practice. Thinning to provide more light may also increase precipitation throughfall, soil water, air temperature, snow cover, and wind speed. A cover of snow may protect plants from cold temperatures. Wet, dense snow, such as often comes early in the winter or late in the spring, can also break stems and uproot forest trees, as can ice. Often damage is caused by a combination of snow and wind (Petola et al. 1999). The likelihood of damage to ponderosa pine plantations from heavy, wet snow is related to elevation, slope, aspect, and the roughness of the canopy—that is, variability in tree height and crown size (Megahan and Steele 1987). The chance of snow and ice damage is higher in dense stands, where trees have small diameters in relation to their height. Snow and wind can also reduce the leaf area of a stand, increasing litterfall and reducing canopy density. Trees killed or damaged by snow or ice may become a habitat for birds or small mammals, or for bark beetle population build up.

Strong winds may affect forests locally, especially after thinning. Wind is often of overriding importance in southeastern Alaska, where windthrow is the most common cause of forest disturbance (Deal et al. 1991; Ruth and Harris 1979). Occasionally there are widespread severe windstorms that have major effects on the forest, such as the 1962 Columbus Day storm in the Pacific Northwest and the 1999 windstorms in northern Minnesota, which caused extensive blowdown.

The effects of wind are strongly influenced by topography and by the density and structure of the vegetation. High wind speeds and turbulence occur near the leeward side of ridgetops (Alexander 1964; Gratkowski 1956). Velocity and eddying are increased where winds are forced through gaps or saddles (figure 3-8). Alexander (1964) found that wind effects are greatest at places where ridges are broken by saddles,

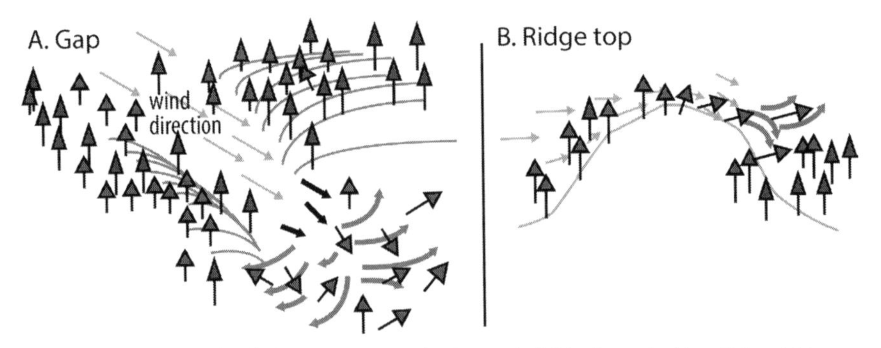

Figure 3-8. Wind speed and turbulence increase on the down-wind side of gaps in ridges (A.) and ridge tops (B.) causing blowing down and damaging tree tops. Bold and twisted lines indicate increased wind speed and turbulence, respectively.

probably because of acceleration and turbulence as the wind passes through the saddle. Windthrow was somewhat greater on moderate slopes (21–40%). Roughness and density of the forest stand canopy increases wind turbulence and causes the wind to eddy into the stand rather than flow over the top of it (Green et al. 1995). Stark (1963) found that over the course of 30 years wind speed within the forest was less than half the speed in adjoining openings.

Climate is also affected by human activity. Air pollutants, especially nitrous oxides, injure the foliage of conifers (especially ponderosa pine) and kill the trees or make them susceptible to mortality from insects and periods of drought (Miller and Millecan 1971; Miller et al. 1963). The worst effects of air pollutants on forests occur where they are trapped by temperature inversions, such as in the Los Angeles basin and the central valleys of California.

In addition, we cannot overlook the silvicultural implications of climate change. Variability in temperature and precipitation are inevitable, but the direction and magnitude of the changes are uncertain. Many forest plants and animals have evolved to adapt to this variability, but the effects of increased variability and rates of change are unknown. Currently, drastic changes in silvicultural practices do not seem warranted. Modification of well-developed practices will likely be sufficient. As we will discuss later, growing mixed species and maintaining genetic diversity in relatively low-density stands may be a reasonable hedge against the uncertainties of future climates.

Forest Soils

Like all soils, forest soils are generally the result of centuries of weathering of parent rock and the accumulation of organic matter and nutrients from plants and animals, as well as inputs from the atmosphere and subsurface water flow. They are a legacy of past geological and ecological processes. Careful implementation of silvicultural practices enables forest soils to continually support these processes and maintain productive capacity. Soil carbon storage can be an important part of C sequestration from the atmosphere.

Soil forming processes are common in all forest soils; however, there is much variation in soil forming factors (Jenny 1941), especially in parent material and consequently in basic soil properties—depth, structure, texture, nutrient availability, and water-holding capacity. In coastal forests, soils are usually deep, and those derived from basalt are cation-rich clays and clay loams. In the Sierra Nevada, the most productive soils are derived from andesites at about 3000 to 5000 ft in elevation, where precipitation is abundant. Less productive soils are often derived from granitic, glacial, and metamorphic parent material. In many forests in the Cascades and the interior of the western United States, soils are formed on parent material derived from recent volcanic activity, which frequently support ponderosa and lodgepole pine forests of low to moderate productivity. These soils are often coarse-textured and contain little organic matter because of their recent origin and the low precipitation east of the Cascades and Sierra Nevada.

Abrupt changes in vegetation are not generally related to changes in soil characteristics. Soil productivity certainly varies by soil type and properties such as soil depth, but the major tree and shrub species occur and grow well in a range of soil types. The exception to this rule is the serpentine soils of northern California and southern Oregon, where mixed-conifer forests change to Jeffrey pine and gray pine, and local populations of *Ceanothus* and other species have become adapted to these soils.

Soils of western forests are generally quite productive. They are formed on rather recent parent material and have a relatively short geologic history, and are often more productive than older soils that have been subjected to long periods of weathering and leaching. In addition, these forests have a short history of intensive human use and in general have not been cleared for agriculture and grazing like many of the forests in the eastern United States and Europe.

Soils support five major ecosystem functions:

(1) They store water and make it available for plant growth.

(2) They accumulate nutrients from weathering, biological fixation, decomposition of organic matter, and the atmosphere, and make them available for plants.

(3) They store carbon in organic matter.

(4) They are the medium for the extension and development of root systems that are the mechanical support for trees and other vegetation.

(5) They are habitat for many organisms that support or interfere with silvicultural objectives.

WATER

The norm in most western forests is summer drought or very low amounts of precipitation from late spring into the fall. But from fall through early spring, forest soils are likely to be filled to field capacity—all the water the soil can hold against the force of gravity—within the rooting zone, as the soil profile is recharged with precipitation. During this period, many sites receive well over 20 in of precipitation. Forest soils generally hold less than 10 in of water against the force of gravity (figure 3-9). Consequently, water flows below ground and into streams. In general, the water available to plants amounts to the water stored in the soil when plants begin to use it in the spring, plus any recharge that occurs from late-spring and early-summer precipitation and subsurface flow. Recharge from subsurface flow may be considerable, but its occurrence is highly variable and unpredictable from site to site. As noted previously, summer fog is common along the coast and can amount to 10–15% of total precipitation in coastal forests where a vegetation canopy is present to condense and funnel the water to the ground. This water may not recharge the soil profile, but it reduces evapotranspiration demand. In most western forests, summer rains are generally too infrequent and drop too little precipitation to recharge much of the soil water and affect plant growth, but in some climates, such as those of northern Arizona and the east central Rocky Mountains, summer precipitation is normal, and some water recharge occurs periodically throughout the growing season.

The ability of forest plants to survive and grow is largely dependent on the amount of water held by the soil. Nutrients are seldom the limiting factor for most native forests. Soil water storage is mainly a function of soil depth (volume of soil), soil texture, organic matter (figure 3-9), and structure. For a given surface area, a deep soil represents

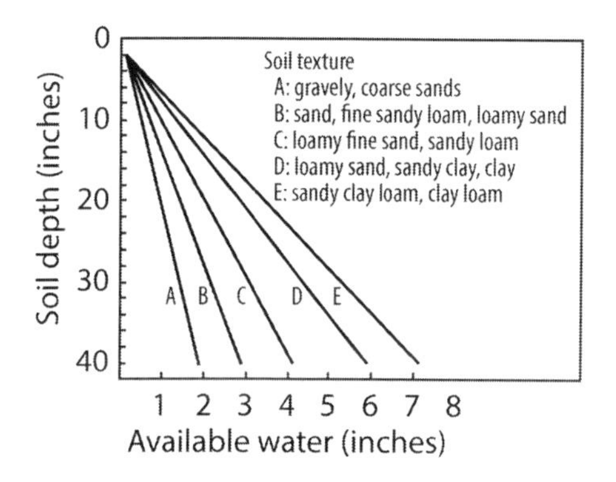

Figure 3-9. Available water holding capacity depends on soil texture and depth. For example when soil C is saturated to a depth of 20 in, it holds approximately 2 in of water at tensions less than 1.5 megapascals that is "available" for use by plants. For soil A only about 1 in of water would be available at this depth. Adapted from Wholetz (1968).

a large sponge with a great capacity to hold water. Soil texture and structure create the size of the pores in the sponge; small pores hold more water than large pores, which allow water to pass through. Pores also affect aeration, which in turn affects the ability of roots to grow, exploit the soil volume, and use the water the soil contains. A heavy-textured clay soil, with very small pores, for example, may hold a considerable amount of water, but much of it may be bound tightly to the soil particles, making it unavailable to plants. Also a high proportion of water-filled pores limits soil aeration and root growth. Sandy soils have large particles and pore spaces and thus hold less water than clay soils and have more air-filtered pores. Clay-loam soils with lots of organic matter have good soil structure and are generally good suppliers of water and nutrients.

Soils with rocks and little organic matter may store less than 1–2 in of water. On the other hand, litter, rocks, and logs on the soil surface act as mulch and reduce evaporation. Reduced evaporation can appreciably increase the water available to plants and other organisms in about the upper 8 in of soil (figure 3-10).

In most western forests, plant growth begins in late spring or early summer. As plants use water, the soil dries and soil water potential (the ease with which plants can extract water from the soil) decreases, making water increasingly less and less available

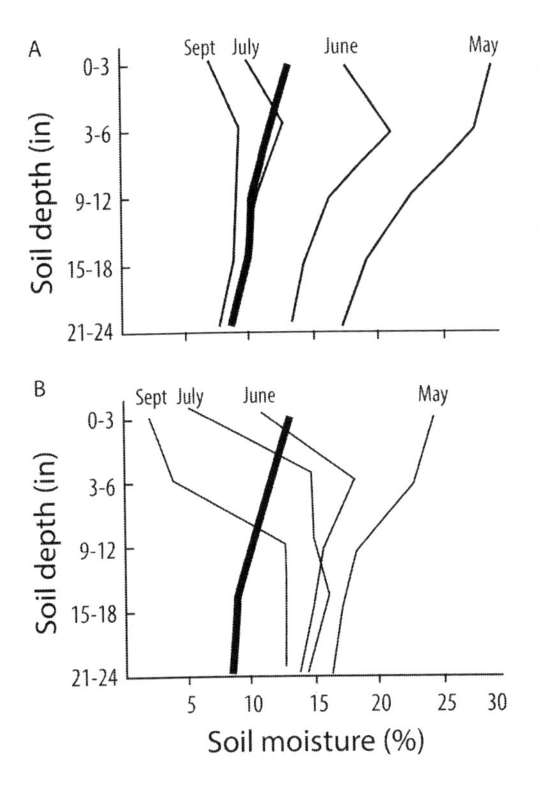

Fig 3-10. Soil moisture depletion during the dry season in the top 24in of soil in a mixed-conifer forest. Bold line indicates soil moisture at permanent wilting point (a tension of –1.5 mega pascals). Forest vegetation was undisturbed in (A).In (B), trenching and killing the vegetation excluded water use except by planted ponderosa pine seedlings .In both (A) and (B), water from 0 to6 in was reduced by evaporation and transpiration. In (A), soil moisture from 0 to 24 in reached permanent wilting point. In (B), soil moisture below about 6 in did not reach permanent wilting point. Adapted from Tappeiner (1967).

to plants (figure 3-10). Soil is at field capacity (held at a tension of about -0.3 megapascals [MPa]) in the spring within a few days after the last rain or after snow has melted, and plants can readily extract water from the soil. As evapotranspiration occurs, soil water is depleted. By midsummer or early fall, without frequent rainfall, soil water is bound tightly to soil particles at a tension of -1.5 MPa, the permanent wilting point. At this tension, water is essentially unavailable for plant growth, and a drought-induced dormancy may occur. Photosynthesis and cell division stop, and leaves of deciduous species may wilt and drop. If temperatures become very high or the drought is prolonged, some plants may die. However, many forest plants can withstand soil water tensions more negative than -1.5 MPa, although their photosynthesis and growth is nil.

Competition has a major effect on water availability to forest plants (See chapter 7). The first year or two after a fire or site preparation, vegetation density is low and there is little competition for soil water. But as plant density increases, so does competition for water. Plants that form dense covers above ground and dense, deep root systems generally obtain a large share of the soil water (Baron 1962; White and Newton 1989). Reducing tree density by thinning or controlling competition by removing competing vegetation are ways of prolonging the availability of soil water to crop trees or shrubs (figure 3-10).

Water availability also depends on plant growth habits. Some plants, mainly grasses and herbs, grow roots early in the spring when soils are cold (<40°F). Thus they can use stored water before it is available to plants such as conifers, whose roots grow at warmer temperatures. Some plants (ponderosa pine, Pacific madrone, manzanita, and *Ceanothus* spp.) have root systems that can penetrate weathered parent material and use water stored there (Zwieniecki and Newton 1994, 1995, 1996a, b). Other plant roots are shallow, grow horizontally and are restricted to the weathered soil mass.

There is evidence that Douglas-fir and ponderosa pine trees may lift or redistribute water from lower soil levels and make it available to plants with their roots in the upper soil layers, including the trees themselves, with their roots in the upper soil layers (Brooks et al. 2002). In effect, the soil-drying trend of the summer drought proceeds but with small diurnal increases in soil water potential. This diurnal increase in water may aid establishment of tree, shrub, or herbaceous seedlings. However, it is unclear what this may mean for silvicultural practices, since these large trees inhibit the growth of smaller trees beneath them (Harrington 2006, Acher et. al. 1998, Zenner et al. 1998).

NUTRIENTS

Soils store nutrients and organic matter from plants, animals, the atmosphere, and the parent material. Most nutrient elements are weathered from parent material and recycled from plants and animals above and below ground. Dead boles, branches, and roots add large quantities of organic matter to forest soils. Nitrogen is added by precipitation and especially by N-fixing plants such as red alder, lupine, and *Ceanothus* spp. Canopy lichens add nitrogen in some older forests. Chemicals from industrial and domestic sources are added via the atmosphere.

The factors that enable soils to store nutrients are nearly identical to those that affect water storage capacity: soil depth, texture, and organic matter content; volume of rocks; and the volume of the soil actually containing roots. Nutrient pools blocked by soil impermeable to roots may not be considered part of the available pool. Therefore, soil structure and permeability that enable the growth of root systems and their uptake of water and nutrients are major considerations in soil nutrient availability in forest stands.

Forest stands can be considered as a series of nutrient pools (figure 3-11; also see Landsberg and Gower 1997 and chapter 9). The largest pool is the mineral soil. The next largest is the forest floor, the site of annual litter deposition and decomposition. The mineral soil and the forest floor together may contain over 85% of the nutrient capital. The crowns of forest trees and the understory vegetation generally contain 5–10%; the boles of the forest trees contain relatively small amounts of the total nutrient capital (<10%). Tree stems are composed mostly of C, and although their biomass can be considerable, their nutrient content is relatively low (figure 3-11). The total amounts vary

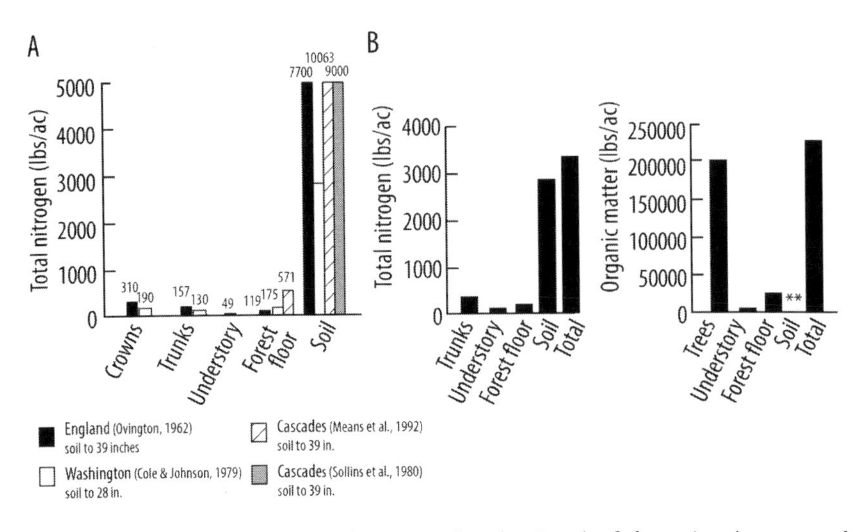

*Figure 3-11. (A) Nitrogen content of soils and vegetation from four Douglas-fir forests (numbers on top of the bars indicate N in lbs/ac),and (B) distribution of biomass and N in a 36-yr-old Douglas-fir stand on site IV glacial out-wash soil in western Washington. (** not available) Adapted from Cole and Johnson (1979).*

from site to site, with greater nutrient capital generally occurring on more productive sites, but the relative distribution of nutrients within stands is similar across many sites.

Nutrient distributions change immediately following fire or harvesting and as a new stand grows. The nutrient pool in the mineral soil remains relatively stable, especially if the top 8–10 in of soil is not removed. After disturbance, there is some loss through leaching; however, the ion exchange capacity of the soil and the rapid regrowth of shrubs, herbs, and grasses generally retain most of the nutrients released from fire or clearcutting.

Information on nutrient pools (figure 3-11) can be used to evaluate the effects of vegetation management activities on nutrient capital. For example, a fire that removed the forest floor would vaporize about 120 to 570 lbs/ac N and cations would be left in the ash on site, assuming it was not removed by water or wind erosion. The more site-specific the information on nutrient distribution, the more accurate will be the estimation of the effects of disturbance. Budget or balance sheets such as these do not account for nutrients available to the plants, and the value of the estimates depends on the laboratory techniques used to extract nutrients. For example, the measure of total N or Ca accounts for the quantity of these nutrients tied up in living and dead tissue as well as in the soil solution. Exchangeable Ca and mineralizeable N are better estimates of what is available to plants. However these estimates vary with the methods used to extract the N and Ca ions from their soil exchange sites. Estimates of total nutrients do not account for future mineral inputs from weathering or from the atmosphere or losses from leaching or erosion. Currently there are no acceptable techniques for estimating plant-available nutrients, but even with these limitations, a knowledge of the nutrient pools is a practical way to assess the potential effects of disturbance to a forest stand (Bengston 1981). For example, Boyle and others (1973) estimated that with whole tree harvest (stems and crowns) of 40-yr-old aspen, Ca would be depleted within nine 30-yr rotations; however, this harvesting regime apparently would not deplete N, P, and K. However, Ca and N inputs from weathering of parent material and the atmosphere were not considered.

Bormann and Gordon (1989) present evidence that forest stands can be self-sufficient in nitrogen. After about 30 years the N levels in the forests they studied returned to preharvest levels even though the stands were clearcut and the relatively nutrient-rich crowns of the trees were removed from the sites. Longer time periods may be required on less productive sites. The amount of N in the forest can also be increased by fertilization and by encouraging the growth of N-fixing plants. N will be conserved by careful use of treatments such as site preparation and prescribed fire, and if harvest rotations are 30 yr or more. More information on nutrient self-sufficiency and availability is needed for Ca and other nutrients.

Nutrient quantity and availability may be viewed from the perspective of both ecological function and production of wood and other plant products. Most forest soils contain sufficient nutrients for trees and other plants to complete their life cycles: to grow, produce seed, and reproduce new cohorts. Many managed and unmanaged western forests have the capacity to produce significant amounts of biomass and support a rich flora and fauna without addition of nutrients. However, fertilizers such as N often temporarily stimulate an increase in biomass production. In a particular stand, nutrients may not be limiting in the ecological sense, if there are sufficient nutrients for a plant to grow and reproduce, but they could be limiting in an economic sense if fertilization would increase commercial yield (see chapter 9).

SUPPORT

A major soil function is the support of forest trees. Trees in Pacific Coast forests reach heights of well over 200 ft and support large crowns that extend to over 50% of their height. These trees require root systems to anchor the tree mass against the forces of gravity, wind, snow, and ice. Trees must be able to develop large, extensive roots both laterally and vertically that can withstand both the tension and the compression exerted by these forces. Gratkowski (1956) found increased windthrow on wet sites and shallow, rocky soil, probably because of restricted root development.

Root systems are a consequence of the genetic potential of the species and soil properties. Some trees produce taproots below the main stem and "sinker" roots extending from horizontal roots down into the soil. Others produce systems of large, spreading shallow roots for support. Apparently all species have many small feeder roots in the upper soil (A horizon), where water and nutrients are most available. Roots that extend through the mineral soil (the B and C horizons) may develop fans of fine roots in localized pockets of nutrients and available water. Discontinuities in soil texture and structure, such as layers of gravel or decomposed root channels, are sites where water and nutrients may accumulate, and concentrations of roots will often develop in these areas. Root development is restricted where accumulation of water leads to anaerobic conditions for much of the growing season. Roots of trees and shrubs pass through the mineral soil and may extend into fractured bedrock several meters below the soil surface. Zwieniecki and Newton (1994, 1996a, b) found roots of whiteleaf manzanita, Pacific madrone, ponderosa pine, and Douglas-fir in fractured bedrock from 0.5 to 3.0 m below the soil surface. Ponderosa pine, manzanita, and Pacific madrone were more efficient at extracting water from the rock than Douglas-fir.

HABITAT

Forest soils support a rich flora and fauna that enable forest growth and development but sometimes cause mortality of trees and other plants. Mycorrhizal fungi provide an extension of root systems that facilitates uptake of water and nutrients by forest plants and may enable them to resist pathogenic fungi. Use of mycorrhizal fungi may become a part of reforestation methods and other silvicultural practices on some sites (Amaranthus and Perry 1987). These fungi may facilitate the growth and survival of some species (Simard et al. 1997); for example mycorrhizae associated with shrubs may aid the survival of young, natural conifer seedlings (Horton et al. 1999). However, the seedlings usually grow very slowly in competition with the shrubs, and it is uncertain if they survive this competition. Considerably more work is needed to establish the importance and use of mycorrhizal fungi in silviculture. Actinomycetes form nitrogen-fixing associations with roots of *Alnus* and *Ceanothus* species and other forest plants. Many invertebrates and fungi are present, especially in the litter layers on the forest floor. Rodents, gophers, mice, shrews, moles, mountain beaver, and other macrofauna are common inhabitants of forest soils, and their life habits—burrowing and the transport and digestion of organic matter—affect soil structure, aeration, water movement, and dispersal of fungal spores. Many of the activities of these organisms are important for maintaining soil structure, aeration, and permeability.

SILVICULTURE AND FOREST SOILS

Silvicultural practices themselves affect forest soils in a variety of ways (Grigal 2000). For example, fire can volatilize nutrients, especially nitrogen and sulfur, and can mineralize cations and make them available to forest plants. Fire and mechanical scarification can stimulate the germination of seed stored in the soil and enable establishment of nitrogen-fixing plants. Disturbance of the organic soil layers prepares a seedbed that favors the establishment of trees, shrubs, and herbs. These disturbances are neither beneficial nor deleterious to forest soils in themselves, but they have the capacity to have major effects on soils and need to be used thoughtfully.

It is important to distinguish between soil disturbance and soil degradation (Senyk and Smith 1989). Soil disturbance includes mixing and redistribution of upper horizons; soil exposure; soil compaction, probably some soil movement or erosion; and nutrient loss. These disturbances may or may not amount to soil degradation, which Senyk and Smith (1989) define as a loss of productivity. Even if soil disturbance is not severe enough to impair soil productivity, however, it can have important effects on other forest resources. For example, low amounts of soil surface movement the first year or two after a fire or site preparation may not reduce site productivity, but may cause unacceptable sedimentation in a stream. The key issue is whether the silvicultural disturbance is sufficiently severe or widespread to cause loss of productivity or adversely affect other resources (Grigal 2000).

It is also important to separate effects of management activities—such as logging methods and road building and maintenance—from silvicultural practices. Roads can be major sources of sediment in streams. They are important infrastructure elements that support silvicultural activities and other forest management objectives, and across large landscapes erosion and sediment yields can often best be controlled by proper road location, construction, maintenance, and traffic use, rather than modification of silvicultural practices. Similarly, choice of logging methods and the timing of logging may have much more important effects on soils in a thinning operation than the thinning itself. Cable systems may cause little soil disturbance in a moderately intensive thinning, whereas the use of ground-based equipment in carrying out the same operation might damage trees and cause soil degradation.

Neither harvesting nor prescribed fire necessarily deplete forest soil nutrient capital. In a summary of worldwide studies on the effects of these practices, Johnson and Curtis (2001) found that on average they had little or no effect on soil N or organic C. Harvesting only sawlogs (the main stems of the trees) in conifer forests actually increased (+18%) soil N and organic C, whereas whole-tree harvesting caused a decrease (-6%) in N. In hardwood forests, organic C and N were apparently not affected by harvesting. Fire generally reduced organic C, but not N, in the A horizon for the first 10 yr after burning. Giardina and Rhoades (2001) reported similar results for lodgepole pine in Wyoming. It is important to know the timing or frequency of treatments when evaluating the effects of fire and harvesting on soils. Powers and others (2005) found that experimental removal of the forest floor reduced C in the upper soil horizons, but not total soil C storage, and had no effect on tree growth for 10 yr. In most silvicultural practices, the forest floor is disturbed, or partially burned, but not removed. Bormann and Gordon (1989) calculated that even with whole-tree harvesting, forests can be self-sufficient for N if rotations are longer than 30 yr. N added from the atmosphere and released from organic matter can replace N removed in tree boles. Also, soil N can increase over pre-harvest levels if N-fixing plants are part of the system.

The effects of harvesting and prescribed fire on soil C are similar to the effects on N. Harvesting sawlogs increased soil C (+18%) when the tops and branches were left on the site, whereas whole-tree harvesting caused a decrease (Johnson and Curtis 2001). These authors report that prescribed fire in Arizona reduced C for 10 yr, after which it returned to pre-burn levels. More information is needed to determine if these generalizations about C and N apply to other nutrients and how they might vary by soil type.

Forest soils sequester considerable amounts of C from the atmosphere. Potential total C storage (net primary production) averaged 1127 Mg/ha in old (150 yr+) coastal to 829 Mg/ha in western Cascade forests, and 195 Mg/ha in ponderosa pine forests in the eastern Cascades. From 15 to 31% of the total C was stored in the soil, to a depth of 1.0m, and 41 to 53% in the trees (Smithwick et al. 2002).

It is difficult to evaluate the effects of soil disturbances on tree growth. In a study of the growth of ponderosa pine on sites that included soils compacted during logging, Helms and others (1986) found that when soils were heavily compacted (that is, at high bulk density), ponderosa pine height growth was reduced by 43% at age 1 yr, and 13% at 15 yr. At 16 yr, less than 10% of the total height was explained by soil bulk density. The depth of the A horizon, shrub density, and the amount of organic matter also affected tree growth. On areas of heavy compaction, such as log landings, soil bulk densities were 1.2–1.3 times higher than those in uncompacted areas. Tree volumes at 16 yr were 70–85% of those on uncompacted areas (Helms and Hipkin 1986). A better understanding of the effects of soil compaction on forest productivity is emerging (Miller et al. 2004). For example, research in northern California has shown that the effects of soil compaction on growth of conifer seedlings can be negative, neutral, or even positive, depending on soil texture and soil water regime (Gomez et al. 2002). Reductions in Douglas-fir growth from soil compaction following ground-based harvesting have been shown to last as little as 2 yr in coastal Washington and up to 7 yr in the Oregon Cascade Mountains (Heninger et al. 2002). Powers and others (2005) report that soils with high water-holding capacity had the greatest reduction in conifer seedling growth following compaction, while on sandy-textured soils growth was generally increased by compaction. Apparently increasing soil bulk density by compaction increased the water-holding capacity of the soil (Ares et al. 2005).

On steep slopes, terracing has been used in site preparation to reduce competition from grasses and sedges and aid survival and growth of ponderosa pine in Montana. Zlatnik et al. (1999) found that terracing involved intensive soil disturbance, especially movement of the soil; however, for 20 yr the ponderosa pine and associated shrubs were larger and produced more volume and understory biomass on the terraces than on the

Table 3-1. Major soil features potentially affected by silviculture practices.

Soil Features	*Importance*
Organic matter, forest floor, and soil A horizons	Improve and maintain soil structure; important for nutrient storage, water storage, water infiltration, habitat for microflora and –fauna.
Bulk density of mineral soil	Compacted soil may inhibit soil gas exchange and root penetration, reduce water infiltration, and increase surface flow.
Large wood and decomposing logs and stumps	Future source of organic matter, habitat for various species and microorganisms, site for establishment of certain plants.

adjoining non-terraced sites. Possibly the terraces reduced erosion and kept water on site, helping to improve tree and shrub growth, and soil C and P, K, pH, particle size, and water holding capacity. This intensive site preparation method is not commonly used, since cheaper, less intensive methods are available; it does however demonstrate the resiliency of forest soils. We question whether the repeated use of terracing would maintain productive soils, especially those that are quite prone to erosion.

It is also important to distinguish between the effects of geological events and silvicultural practices. Sedimentation in streams can result from erosion of stream banks, and from floodplains during high water flow. Deep-seated mass movement (debris flows) and landslides can occur in both managed and unmanaged forest stands. Mass movement is important in some streams because it delivers gravel and logs to the stream that eventually form pools and gravel beds that are important for fish habitat. But it can also deliver sediment to the stream and degrade aquatic habitats. Mass movement and erosion are normal processes. In fact, in the long term they may contribute to aquatic habitat (Reeves et al. 1995). The issue with respect to forest management is not to prevent these processes but to keep their frequency and severity within acceptable limits.

Some of the major soil properties that can be affected by silvicultural practices and their importance are shown in table 3-1. Research has shown that these variables are all important parts of soil processes and productivity, but we are just learning to predict their effects on forest growth. There is little information that indicates when, for instance, loss of organic matter, soil compaction, and reduced porosity decrease soil productivity. Silvicultural practices cannot avoid disturbing these soil features (Grigal 2000). As Bengston (1981) argues, replacement of nutrients by fertilization, in addition to being costly, is likely to be only about 50% effective, because nutrients from fertilizers are not completely used. Also C is sequestered from the atmosphere in forest soils. Therefore it is wise to protect the soil during silvicultural and other forest management operations.

Insects, Fungi, and Mistletoes

Insects, fungi, and mistletoes are diverse groups of organisms common in all forest stands. They are important components of the forest ecosystem because they break down dead plant parts and facilitate nutrient cycling. Some species of fungi form mycorrhizal associations with forest plants. Their fruiting bodies provide food to many species of animals; some species of fungi are harvested commercially. But fungi are also pathogenic to trees of all sizes and species (table 3-2) and to other forest plants as well. Insects aid in the breakdown of organic matter and recycling of nutrients, pollinate flowers, and provide food for many species of birds and mammals. However, at high levels, they can seriously affect forest management goals. In this section we briefly describe some of the more common insects, pathogenic fungi, and mistletoes and provide references to the literature (tables 3-2 and 3-3).

The effects of insect pests and pathogens are best evaluated from the perspective of forest management objectives. At low levels they may be viewed as positive or having little effect; resulting dead or weakened trees provide snags or logs on the forest floor that add organic matter to the soil, serve as habitat for wildlife and other fungi and insects, and provide a substrate for regeneration of some species of vascular plants. Depending on cost and accessibility, many killed or weakened trees are salvaged in wood production forests. However, epidemic levels of pathogenic organisms are generally unacceptable and considered to be a sign of poor forest health. Spruce budworm, mountain pine beetle, and Swiss needle cast epidemics have caused widespread mortality and growth reduction in many stands, and pine beetle mortality has dramatically increased the potential for severe fire.

Pathogens and insects are interactive in causing tree mortality, especially in dense stands and during periods of drought. Insect activity in ponderosa pine stands may be initiated on trees weakened by root disease and therefore less able to compete for water in dense stands or during drought. In this case, insects may be indicators of other factors affecting tree vigor. At the epidemic level, insects may kill trees outright.

It is not possible to predict when or where outbreaks of insects and pathogens will occur. Most of them are species specific, however. Growing mixed-species stands and avoiding dense stands on dry sites are important ways to provide some resistance to pathogens and insects and to preserve options for forest stands when outbreaks occur.

MISTLETOES

Mistletoes are vascular plants that grow in the crowns of most conifers. They reduce foliage density (thereby reducing vigor and stem growth) and weaken stems, making their hosts susceptible to breakage from wind and heavy snow or ice. It is difficult to recognize the incipient stages of mistletoe in the crowns of large trees. However, as mistletoe plants grow large, “brooms” or clumps of conifer foliage are formed in tree crowns. Trees weakened by brooms grow slowly and are susceptible to bark beetles. Mistletoe can spread from an overstory of infected trees to saplings in the understory. Thus mistletoe infestations can have major impacts on multiple-story stands of the same species. However, some level of mistletoe may be desirable because brooms on large trees may provide nest platforms for canopy-dwelling wildlife (Parks et al. 1999).

ROOT DISEASES

Root diseases are often difficult to recognize until trees die. *Heterobasidium annosum* and *Phellinus* often spread concentrically root-to-root through a stand, forming gaps in the canopy. Shrubs (vine maple), hardwoods (bigleaf maple) and sometimes conifers (western hemlock) that are resistant or immune to the disease often invade these gaps (Thies and Sturrock 1995). Several dead trees surrounded by trees with sparse foliage often indicate a root disease center. Root diseases can persist in roots for decades and can infect older trees, poles, and saplings. They are generally species-specific; therefore, thinning and reforestation to favor resistant or immune species are ways to reduce the effects of root pathogens. Trees that die from root diseases provide snags and logs for habitat. Douglas-fir killed by *Phellinus* may not provide usable snags for long because the roots are severely decayed. Gaps that form in the forest canopy following tree death provide growing space for conifers and hardwoods immune to the root diseases, increasing species diversity and complexity of stand structure.

RUSTS AND FOLIAGE DISEASES

Rusts and foliage diseases have the same potential for widespread impact in western forests as chestnut blight and Dutch elm disease in eastern forests. Blister rust, introduced from Asia in the 1910 (Hummer 2000), has had a major impact on sugar pine, western white pine, eastern white pine, and whitebark pine throughout the range of these species. It appears that large trees are able to survive infection for decades, but smaller seedlings, saplings, and poles are readily killed as cankers grow from branches into the bole and girdle it. Blister rust resistance is under genetic control, and rust-resistant planting stock is being developed. An intriguing possibility for encouraging resistance in the native population is to select large trees that are resistant or have low levels of infection and encourage natural regeneration of their progeny. This would require using small-scale seed tree or shelterwood regeneration methods, including site preparation and shrub control around resistant trees (Hoff et al. 1976).

Swiss needle cast of Douglas-fir is caused by an endemic fungus, *Phaeocryptopus gauemannii*, that until recently had never caused much concern. However, Swiss needle cast has caused significant growth losses and declining health in Douglas-fir plantations throughout north coastal Oregon, as well as less severe damage in Washington and south coastal Oregon (Hansen et al. 2000; Filip et. al 2000a). The disease causes

Table 3-2. Examples of major diseases of western conifers. Also see Scharpf (1993).

Species Affected	*Comments*	*References*
Dwarf Mistletoes (Arceuthobium spp.)		
Mistletoes		
All conifers	Locally important in the Sierras, Cascades, and east; easily spread from overstory to understory; weaken tree crowns; may form nesting sites.	Hawskworth and Wiens (1996); Parks et al. (1999); Parmeter (1978); Roth (2001); Scharpf and Parmeter (1976)
Root Diseases		
Heterobasisium annosum		
All conifers, Pacific madrone	Locally important, especially in true fir and ponderosa pine; may predispose stand to bark beetles; shortens rotation lengths; can be spread by thinning.	Bega (1978); Filip et al. (1992); Otrosina et al. (1989); Sullivan et al. (2001)
Armillaria (shoestring)		
Most conifers, hardwoods	Spreads from dead/dying hardwoods to conifers via root contact.	Bega (1978); Filip (1986); Filip et al. (1989a, 1999); Roth et al. (2000)
Leptographium (Verticicladiella) (black stain)		
Douglas-fir, ponderosa pine	Locally important, enters young trees through stem wounds; control timing of thinning.	
Phytophthora sp.		
Port-Orford cedar, yew	Important within Port-Orford cedar stands; avoid transporting spores from infected stands; spores transported by subsurface water flow.	Zobel et al. (1985)
Phellinus weirii (laminated root rot)		
Douglas-fir, Western hemlock	Important in coastal and western Cascade Douglas-fir forests; common in 30+yr stands; makes openings in canopy 0.5–1.0+ ac.	Thies and Nelson (1997); Thies and Sturrock (1995)

Table 3-2 cont.

Species Affected	*Comments*	*References*
Rusts		
Cronartium sp.(blister rust)		
Sugar pine, white pine	Extremely important throughout western mixed conifer forests; kills seedlings, saplings, and small poles; encourage natural regeneration from seed trees with low incidence of infection; disease resistance genetically controlled.	Kinloch et al. Western (1992)
Foliage Diseases		
Elytroderma		
Ponderosa pine, Lodgepole pine, Jeffrey pine	Attacks mainly poor-vigor trees around the edges of lakes and meadows; spreads from needles into twigs.	Scharpf and Bega (1981)
Phaeocryptopus gaeumannii	Swiss needle cast	
Douglas-fir	Currently important in young (15–40-yr-old) stands in western Oregon; impact on stand growth reduced in mixed-species stands.	Filip et al. 2000a
Stem Decay Fungi		
Eichinodontium, etc.		
All species of conifers/ hardwoods	Enters trees through wounds from branch stubs, sunscald, fire, frost, logging, insects, animals, broken tops, etc.; predisposes trees to wind fall and stem break; reduces wood quality and merchantable volume	

premature loss of older cohorts of needles, seriously reducing total foliage mass, tree health, and tree growth. The disease appears to fluctuate in severity with annual variation in climatic variables; mild winter temperatures and late spring precipitation tends to favor fungal development and spore germination.

STEM DISEASES

Stem decay fungi are generally considered of minor importance because they do not cause significant mortality and their effects are not easily seen. These fungi enter the stem through dead branches or through wounds on the stem from fire, sunscald, frost cracking, windthrow, and logging. However, they can be quite important economically, especially after fire and thinning in true fir and western hemlock stands. Significant amounts of decay can occur in the bole within 10–20 yr of wounding. Stem decays are important in the decomposition of snags and logs, enabling them to become habitat for a large number of wildlife species. For example, wood-decaying fungi form hollow logs and trees that provide den or nesting sites.

BARK BEETLES

Bark beetles are most common in ponderosa and lodgepole pine, true fir, and mixed-conifer forests (table 3-3). They are always present in scattered, weakened trees, recent windthrow, etc. They are often found in trees infected with root pathogens such as *H. annosum* or *Armillaria*. Occurrences such as these are sometimes thought to indicate poor forest health, but they do not. Populations may reach epidemic dimensions during periods of drought, when many trees are killed on shallow soils, on south slopes, and especially in dense stands. Populations can build up after extensive fires that leave many acres of trees freshly killed or severely weakened by crown and stem damage. Large concentrations of fresh slash, which provide good habitat for bark beetles, can result from logging or precommercial thinning; however, cutting and scattering slash and timing thinning to promote drying of the wood can reduce bark beetle use. Most severe outbreaks of bark beetles occur in eastside lodgepole pine and ponderosa pine forests and on dry sites in mixed-conifer forests of southwestern Oregon and California.

DEFOLIATING INSECTS

Defoliating insects have recently been a major concern in mixed-conifer stands in eastern Oregon. Tussock moth and spruce budworm have attacked Douglas-fir, white fir, grand fir, and Engelmann spruce in pure stands and where these species were mixed with ponderosa and lodgepole pine (table 3-3). Fire control over the past 50+ yr had enabled these species to establish dense multiple layers beneath the pine, structures that are ideal insect habitat because larvae can disperse downward through a continuous canopy of susceptible species. In addition, recovery of trees and their ability to produce

Table 3-3. Examples of major tree-damaging insects of western forests.

Species Affected	*Comments*	*References*
Bark Beetles		
Dendroctonous sp. bark beetles		
Ponderosa pine, sugar pine Lodgepole pine, Jeffrey pine, Douglas-fir	Generally endemic, can become epidemic during several years of drought; populations can increase in trees weakened or killed by fire, windthrow, or root diseases, tops broken from snow, and in logging slash; stands with low densities of vigorous trees are resistant.	Furniss and Carolin (1977); Miller and Keen (1960); Sartwell and Stevens (1975)
Scolytus sp. (fir engraver), *Ips* sp. (pine engraver)		
White and red fir, Ponderosa pine, sugar pine, Lodgepole pine, Jeffrey pine	Trees weakened from high stand densities and root diseases are susceptible. Stands with low densities of vigorous trees are resistant.	Ferrell and Hall (1975); Ferrell et al. (1994); Wickman and Eaton (1962)
Defoliators		
Orgyia sp. Tussock moth, *Choristoneura occidentalis* western spruce bud worm		
White fir, interior Douglas-fir, Grand fir, Engelmann spruce	Outbreaks in eastside forests occur in dense stands or understories of susceptible species; no direct silvicultural control; maintaining ponderosa and lodgepole pine in stands helps reduce overall effects.	Brooks et al. (1987); USDA Forest Service (1978); Wickman (1988)
Lymantria dispar Gypsy moth		
Douglas-fir	Potentially a very destructive insect; has occurred in young Douglas-fir forests; common in eastern forests; has migrated west.	
Wood Borers		
Melanophila sp. Flathead borers		
Ponderosa pine, sugar pine, Jeffrey pine, Douglas-fir	Generally considered to be secondary insects that attack weakened or dead trees.	Ferrel and Hall (1975)
Tetropium sp. Round head borers		
True firs	Generally considered to be secondary insects that attack weakened or dead trees.	
Eucosma sp. Tip moth		
Ponderosa pine, Jeffrey pine	Kills or shortens growth of terminal leaders; causes forking of seedlings, saplings, and poles; restricted to eastside forests.	Stoszek (1973)
Cone and Seed Insects		
Conopthorus, Dioryctria, Barbara, Megastigmus, Contarinia sp.		
All major tree species	Can destroy large proportions of the cone/seed crop, especially in years of small crops; generally not important in years of moderate to high seed production.	Keen (1958)

new foliage was affected by several years of drought. Consequently, dead true fir and Douglas-fir are mixed among live lodgepole and ponderosa pine over many thousands of acres of eastside forests, creating the potential for severe fires. Reducing the density of true fir and Douglas-fir in mixed-conifer stands and growing vigorous ponderosa and lodgepole pine at low to moderate densities might reduce the impact of defoliators.

CONE AND SEED INSECTS, WOOD BORERS

Cone and seed insects can have major effects on seed quality and abundance, especially in years of low seed production, and thereby impact natural regeneration, especially of ponderosa pine and Douglas-fir. Woodborers can be locally important, generally reducing the growth of seedlings and saplings and killing larger trees weakened by other causes.

Fire

Fire was an important disturbance in presettlement forests of western North America (Agee 1993; Arno 1976; see chapter 11). It occurred in all major forests, although the frequency varied from very short intervals in ponderosa pine forests at lower elevations and on drier sites (< 10 yr) to well over 100 yr at moist coastal and high elevations (table 3-4). The effects of fire were related to its frequency and intensity. In forests where fire occurred frequently and reduced fuels to low levels, fire severity was low. Although fires damaged and killed small trees and understory vegetation, they probably had less effect on large trees than today's fires do. Large ponderosa and sugar pine and Douglas-fir have thick bark that insulates their stems from surface fires with low flame heights, which move rapidly and release little heat. However, the relatively recent increase in fuels in these forests increases the frequency of intense fires that kill the cambium and buds in the crown.

Today's fire regimes and the potential effects of fire are different from those in pre-European-settlement forests, especially in forests at lower elevations and on dry sites where fires once burned frequently. The policy of fire control implemented in nearly all forests throughout the West (Donovan and Brown 2007 has enabled shade-tolerant species to reproduce and changed the typical stand structure from a relatively wide-spaced, single-layered forest to a dense, multiple-layered forest. These multiple layers create fuel ladders that extend from the ground to the crowns of the larger trees. High stand density predisposes these forests to insect attacks, producing more, dry, dead fuel from insect-killed trees. Logging slash has also added to the fuel and flammability on many sites. Finally, the potential for fire ignition is much higher because humans are much more widespread through the forest than in earlier eras (Duane 1996). The wildland–urban interface (WUI) is a rapidly growing phenomenon, as cities and recreation areas continue to expand into forests. Thus, the potential for fire and its consequences are rapidly changing and diverging from those of pre-European-settlement forests (Stephens et al. 2014). Silvicultural practices, including prescribed under-burning and thinning, clearly have the potential to reduce future fire severity.

Table 3-4. Natural fire intervals in common forest types.

Forest Type	*Interval (yr)*	*Region*	*Reference*
Spruce Hemlock Redwood	230 95–145 100 (*), 130–150 (**)	Coastal/moist	Fahnestock and Agee (1983); Morrison and Swanson (1990); Teensma (1987)
Silver fir	235–545 108–192	Mt. Rainier Central Cascades	Agee (1990); Hemstrom and Franklin (1982); Morrison and Swanson (1990)
Red fir	40–42 65 42–39	NE California Southern Sierra Nevada Southern Cascades	Chappell (1991); McNeil and Zobel (1980); Pitcher (1987); Taylor and Halpern (1991)
Douglas-fir/ hardwoods	18–20	Siskiyou mountains	Agee (1991); Sensenig (2002); Sensenig et al. (2013)
Douglas-fir forests (dry)	7–11 10 10–24	Wenatchee, WA Blue Mountains, OR Okanagon, WA	Hall (1976)
White fir	9–25 9–42 43–61	Southern Cascades Southern Cascades Siskiyou mountains	Agee (1991); Bork (1985); McNeil and Zobel (1980); Sensenig et al. (2013)
Grand fir	33–100 25–75	Warm Springs, OR Montana	Agee (1993); Antos and Habeck (1981)
Ponderosa pine	4 2–6 11–16 16–38	Arizona Arizona Warm Springs, OR Pringle Butte, OR	Bork (1985; Dietrich (1980); Savage and Swetnam (1990); Weaver (1959, 1961)
Lodgepole pine	60 39 60	Crater Lake, OR Crater Lake, OR Fremont Forest, OR	Chappell (1991); Stuart et al. (1989)
Ponderosa pine-mixed conifer, mixed conifer	8–10 5–11	Southern Sierra Nevadas	Caprio and Swetnam (1996); Kilgore and Taylor (1979)
Giant sequoia–mixed conifer	14–32	Southern Sierra Nevada	Caprio and Swetnam (1996)
Ponderosa pine at 3000 ft to true fir at 8000+ ft.	6–11 yr (ponderosa pine) to 30–40 (true fir)	Northern Rocky Mountains	Arno (1976)

Vertebrates

Every stage of stand development and every type of stand structure provides habitat for a large variety of animals. Forest management objectives frequently call for providing habitat for selected species or for maintaining a broad range of habitats for many species. Silvicultural practices will often be the major means of meeting these objectives. Readers are referred to Black (1992), Brown and Mandery (1962), and to local expertise for species and site-specific information.

Animals are likely to have the greatest effect on the development and management of stands of young trees, especially at the time of regeneration and seedling establishment (table 3-5). Squirrels consume large quantities of seed; they also cut and cache cones and significantly reduce seed supply. Chipmunks, mice, and voles consume seed as it overwinters (Lawrence and Rediske 1962; Tevis 1956). On some sites, rodents and birds cache seed and may aid in the dispersal of seed and success of seedling establishment (Vander Wall 1992, 1995).

Rodent populations generally expand in response to the abundant covers of grasses, herbs, and shrubs that develop after fire or logging as planting or natural regeneration occurs. Rodents eat phloem, foliage, and roots of conifer seedlings and other plants. At high elevations, where shrubs and herb cover are sparse during the winter, rodents, especially gophers, may burrow through the snow to forage on conifer stems and foliage. Browsing of new plantations exerts a major impact on seedling growth and survival. Protecting seedlings from browsing can be costly (Brodie et al. 1979).

Deer and elk may also have considerable impact on natural and planted seedlings in plantations and in the understory. One important reason to release conifer seedlings from competing vegetation is to allow them to grow rapidly to a size at which they will not be affected by browsing. Browsing heavily impacts the growth of red alder and bigleaf maple seedlings (Fried et al. 1988), as well as those of Pacific madrone and California black oak. Repeated browsing on the tops of these trees in the understory of conifer stands often keeps them from growing into saplings and poles, greatly affecting stand structure and species composition. Deer and elk also browse on planted and natural conifer regeneration, especially Douglas-fir, often ponderosa pine, less frequently on true firs and western hemlock.

Bears, porcupines, beavers, and squirrels can also affect growth, density, and species composition of a stand. Bears strip the bark from the lower boles (4–6 ft) to eat the phloem tissue of young trees 5–10 in dbh. Significant bear damage has been reported in young redwood, in Douglas-fir plantations in western Oregon and Washington, and in young larch stands in the northern Rocky Mountains (Black 1992). Porcupines feed on the phloem in the upper crowns of trees 8–14 in dbh, killing treetops and also girdling stems of smaller trees near the ground. They browse on nearly all conifers but are most common in younger ponderosa pine forests, especially near water, on edges of

Table 3-5. Vertebrates that affect silviculture and management of stands in western forests and selected references. (See Black [1992] for further references to the literature and discussion on effects of animals on silvicultural practices.)

1. Seed Eaters	*Animals affecting young stands and regeneration*	*References*
Shrews (*Vorex* sp.); mice (*Peromyscus* sp.); chipmunks (*Tarnius* sp.); ground squirrels (*Spermophilus* sp.); various birds, such as sparrows (*Zonotrichea* sp.), crossbills (*Loxia* sp.), dark-eyed junco (*Junco* sp.)	Eat seed on trees and on the ground; may also cache and disperse seeds	Shearer and Schmidt (1971); Sullivan, 1979
2. Seedlings, saplings, small shrubs	*Foraging in young stands*	*References*
Voles (*Phenacomys* sp., *Clethrinomys* sp., *Microtus* sp.)	Girdle roots and stems; live below ground; may feed under the cover of a thatch of leaves, twigs, and snow; affect most conifers, hardwoods, and shrubs.	Hooven and Black (1976)
Pocket gophers (*Thomomys* sp.)	Feed on roots; clip stems and branches and girdle stems; feed below ground or under a cover of snow; feed on seedlings and small saplings; most effects at high elevations and in eastside pine forests.	Crouch (1971, 1979); Howard and Childs (1959)
Mountain beaver (*Castor canadensis*)	Eat foliage from shrubs, hardwoods, conifer seedlings and saplings; found in draws and wet areas; most important in young Douglas-fir stands, especially near streams.	Martin (1971); Neal and Borrecco (1981)
Deer (*Odocoileus* sp.), Elk (*Cervus* sp.)	Browse shrubs, forbs, leaders, and branches of seedlings and saplings of hardwoods and conifers; species preference and time of browsing varies among forest types; browsing increases on fertilized seedlings and with access and proximity to cover (edges).	Brown and Mandery (1962); Gourley et al. (1990); Hanley (1983); Witmer (1982)
3. Larger trees (saplings and poles)	*Girdling and stripping bark from trees (>8 inches dbh)*	*References*
Porcupines (*Erethizon* sp.)	Girdle stems to feed on phloem, usually at the top of poles and saplings; common in pine forests near meadows, springs, and rock outcrops; prefer trees in low-density stands; will also feed on stems of saplings.	Eglitis and Hennon (1997); Smith (1982); Sullivan et al. (1986)
Black bears (*Ursus* sp.)	Feed on inner bark on phloem of young, rapidly growing trees; most common in thinned Douglas-fir stands 15–20 yr old and larch; may prefer trees in fertilized stands; prefer dominant/codominant trees.	Lindzey and Meslow (1977); Mason and Adams (1989); Nelson (1989); Piekielek (1975)
Tree squirrels (*Tamiasciurus* sp.)	Strip thin bark and feed on branches of nearly all western conifers; often active in lodgepole pine; thinning stands may reduce population densities; also cut and cache cones.	Sullivan and Sullivan (1982); Sullivan and Vyse (1987)

meadows and grassy areas. Tree squirrels also feed on the bark of stems and branches of conifer trees, especially lodgepole pine, as well as eating cones and seeds.

Examples of Evaluating Forest Stands As Ecosystems

Following are three examples of how evaluating stands using the environmental variables and the discussion of them presented above provides a basis for developing a silvicultural prescription. In addition to evaluating these variables at the stand level, it is often important to look beyond the stand for variables such as seed availability, potential fire, wind, etc. These examples assume that the landowner's objectives and the relevant forest resource values are known. The task is to develop a silvicultural prescription that best meets the objectives.

Case 1: A 10–15-yr-old mixed-conifer plantation on public land

A 10–15-yr-old mixed-conifer plantation that is quite dense (with both conifer saplings and hardwood sprout clumps) on public land.

History: The stand was planted to ponderosa pine; white fir and Douglas-fir regenerated naturally, and hardwoods sprouted.

Management objectives:
- Develop a multi-story, mixed-species stand for old-forest characteristics.
- Produce commercial wood.
- Retain a hardwood component in the stand, but prevent widespread hardwood overtopping of ponderosa pine and Douglas-fir.

Evaluation of the stand and similar nearby stands suggested that the following variables were likely to be important:
- Hardwoods overtopping conifers.
- Root disease.
- Insects, including bark beetles, in the slash if the stand is thinned and during periods of drought as density increases.
- Heavy, wet snow.
- Periodic drought and moisture stress on the southerly slope.
- Fire from heavy fuel loads in surrounding stands, and from slash if the stand is thinned.

Response to these variables might be:
- Thin as soon as is practical to keep trees from becoming susceptible to breakage from snow.
- Thin during late spring or early summer so that slash will dry out and not become habitat for insects.

- Thin to prevent stand from becoming dense and susceptible to chronic insect attacks before it is 30–50 yr old, when commercial thinning can be done to control density.
- Evaluate the cost and effectiveness of reducing thinning slash and other fuels; possibly plan for underburning in 10–15 yr.
- Favor a mixture of ponderosa pine and Douglas-fir that can tolerate underburning. Douglas-fir is less susceptible to bark beetles than ponderosa pine.

Case 2: A dense, productive 40–50-yr-old Douglas-fir/hemlock stand

Current relative density is high: 0.55 to 0.60 (of a maximum of 1.0; see chapter 6). Basal area is nearly 200 ft^2/ac, increasing at a rate of 3–4 ft^2/ac/yr; live crown ratios of the dominant and codominant trees are less than 40% and are likely to drop to 30%. The objectives are to produce yields of timber, and at the same time grow the stand on a rotation of 100 yr or more, and produce large (30 in +) trees for green tree retention/wildlife habitat.

Important environmental variables are likely to be:
- Root disease (*Phellinus weirii*).
- Swiss needle cast infection.
- Wind.

Possible responses to these variables are:
- Because of the high density and increasing basal area, thin the stand soon to develop wind-resistant trees with large crowns.
- To minimize wind effects, thin lightly or not at all on exposed windward edges of the stand. Favor largest trees most likely to be windfirm. Thin lightly initially to develop wind firmness, and then plan for additional thinning in 10–15 yr.
- In response to potential disease, leave 30% or more of the basal area in western hemlock and western redcedar, where they occur.
- Hemlock and cedar are immune to Swiss needle cast; hemlock is resistant to Phellinus, and cedar is immune. Return to the stand in about 10 yr to evaluate the potential effects of the diseases.
- Remove some infected Douglas-fir with declining crowns to capture future mortality losses, leave some for logs on the forest floor and snags.

Case 3: Regeneration of a true fir stand

The objective is to harvest timber and to quickly and cost-effectively regenerate a true fir stand. Important environmental variables are likely to be:
- Gophers.
- Cold air and frost pockets.
- Presence of grasses shrubs and lodgepole pine seed sources.

Responses to these variables might be:

- Leave advance regeneration (1–4 m tall), which is likely to be less susceptible to damage from gophers and frost than newly planted seedlings.
- Plant large seedlings as soon as possible before gopher populations expand in the invading shrub and grass cover.
- Plant lodgepole or Jeffrey pine, which are cold tolerant and will grow rapidly, to protect smaller, slower-growing, natural true fir seedlings from frost.
- Leave large true fir seed trees to provide seeds that will germinate to augment regeneration in areas of gopher and frost-related mortality; these trees may also provide shelter to reduce frost potential.
- Evaluate the possibility of using the shrub cover to protect seedlings from frost, realizing that it will likely delay growth of established seedlings by several decades.

Disturbance, Succession, and Stand Development

EFFECTS OF DISTURBANCE

Forests are long-lived, and it is likely that both natural and managed forests will experience unexpected disturbances. Oliver and Larson (1996) provide a thorough summary of disturbances that affect forest development and succession, including fire, wind, floods, landslides, erosion, avalanches, glaciers, volcanoes, ice storms, and biotic agents (herbivory by mammals and insects, pathogens). Time and severity of disturbance are only roughly predictable. For example, windstorms and windthrow are quite common in the eastern and southeastern lower 48 states and in southeastern Alaska (Deal et al. 1991), but they occur infrequently in the west, an exception being the well-known severe windstorm that occurred throughout the coastal and western Cascade forests of the Pacific Northwest and in the Sierra Nevada on Columbus Day, 1962.

Large fires occur nearly every year, and lately have occurred with increasing frequency in western forests. The effects of these disturbances are often interactive. For example, windthrow may make a forest susceptible to insects that breed in trees killed or weakened by the wind, and then later to fire as dead trees dry out and become flammable.

In their effects on stand dynamics and plant succession, silvicultural practices are quite similar to the natural disturbances listed above. Thinning, single tree selection, and group selection methods make gaps in the canopy like those made by root diseases, snow, ice storms, or isolated windthrow. Like severe fire, clearcut and shelterwood regeneration methods followed by site preparation kill trees and shrubs and leave a site with bare mineral soil. However, removing trees changes microclimate, habitats, and fire potential from what they would be after a severe fire. Low-severity fires are similar to thinning and to prescribed burning in the understory; in either case, new seedlings or sprouting trees and shrubs can revegetate the site. From an ecosystem perspective, however, natural disturbances differ from those that result from silvicultural treatments. Natural fire and windthrow leave dead wood on the site that can provide habitat for animals, recycle nutrients, and make the stand more susceptible to fire or insects.

From the perspective of plant succession and forest stand dynamics, the effects of disturbance are more important than the types of disturbance. For example, severe fires, avalanches, floods, and landslides all expose bare mineral soil, a substrate that favors the establishment of seedlings of many species. Similarly, tree seedlings, saplings, and shrubs in the understory are likely to grow rapidly in response to the death of trees, whether caused by wind, ice, or insects, because all these disturbances increase the availability of light, water, and nutrients to the understory. The severity of disturbance is also important, of course. Intense fire may consume organic soil layers and kill overstory trees on some sites within a burn; on other sites it may leave the overstory undamaged and a mosaic of dead and living shrubs and saplings in the understory. The response of vegetation to disturbance depends not only on the severity or the result of the disturbance but also on the plants and propagules (seed and vegetative buds) present on the site or those that reach the site (mainly from wind-blown seed) after the disturbance. For example, light to moderate disturbance of the overstory favors growth of plants already on site, including overstory trees, seedlings, and shrubs. Disturbances that kill the tops of shrubs and hardwoods result in the release of buds on the bases of stems, in burls below ground, and in rhizomes and roots. This type of response is quite predictable because the size and density of the growth are related to the size and cover of shrubs or the numbers and sizes of trees before the disturbance (Harrington et al. 1991). Disturbances that expose mineral soil promote regeneration from seed that is either stored or produced on site or comes in on the wind.

PLANT SUCCESSION AND STAND DEVELOPMENT

Understanding the theory of secondary plant succession is important for the practice of silviculture. This theory provides a basis for evaluating how forest vegetation may respond to disturbances and for formulating silvicultural practices and systems to control vegetation development.

Change in vegetation is classified as either primary or secondary succession. Primary succession occurs after events that remove an entire biotic community or leave a new geologic surface, such as island formation from volcanic activity, glacial retreat, and major landslides and avalanches. Soil formation is usually part of this process. Silvicultural practices are normally involved with secondary succession, which follows such major disturbances as severe fire, insect outbreaks, windthrow, and timber harvest by clearcutting. Most of these disturbances affect vegetation, but they generally have little effect on soil productivity.

Theory suggests that several patterns of secondary succession can be generalized to most forest types (Kimmins 2004; Oliver and Larson 1996). One is "relay floristics," in which one community of plants becomes established and modifies the environment so that other species invade and replace the first community. Another pattern of succession is termed "initial floristics." This type of succession is common in many forests. After a disturbance, many species are already on the site as vegetative buds or seed and others invade almost immediately. Because of differences in density, growth rates, potential size, longevity, and time of establishment, some species remain on the site for a relatively short time, whereas others become dominant and may maintain their dominance for long periods. For example, western conifers such as ponderosa pine and Douglas-fir can grow rapidly when they are young and, once in a dominant position, may remain on the site for 300+ yr, or until a severe disturbance occurs. However, if shrubs or hardwoods are present and initially dominate the site, those conifers may be excluded and succession will take an entirely different path.

Competition or competitive exclusion is also part of forest succession. Plants interact by altering the environment of other plants on the same site. This alteration can be beneficial (for example, by providing shade for seedling establishment), neutral, or deleterious for other species. Some plants rapidly produce cover using large amounts of site resources (water, light, nutrients), thus preventing or substantially slowing the growth of other plants. Understanding the effects of competition is important for projecting the future species composition of a forest stand and its rate of growth and development, both of which are major concerns in the practice of silviculture (Walstad and Kuch 1987; see chapter 7).

Plants that occupy a site after a disturbance are often present on the site before the disturbance. This is nearly always the case following less severe disturbances such as mortality of individual trees that causes small gaps in the canopy, but it is also true after severe disturbances. Many species of shrubs (salal, salmonberry, manzanita) and trees (aspen, Pacific madrone) sprout from buds below ground after their tops are killed, ensuring their carryover from one stand to another. Some species persist as buried seed that remains viable long after the plants that produced it have died. Plants that sprout from buds or germinate from seed stored in the forest floor are often major parts of the initial floristics after disturbance, and they are also often part of the process of succession in later stand development.

IS SUCCESSION PROBABILISTIC OR DETERMINISTIC?

For a particular site, succession may have quite different pathways and outcomes. It is the role of silviculturists to predict succession and, if needed, manage it for the desired outcome. In some cases, vegetation that develops after disturbance depends on the chance occurrence of trees and shrubs (figure 3-12). Two cases illustrate these points.

In many forests throughout the West, fire, logging, and site preparation stimulate germination of buried seed of *Arctostaphylos* spp. *Ceanothus* spp., *Prunus* spp., and other shrubs and trees. Seedling density may be quite high (>10,000/ac) and a dense cover of shrubs develops within 2–4 yr (Hughes et al. 1987). The cover may be a combination of sprouts and seedlings, because these species can sprout from the base when their tops are killed. If conifer seed is present and germinates along with the shrubs, then a stand of conifers will develop along with the shrubs. The conifers will eventually overtop the shrubs, which will die if conifer density and canopy cover become great enough to shade them out. But if few conifer seeds are present immediately after the disturbance, a dense community of shrubs will develop that can persist for many years if no further disturbance occurs. If a seed source of a shade-tolerant species such as white fir is present, conifer seedlings may become established beneath the shrubs and very slowly overtop them and shade them out (Conard and Radosevich 1982a). This process, which might take 50–60+ yr, will result in a stand of white fir. In that hypothetical case, the initial floristics determined the early species composition: shrubs or shrubs and trees. Later on, the cover of shrubs facilitated the establishment of a pure stand of fir because it provided an environment that excluded other conifers, probably by a combination of

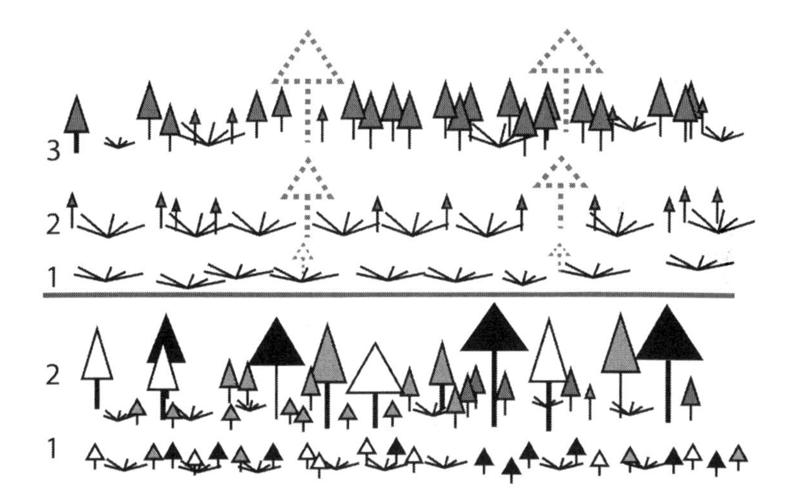

Figure 3-12. Two possible pathways of succession in a mixed conifer forest, depending on the timing of seed fall following disturbance. Theoretical succession in mixed conifer stands following disturbance.

Top: 1. Dense shrub community develops from a seed bank. Few conifer seed available, possibly a few shade-intolerant species (dashed lines) are established along with the shrubs. 2. Shade tolerant conifers are eventually established beneath the shrubs and overtop them. 3. Shade tolerant conifers (grey crowns) suppress shrubs and form a dense, nearly pure even-age stand (Conard and Radosevich 1982a).

Bottom: 1. Shrub seed present, and conifer seed available within a year or two following disturbance. Shade-intolerant (white and black crowns) and shade tolerant (grey crowns) conifers grow with the shrubs and suppress them. 2. A mixed stand forest results. Shade tolerant species become established in the understory.

reducing light intensity, consuming the soil water, and providing cover for seed-eating rodents. The shrub cover also prevented rapid growth of fir seedlings. Thus the species composition and density of the stand at 80–100+ yr results from establishment and growth that occurred within a few years after disturbance (initial floristics) along with the later arrival of seed and the cover provided by the shrubs (relay floristics).

After a recent, severe fire in these forests, planting established productive ponderosa pine or mixed-conifer stands of ponderosa pine, Douglas-fir and white fir within 16 years (Zhang et al. 2008). Non-reforested sites are covered with shrubs and some sprouting hardwoods and will slowly become pure fir stands as described above. Thus succession to mixed-species forests of trees following a severe fire can be facilitated by reforestation and other silvicultural practices. Thinning in the next 10 to 20 yr would likely be desirable to reduce probable mortality from mountain pine beetles and thus decrease the potential for severe fire in these young forests.

Another case is that of conifer/red alder establishment after disturbance from fire, logging, or mass movement in the Coast Range forests of Oregon. Disturbance plus early establishment of conifer seedlings generally results in a stand of Douglas-fir with some hemlocks and western redcedar. This community can persist for several centuries in the absence of severe disturbances. However, red alder can also readily regenerate on these sites, especially near the coast on northerly slopes and on disturbed soil (Haeussler et al. 1995). Red alder seed is often plentiful and is readily blown by the wind, and red alder seedlings can grow rapidly and overtop conifers, which subsequently become suppressed and die beneath the alder. Thus a stand of red alder will occur if red alder seedling density is high. A salmonberry shrub understory, 6–12 ft tall, often becomes established beneath the alder (Hibbs and Giordano 1996). It is doubtful that red alder facilitates salmonberry establishment, because it can form dense covers in the open. Red alder is relatively short-lived, and stands begin to deteriorate after about 60 yr. As the red alder canopy dies, a salmonberry community often occupies the site. There are no reported examples of salmonberry communities succeeding to conifers. The current hypothesis is that salmonberry can persist for long periods, 50+ yr (Tappeiner et al. 2001). Therefore the initial floristics (conifers or red alder) largely determines whether the stand will become a conifer stand or a red alder stand that is replaced by salmonberry after 60+ yr.

Succession can also be quite deterministic. For example, stands of aspen, tanoak, and Pacific madrone readily sprout after their tops are killed (Harrington et al. 1994). Many seedlings of lodgepole pine and western hemlock are often present in the understory of conifer stands (Bailey and Tappeiner 1998), and they begin to grow when the overstory trees are logged, blown down (as with hemlock), or killed by insects (as with lodgepole pine). Thus in the case of lodgepole pine, succession is likely to produce stands very similar to the previous stand, at least as far as the major tree species are concerned, while hemlock will eventually replace Douglas-fir. The role of silviculture is to ensure that stand development is deterministic when a stand with known species composition and density is the goal. For other objectives, the goal of silviculture is to point out the most likely courses succession will take and the consequences of not managing vegetation after disturbance. From a silvicultural perspective, it is important to evaluate vegetation within one or two growing seasons after disturbance to determine which scenario is likely to occur and then to use silvicultural practices to alter succession, if necessary. Relevant practices might include planting, controlling shrub and hardwood density, and cutting hardwoods during precommercial thinning.

OVERSTORY AND UNDERSTORY DEVELOPMENT

We will discuss two general patterns of overstory development: a) following a major stand-replacement disturbance when the new stand begins with the establishment of a new cohort of trees, and b) when new cohorts of trees are established among older ones following a light to moderate disturbance. Overstory development generally begins when a site has experienced a severe disturbance and is beginning to be revegetated by sprouting plants and natural or planted tree seedlings. However, development also includes periodic disturbance by wind, low-severity fire, and other causes.

The early stage of development following a severe large-scale disturbance is called stand initiation (Oliver 1981; Oliver and Larson 1996) and is characterized by active regeneration and readily available resources for tree growth. Establishment of this new cohort may occur promptly in plantations or with species like western hemlock and lodgepole pine, which regenerate readily from seed or advanced regeneration, and with aspen, which regenerates vegetatively from root suckers. Winter and others (2002) and Dowling (2003) report recruitment of Douglas-fir lasting over 30 yr, and McCaughey and others (1991) found that regeneration was still occurring 20 yr after shelterwood and clearcutting in the Rocky Mountains, and that seedling ages on several well-dispersed sites ranged from 2 to 20 yr. As the trees grow, their crowns and roots expand. As resources become limited, there is strong competition among trees, and self-thinning occurs. This stage is called the stem-exclusion (Oliver and Larson 1996) or competitive exclusion (Franklin et al. 2002) stage. The cover of trees and shrubs in the understory decreases markedly in this stage.

As tree growth and self-thinning proceed, gaps in the canopy are formed from the deaths of larger trees, the breaking of branches from ice and snow, and the clashing of tree crowns during windstorms. These gaps permit trees and shrubs to become established in the understory, a stage of stand development that is called understory reinitiation. As stand development continues, trees get larger, small-scale disturbances continue, and the understory grows. Eventually, the stand develops a complex structure that is often described as old growth (Spies and Franklin 1991, Poage and Tappeiner 2005).

Such a process is typical of the development of even-aged stands. It represents stand development on many sites—for example, plantations of Douglas-fir, ponderosa pine, and mixed conifers. It is therefore a useful scenario for explaining stand dynamics and development in many forests. However, it is important to recognize (as do Oliver and Larson 1996) that there can be important variations in this scenario. Following are some common variations, though this is certainly not an exhaustive list:

• The rate of stand development varies with species and site and proceeds rapidly with fast-growing species on productive sites.

• The rate of stand development depends on tree density. Self-thinning may not occur if trees are widely spaced. In this case, although there may be inter-tree competition, it is not enough to trigger self-thinning.

• Stands may stagnate if they are very dense, thus preventing or substantially delaying self-thinning and stem exclusion. This occurs with shade-intolerant species (lodgepole and ponderosa pine) on sites of low productivity. In these cases self-thinning may not occur, and a disturbance (ice or snow breakage, insects, thinning) may be needed to enable stand development to continue.

• Unless they are growing at high densities, some tree species may not have dense enough crowns to shade out understory species; for example, ponderosa pine with an understory of manzanita and *Ceanothus*, red alder with salmonberry, and Douglas-fir with a vine maple understory.

• Understory shrub layers may be so dense that understory reinitiation does not lead to the multilayered stands typical of old growth. Canopy gaps can lead to establishment and growth of shrubs as well as trees.

• Stand dynamics can vary considerably within a forest stand because of differences in initial stand density, species composition, and disturbance patterns.

• Establishment of reproduction may occur over several decades. Trees established earlier may become dominant and grow more rapidly than those established later. Alternatively, the species established first may occur at low densities, allowing the slower-growing, later-established species to catch up and reach dominant positions in the canopy.

• Two-storied stands may result from the differences in growth rates of different species even though they became established at nearly the same time—for example, Douglas-fir will likely grow faster than western redcedar or western hemlock, with these more shade-tolerant species forming a second layer.

Disturbance has been common in western forests, especially when there were frequent low-severity fires. Not surprisingly, stand dynamics may differ considerably on sites subject to chronic or fairly regular disturbance. For example, even relatively severe disturbances may not kill all overstory trees on a site. Trees of varying species, sizes, and ages may remain after what was, for the most part, a moderately severe disturbance.

The average variation in tree ages in western Oregon old-growth stands was over 170 yr, with a range from 50 to more than 300 yr (Poage and Tappeiner 2002; Tappeiner et al. 1997). Thus, the current stand was not initiated by a single disturbance, but probably resulted from trees becoming established after several disturbances that occurred after long disturbance-free intervals. Sensenig et al (2013) found a broad range of tree ages in old mixed-conifer and Douglas-fir stands in southwestern Oregon. In the 18 stands he studied, fire occurred during about 70% of the decades from 1700 until 1900 (figure 3-13). Trees became established during these two centuries, but there was no evidence that establishment was better or worse during decades when fire occurred as opposed to decades when it did not. Establishment times varied from site to site, but in general no more than 20% of the trees were established in any decade. Apparently fire burning at relatively low intensities was a frequent agent in these forests and probably killed some trees while it prepared a seedbed for the establishment of others. Thus, unlike the scenarios described above, disturbance and recruitment were ongoing processes. The natural dynamics in forests with chronic disturbance seem to have consisted of an ongoing recruitment of trees, variable-sized stand structures, and no clear beginning or ending as a result of severe disturbances. This pattern of disturbances is quite different from the one described above and resembles the stand dynamics that would occur with uneven-aged management.

The examples of natural stand development discussed above suggest that a wide range of possible silvicultural systems and practices can be used to develop a variety of stand structures in western forests. Natural stand development in these forests has ranged from a single cohort of trees with a narrow range of tree sizes and ages established after a single severe disturbance (similar to even-aged silvicultural systems)

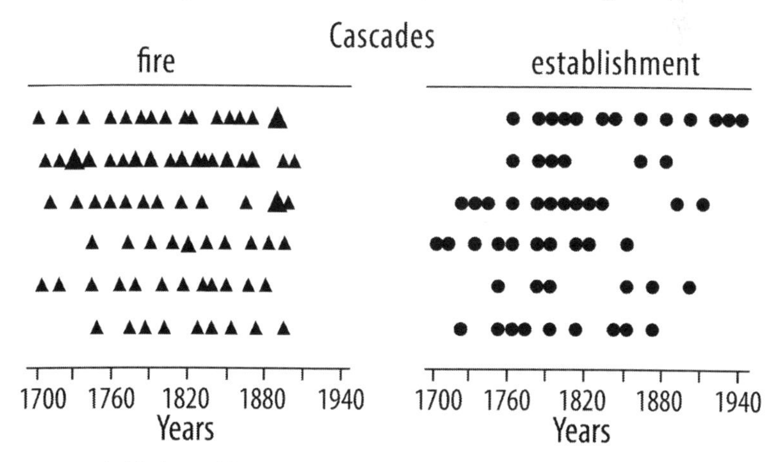

Figure 3-13. Probable dates of fire occurrence and conifer tree establishment in six 20-acre stands in southern Cascades of Oregon (adapted from Sensenig et al. 2013).) Regeneration of conifers occurred during periods of fire, but there was no clear relationship between estimated dates of fire and tree establishment. There were few fires and little tree establishment detected for the 20th century.

to stands with a range of tree sizes and ages that became established during periods of frequent disturbances (similar to uneven-aged silvicultural systems). The results of partial cutting in old forests in southeastern Alaska (Deal and Tappeiner 2002) suggest that this practice resembled chronic disturbance from wind (Deal et al. 1991) and that it would retain stand structures similar to those that result from wind disturbance. Similarly, in ponderosa pine forests in northeastern California, harvesting trees can produce structures similar to those that fire produced in the past (Dolph et al. 1995).

Effects of Silvicultural Treatments on the Physiology and Morphology of Trees

Silvicultural treatments such as thinning, clearcutting, vegetation management, etc., increase the availability of water, light, and nutrients. Treatments that provide shade (shelterwood and group selection methods) reduce temperature extremes and increase relative humidity. Changes in resource availability modify tree and shrub physiology, and their allocation of C to leaves, woody tissue, or roots (Goldberg 1990; figure 3-14). Note that some of the relationships shown in figure 3-14 operate in both directions. For example, changes in physiology (e.g., increased transpiration) and morphology (e.g., increased leaf area/plant) will increase the rate of water depletion, and hence reduce its availability.

In the past several decades, portable instruments such as pressure chambers (to measure water potential in plants), infrared gas analyzers (for measurement of CO_2 exchange), and soil-water and temperature sensors have enabled an understanding of some of the basic physiological responses of trees to silvicultural treatments; however, there is still much to learn. These studies have generally focused on tree water use, photosynthesis, leaf production, and stem growth (table 3-6). Only a limited understanding exists of the effects of treatments on tree roots and the effects of fertilization on tree physiology. Here we provide a brief overview of the better-understood physiological and morphological responses of trees to silvicultural treatments. We discuss the resulting changes in microclimate, resource availability, tree physiology, and morphology responses. For more complete discussions of this area of forest science, see Cannell (1982), Cannell and Last (1976), Hinckley et al. (1978), Kozlowski et al. (1991), and Waring and Schlesinger (1985).

TEMPERATURE

Silvicultural treatments that reduce stand density and canopy coverage usually modify the thermal regime of the air, forest floor, and soil by increasing average temperatures, but they also increase temperature extremes (figures 3-4, 3-5). Temperatures and vapor pressure deficits under a forest canopy are increased by reducing canopy or reduced by

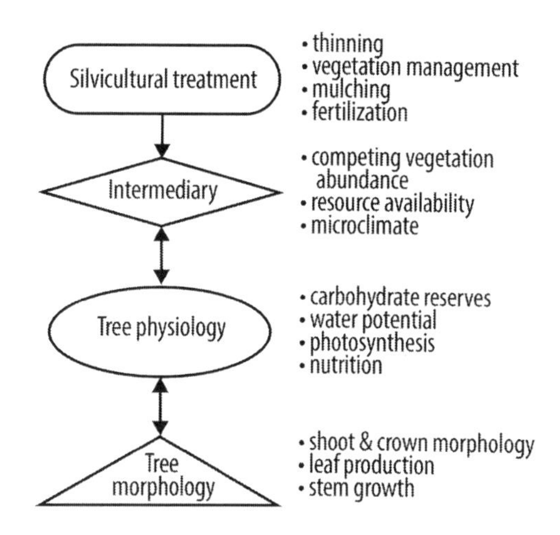

Figure 3-14. Conceptual diagram of the influence of a silvicultural treatment on tree responses. The treatment affects an intermediary (like soil water, or temperature) which subsequently influences tree physiology, enabling changes in tree morphology. Adapted from the concepts of Goldberg (1990).

increasing canopy density. Barg and Edmonds (1999) found that summertime maximum air temperatures in clearcut and green tree retention (20–30 trees/ha) treatments were 5–8°C higher than in intact forest. A heavy thinning of 25-yr-old Douglas-fir increased soil temperature by 0.5–1°C, with differences declining over time (Strand 1970). In an intensive study of microclimate within an Engelmann spruce-subalpine fir forest, Spittlehouse et al. (2004) found that most of the changes in microclimate associated with proximity to a forest opening occurred within one tree height of the canopy edge. Soil and air temperatures varied with opening size and were 2–4°C higher in large (10-ha) openings than in intact forest (Spittlehouse et al. 2004).

LIGHT

Forest cover has a major effect on the quality and quantity of light. Drever and Lertzman (2003) found that light transmission increased with increasing levels of overstory removal of Douglas-fir. Light was highly predictable from the summed heights of retained trees (r^2=0.99). Shrubs and hardwoods at 10+ yr of age can have stand basal areas nearly the same as those of fully stocked stands of conifers at 20+ yr, and their canopies are able to fully shade the soil because they can support leaf area indices of ±3–7 m^2/m^2 (Harrington et al. 1984, Hughes et al. 1987). Furthermore, these species are often evergreen and thus have lower light levels year round. For example, only 20–38% of PAR (photosynthetically active radiation—the wavelengths plants use to photosynthesize) penetrates tanoak sprout-clump canopies (Harrington et al. 1994).

Removing overstory trees can change not only the light intensity, but also its spectral quality, and this can influence development of tree regeneration (Atzet and Waring 1970). Changes in spectral quality within a stand due to crowding have been shown to

Table 3-6. Selected examples of morphological and physiological responses of trees and shrubs to dilvicultural treatments.

Treatment and species	*Response(s)*	*Reference*
Precommercial thinning of Douglas-fir saplings	Increase of buds on terminal and lateral leaders	Maguire (1983)
Release of Douglas-fir seedlings from shrubs and hardwoods	Increase in bud numbers on the leaders; later growth related to number of buds after release	Tappeiner et al. 1987b
Release of Douglas-fir saplings from hardwoods	Increase in bud number and size and numbers of needle primordia per bud and needle size	Harrington and Tappeiner (1991)
Release of ponderosa pine seedlings from shrubs	Increase in fascicle number and needle length; later growth of seedlings related to fascicle number	McDonald et al. (1992)
Thinning ponderosa pine stands 43 yr old	Needle fascicles longer and heavier in the thinned stand 8 yr after thinning	Wollum and Schubert (1975)
Thinning lodgepole pine about 50 yr old	Soil moisture depletion decreased; needle water potential and predicted photosynthesis increased after thinning compared to the control	Donner and Running (1986)
Removal of salal understory from a Douglas-fir stand	Soil water potential, stomatal conductance, and photosynthesis increased	Price and Kelliher (1986)
Thinning red pine stand 18 yr old	Increased soil moisture throughout the summer; increased water potential in the needles on 4 of 10 measurement days	Sucoff and Hong (1974)
Thinning a Douglas-fir stand in France 19 yr old	Soil water potential, predawn water potential in the foliage, and sapwood area increased in the thinned stand compared to the control	Aussenacand Granier (1988)
Douglas-fir stand thinned and fertilized 24 yr old	Water potential in the soil and predawn and early morning water potential in the trees increased during the summer for a 10-yr period after thinning; fertilization increased leaf area by 50% but had little effect on soil water	Brix and Mitchell (1986)
Release of Douglas-fir seedlings from Pacific madrone	Moisture stress of Douglas-fir during the summer drought was positively related to hardwood leaf area; soil water potential consistently increased where hardwoods were controlled	Pabst et al. (1990)
Release of Douglas-fir saplings from Pacific madrone	On sites with little madrone, water depletion from June through September was less than on untreated plots; predawn water potential of Douglas-fir was greatest where madrone was removed	Wang et al. (1995)

Table 3-6 cont.

Treatment and species	*Response(s)*	*Reference*
Removal of shrubs and herbs from a 5-yr loblolly pine plantation during the third yr of a dry period	Removal of arborescent vegetation increased xylem water potential; additional increase occurred with removal of herbs; cations were generally more available in the soil solution when shrubs and herbs were removed	Carter et al. (1984)
Removal of manzanita and canyon live oak from among suppressed Douglas-fir saplings	Complete removal of shrubs and hardwoods increased water potential in the soil	Hobbs and Wearsler Wearstler (1985)
Ponderosa pine and green leaf manzanita seedlings "replacement series"	Leaf water potential for pine decreased throughout the summer especially with increasing density of manzanita; manzanita water potential decreased during the summer but was not affected by tree or shrub density	Shainsky and Radosevich (1986)
White fir seedlings and saplings under manzanita, snowbrush, and chinkapin	Soil water tension was least in the treatment with least shrub cover; after 4 yr white fir leader growth increased 200% in treatments with less soil water tension and 20% shade; growth increase was less where there was less shade	Conard and Radosevich (1982a)
White fir seedlings and saplings under manzanita and snowbrush	Photosynthesis of ceanothus was generally 1.5–2.0 times greater than that of white fir and manzanita; water stress reduced photosynthesis of white fir more than that of the shrubs	Conard and Radosevich (1982b)
Douglas-fir saplings and Ceanothus	Soil water potential greater where shrubs and/or herbs were removed; complete removal of vegetation increased predawn water potential in Douglas-fir	Petersen et al. (1988)
Site preparation with three levels of shrubs and planted ponderosa pine, sugar pine, and white fir	No difference in incident radiation among treatments; intensive treatments increased soil water potential compared to less intensive treatments and the control; predawn and midday water potential and growth of all conifers increased as shrub volume decreased	Lanini and Radosevich (1986)
Douglas-fir saplings with varying amounts of overtopping tanoak or Pacific madrone	Hardwoods reduced photosynthetically active radiation by 20–38% of full sunlight and reduced soil moisture; throughout the summer temperatures were elevated and relative humidity reduced in the hardwood canopy; consequently Douglas-fir photosynthesis was least where there was a conifer cover, especially during the late summer	Harrington et al. (1994)

Table 3-6 cont.

Treatment and species	*Response(s)*	*Reference*
Thinning and fertilization of Douglas-fir 53 yr old	Effects were apparent 20–30 yr after treatment; thinning increased leaf area/tree; fertilization increased both stand and tree leaf area; both treatments increased the rate of stem-growth per unit of leaf area	Binkley and Reid (1984)
Ponderosa pine saplings treated for insect control, fertilizer, and shrub and herbaceous plant control	Control of vegetation (increase in available soil water) increased tree growth the most, followed by fertilizer; little fertilizer effect without first controlling vegetation	Powers and Ferrell (1996)
30-yr old Douglas-fir fertilized with 0–869 Kg/ha of N	Higher rates of photosynthesis for 1 yr after fertilization	Brix (1986)
24-yr-old Douglas-fir thinned and fertilized	Treatment doubled growth of individual trees for 7 yr when applied separately but quadrupled it when applied together; growth per unit of leaf area increased for 3 yr and then was the same as the control; treatment caused a marked increase (2–3+ times) in leaf area per tree.	Brix (1983)
Fertilization and thinning of Douglas-fir 24 yr old	Sapwood area, leaf area, and radial growth of individual trees increased	Brix and Mitchell (1980, 1983)
Fertilization of Douglas-fir 24 yr old	Photosynthesis for current shoots the yr of treatment increased as did 1-yr-old shoots the second season	Brix (1971)
Two applications of N to Douglas-fir 20 yr old	Needle length and width, number of branches, and leaf area per shoot increased; rates of photosynthesis did not increase	Brix and Ebell (1969)
Thinned Douglas-fir 38 yr old	Short-term increases of photosynthesis of codominant and suppressed trees but not of dominant trees	Helms (1964)

stimulate short-term increases in the growth of Douglas-fir seedlings (Ritchie 1997). Increases in the ratio of red/far red light from an opening in the forest canopy can stimulate germination of red alder seed, facilitating the species' establishment in disturbed habitats (Haeussler and Tappeiner 1993).

Mailly and Kimmins (1997) found that Douglas-fir needs at least 40% full sun to ensure survival and continued morphological development, although its maximum growth rate occurs under full sun conditions (Drever and Lertzman 2001). In contrast, western redcedar requires only ±10% full sun to survive (Wang et al. 1994), and it approaches its maximum growth rate at approximately 30% full sun (Drever and Lertzman 2001). Following a series of studies across different sites, Carter and Klinka (1992) formulated the general hypothesis that the light required by Douglas-fir, western hemlock, and western redcedar to attain a given rate of height growth increases as soil moisture availability on the site increases. In other words, as available water increases, light may become increasingly important.

WATER

Soil water availability increases when vegetation density is reduced. Soil water is generally measured as the proportionate weight of water per unit dry weight of soil (gravimetric content) or as the proportionate volume of water per unit volume of intact soil (volumetric content). As the water content in the soil decreases from evapotranspiration the water pressure potential decreases; that is, it takes more energy for plants to extract water from the soil. Thus water becomes less available to plants (figure 3-10). As we discuss below, at soil water tensions of –1.5 MPa, which are common during the summer, it is difficult for plants to extract water from the soil, and—as in the soil—the water in the plants is held at high tensions.

Adams et al. (1991) studied the long-term changes in soil water after clearcutting old-growth Douglas-fir. In the first yr after harvest, the top 120 cm of soil had 10 cm more water than adjacent intact forest. However, by the fourth yr after harvest, the clearcut area had 2 cm less soil water than the intact forest because of rapid revegetation and use of soil water by fireweed, snowbrush, and vine maple. Thinning a forest stand has been shown to increase the soil water potential (i.e., values became less negative), indicating greater soil water availability to lodgepole pine in Montana (figure 3-15; Donner and Running 1986); Douglas-fir in France and the Pacific Northwest (Aussenac and Granier 1988; Brix and Mitchell 1986); and red pine in Minnesota (Sucoff and Hong 1974).

Soil water is a critically important resource for which forest plants compete on many sites in the summer-dry climate of western forests. Competition is especially common between small trees and shrubs (Powers and Reynolds 1999; see chapter 9) and severe competition for soil water can result from as little as 20% shrub cover

(Oliver 1984; Shainsky and Radosevich 1986; White and Newton 1989). Ericaceous shrubs, such as manzanita, can deplete soil water content far below what conifers can extract (Zwieniecki and Newton 1996a, b). Controlling competing vegetation can influence not only the availability of soil water (Harrington et al. 1994; Hobbs and Wearstler 1985; Pabst et al. 1990; Petersen et al. 1988; Wang et al. 1995), but also the availability of soil nitrogen (Cole and Newton 1987; Harrington 2006; Roberts et al. 2005; Rose and Ketchum 2002).

PLANT-WATER RELATIONS

The water in a plant, like the water in the soil, is under tension. This tension results from the combined effects of evaporative demand from the atmosphere and the tension by which water is held within the soil. Water stress of plants and soil is measured as a pressure potential, where a value of −1.5 MPa or −15 atmospheres is generally associated with the wilting of an herbaceous plant such as a sunflower. The wilting point occurs when the turgor of vascular tissue within a plant will not recover without the addition of soil water. Plant water potential, also called xylem pressure potential, is measured by inserting an excised plant part into a strong sealed chamber that is then pressurized with an inert gas, usually N. The pressure required to force water to exude from the cut surface of the vascular tissue, expressed as a negative value, is the xylem pressure potential (Cleary and Zaerr 1980). Highly negative values (e.g., <−1.5 MPa) indicate that considerable tension of water exists in the plant's xylem, just as low availability of soil water is measured as the tension (negative pressure) of water within the soil. Small negative values for water potential (−0.5 MPa) in plants generally indicate that soil

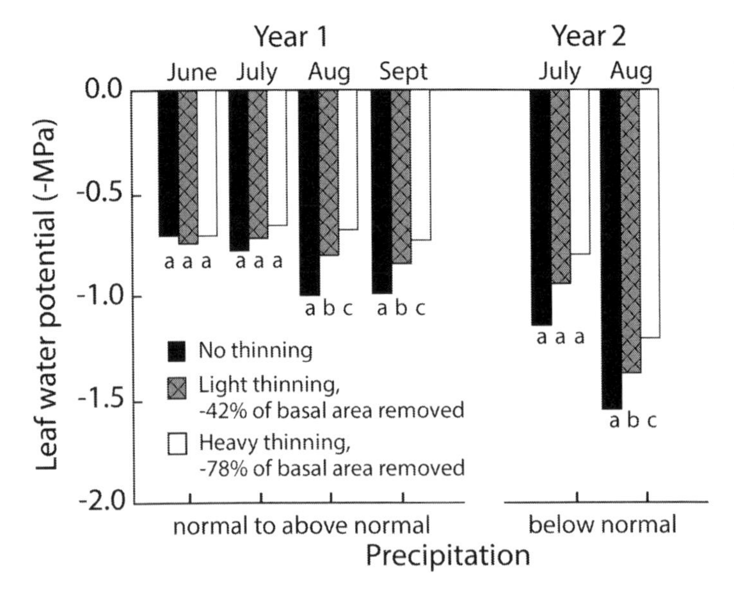

Figure 3-15. Effect of thinning on leaf water potential in lodgepole pine in years with normal and below normal precipitation. Means with different letters are significantly different (95% confidence level). Adapted from Donner and Running (1986).

water is readily available and that they can open their stomates and photosynthesize—assuming that light, temperature, and other variables are not limiting.

When soil water is plentiful, plant water potential is generally high at predawn, morning, and midday. Donner and Running (1986) showed that thinning lodgepole pine increased the leaf water potential in July and August, thereby extending the growing season. Similarly, after controlling the density of Pacific madrone, the predawn and midday water potential in Douglas-fir were strongly related to soil water potential. Predawn water potential in Douglas-fir (growing among the madrone) increased from –1.8 to >–0.05 as soil water potential increased from –2.0 to >–0.3 MPa (Pabst et al. 1990). However, midday water potential values are often a function of the current vapor pressure deficit (i.e., evaporative demand of the air) and may be greater on north slopes or in the shade than on exposed sites or south slopes. In northern Minnesota, where summer rains are common, Sucoff and Hong (1974) found that plant water potential in thinned stands was greater than that in unthinned stands on four of the ten measurement days. They estimated that water potential in the foliage of trees in thinned stands was greater than that in unthinned stands on only 20% of days in the growing season. Nevertheless, trees in the thinned stands grew 36% faster than those in the unthinned stands. Similar results were obtained when forbs, grasses, shrubs, and sprouting hardwoods were removed around conifer seedlings and saplings. Both soil water potential and Douglas-fir water potential increased after reducing the density of Pacific madrone (Pabst et al. 1990; Wang et al. 1995), tanoak (Harrington et al. 1994), snowbrush ceanothus (Petersen et al. 1988), canyon live oak, and manzanita (Hobbs and Wearstler 1985). Water potential increased in ponderosa pine, white fir, and sugar pine following complete removal or reduced density of manzanita (Conard and Radosevich 1982a, b; Lanini and Radosevich 1986; Powers and Reynolds 1999; Shainsky and Radosevich 1986).

The ability of a plant to withstand drought is often measured as threshold relationships of photosynthesis, conductance, survival, or growth with plant water potential. For example, Cleary and Zaerr (1980) found that photosynthesis declines gradually to zero as water potential of Douglas-fir seedlings decreases from –1.0 to –3.0 MPa, whereas it declines abruptly to zero as water potential of ponderosa pine seedlings decreases from –1.0 to –2.0 MPa. Sands and Correll (1976) observed that growth and metabolic reactions of radiata pine became inhibited as water potentials reached –1.0 and –1.5 MPa, respectively. Relative growth rate of red alder saplings declined abruptly at water potentials of only –0.3 to –0.4 MPa induced by variable stand densities. Photosynthesis of red alder essentially stopped as water potential dropped below –1.2 MPa (Hibbs et al. 1995a). Some evergreen species are able to withstand extremely negative water potentials before their stomata close, such as –3.4 MPa for Pacific madrone (Harrington et al. 1994; Pabst et al. 1990).

PHOTOSYNTHESIS

The net result of photosynthesis is the production of carbohydrates that supply the energy for physiological processes and provide the raw materials for production of leaves, roots, stemwood, cones and seeds, and defensive chemicals. Typically, photosynthesis is measured by placing one or more individual leaves or shoots of a plant into a sealed, clear plastic chamber (cuvette) that has a constant rate of air flow entering and leaving. An infrared gas analyzer connected to the cuvette detects changes in concentration of carbon dioxide. When net photosynthesis is occurring (i.e., uptake of carbon dioxide exceeds output from respiration), the CO_2 concentration decreases and photosynthesis rate can be expressed as CO_2 uptake per cm^2 of leaf area per second ($\Delta CO2/cm^2/sec$). Often photosynthesis rate is measured under fully illuminated conditions to determine a potential rate of photosynthesis given no limitations in light availability.

Measurements of photosynthesis indicate whether growth responses are more attributable to shifts in the physiology or the morphology of a tree. Although any additional accumulation of C in a tree resulting from a silvicultural treatment must be a consequence of enhanced photosynthetic activity, it can be difficult to identify when and how it occurred. For example, increases in leaf area after thinning, combined with little or no increase in photosynthesis rate ($\Delta CO2/cm^2/sec$), can drastically increase the biomass accumulation of a tree (Helms 1964, Brix 1971). Samuelson (1998) found that rapid biomass accumulation of seedlings of loblolly pine and sweetgum from fertigation (irrigation plus fertilization) was primarily a result of increased leaf area rather than increased potential rate (light-saturated) of photosynthesis. Munger et al. (2003) found a similar result for loblolly pine that received repeated applications of fertilizer. In contrast, repeated fertilization on an extremely nitrogen-poor site in the North Carolina Sand Hills caused the potential rate of photosynthesis ($\Delta CO2/cm^2/sec$) of loblolly pine to increase by up to 20% during the first two years of treatment (Murthy et al. 1996). In this latter example, the potential rate of photosynthesis was correlated with leaf nitrogen concentration (i.e., nitrogen availability was limiting the potential rate of photosynthesis), whereas in the previous examples (Samuelson 1998, Munger et al. 2003) it was not.

One of the factors driving photosynthesis is the amount and activity of chlorophyll in the leaves. In a study of 2-yr-old seedlings of Douglas-fir, ponderosa pine, western hemlock, and western redcedar, Kahn et al. (2000) found that chlorophyll concentration increased while chlorophyll fluorescence (a measure of the ability of chlorophyll to capture light energy) decreased with increasing shade. This result suggests that limitations in photosynthetic capacity of shade-grown seedlings may be attributed to changes in chlorophyll amount and activity that can develop within a relatively short time.

Research on climate change has prompted studies to see how trees will respond to forecast increases in CO_2 concentrations. Doubling the CO_2 concentration stimulated 83–91% increases in potential photosynthesis of loblolly pine (Murthy et al. 1996). Hibbs et al. (1995) found similar responses for red alder and inferred that the higher CO_2 concentration may have allowed the species to photosynthesize at lower water potentials, making it somewhat more tolerant of an increasingly droughty environment.

NUTRITION

Often, but not always, the application of fertilizers improves tree nutrition, measured as the foliar concentration of specific nutrients and stem growth (See chapter 9). These increases can be short-lived because the rapidly expanding crown of the tree redistributes the additional nutrients, resulting in a dilution of their concentration within the foliage. To avoid this dilution effect and sustain tree growth responses to enhanced nutrition, some studies use repeated nutrient amendments during the growing season with fertilizer rates increasing in proportion to the biomass of trees being cultivated (Powers and Reynolds 1999; Timmer 1997). In general, peak N demands occur at crown closure, when stand leaf area reaches its maximum value. One of the effects of improved nutrition is often an increase in the leaf area of trees (Binkley and Reid 1984, 1985) and photosynthesis (Brix 1981, 1983).

LEAF AND CROWN MORPHOLOGY

A variety of leaf responses have been observed following increases in resource availability associated with a silvicultural treatment. Typically, the weight and number of foliage components increases as tree vigor increases. For example, after thinning a 43-yr-old ponderosa pine stand in Arizona, Wollum and Schubert (1975) found that needle weight and fascicle number per shoot increased. Similar responses have been observed for ponderosa pine seedlings. After shrubs were removed, ponderosa pine seedlings grew more needle fascicles on their shoots, and the needles were longer (McDonald et al. 1992). The number of fascicles and shoots was strongly related to growth of the pine seedlings 3–5 yr later (McDonald et al. 1992).

Douglas-fir seedlings and saplings increased the number and sizes of interwhorl buds within 1–2 yr of precommercial thinning and shrub control. In all cases there were more interwhorl buds on the leader of the main stem and primary branches of treated trees than on untreated trees (Harrington and Tappeiner 1990; Maguire 1983; Tappeiner et al. 1987b). In addition, the branch density increased after thinning and fertilizing Douglas-fir (Brix 1981). The branches that come from interwhorl buds make up nearly 90% of the foliage of Douglas-fir (Jensen and Long 1983). After treatment, buds are not only more abundant, they are larger and contain more needle primordia (sites of needle initiation), and the resulting needles are larger than those of untreated trees (Harrington and Tappeiner 1991). Increase in bud number and size are the precursors of a dense crown. As in ponderosa pine, the stem growth of seedlings for 3–5 yr after treatment was strongly related to the number of buds produced the first growing season after treatment (Tappeiner et al. 1987b); however, if shrubs reinvaded the Douglas-fir stand, this potential was not realized and some seedlings died. Changes in bud size and density are precursors of greater crown development and, consequently, water movement. Thinning in Douglas-fir stands 24-53 yr old increased the leaf area per tree of Douglas-fir (Binkley and Reid 1984; Brix and Mitchell 1983). Sapwood area per tree (the conducting tissue for water and nutrients), which is related to leaf area (Waring 1983), has been shown to increase following thinning (Aussenac and Granier 1988, Binkley and Reid 1984, Brix and Mitchell 1983). Sapwood is the conducting tissue for water and nutrients, so it is functionally related to the total leaf area of the tree (Büsgen and Münch 1926; Waring 1983; chapter 5)

Conditions within the foliage may change as more resources become available following treatment. The surface area-to-weight ratio (specific leaf area) of foliage decreases with increasing availability of light, as found in seedlings of Douglas-fir (Del Rio and Berg 1979) and western redcedar (Wang et al. 1994), probably because of an increase of carbohydrates in the foliage. Elevated CO_2 concentrations caused the specific leaf area of Douglas-fir seedlings to decrease due to increased levels of sugars and non-structural carbohydrates (those available for plant growth) (Olszyk et al. 2003). However, foliage responses can be subtle and result in changes at the cellular level. Following partial canopy removal, stomatal density of foliage of subalpine fir and Engelmann spruce seedlings increased, but overall growth was not affected during the 4 yr after treatment (Youngblood and Ferguson 2003).

ROOT SYSTEMS

Of all tree morphological variables, least is known about root responses to silvicultural treatments. The size and extent of root systems are illustrated by a study of tree water use (Ziemer 1968) around a single sugar pine tree (28 in dbh) on a deep soil. It was isolated by removing 88% of the trees larger than 3.5 in dbh from around it. The zone of greatest soil moisture depletion occurred about 10 ft from the tree at a depth of 8–13 ft, and the zone of soil moisture depletion from the tree extended about 20 ft away from it to a depth of 15 ft.

Douglas-fir and ponderosa pine trees may lift or redistribute water from lower soil levels and make it available to plants, including the trees themselves, with their roots in the upper soil layers (Brooks et al. 2004). Although this does not prevent the summer soil-drying trend, it causes small diurnal increases in soil water. It is not yet clear what this water transfer may mean for silvicultural practices such as thinning or reforestation.

The density of roots below ground is somewhat related to the abundance of tree biomass above ground. Strand (1970) found that heavy thinning of Douglas-fir resulted in a lower density of fine roots within the soil, particularly at 60–80 cm depth. Ahrens (1990) found that sprouting tanoak retained most of the original root system through the first 4 yr after cutting the stem of the parent tree. Thus, the massive root system of a tanoak sprout clump is able to tap a large volume of soil and therefore be highly competitive with the much smaller root system of Douglas-fir seedlings.

Root system size and surface area are related to resource availability. Kahn et al. (2000) found that root volume of four conifer species decreased with increasing shade, suggesting that seedlings grown in the shade of an evergreen competitor will have very little opportunity to advance their root systems into areas of greater resource availability. Similarly, Ludovici and Morris (1996) found that low availability of soil water reduced root surface area of loblolly pine and sweetgum by 28% and 18%, respectively, compared with well-watered seedlings. However, the root surface area of both species increased dramatically in the presence of localized pockets of increased N availability that were included in the experiment. Sweetgum was the more effective competitor for N with a low availability of soil water.

SUMMARY

Forest trees and other plants integrate the use of water, light, temperature, and nutrients as they grow and maintain themselves in forest ecosystems. In western forests with a pronounced summer drought, water is often the limiting resource, but this may vary with the time of year. In western Pacific Coast forests, winter temperatures are often mild (figure 3-4), and evergreen trees can photosynthesize and produce net carbohydrates year-round (Emmingham and Waring 1977; Helms 1964; Waring and Franklin 1979 a, b). In these forests, although water limits photosynthesis during the drought, temperature and light can limit photosynthesis during the wet winter and early spring (Harrington et al. 1994). Similarly, soil nutrients may limit tree growth below its potential, but fertilization will not increase tree growth if soil water is not available.

Silvicultural practices usually cannot focus on favoring or limiting a single resource that affects plant growth. Thinning and regeneration harvesting, and disturbances such as fire or windthrow, affect several resources simultaneously. Opening a forest canopy could provide light and soil water that favor some species while decreasing winter temperature minimums to the detriment of other species (or size classes of the same species). Therefore, silviculturists must be aware of how plant resources and microenvironments are affected by their practices and how they interact.

REVIEW QUESTIONS

1. What is a forest stand from a silviculture point of view? How does it vary from an ecological viewpoint? Why does a forest stand, defined from this point of view, inherently have lots of variability?
2. Why is it important that forest stands be relatively independent of each other? Discuss.
3. What is stand structure, and how is it related to silvicultural systems? What are the biotic and non-biotic components of stand structure?
4. Why is stand structure the basis of silvicultural prescriptions?
5. How is stand structure related to the forest ecosystem?
6. What are the components of a forest ecosystem from a silvicultural point of view?
7. Give an example of how the components of a forest ecosystem might interact to cause a change in a forest stand.
8. How might you use an "ecosystem check list" to evaluate a silvicultural treatment? Give an example.
9. How might broad-scale climates in western forests affect the practice of silviculture? What features of the climate do most western forests have in common?
10. Describe how microclimate varies by slope and aspect.
11. How and why does forest cover affect radiation and soil and air temperatures?
12. Where is wind damage most likely to occur in a forest? Explain why.
13. How does the water holding capacity of a soil vary with soil texture and depth?
14. What is meant by the permanent wilting point of a soil? Does it apply to forest plants? Explain.
15. Using a graph, describe how the top 24 in of a soil would dry out during the summer drought in (A) a forest stand, and (B) an area with very little vegetation. Show the effects of transpiration and evaporation and indicate permanent wilting point. Using this graph, explain how practices like thinning and site preparation and leaving shelterwood trees affect soil water.
16. How would 1 in of precipitation affect soil water during the summer drought?
17. How do rocks in a soil profile affect water-holding capacity of a soil?
18. How is biomass distributed in a forest?
19. Using N as an example, how are nutrients distributed in a forest? Explain how silvicultural practices may affect the nutrient distribution in a forest.
20. How can silvicultural practices affect forest soil properties? Do silvicultural practices lower the productivity of soils? Explain.
21. Explain how competition is part of secondary succession and stand development.
22. What do we mean by succession being deterministic or probabilistic? Give an example using shrubs/conifer succession in eastside forests or alder/conifer development in the Coast Ranges of Oregon or Washington.

23. What are the four stages of forest stand development outlined by Oliver and Larson? Describe each stage. Discuss how these stages might be altered by silvicultural practices.
24. Describe a course of stand development that might occur in ecosystems where there are chronic disturbances.
25. In what ways do silvicultural practices change the operational environment of forest plants?
26. Explain how changes in the physiology and morphology of a plant in a stand that has been thinned or had shrub cover reduced might compare to those in an untreated stand.

CHAPTER 4
Ecology of Shrubs, Hardwoods, and Herbs

Introduction

Most of silviculture is focused on the reproduction and growth of trees, because they are the dominant part of the forest structure and because of their commercial value. However, an understanding of the ecology of shrubs is also important to the science and practice of silviculture. These species affect stand development and growth and they form dense understories, making them an important part of forest structure. They respond to silvicultural practices, and they are a component of stand development whether or not their growth and cover is a goal of stand management (Dyrness 1973; Halpern and Spies 1995; Schoonmaker and McKee 1988). Whether it is intended or not, management of understory species occurs in the process of managing the overstory. This chapter provides background on the ecology of forest vegetation and examples of how it is related to silviculture.

Forest vegetation is a term that generally includes forbs (broadleaf herbaceous species), graminoids, shrubs, and often non-commercial hardwoods. **Vegetation management** has become the science and art of controlling plants that compete with young trees, particularly during the first 10+ yr of response to site preparation or severe fire. However, in this chapter we discuss the autecology of these plants in the understory of forest stands and the dynamics of their invasion and growth following a major disturbance to the forest. We include hardwoods in this discussion because many western hardwoods grow both in the understory of conifer stands and as sprouts on recently disturbed sites. Understanding of the ecology of these species enables silviculturists to incorporate them into silvicultural systems (Wagner and Zasada 1991).

Plant Strategies

Once they are established, plants have different strategies for maintaining themselves in communities and interacting with other plants. Grime (1981) recognizes three typical strategies: competitors, stress tolerators, and ruderals. These categories are useful for understanding the ecology of forest plants, although as Grime suggests, it is sometimes difficult to place plants in a particular category.

Competitors are plants that grow rapidly, quickly occupy a site and—because of their longevity, size, and rapid growth—may preclude the establishment of other species. They are generally the major species on a site for several decades or more. For western forests, examples of these species include shrubs (ceanothus, manzanita), hardwoods (aspen and red alder, Pacific madrone), and conifers (Douglas-fir, redwood, lodgepole and ponderosa pine, larch). Species such as Douglas-fir maintain dominance on a site for hundreds of years. Similarly, on dry sites there are 50+-yr-old communities of whiteleaf manzanita and other shrubs with no tree species in the understory. There is no indication that these communities will be replaced in the near future without a major disturbance such as fire.

Stress tolerators are plants that can persist in the understory in low light, low moisture environments, and grow very slowly; however, when the overstory is disturbed by wind, disease, or logging, they can respond to the new environment and grow vigorously. These species include shrubs (salal, vine maple, Oregon grape, salmonberry, bearclover, and *Vaccinium* spp.), hardwood trees (bigleaf maple, tanoak) and conifers (seedlings and saplings of western hemlock, white fir, and Douglas-fir, and ponderosa pine on dry sites), and herbs (twinflower, bunchberry, vanilla leaf, trailing blackberry, etc.). Stress tolerators can assume dominance when they are released from the overstory. Douglas-fir may be classified as a stress tolerator on dry interior sites where it can persist in the understory. However, on productive coastal sites, it is clearly a competitor.

Ruderals are adapted to disturbances. They are annuals and biennials that quickly complete their life cycles and produce seed for a new generation. They are most common in the early stages of succession and stand development (Halpern et al. 1997). They include both natives and non-natives such as thistles, grasses, foxglove, fireweed, Scotch broom, etc.

These categories are useful in describing how plants maintain themselves in forest ecosystems, but they do not explain how species interact. For example, when salal, salmonberry, manzanita, and tanoak persist in the understory, they are probably competing to some degree with the overstory trees for water (Price et al. 1986). Also, these species sprout vigorously when the overstory is removed, and they compete with other plants in the early stages of succession and stand development, in the same way as species classified as ruderals and competitors.

Roles of Shrubs, Hardwoods, Herbs, and Ruderals in Forest Stands

Shrubs, hardwoods, herbs, and ruderals are important components of forest stands. Shrubs and herbs provide browse for ungulates (Hanley 1983), substrate for prey, and hiding cover and nesting sites for many mammals. Shrub layers are important for bird habitat because they provide cover and nesting sites (Hagar et al. 1996; Hayes et al. 1997). The life cycles of certain insects that are prey for birds are closely tied to particular species of understory plants. Some shrubs and hardwoods are substrates for lichens and bryophytes (Muir et al. 2002).

Depending on their density, understory plants cycle considerable amounts of nutrients (although generally less than the overstory trees), especially cations, and they may cause a more rapid turnover of the organic matter in the forest floor (Fried et al. 1989; Tappeiner and Alm 1975). Total litterfall of understory shrubs, herbs, and hardwoods and conifers is greater than under conifers alone. However, this litter is rapidly recycled and the forest floor weights per unit area are less under shrubs and conifers than under conifers alone (Fried et al. 1989). After a severe disturbance, plants that rapidly revegetate the site use and retain nutrients that might otherwise be leached below the rooting zone and become unavailable to forest plants. Some species (e.g., snowbrush, deerbrush, bitterbrush, and alder) fix N from the atmosphere; *Ceanothus* species are well known for this ability. Populations of *C. velutinus and C. sanguineus* 0 to 33 yr old are reported to fix N at rates from 0 to nearly 110 lbs/ac/yr (Binkley and Husted 1983; Youngberg and Wollum 1976; Zavitkowski and Newton 1968); the reasons for the wide range in the rates of fixation are not known. Over a 35-yr period in a ponderosa pine forest on a sandy-loam pumice soil, there was more soil N in the organic layers (about 161 vs 236 lbs/ac) and the upper 0–4 cm of mineral soil (234 vs 336 kg/ha) in areas where there were understory shrubs (greenleaf manzanita, bitterbrush, and snowbrush) than in areas where shrubs were removed (Busse et al. 1996). Microbial biomass was also greater where shrubs were present. However, the presence of understory shrubs on the site reduced conifer growth for at least two decades (Barrett 1982). Thus, although it may be important to control these species to establish conifer reproduction, it may be equally important not to eliminate them from a site altogether.

Some shrubs and hardwoods have commercial value, while many are exotic (invasive plants). For example, the barks of yew and cascara have medicinal value, and the foliage of sword fern and salal is commercially harvested for use in floral arrangements. Vine maple is a substrate for commercial crops of moss as well for a variety of native lichens and other bryophytes; it also provides aesthetically pleasing fall color in western forests. Stands of red alder and cottonwood are grown commercially. Large logs of California black oak and bigleaf maple yield high-quality wood, and tanoak has been used for industrial wood as flooring and pallets. Invasive plants include Himalayan blackberry, Scotch broom, and English holly, all of which are common shrubs; as well as many grasses and forbs: oxeye daisy, foxglove, several thistles, false brome, etc.

Natural Seedling Establishment and Shade Tolerance

The natural establishment of shrub and herb seedlings is similar to that of conifer and hardwood trees. The seeding characteristics of some common shrubs and hardwoods are summarized in table 4-1 (see chapter 9 for a discussion of natural regeneration from seed). Important steps in the natural regeneration of all forest plants include seed production and dispersal, overwintering of seed, germination of seedlings, and early survival and growth. Lack of seed production, loss of seed to predators, and limited dispersal of seed are all impediments to regeneration. After germination, predation by insects and pathogens, high soil temperatures and loss of water from the upper soil layers, frost heaving, litterfall, and other factors also limit herb, shrub and hardwood regeneration (Nord 1965; Tappeiner and Alaback 1989; Tappeiner and Zasada 1993). Although the steps involved in regeneration are the same for all shrubs and hardwoods, there are substantial differences among species in their natural regeneration processes.

As with conifer seedlings, certain species of shrubs are shade tolerant and become established in the understory of forest stands. Salal, *Vaccinium* spp., vine maple, tanoak, and bigleaf maple are shade tolerant and may become established and persist in relatively dense conifer stands, whereas Pacific madrone, salmonberry, and red alder seedlings will die within 1–2 yr on the same sites (Hauessler et al. 1995; Tappeiner and Alaback 1989; Tappeiner and Zasada 1993). Many shade-tolerant species produce large seed (bigleaf maple, Oregon grape, vine maple, tanoak). Consequently, their seedlings are large and tolerate shady environments, and they are resistant to mortality from pathogens, drought, litterfall, and invertebrate predation. Survival of small-seeded, shade-tolerant conifers such as western hemlock and Sitka spruce (Harmon and Franklin 1989) and shrubs such as salal and huckleberry is often greatest on decomposing wood, but not restricted to this substrate (Huffman et al. 1994; Tappeiner and Zasada 1993). Undoubtedly, animals are important for seed dispersal of understory species, whose seed is produced near the ground (<5 ft). Animals often carry seed to snags more than 30 ft tall, well above its source. Bitterbrush seed is cached or "planted" by animals, and small clusters of seedlings germinate in abandoned caches. Caching aids seed dispersal and possibly seedling survival and establishment as well (Nord 1965; Vander Wall 1994, 1995).

Generally the rates of mortality are high in the first 5+ yr after germination, especially for small-seeded, herbaceous species (Tappeiner and Alaback 1989; Tappeiner et al. 1986, Tappeiner and Zasada 1993). Once established, shade-tolerant species (Oregon grape, salal, *Vaccinium* spp.) often grow quite slowly, especially in dense conifer stands, and are regarded as stress tolerators (Grime 1981) because they can persist in low-light environments for many years. They form understory seedling populations, or seedling banks (Grimes 1981), similar to those of advance regeneration of shade-tolerant conifers (table 4-1). If the overstory is disturbed, the rate of seedling growth and vegetative spread by rhizomes increases. Blueberry seedlings in Alaska were only 4 in tall after 6–10 yr and were just beginning to produce rhizomes, but after opening of the forest canopy by windthrow, their growth rate tripled in 2 yr (Alaback and Tappeiner 1991).

Shade-intolerant species often have seed banks that readily germinate after fire, logging, or windthrow. Snowbrush, deerbrush and redstem ceanothus, greenleaf and whiteleaf manzanita, pin cherry, gooseberry, salmonberry, and other species store seed in the forest floor, even in dense conifer stands (Halpern 1999) and large numbers of seedlings germinate following severe disturbances to the tree canopy and forest floor (Hughes et al. 1987; Quick 1956, 1959). Unlike species adapted to growing in the understory, seedlings of these species grow rapidly and can form dense covers within 5–10 yr. Two yr after clearcutting, 20,000 seedlings/ac of whiteleaf manzanita, deerbrush, and snowbrush were found on sites in southwestern Oregon and northern California (Cronemiller 1959; Gratkowski 1961). At 10 yr, leaf area was $2m^2/m^2$ for whiteleaf manzanita and deerbrush and $4m^2/m^2$ for varnishleaf ceanothus. Cover for all species was nearly 90% of maximum (Hughes et al. 1987). Within 10–15 yr the basal area of shrub stems at 6 cm above the ground was about 110–130 ft^2/ac for whiteleaf manzanita and deerbrush and more than 170 ft^2/ac for varnishleaf ceanothus (Hughes et al. 1987). Normally, Douglas-fir in stands grows basal areas (at a height of 4.5 ft) of about 87 ft^2/ac at 20 yr (site index 130 ft/100 yr, McArdle et al. 1961). Thus the basal area (or cross-sectional area of stem wood near the ground) of young, regenerating seedling shrub communities can quickly exceed that of conifer stands at young ages. These dense young shrub communities often inhibit the establishment of conifers. At older ages, of course, the basal area and biomass of the conifers will greatly surpass that of the shrubs.

Both shade-tolerant and -intolerant species are likely to regenerate from seed after fire, wind, thinning, or harvesting reduces the density of the canopy. The seedling establishment of both shade-tolerant (salal, vine maple, *Vaccinium*, tanoak, bigleaf maple) and shade-intolerant species (salmonberry, Pacific madrone) were greater in thinned stands or partial shade than in unthinned stands (Huffman and Tappeiner 1997; Huffman et al. 1994; Tappeiner and Zasada 1993; Tappeiner et al. 1986). Even the survival of red alder, a very shade-intolerant species, is aided during its first 2 yr by partial shade, which protects seedlings from frost and high soil-surface temperatures (Haeussler et al. 1995).

Vegetative Reproduction

Many forest plants reproduce vegetatively. They sprout after their tops are killed, and their extension of rhizomes, roots, and above-ground stems enables many species to colonize and persist in the understory of forest stands. Table 4-1 shows examples of species with the ability to reproduce vegetatively. Even very young seedlings have the capacity for vegetative reproduction. Beaked hazel seedlings 3 yr old and 3 in tall sprouted after a fire killed their tops (Tappeiner 1979), and we have seen tanoak seedlings 3 weeks old sprouting after their tops were cut.

Vegetative reproduction includes (1) sprouting of new aerial stems from the stem base or from burls (lignotubers below ground, figure 4-1); (2) the production of ramets (new independent plants) from stolons, rhizomes, or roots; and (3) layering (rooting) of aerial stems. For example, greenleaf manzanita, deerbrush, bigleaf maple, and tanoak produce buds along the aerial stem and below ground at the base of the stem that sprout vigorously below the point where a stem is killed (Ahrens 1990; Harrington et al. 1984). Tanoak and greenleaf manzanita form burls at the bases of their stems. Burls are the site of prolific sprouting when the tops of the parent plants are killed. In the case of tanoak, burls may be only 0 –1 in in diameter on seedlings 5–15 yr (Tappeiner and McDonald 1984), but gradually they produce large, multi-stemmed trees that can have several burls more than 50 cm in diameter at their base. Over time the tissue between the burls may decay and the aerial stems become independent trees as the clone breaks up.

Hardwood seedlings often become seedling sprouts. In tanoak the original stem dies within 6–12 yr after germination and is replaced by new sprouts from a small burl (McDonald and Tappeiner 2002; Tappeiner and McDonald 1984). Stems die and are replaced by new stems. Seedling sprouts often have three or more stems, and often the largest is the youngest.

Many forest plants are rhizomatous; that is, they produce underground stems called rhizomes (figure 4-2). By 6–10 yr seedlings begin to produce rhizomes at their bases. As rhizomes grow, aerial stems and taproots form ramets, which can become independent, yet genetically identical to the parent clone. Rhizomes contain many dormant buds. In western Oregon forests, clones of salmonberry and salal have networks of rhizomes extending over an area 15 ft in diameter (Huffman et al. 1994; Tappeiner et al. 1991).

Figure 4-1. Sprouting from a large tanoak tree about 24 in diameter, 2 yr after clearcutting. The clump of sprouts is > 6 ft in diameter.

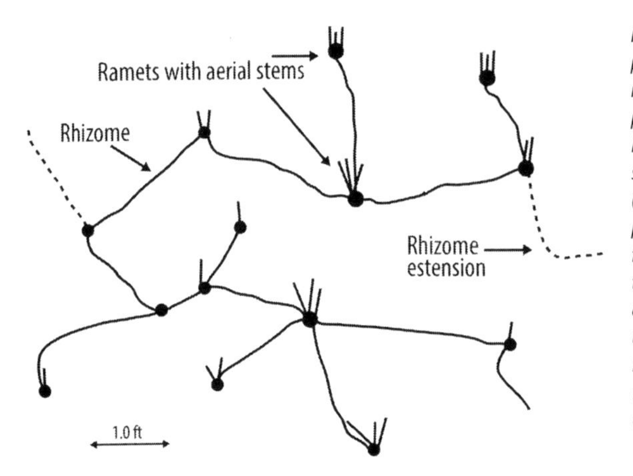

Figure 4-2. Top view of a clonal plant showing rhizomes connecting ramets (black circles). Ramets produce new rhizomes (dashed lines) that increase the horizontal spread of the clone and aerial stems (black vertical lines), that and can produce new clones when separated from the rhizomes. This view is typical of salmonberry (Tappeiner et al. 1993), salal (Huffman et al. 1997), bear clover, and other shrubs and herbs. Several clones may merge to form a dense cover in the understory of forests and in open conditions.

Rhizome growth of Oregon grape in Oregon and beaked hazel in northern Minnesota is much more restricted, and clones are smaller (<6 to 10 ft in diameter) (Huffman and Tappeiner 1997; Tappeiner 1971).

The actual size and age of older clones often cannot be determined. Frequently, the clone origin (the ortet) and the initial aerial stems and rhizomes die or are destroyed by fire or fungi. Rhizomes may decay and become detached from the ortet and from other rhizomes; therefore clones are fragmented into genets.

Clone size and vigor are related to stand density. Large clones with long rhizomes, rapid rhizome growth, and numerous ramets are found on disturbed sites and in low-density stands. In dense conifer stands, clone extent is smaller and rhizome growth and numbers of stems less than on more open sites (Huffman et al. 1994, Tappeiner et al. 1991). Thus, there is a strong inverse relationship between overstory density and clone size and vigor—the production of aerial stems and rhizomes.

Aspen forms clones in much the same way as rhizomatous shrubs, except that aerial stems (suckers) sprout from its roots. Aspen clones may cover tens of hectares or more. They can be identified by leaf phenology, with clones on the same site differing in bud break in the spring or fall coloration. Dye injected into an aspen tree may be transmitted to many other trees, suggesting that clones remain intact via the root systems for several decades (Debyle 1964; Kemperman and Barnes 1976).

Many herbaceous plants (twinflower, bunchberry, foam -flower, and vanillaleaf) are also clonal, with rhizomes or stolons (stems that grow on the soil surface). Some species—goldthread and foam flower, for example (Tappeiner and Alaback 1989)—produce small, compact clones, only 5–25 in in diameter, and annual rhizome extension may be only 0.5–1 in. Other species, such as twinflower, bunchberry, rose, and trailing blackberry, may extend rhizomes (or stolons) 2 ft or more per year. As part of a clone dies from being shaded by taller shrubs and trees or from being covered by leaf litter, its stolons or rhizomes extend into favorable environments. These species seem to wander through the understory (Tappeiner and Alaback 1989). Some species, such as trailing blackberry and poison oak, can become vines (epiphytes) on tall shrubs and trees.

Layering is a method of vegetative reproduction in which aerial stems or branches come in contact with the forest floor and produce new roots and aerial stems (O'Dea et al. 1995; Roorbach 1999). The rooted stem may form a new ramet or an independent plant. Layering is common in vine maple and devil's club. Falling branches and trees or logging slash can land on the aerial stems, pinning them to the ground and causing them to layer. In a study in western Oregon, more than 90% of all layered vine maple stems had been pinned to the forest floor by fallen branches and trees (O'Dea et al. 1995). When aerial stems attached to the parent plant are held in contact with the soil, they root; stems severed from the parent, on the other hand, do not (O'Dea et al. 1995). Successive pinning of layered, rooted vine maple stems can produce clones more than 150 ft in extent. Vine maple cover can be increased during thinning or regeneration harvest by leaving slash on the stems. Cutting or burning the aerial stems and thereby confining vegetative reproduction to small clumps of sprouts at the bases of the dead stems can decrease cover.

Devil's club, a shrub that grows on moist sites, layers in a manner similar to vine maple (Roorbach 1999). However, the weight of its branches alone enables it to contact the forest floor, and it layers without being pinned by fallen limbs. Western redcedar, Engelmann spruce, white spruce, and tanoak also layer, probably when their branches are pushed to the forest floor by snow or fallen branches.

Sprouting Vigor

The vigor of vegetative reproduction (the number of sprouts and the rate of height and diameter growth) varies with species, time of top kill, and size of the sprouting plants. For example, salmonberry sprouts reached 2.0 to 6.0+ ft in the second growing season after burning (Zasada et al. 1994), whereas salal stems of about the same size rarely exceeded 2.0 ft in the same period (Huffman et al. 1994). Killing the tops of salmonberry caused initiation of new aerial stems and rhizomes. Tappeiner et al. (1991) reported that the density of new salmonberry rhizomes was 15–20 times greater in areas where a fire had killed the overstory 2–3 yr previously than in nearby undisturbed populations. Cutting or burning of stems in the spring, just after the leaves are fully developed, usually reduces sprout vigor, probably because carbohydrate levels are low at this time. Zasada et al. (1994) found that bud activity on salmonberry rhizomes in May, just after leaf development, was three times less than it was during July–September, when leaves were fully expanded and non-structural carbohydrate was at its lowest level for the year. Vegetative reproduction and sprout vigor of salmonberry growing in the understory of conifer stands is generally much lower than it is on salmonberry growing in the open (McDonald and Tappeiner 2002; Tappeiner et al. 1996; Tappeiner et al. 2001).

The sprouting potential of hardwood trees is related to tree size. The diameter of Pacific madrone and tanoak sprout clumps (figure 4-1) after cutting, and their leaf area, biomass, and stem number, are all strongly related to the size (dbh) of the parent tree stems (Harrington et al. 1984; Tappeiner et al. 1984). The size of tanoak, madrone, and chinkapin sprout clumps as long as 16 yr after cutting or burning can be predicted from the sum of the basal area of the stems of the parent trees at the time of cutting or burning. Thus, the cover of these species following top kill can be predicted from a stand table if the number of trees in each diameter class is known prior to cutting (Harrington et al. 1991, 1992).

The sprouting of bigleaf maple is related to the height as well as the diameter of the stumps of cut trees. This species can sprout above ground, from buds on the undamaged stem, as well as from below-ground buds at the root collar or from burls (Tappeiner et al. 1996). Sprout clump size was less when the parent tree was cut 1 or 2 yr before harvesting a conifer stand vs being cut at the same time as the conifers (Tappeiner et al. 1996).

Red alder's ability to sprout decreases as trees age. Like salmonberry, red alder sprouts less in the late spring and summer, after leaf production, than during the rest of the year (Harrington 1984, DeBell and Turpin 1989).

Effects of Stand Development and Succession

The previous discussion emphasized that regeneration from seed varies considerably among species, with some species favored by disturbance to the forest canopy and floor, and others more suited to undisturbed environments. Vegetative reproduction is common to nearly all shrubs and hardwoods, although some sprout much more vigorously than others (table 4-1). All are adapted to disturbance and are part of the early-successional vegetation, along with plants that invade from seed banks (Gratkowski 1961; Morgan and Neuenschwander 1988) and from windblown seed. Species like vine maple and salal as well as hardwoods can persist from one stand to the next following disturbance. After disturbance they are part of early serial communities and then may become part of the understory in the next stand.

As tree canopies develop, many species are shaded out (Schoonmaker and McKee 1988). The rate of shading out for each species is related to its shade tolerance and to the overstory species that are present and their density. Shade-tolerant plants such as Oregon grape and vine maple are better adapted to the stem-exclusion stage, when the overstory is dense. Salmonberry will be nearly excluded from dense, young conifer stands at the stem-exclusion stage (Oliver and Larson 1996), but it will develop a dense vigorous understory beneath alder (Tappeiner et al. 1991). Vine maple, Oregon grape, and salal will survive in dense Douglas-fir stands, but they may die if species with dense crowns (high leaf area), such as western hemlock, are present in the overstory. Dense self-thinning stands may be an opportunity to control unwanted species, including exotics. Low-intensity prescribed fire and low concentrations of herbicides can be very effective on plants under severe competition from the overstory. More intensive treatments would often be needed after the overstory is thinned or removed (Tappeiner 1979). Thinning that maintains an open stand will prolong the presence of shrub species and possibly allow shrubs to persist that would otherwise have been shaded out. In that case, hardwoods such as bigleaf maple, tanoak, and madrone, will become part of the main canopy, especially if they are sprouting from large trees. The conifers will overtop them at 10–40+ yr, and then, depending on conifer density, the hardwoods may either be shaded out (as with Pacific madrone, Oregon white oak, and California black oak) or form a second canopy (as with bigleaf maple and tanoak).

Red alder, which is shade intolerant and a prolific seed producer with rapidly growing seedlings, often produces pure, even-aged stands. Shade-intolerant species that produce seed banks (manzanita and *Ceanothus* and *Prunus* spp.) will be shaded out when overtopped by Douglas-fir and true fir, but may persist under open ponderosa pine stands (Oliver 1984, 1990). These species will have established a seed bank before they are shaded out.

After the stem-exclusion stage, shade-tolerant shrubs and herbs begin to reproduce in the understory, both by the establishment of seedlings and by the sprouting and growth of remaining clone fragments. The stem-exclusion stage is probably a critical point in stand development for many plant species because the understory is sparse or patchy, with little competition among understory plants. Salal, Oregon grape, California hazel, vine maple, bigleaf maple, and huckleberry seedlings will become established in the understory at this stage of stand development (Tappeiner and Zasada 1993), along with shade-tolerant conifer seedlings. Seed production of understory plants increases as stands are thinned, providing seed for plant establishment and mast for birds and small mammals (Wender et al. 2004).

Thinning, if it occurs before a dense understory develops, appears to favor establishment of understory shrubs and herbs and overstory trees alike (Bailey and Tappeiner 1998; Bailey et al. 1998). Heavy thinning of the overstory plus soil disturbance may encourage establishment of some early-successional species such as salmonberry, ceanothus, and manzanita. In most instances, soil water seems to be the main factor in increasing understory plant biomass, but several factors are important in increasing the diversity of species. On dry sites in a ponderosa pine forest, soil water was found to be the major variable affecting aboveground understory biomass (Riegel et al. 1992). However, on the same site, an increase in the diversity of understory species was related to increases in multiple variables (light, air and soil temperatures, soil water potential,

Table 4-1. Characteristics of vegetative reproduction and seedling establishment of selected shrubs and hardwoods (1000s of seed/lb). (From Schopmeyer 1974).

Species and Habitat	*Vegetative Reproduction*	*Reproduction from Seed*
Acer macrophyllum: **Bigleaf maple.** Understory tree in older conifer stands, sprout clumps in plantations and burns; overstory tree in young conifer and riparian stands	Sprouts along main stem when overstory is removed and at base when main stem is cut; may form large burls at ground surface or below; vigorous sprouts exceed 9 ft in height in 2–4 yr (Tappeiner et al. 1996)	Seedlings common in understory, especially after thinning; seedling banks; seedlings heavily browsed, slow growth (2.7–10.7/lb) (Fried et al. 1988, Tappeiner and Zasada 1993)
Acer circinatum: **Vine maple**. Large shrubs 12+ ft long; drooping branches are common in the understory of old- and young-growth conifer stands; sprout clumps in burns and clearcuts	Sprouts along main stem after layering; layers readily when stems pinned to ground; sprouts vigorously from base when stems are killed; clones more than 100 ft in extent (O'Dea et al. 1995)	Seedlings in understory favored by thinning, generally low numbers of seedlings; seedling growth is slow (3.5–5.5/lb) (O'Dea et al. 1995, Tappeiner and Zasada 1993)
Alnus rubra: **Red alder.** Early successional tree, forms dense stands on disturbed coastal sites and in riparian areas	Sprouts mainly at base of stem; young trees have high sprouting capacity; older trees (5+ yr and 6 in dbh) sprout with low vigor (Harrington 1984, DeBell and Turpin 1989)	Abundant seed, seedlings established on open, disturbed, moist sites on north slopes; seedlings grow rapidly (3+ ft in 3 yr, 380–1100/lb) (Haeussler et al.1995)
Arbutus menziesii: **Pacific madrone.** Sprout in burns and clearcuts; understory tree in open conifer stands and overstory tree in young conifer stands and pure madrone stands	Burls below ground; sprout growth vigorous and positively related to size of parent tree (Harrington et al. 1984, Tappeiner et al. 1984)	Small seed in berries, dispersed by animals; clumps seedlings favored by disturbed forest floor and part shade; reproduce in open stands after fire or thinning (197–320/lb) (Tappeiner et al. 1986)
Arctostaphylos patula: **Greenleaf manzanita**. Dense cover 3–6 ft tall; early-successional species; common as seedlings and sprouts in burns and clearcuts and older shrubs in the understory of open ponderosa pine stands	Sprouts from below ground, burls; populations with burls sprout vigorously, those without burls sprout with low vigor	Store seed in the soil/ forest floor; 10,000s per ac germinate after fire or regeneration harvest; form dense cover in 3–4 yr (18–27 seed/lb) (Hughes et al. 1987)
Arctostaphylos viscida: **Whiteleaf manzanita.** Dense cover 3+ ft tall; early-successional species; common as seedlings in burns and clearcuts and older shrubs in the understory of open ponderosa pine stands	Sprouts at stem base; sprouts with low vigor, may not sprout if entire top is killed	Seed stored in the soil/ forest floor; 10,000+ per ac germinate after fire or regeneration harvest; form dense cover in 3–4 yr. (Hughes et al. 1987)

Table 4-1. cont.

Species and Habitat	*Vegetative Reproduction*	*Reproduction from Seed*
Berberis nervosa: **Oregon grape.** Understory dense compact shrub 2+ ft tall in understory of old and young Douglas-fir forests; persists after disturbance as dense cover (generally < 1 ft) in burns and clearcuts	Sprouts from rhizomes 1-8 in below ground and base of aerial stems; clones are compact; clone density developed from rhizomes (Huffman and Tappeiner 1997)	Seedlings common in understory of conifer stands, coalesce to form dense mats; seedlings germinate/ survive on undisturbed soil (30–57/lb) (Huffman and Tappeiner 1997)
Ceanothus cordulatus: **Buckbrush.** Slow growing; stems 3-6 ft tall; early-successional moderate	Low sprout vigor, growth	Reproduces from seed stored in the forest following fire or mechanical disturbance; seedling growth is
Ceanothus integerrimus and *velutinus:* **Deerbrush and snowbrush or varnishleaf.** 6–9ft tall, rapidly growing early-successional species	Numerous sprouts from base of large shrubs and seedlings; sprouts grow rapidly from base of stems of large shrubs	These species store seed in forest floor; seedlings rapidly germinate and grow to form dense cover 3–4 yr after disturbance, especially fire; deer- and snowbrush seedlings grow rapidly (58–152/lb) (Cronemiller 1959; Hughes et al. 1987; Quick 1956, 1959)
Ceanothus prostratus: **Mahala mat.** Low-growing shrub 1.5 to 3.0 in+ tall that forms mats on open sites	Spreads by rhizomes that grow on or just below the soil surface	Stores seed in the forest floor that germinates in the area of former mats after fire; seedlings grow together to form dense mats; mats are microsites that favor natural conifer regeneration (37–47/lb)
Chamaebatia foliosa: **Bearclover.** Dense cover 15–20 in tall in the understory of open pine and mixed conifer stands	Rhizomes 2–25+ in below ground; stems 5–30+ in tall; sprouts vigorously from rhizomes as deep as 15+ below the soil surface after tops are killed	Seedlings small and grow slowly in partial shade; little rhizome development after 10 yr
Corylus cornuta: **Hazel.** Compact, multi-stem shrubs; scattered in understory 6–12 ft tall; sprout clumps in clearcuts and burns	Base of stems; rhizomes; compact clone with multiple stems 1–10 ft long; moderate sprout vigor	Low rates of seed production; large seed dispersed by animals; occasional seedlings in understory (0.4–0.7/lb)

Table 4-1 cont.

Species and Habitat	*Vegetative Reproduction*	*Reproduction from Seed*
Gaultheria shallon: **Salal**. Forms dense cover 2–6ft tall in understory of old and young Douglas-fir forests; persists after disturbance as dense, short cover (generally <3 ft) in burns and clearcuts	Sprouts from rhizomes 2–20 in below ground; vigorous rhizomes may extend 3+ ft/yr and form dense canopy; 100,000+ buds/ac on rhizomes; high sprout density but slow height growth when tops are killed; clones can exceed 30+ ft in extent; main stems that die occasionally are replaced by new sprouts from rhizomes (Huffman et al. 1994, Tappeiner 2001)	Regular annual seed production; many seeds/ berries; animal dispersed; seedling establishment common on decaying wood in the understory; seedlings very small, <1 in at 2–3 yr (2600–3800/ lb;) (Messier and Kimmins 1991, Huffman et al. 1994, Tappeiner and Zasada 1993)
Holodiscus discolor: **Oceanspray.** Compact multi-stem 6–12 ft tall; scattered in understory; sprout clumps in clearcuts and burns	Compact clone with multiple upright stems 3–10 ft+ long; sprouts with moderate vigor from buds at the base of the stems	Light, windblown seed; seedlings established on shady disturbed sites grow slowly (5340/lb)
Lithocarpus densiflora: **Tanoak**. Understory tree seedlings; sprout clumps in burns and clearcuts; forms a secondary canopy in old conifer stands	Burls below ground; dense sprout clumps 6-10 ft tall by 6 ft diameter produced 4–5 yr after top kill; sprout clump size strongly, correlated with parent stem basal area (Ahrens 1990, Harrington et al. 1992 and 1984, Tappeiner et al. 1984)	Animal-dispersed seed; slow growth 4-8 in in 8–12 yr; seedling banks common in understory of conifer and hardwood stands; seedlings sprout after tops are killed (0.06–0.11/lb) (Tappeiner and McDonald 1984, Tappeiner et al. 1986)
Oplopanax horridum: **Devil's-club.** Tall spreading shrub 2–3 m; occurs in understory or in openings on wet sites	Commonly spreads by layering after branches contact the ground	Large seed in berries (Roorbach 1999)
Populus tremuloides: **Quaking aspen.** Early-successional tree; forms dense groups or small stands in openings in conifer forests or along meadow edges	Root suckers; dense stands of root suckers result from top kill of trees; individual aspen clones may extend more than several ac (Kemperman and Barnes 1976)	Seed easily wind dispersed; requires disturbed, moist sites for seedling establishment (2500–3000/lb) (Zasada 1986)
Prunus spp.: Early-successional trees and shrubs; generally scattered, overtopped by conifer in 20–30 yr	Rapid sprout growth from stem base when top is killed	Buried seed that germinates after disturbance; seedlings grow rapidly but are often browsed (2.8–14.0/lb)

Table 4-1 cont.

Species and Habitat	*Vegetative Reproduction*	*Reproduction from Seed*
Pteridium aquilinum: **Bracken fern.** 2–6+ ft tall; occurs in understory and in open after disturbance	Rhizomes appear to spread rapidly after disturbance; spreads rapidly after thinning (Bailey and Tappeiner 1998)	Produces spores but the germination of spores and gametophyte generation rarely seen in the field
Quercus sp.: **California black oak** and **Oregon white oak**. Occur in pure stands or in open conifer stands as either understory or overstory trees; sprout clumps after fire or clearcutting	Sprout along main stem when overstory is removed and from the stump when main stem is cut; vigor of stump sprouts related to tree size (McDonald 1978a)	Seed, dispersed by animals, cached; seedlings plentiful in the understory of conifer stands and in open sites; seedlings grow slowly (0.05–0.18/lb) (McDonald and Tappeiner 2002)
Rubus parviflorus: **Thimbleberry.** Forms dense cover 2–6 ft+ tall on open sites	Rhizomes spread rapidly and form dense cover of stems 4 ft tall	High rates of seedling establishment after disturbance; rarely found in dense understory
Rubus spectabilis: **Salmonberry.** Dense stems; 6–12+ ft tall under alder and in open; generally low, sparse cover in conifer stands	Produces rhizomes and aerial stems; rhizomes contain 100,000s buds/ac; dense sprouts 6+ ft tall in 2 yr after cutting or fire; maintains cover by annual production of sprouts that replace the few that die (Tappeiner et al. 1991, 2001; Zasada et al. 1994)	Seed dispersed by animals; probably develops a seed bank; some seedlings germinate on undisturbed sites; high rates of germination after disturbance; seedlings grow 8+ in tall in 3–4 yr in the open (143/lb) (Tappeiner and Zasada 1993)
Symphoricarpus albus: **Snowberry.** Compact shrub 2-4ft tall; open stands	Rhizomes spread and produce burls and aerial stems	(54–113/lb)
Vaccinium spp.: **Red huckleberry/ blueberry.** Understory shrub 2–6ft tall	Rhizomes grow slowly; limited sprouting; sprouts generally fairly short stems (<3 ft)	Frequent, plentiful berry production, especially on open sites; seedlings germinate and become established on dead wood in understory (2400–3200/lb) (Minore et al. 1979; Wender et al. 2004)

pH, and N) following reduction in overstory canopy and root density (Riegel et al. 1995). On more productive sites with denser overstory canopies, light alone might be the primary variable controlling both growth and diversity of understory species.

Persistence of Shrub Communities

Shrub communities in the understory of forest stands can be persistent. As we have discussed, covers of manzanita and ceanothus may delay development of a conifer stand for well over 40 yr. Overtopping by conifers will kill these shrubs, but their reproductive potential persists as a seed bank in the forest soil that will germinate after disturbance.

Rhizomatous shrubs often become established from seed in the understory of conifer stands. This behavior is typical of stress tolerators (Grime 1981), which persist and grow slowly at low levels of light and soil moisture. Disturbance to the overstory will produce increases in the growth rates and densities of aerial stems and rhizomes of stress tolerators. Unlike early-successional seed bank species, their seedling growth rates are slow. However, like early-successional shrubs, once a dense cover of a clonal species is established, it may be quite persistent. Rhizomes of shrubs grow among each other, and the clones coalesce to form a dense mat below the ground. In open conditions, more than 33 ft of salal rhizome for each ft^2 of surface area can be present (Huffman et al. 1994; figure 4-2). Salal seedlings are rare; it appears that this species form dense covers by rhizome expansion and aerial stems above ground. For Oregon grape and hazel, the extent of rhizomes and clones is smaller than for salal, salmonberry, and bearclover. However, those species apparently establish seedlings more readily than does salal (Huffman and Tappeiner 1997; Tappeiner 1971).

As the rhizomes of shrubs expand, they produce a bud bank below ground, enhancing their persistence. Zasada et al. (1994) and Huffman et al. (1994) estimated that dense populations of salmonberry and salal produce one bud for every 1–2 in of rhizome. Rhizome density of salmonberry ranged about 1–5 ft of rhizome per ft^2 of land surface area; for salal, the range was about 3.5–40 ft of rhizome per ft^2 of land surface area (Tappeiner et al. 2001). Thus there are potentially hundreds of thousands of buds per ac in the rhizome bud bank beneath shrub communities.

This bud bank is the source of community stability for two reasons. First, after a disturbance that kills the above-ground stems, many of the buds in the rhizomes are released, producing new aerial stems at a density that generally exceeds that of the stems that were killed (figure 4-3). Second, even without disturbance, some new aerial stems sprout annually from the rhizomes. For example, salmonberry populations produced 1+ new stem per ft^2 annually over an 8-yr period (Tappeiner et al. 2001). These stems appeared in the spring, and most died or were browsed by winter; however, when a large aerial stem died, it was replaced by one of the new stems. Thus, annual sprouting maintained a closed shrub canopy and probably kept the shrub layer from being

Figure 4-3. Salal (A) and salmonberry (B) communities. Bundles of rhizomes were excavated from 11 ft2 plots in the salal and 44 ft2 plots in the salmonberry communities like those in the background in A and B. The salal was in the understory of a 50–60-yr-old Douglas-fir stand; the salmonberry was in a 30–40-yr-old red alder stand. Rhizome lengths of more than 30 ft/ft2 for salal and 3 ft/ft2 for salmonberry are common (Huffman et al. 1995, Tappeiner et al. 1992). The salmonberry population in C developed under an alder stand that was cut and burned about 5 yr previously.

replaced by other species. In a 19-yr study of understory hazel populations, Kurmis and Sucoff (1989) found that stem density varied by several thousand stems/ha during that period, but the number of stems after 19 yr was nearly the same as it was in year 1—even though the age of the oldest stem was only about 15 yr. The whole population of aerial stems turned over, but the shrub cover remained relatively stable.

Thus it appears to be a general characteristic of clonal populations of shrubs that they do not self-thin as even-aged populations do. Rather, they are uneven-aged populations of aerial stems in which annual sprouting replaces the stems that die, thereby maintaining the population indefinitely (Balough and Grigal 1987; Kurmis and Sucoff 1989; Huffman and Tappeiner 1997; Huffman et al. 1994; Tappeiner et al. 1991, 2001). Although a major disturbance will stimulate a dense reinitiation of aerial stems, such a disturbance is not necessary for shrub communities to maintain a dense canopy.

These dense, persistent communities have an important influence on forest succession by inhibiting the establishment of trees and other species. If a gap occurs in the overstory, the shrub density often increases and interferes with tree establishment. Intense fire, herbicides, or mechanical disturbance might be needed to kill rhizomes, reduce sprout density, and enable the establishment of other species.

REVIEW QUESTIONS

1. How are shrubs and hardwoods important to forest ecosystems?
2. Outline Grime's classification of species strategies. How do plants with these three strategies grow at different stages of stand development?
3. What is layering? Explain how vine maple layers. How can vine maple layering be encouraged or reduced?
4. How can shrub communities (for example, salal, salmonberry, manzanita, and ceanothus) maintain themselves in forests?
5. What are seed banks, how are they formed, and how can you predict that a seed bank is likely to be present? How can you predict sprouting vigor?
6. What are bud banks and where do they occur? Why are they important in the practice of silviculture? How can you predict that a bud bank is present?
7. Explain how the vigor of sprouting is related to shrub and hardwood size and vigor.
8. What is the importance of reproduction by seed vs vegetative reproduction for clonal forest shrub?
9. What are rhizomes, burls?

CHAPTER 5

Growth of Forest Trees and Stands

Introduction

Understanding the growth of trees and forest stands is basic to the practice of silviculture. Thinning, density of planted or natural seedlings, competition, fertilization, and prescribed fire all affect the density of a stand, tree growth, and stand development. It is growth that drives stand dynamics and development and leads to changes in tree dimensions and stand structure. In the first part of this chapter we present basic concepts of tree and stand growth; in the second part we discuss how these concepts can be applied to the practice of silviculture.

Some of the information we present provides a background for better understanding tree and stand growth, but not all of it is developed enough to be applied directly to the practice of silviculture. Many important concepts of tree and stand growth have been discovered via intensive research. For example, we know that the growth of trees varies along the stem in relation to variables such as the force of the wind and the position and size of the crown. We also know that the growth of dominant and codominant trees is affected by stand density and wind on some sites. It is not currently possible, however, to reliably predict the effect of these variables on stand growth. An appreciation of these concepts will, we hope, stimulate development of methods by which to apply them to the management of forest stands in the future.

Growth of Forest Trees

TIMING OF GROWTH

Most forest trees begin shoot growth in late winter or spring. At that time, stem units (precursors for stem, foliage, and reproductive parts) formed in the buds during the previous growing season expand and produce new leaves and shoots—leaders—on the branches and main stem (Daniel et al. 1979). Shoot growth of Douglas-fir begins in early spring in western Oregon and stops in August, whereas cambium growth continues into October. Emmingham (1977) found that the growth rate of trees from coastal seed sources was greater than that of trees from interior sources, but timing of growth was nearly the same for both. Height growth has been shown to be under genetic control to some degree. Twenty-year height growth of ponderosa pine seed sources, planted on the same site, was greater for seed sources from mid-elevation than from of high-elevation seed sources (Conckle 1973; Kitzmiller 2005).

Most western conifers and hardwood trees usually have only one flush of growth during most growing seasons. Their growth is largely determinate; that is, the number of cells in the bud set during the previous growing season determines the shoot elongation in the current season. Ponderosa pine and Douglas-fir may initiate a second flush of stem growth late in the season (known as lammas growth) on sites where a combination of deep soil and lack of competition ensure plentiful soil moisture (Roth and Newton 1996a, b). Second flushes of growth are usually not as large as the initial ones. Some species (Monterey and knobcone pine) exhibit determinate/free growth—that is, growth is determined by the cells fixed in the bud during the current growing season (Daniels et al. 1979), but new cells may be formed for additional shoot growth. Other species (western redcedar and alder) exhibit indeterminate growth—that is, their shoot growth depends largely upon current season growing conditions.

HEIGHT GROWTH

Trees can grow tall on productive sites where generally more water and possibly more nutrients are available for tree growth, and where the soil depth enables a large rooting volume. As trees become tall, their height growth slows, because it becomes increasingly difficult for them to move water from the soil to their tops against the force of gravity and internal resistance to water movement (Ryan and Yoder 1997). Thus the tallest trees occur on sites with deep soils where soil water is optimal (Waring and Franklin

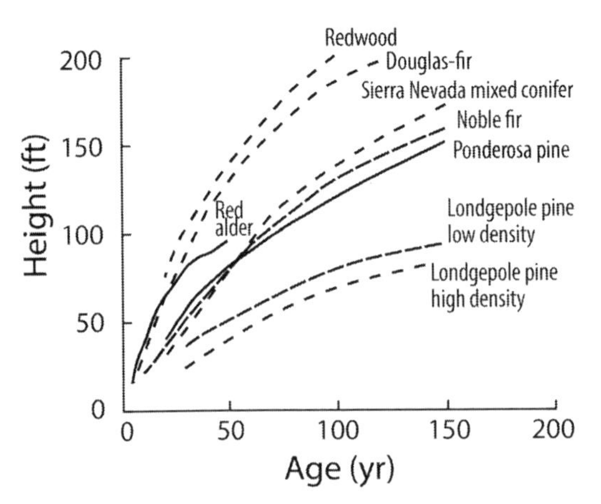

Figure 5-1. Height growth curves of selected western conifers on productive sites. Redwood: SI 200 ft @ 100 yr (Lindquist and Palley 1963); Douglas-fir: SI 120 ft @ 50 yr (King 1966); Red alder: SI 65 ft @ 20 yr (Harrington and Curtis 1968). Noble fir: SI 130 ft @ 100 yr (Herman et al. 1978); Lodgepole pine: SI 80 ft, crown competition factor <125 and 300 (Alexander 1967); Sierra Nevada mixed conifer: SI 80 ft @ 50 yr (Dunning and Reinecke 1933); Ponderosa pine: SI 120 ft @ 100 yr (Meyer 1938). *SI=site index, height of the dominant and codominant trees at the base ages of 20, 50, or 100 yr*

1979a, b). Height growth is most rapid within the first 20–40 yr for trees growing free of competition from shrubs or adjoining trees (figure 5-1). Each tree species has its inherent height-growth pattern that is modified by the effects of competition from shrubs and adjoining trees, as well as environmental effects such as soil conditions, wind, relative humidity, snow and ice storms, and insects.

Rates of height growth and the general shapes of the height/age relationships of dominant and codominant trees may vary with soil type and elevation. Carmean (1956) found that trees on gravel soils grow more rapidly at early ages than trees on soils derived from basalt—a 30 ft height difference at 50 yr. However, after 50 yr, height growth slowed for trees on gravel soils, and at 100 yr, dominant and codominant trees on the two soils were the same height. At 100 yr, height growth for trees on basalt was continuing; it had stopped for those on gravel soils. Pattern of height growth was related to root development and soils. Ponderosa pine height growth increased when roots grew through the coarse-textured pumice layers into finer-textured lower layers (Hermann and Petersen 1969). Douglas-fir height growth varies by elevation and from the Cascades to the Coast Range (Curtis et al. 1974).

Height-growth patterns of red alder, Douglas-fir, noble fir, and redwood are distinctly different (figure 5-1). Red alder grows more rapidly than most conifers for its first 10–20 yr but much more slowly thereafter (Harrington and Curtis 1986). Redwood's early height growth is rapid because it sprouts from stumps and uses reserves stored in the roots. Like Douglas-fir, it can maintain relatively rapid height growth at older ages, especially on productive sites. Noble fir, like other true fir species in the West, grows slowly during its early years but sustains height growth well beyond 100 yr. It grows only to about 80 ft at 50 yr, compared to nearly 130 ft for same-aged Douglas-fir, but it grows to 160+ ft by 150 yr (figure 5-1; Herman et al. 1978). In Sierra Nevada mixed-conifer forests, dominant trees of the principal species (ponderosa pine, Douglas-fir, white fir) can grow in height at the same rate (Dunning and Reineke 1933). Lodgepole pine has a much slower growth rate and does not become as tall (often <100 ft) as other conifer species (figure 5-1).

Height-growth rates are often determined by reconstructing past height-growth patterns of dominant and large codominant trees (King 1966). This is done either by measuring the height to whorls of branches formed in successive years or by sectioning trees to determine ages at successive heights. Dominant and large codominant trees are used to determine potential height growth for a species because, compared with intermediate or small codominant trees, their growth rate is less likely to be affected by competition from surrounding trees.

Height-growth curves for dominant and codominant trees represent the potential height growth for a species on sites of a given quality. These curves do not usually account for the effects of early competition on tree growth. However, during stand

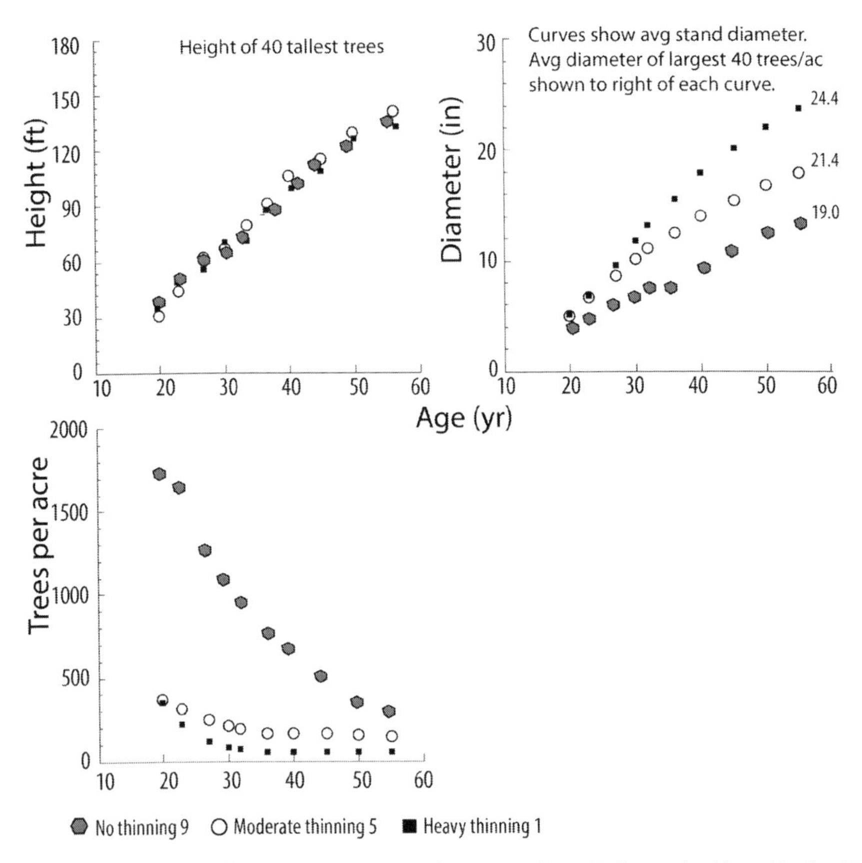

Figure 5-2. Height of the tallest 40 trees/ac, average diameter, and trees/ac in stands with no thinning (9), moderate thinning (5), and heavy thinning treatment (1). Adapted from Marshall and Curtis (2002).

establishment, competition from shrubs, sprouting hardwoods, and other trees occurs on many sites, reducing height growth for several decades. Oliver (1990) found that dominant and codominant ponderosa pine growing with shrubs in the understory averaged 6 ft less in height at 30 yr than those growing where shrubs had been removed. Similarly, competition from shrubs and sprouting hardwoods caused an approximate 6 ft reduction in the cumulative height growth of 12-13-yr-old Douglas-fir in southern Oregon Harrington and Tappeiner 1997) and a 16 ft reduction by the time the same trees were 24-25- yr old (Harrington and Tappeiner (2009). This early reduction may be overcome if the competition is removed silviculturally or shaded out when the conifers overtop it. Douglas-fir and white fir growing under an overstory grew an average of 25 cm/yr in height; growth increased to 60 cm/yr 10 yr after the overstory was removed (Tesch and Korpella 1993). Similarly, white fir growing in a brushfield of manzanita and ceanothus may reach heights of only 4–6 ft after 30+ yr, and then grow 2+ ft/yr in height after overtopping the shrubs or being released from them. It appears that early suppression of height growth will increase the time it takes trees to reach their potential height, but it will not prevent them from reaching it.

HEIGHT GROWTH AND STAND DENSITY

Assuming the same age and site quality, average height is always less in dense stands than in more open stands, because the former have more trees in the lower crown classes. Unlike diameter, the height of the largest trees is often independent of stand density. In coastal Douglas-fir forests, there was no practical difference in the height of the 40 largest trees/ac at densities ranging 50–500+ trees/ac, although there were large differences in diameter in these stands (figure 5-2; Marshall and Curtis 2002).

However, stand density may affect the height growth of dominant and codominant trees of shade-intolerant species on sites of low productivity. It is often assumed that, because their crowns are in the main canopy, the height growth of these trees is independent of stand density. However, high stand densities can reduce height growth of dominant and codominant lodgepole and ponderosa pine (Alexander et al. 1967; Holmes and Tackle 1962). Similarly, the height growth of Douglas-fir on lower site classes was reduced at higher stand densities (Curtis and Reukema 1970), although part of the reduction was likely caused by soil characteristics (Miller et al. 2004).

When stands of trees are so dense and uniform that height growth is strongly inhibited, overall stand growth can stagnate. In the case of lodgepole and ponderosa pine growing on low-productivity sites, high stand density can reduce the growth of even the largest trees to only a few cm/yr. When height growth is markedly slowed, trees do not differentiate into crown classes, and little suppression mortality (self-thinning) occurs. Some disturbance or thinning is probably needed to overcome stagnation.

In the first 5 yr after planting, growth of Douglas-fir actually increases with planting density (Cole and Newton 1987; Scott et al. 1998), after which it decreases. Increased growth with density occurs at very high densities (550–1200 trees/ac) but only for 2–8 yr after planting (Woodruff et al. 2002). Because of its short duration and the high densities at which it occurs, this growth increase is not likely to have practical application in silviculture; precise timing of precommercial thinning to prevent a reduction in crown size and diameter growth would be needed to take advantage of this brief acceleration. Apparently, this transient stimulus of growth in both height and stem diameter is triggered by changes in the ratio of red to far-red light within the stand (Ritchie 1997). Transient growth increases from high-density plantings have also been observed in red alder (Knowe and Hibbs 1996), loblolly pine (Pienaar and Shiver 1993), and eastern cottonwood (Krinard 1985).

DIAMETER (XYLEM) GROWTH

As buds expand in the spring, they produce growth regulators that migrate from the buds into the branches and main stem. These regulators initiate cambial activity and growth of phloem and xylem (Daniel et al. 1979; Zimmerman and Brown 1971). Phloem is the tissue just beneath the bark that transports photosynthates downward from the foliage. Xylem, which forms the main part of the stem, transports water and nutrients and provides the structure that, along with the roots, supports the crown of the tree.

In conifers, xylem cells produced early in the spring—early wood—are large and have thin walls. Cells produced later, after foliage development stops—late wood—are smaller and have thick cell walls. In some hardwoods, such as *Quercus* spp., early wood contains large vessels and late wood consists of small vessels and fibers. These species are called ring-porous species. In other hardwoods, such as *Alnus* and *Populus* spp., vessels and fibers occur throughout the ring, making them diffuse-porous species. In most conifers and ring-porous hardwoods, it is easy to count the annual rings, especially when growth is relatively rapid (i.e., the tree has >10–15 rings/in). In contrast, the annual rings in diffuse-porous species are difficult to distinguish from one another.

In very dense stands, resources for xylem growth may be so limited that xylem rings are produced only in the crown, not at the base of the tree. Cessation of xylem growth is generally synchronous with the late-season decline in soil water in most western forests. Xylem growth increases under favorable moisture conditions and decreases during periods of drought, especially in dense stands.

Sapwood, the moist outermost band of xylem tissue around the stem of the tree, transports water and nutrients from the roots to the crown. Within the crowns of young trees the stems are often nearly all sapwood. Because water must move through it, the amount of sapwood in the stem is related to the total amount of foliage in the crown (Büsgen and Münch 1929; Grier and Waring 1974). Trees with large bands of sapwood (high sapwood basal area) are generally more vigorous than trees on the same site with smaller bands. Tree vigor can be evaluated by measuring volume, basal area, or diameter growth in relation to sapwood area (Mitchell et al. 1983; Waring et al. 1980).

Wood strength is related in part to the density and micro-anatomical properties of xylem cells and the number and size of knots—the remains of branches in the xylem. Wood density is determined largely by the proportion of late wood (with small, thick-walled tracheids) to early wood (with large tracheids). The density of wood varies with its position in the tree. Cell walls are thicker, cells are smaller, and wood is generally denser near the base of the tree and near the bark than farther up the tree and near the pith (Daniel et al. 1979; Zimmerman and Brown 1971).

Wood density is more closely related to location in the tree than to the tree's radial growth rate. Wood produced in the stem near the pith—the center of the stem—is generally quite different from wood produced near the bark. The wood produced from the pith to 15–20 yr (rings) out is called juvenile wood. It is characterized not only by large, thin-walled tracheids with low specific gravity, but also by low cellulose content and high microfibril angle. Juvenile wood shrinks more than mature wood when it is cut into boards. Trees grown at high densities when they are young, or grown in an understory and then released, may have small cores of juvenile wood, while trees growing rapidly when they are young will have large cores of juvenile wood; however, the rapid growth rate will also result in high rates of production of mature wood after the trees are about 15–20 yr old. Xylem growth in mature wood may produce wide rings, but this will not usually affect wood density because in general the proportion of late wood to early wood increases as the tree grows its annual ring. In the past, low wood quality was associated with wide rings, but in those cases wood with wide rings was actually juvenile wood, because the widest rings typically occur in the first 10+ years from the pith. As Megraw (1986) notes, juvenile wood is a convenient term to describe the wood produced near the pith; however, the duration of the juvenile period varies among tree species and stand conditions. In Douglas-fir, the complete transition between juvenile and mature wood might last several decades.

DIAMETER GROWTH AND STAND DENSITY

Diameter growth in young stands is often affected by competition from shrubs, grasses, and overtopping trees (White and Newton 1989, Harrington and Tappeiner 1997) and by competition from adjoining trees in older stands (figure 5-2). As trees are released from competition, crowns expand and diameter growth increases. Diameter growth is strongly and inversely related to stand density (figure 5-2). For example, average diameter of Douglas-fir at 55 yr ranged from 24 in under heavy thinning to 13.5 in with no thinning (Marshall and Curtis 2002). Part of the change in diameter is the arithmetic result of removing the smaller trees during thinning, but more is due to growth acceleration in the less dense stands. In the heaviest thinning, diameter growth was 0.53 in/yr at 40 yr and 0.32 in/yr at 55 yr. In the unthinned control, these values were 0.15 and 0.13 in/yr, respectively. It is important to note that not only average diameter, but also the diameter of the largest trees, is related to density. Thirty-one years after thinning, diameter range for ponderosa pine at 140–150 trees/ac was 11.9–12.8 in, whereas the range for trees growing at 380–480 trees/ac was only 7.7–8.2 in (Cochran and Barrett 1993).

As mentioned above, individual trees have smaller crowns in dense stands, and thus each tree has less capacity for diameter growth than trees with larger crowns in less dense stands. The total stand leaf area may be similar in two stands but the amount of leaf area per tree varies with the number and species of trees present.

DIAMETER GROWTH AND STEM FORM

The rate of diameter growth—that is, xylem ring width—and stem cross-sectional area growth are also related to crown size and position of the tree within the canopy (Kershaw and Maguire 2000; figure 5-3). Trees with large crowns and high amounts of leaf area per tree have greater diameter growth rates and larger stems than trees with small crowns. Foliage mass is often greatest in mid-crown, but relatively more foliage occurs lower on the stem in dominant trees, and higher on the stem in trees with short crowns (Maguire and Bennett 1996). In open-grown trees, maximum diameter and stem-area growth occur lower on the stem than in stand-grown trees (Duff and Nolan 1953; Larson 1963). As crowns close during stand development, the point of maximum growth shifts up the stem (figure 5-3). Maximum diameter growth often peaks in the crown, whereas maximum stem-area growth may peak in the crown or below the crown. Suppressed tress with very small, receding crowns may not produce xylem rings at breast height and the lower parts of the stem (Gray 1956; Larson 1963). Stem-area growth rates can remain constant for many years in old trees that apparently once had big crowns and were the larger trees in the stand (Poage and Tappeiner 2002).

Stem cross-sectional increment is often assumed to increase with distance from the tip of the tree to crown base, but to remain constant below crown base (Pressler's hypothesis; see Assmann 1970). In a study of the stem-area growth of two shade-tolerant species from New England and the Pacific Northwest, Kershaw and Maguire (2000) found that stem cross-sectional area growth generally increases from the top of the tree down as foliage or crown area increases. The increase was not linear, however; it leveled off with increasing distance from the top of the crown. Stem-area growth below the crown did not conform well to Pressler's hypothesis, and probably varies by species, age of the tree, crown size, and distance from the top of the crown.

Several theories have been proposed to explain variations in stem form in trees. Open-grown or large-crown dominant and codominant trees have more taper to their stems than smaller trees because they maintain relatively rapid rates of growth on the lower part of the stem. Taper is less in stand-grown trees with small crowns because the growth in the upper part of the stem, near the base of the crown, is greater than the growth in the lower part of the stem.

Butt-swell at the very bottom of the tree, starting at or below the surface of the ground is also part of stem form. Butt-swell is most apparent in western hemlock and Sitka spruce. Trees with large crowns have the most butt-swell. It is apparently a mechanism that helps support the tree, but how it develops is not well understood (Larson 1963).

The mechanical stresses of wind have been shown to affect stem form. Trees must support themselves against the forces of gravity, wind, and ice and snow on their crowns, as well as the weakening of their stems and roots from pathogens or other mechanical or biological stresses. Jacobs (1954) guyed trees to reduce swaying from the

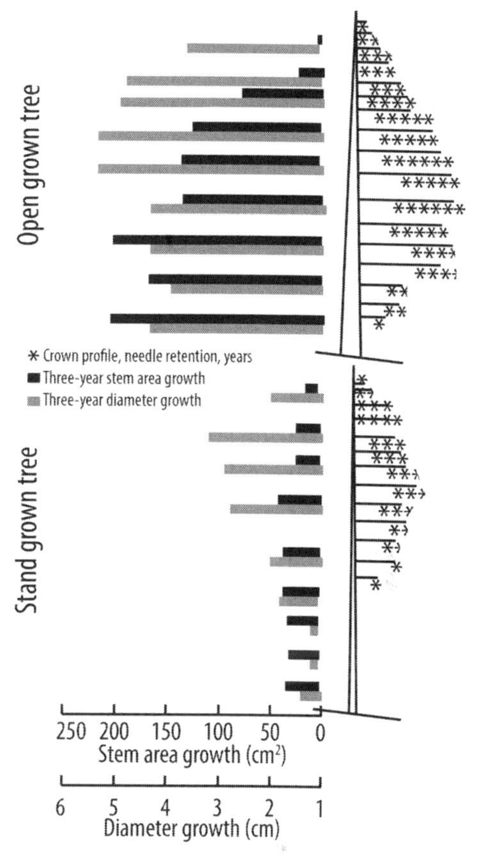

Figure 5-3. Three-year cross-sectional area and diameter growth between every third whorl of branches in an open-grown and a stand-grown Douglas-fir. The profile of the crowns, including branch length and needle retention, is shown to the right. The trees were about 30 yr old and 60 ft tall. The open-grown tree has a larger crown, longer needle retention, and greater radial and stem-area growth than the stand-grown tree. In the stand-grown tree, the stem area and radial growth decrease below the crown.

wind and found that diameter growth increased above the support, whereas the diameter of unsupported control trees did not increase. Thus, allocation of growth can be at least partly determined by forces on the stem. When the guys were removed, the trees were no longer stable and were blown down. Following thinning, stem form changes as stem diameter increases relative to height (height-diameter ratios decrease) and this response helps to reduce wind damage (Mitchell 2000; Meyer 1963; see chapter 8).

Trees with low height–diameter ratios (trees with taper) are known to resist the forces of wind, ice, and snow better than those with large height–diameter ratios (Mitchell 2000; Wilson and Oliver 2000; Wonn and O'Hara 2001). Working with nine North American conifers, Dean et al. (2002) concluded that stem diameter is proportional to the bending moment at the place on the stem to which the force is applied. Thus stem taper maintains a relatively uniform bending curvature that supports the stem against mechanical forces.

As trees are subjected to the forces of wind, ice, and snow, or to loss of mechanical support when weakened by decay fungi, they tend to maintain themselves in an upright position by forming reaction wood along the stem. Formation of reaction wood is a growth response of the xylem caused by the movement of growth regulators to the affected zone of the stem (Daniels et al. 1979; Zimmerman and Brown 1971). Apparently, the reaction is not caused by the force itself but by the effect of gravity on the distribution of growth regulators after the stem bends. In conifers, reaction wood is often termed compression wood because the reaction occurs on the side of the stem opposite to the stress, causing increased xylem growth in that part of the stem. In hardwoods, it is called tension wood because the reaction occurs on the same side of the stem where the stress is applied. In either case, the result is a thickening of the stem from an increase in xylem growth on one side. Reaction wood increases shrinkage and warping in lumber and causes significant reduction in wood quality. Western hemlock apparently responds to wind forces by developing flutes at the base of the stem. Julin et al. (1998) found that flute depth was greatest at the base of the tree and decreased with height and that, at a given height, flute depth increased with age. When these researchers guyed trees to support them against wind, the rate of flute deepening decreased. Although theories involving nutritional gradients in the crown and the effects of mechanical forces such as wind and bending help explain variation in stem form, Larson (1963) suggests that a better understanding of hormone production and distribution in trees will provide a physiological explanation of how it occurs.

Crown and Canopy Structure and Dynamics

Crown and canopy structure determine the value of a stand for many forest resources. In this book, crown is defined as the live branches and foliage of individual trees, and canopy is the stand-level composite of individual tree crowns. The amount and spatial arrangement of foliage controls the interception of solar radiation, the photosynthetic potential of the canopy, and the net primary productivity (NPP) of the stand. The canopy is primary habitat for many species of wildlife and for algae, lichens, and arthropods. It regulates precipitation throughfall and microclimate within the stand, and consequently influences factors such as water yield, fire behavior, and the composition and development of the understory vegetation.

Crowns are most often characterized by length, width, percentage of the tree stem they cover (live crown ratio), and height above ground. Total leaf area, nutrient content, individual branch size and distribution, and foliage length are also important descriptors. Information derived from needle characteristics, such as nutrient concentrations and photosynthetic rates within crowns, may also be of value for predicting or understanding responses to treatment and for diagnosing certain types of stresses or diseases. The most commonly measured—and probably the single most important—crown dimension is live crown length, or the length of the stem containing contiguous live branches. Despite some ambiguity in defining crown base, or the lower limit of live foliage, the measure has proved quite valuable for estimating tree growth in unmanaged stands and predicting likely response to thinning. Most trees have uniform crowns with few gaps at young ages, but very patchy crowns when they are older. Crowns become irregular as trees lose branches from wind, pathogens, and other stressors. In addition, the crown is not always evenly distributed along the stem; sometimes it will begin farther up on one side than on the other due to irregular spacing between trees. Epicormic branching following a disturbance and mistletoe infections may cause additional variability within tree crowns. This phenomenon also occurs in old-growth trees that develop fan-shaped branches at mid-canopy levels.

The typical way to determine the location of the crown base is to imagine filling in holes in the upper crown with branches from the bottom of the crown, until the bottom is even around the stem (referred to as the compacted crown ratio, Monleon et al. 2004). Often, more objective measures of crown length are used. Defining the crown base as at the lowest live whorl works well in stands with relatively uniform spacing of trees. Lowest contiguous live whorl has been defined as the first live whorl above the first dead whorl encountered from the tip of the tree (Maguire and Hann 1989).

Other definitions of crown base have been based on number or geometry of live branches in whorls near the bottom of the crown. Full crown base has been defined as the lowest whorl with live branches around at least 3/4 of the stem circumference (Maguire and Hann 1989) or at the lowest whorl with live branches in at least three of the four quadrants around the stem (Curtis 1983). Live crown length is often expressed as live crown ratio, or the ratio of live crown to total tree height. Ferrell (1980, 1983) characterized the entire crown by combining evaluations of live crown percent, crown raggedness (holes throughout the crown from dead branches), top condition (pointed or rounded), and branch angle. Crown width is a frequently used crown dimension (Moeur 1981), primarily because of its relationship to crown projection area (the projection of the crown cross-sectional area on the surface of the ground) and canopy cover. The latter is increasingly related to habitat suitability for many wildlife species. Crown width is usually based on the length of the longest branches. Estimation of crown widths on stand-grown trees, sometimes referred to as largest crown width, has been made possible by reducing maximum crown widths of open-grown trees (Paine and Hann 1982) proportionally to reductions in crown ratio (Hann 1997). The full crown profile has occasionally been measured to characterize competition for aerial growing space, for assessment of potential wind damage, to depict the three-dimensional shape of crowns, or to estimate the surface area of the crown (Dubrasich et al. 1997). Also, the ratio of crown width to crown length has been discussed as a criterion for identifying crop ideotypes and for depicting the packing of trees in a stand and the degree of crowding or inter-tree competition.

The growth rates of individual trees are related to gross crown dimensions. However, in addition to crown size, foliage density and quality are also important.

Trees with small crowns may become quite vigorous and accelerate their growth rates when released from a cover of shrubs or from a dense overstory, for example. Trees with the same-sized crowns may differ in foliage density and capacity for photosynthesis, characteristics that are not included in gross crown dimensions. Ferrell (1980, 1983) has developed a method to adjust for gross gaps or raggedness in a crown that improves estimates of tree vigor, but this method does not account for foliage density. Consequently, trees with the same gross crown dimensions may have different growth rates. Theories involving nutrition and water transport argue that trees with large crowns need greater sapwood area to move water to the foliage than do trees with small crowns (Waring 1983; Waring and Schlesinger 1985). Therefore, the larger-crowned trees have stems that contain more sapwood. Similarly, trees with the same-sized crowns may have different foliage densities. Thus, sapwood area may prove to be a more reliable way to assess tree vigor than measures of crown size. Maguire and Kanaskie (2002) have proposed the ratio of live crown length to sapwood area as a crown density or sparseness index that correlates closely with the growth rates of individual trees.

Shrub leaf area is strongly related to shrub stem diameter or basal area (Hughes et al. 1987). In the future, it may be possible to use sapwood area or leaf area as a measure of stand density and plan silvicultural treatments or evaluate site productivity based on leaf area for various combinations of sites and species (O'Hara 1996).

Crown dimensions change as stands develop. As trees grow in height, their crowns recede as the lower branches die when light levels in the lower crown fall below the light compensation point (the point where respiration is balanced by photosynthesis). This type of branch mortality occurs on all trees, not just smaller ones being overtopped (figure 5-4). Thus the larger trees in dense stands will have smaller crowns than trees of the same height in less dense stands. In addition, trees growing in dense stands on poor sites are likely to have smaller crowns than trees in stands of the same density on productive sites, because stand density may limit height growth as well as crown growth (Curtis and Reukema 1970).

Growth of Forest Stands

LEAF AREA OF FOREST STANDS

The growth of forest stands depends upon the amount of solar energy captured by the forest canopy for converting CO_2 to carbohydrates, along with soil water and nutrient availability, climate or microclimate, etc. Each species (genotype) or combination of species and site has the potential to display a certain amount of leaf area or leaf area index (LAI)—the surface area of leaves per unit area of ground surface (ft^2 of leaves/ft^2 of ground, or m^2/m^2)—and produce biomass. This potential is influenced by factors such as herbivory, pathogens, and disturbances such as ice storms, etc., and also by silvicultural treatments such as thinning and fertilization.

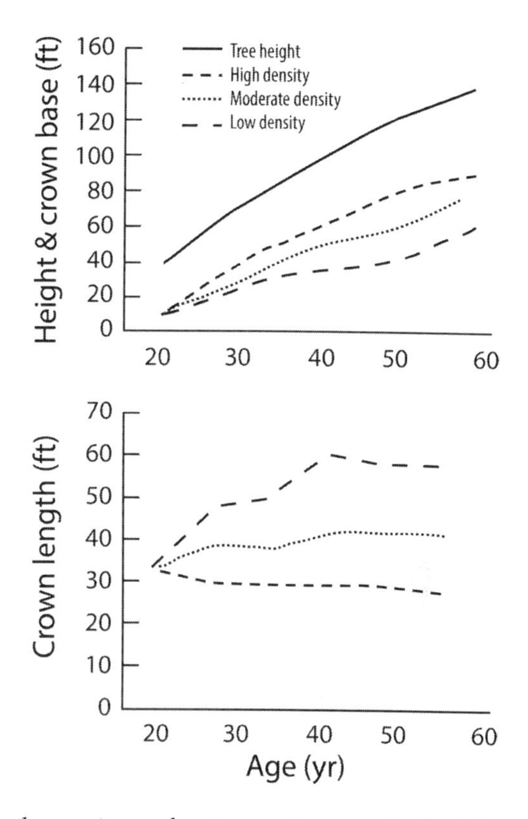

Figure 5-4. Tree height and height to the base of the live crown (upper), and crown length (lower) of the 40 largest Douglas-fir trees per acre growing at three densities. Tree height was the same in the three densities. Adapted from Marshall and Curtis (2002).

As forest stands grow and canopies close, a site reaches its carrying capacity for foliage (Long and Smith 1984; figure 5-5). Once the canopy closes, the amount of foliage (its weight or leaf area) is relatively constant—not absolutely constant, but it is likely that leaf area increases as crowns close and then fluctuates as crowns decrease and recover from disturbances and weather extremes.

The rapid volume growth in young stands is probably associated with a period of rapidly expanding tree crowns and stand leaf area. As trees get larger, inter-tree competition occurs and, depending upon stand density, smaller trees may die. The canopy of foliage in even-aged stands moves further from the ground, but the total amount of foliage remains about the same—in a sense, the leaf area is shifted to larger trees as smaller ones die. Later in stand development, leaf area decreases from wind sway and crown abrasion. In addition, large trees die and gaps appear in the canopy. If remaining trees are vigorous, their crowns expand, leaf area increases, and gaps close.

The total leaf area of a stand indicates the stand's capacity to absorb solar radiation and exchange gases; hence, it is a major determinant of gross primary productivity (GPP) and NPP. Kaufman and Ryan (1986) computed an index of potential radiation absorption as the product of leaf area and intercepted direct-beam radiation. This index was highly correlated with volume growth of individual trees of lodgepole pine,

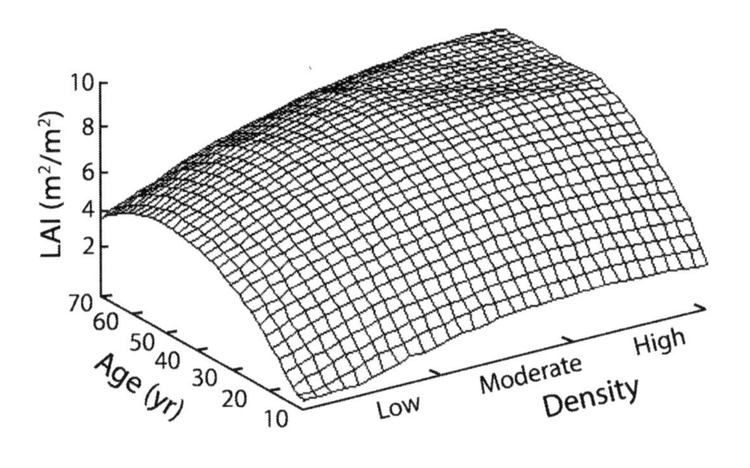

Figure 5-5. Theoretical diagram of leaf area in stands of various ages and densities. Also see Long and Smith (1984).

Engelmann spruce, and subalpine fir. On a regional scale, Kira (1975) demonstrated a strong correlation between NPP and the product of leaf area index and growing season length (LAD, or leaf area duration).

Pacific Coast coniferous forests have some of the largest LAIs of any forests in the world, reaching as high as 20 m^2/m^2 in some species (Tadaki 1966; Waring and Schlesinger 1985). The age at which LAI peaks in even-aged conifer stands varies by site and stand density (figure 5-5, Kimmins 1987). Ford (1982) found that in Sitka spruce LAI peaked at 10–11m^2/m^2 at 16 yr. In one study of Scots pine LAI peaked at 7m^2/m^2 in yr 16 (Albrektson 1980), but another study found a peak of 11m^2/m^2 at yr 20 (Ovington 1962).

Some species do not have a well-defined peak in LAI, but rather seem to climb gradually to an upper asymptote; this has been documented for Japanese red pine (Kira and Shidei 1967; but see also Hatiya et al. 1989 and Osawa and Allen 1993) and lodgepole pine (Pearson et al. 1987). LAIs for various genera range from 0.7m^2/m^2 for a 115-yr-old Mexican pinyon-alligator juniper woodland to 20.2 m^2/m^2 for a 121-yr-old stand of shade-tolerant western hemlock-Sitka spruce. The maximum LAI attainable at a given site appears to depend on light (Kimmins 1987; Monsi et al. 1973), water balance (Gholz 1982), and nutrient status (Brix and Ebell 1969).

In addition to the described general relationship between leaf area and GPP as a stand develops, other attributes of the forest canopy also influence patterns in GPP and/or NPP. These include spatial distribution of the foliage (Whitehead 1986), range in specific leaf weight (weight per unit area of foliage, Oren et al. 1986), and concentrations of N (Russell et al. 1989; Thompson and Wheeler 1992), chlorophyll (Thompson and Wheeler 1992), and photosynthetic enzymes (Van Keulen et al. 1989) in the foliage. For a given LAI, some researchers have speculated on a growth advantage for stands with a wide vertical distribution of foliage or trees with long crowns. This may be due to better penetration of light (Jahnke and Lawrence 1965), better facilitation of gas exchange (Raupach 1989), and/or lower structural and maintenance investment in branch wood (Kärki 1985). However, for the stand to achieve the density required for a high rate of growth (and because of inter-tree competition in dense stands), there will be some shortening of the crowns.

LEAF AREA AND PHOTOSYNTHESIS

GPP is determined by the process of photosynthesis at the stand level, and is thus dependent on the amount of leaf area. However, the rate of photosynthesis per unit of leaf area is determined by within-tree variables such as leaf water potential, concentrations of photosynthetic enzymes, leaf age, etc. Helms (1964) and Woodman (1971) found that, for Douglas-fir, very young and very old foliage had lower or more erratic rates of photosynthesis than foliage 1–2 yr old, and photosynthesis varied by crown position. This variation may have resulted from differences in leaf age in different parts of the crown, and differences in levels of shade. Environmental conditions such as temperature, relative humidity, light levels, and ambient CO_2 also affect rates of photosynthesis throughout the crown. On sites where these variables are nearly optimal throughout the growing season, such as Pacific Northwest coastal forests (Waring and Franklin 1979b), photosynthesis and GPP will be high.

GROSS AND NET PRIMARY PRODUCTIVITY AND BIOMASS

As stands of trees grow, they accumulate C in plant structural material and use it in physiological processes (figure 5-6). Organic dry matter contains about 50% C. Along with accumulation of C from the atmosphere, plants accumulate nutrients, mainly through their roots.

Waring et al. (1998) found that on nine temperate forest types NPP averaged 47% (±4%) of GPP. They provide a breakdown of NPP and GPP for a Sitka spruce-western hemlock stand in western Oregon (table 5-1). Thus nearly half of the energy that a

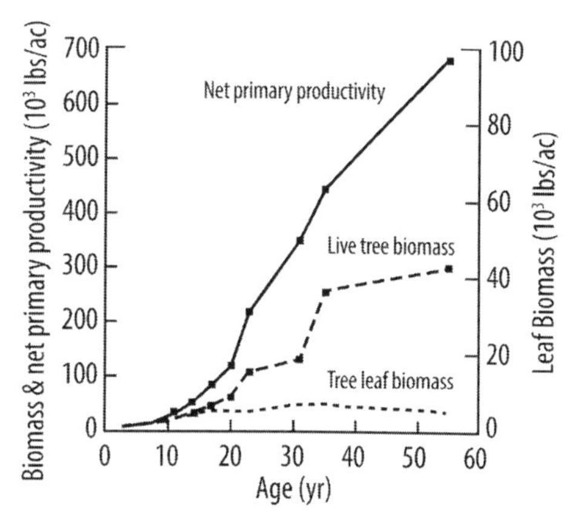

Figure 5-6. Net primary productivity, total biomass, and total leaf biomass in Pinus sylvestris plantations (Ovington 1962). See text for definitions.

Table 5-1. Respiration of CO_2 in relation to the growth and maintenance stand components (Waring et al. 1998).

Stand component	*Respiration CO_2 ($g/m^2/yr$)*
NPP above ground	525
NPP below ground	156
Total NPP	681
Growth respiration above ground	-131
Maintenance respiration, leaves	-97
Maintenance respiration, stems and branches	-333
Total respiration, roots	-156
Total respiration, stand	-717
GPP=NPP+total respiration	1398
NPP/GPP	49%

stand "fixes" is consumed in respiration and other physiological processes. The biomass of living plants and the volume of timber in a forest stand (what one sees or measures) represent an even smaller proportion of cumulative GPP than does NPP (figure 5-6). Thus the amount of GPP that goes into the production of usable biomass in terms of timber production, browse, etc., is a very small proportion of the total stand production. A major reason for the difference between NPP and biomass is the shedding and decay of plant parts or litterfall. Fallen and decomposed tree stems, leaves, branches, cones, pollen, and roots are a large part of NPP, but are not usually measured in the standing biomass of trees and understory vegetation. Overstory trees generally contain 85–95% of the living biomass depending on tree size or age (Ovington 1962) and stand density. However, understory vegetation, which is generally a small part of the total biomass, is often important for many ecosystem functions including nutrient cycling (Tappeiner and John 1973), forest stand dynamics (George and Bazzaz 1999; Maguire and Forman 1983), and wildlife habitat.

Net primary production removes C from the atmosphere and stores it in forest soils and vegetation. Smithwick and others (2002) estimated that total C storage in old western forests ranged from 1127 Mg/ha in coastal hemlock to 195 Mg/ha in interior ponderosa pine forests. From 41 to 53% of the C was stored in trees, mostly in stem wood. In old forests the amount of C added to the forest is about equal to the amount lost by respiration of dead biomass. Acker and others (2002) report that 29-yr-old Douglas-fir stands in the western Cascades add 7 Mg/ha/yr of C to forests, 200-yr-old stands 4-5Mg/ha/yr, while stands at 400 yr add no C. The C lost from stands of these ages was 0.1-0.3, 1-2, and 2-6 Mg/ha/yr, respectively (Acker et al. 2002). Thus the volume growth of wood in forest stands is closely related to rates of C fixation.

STAND VOLUME GROWTH

An important characteristic of many western forest species is that they are productive over a broad range of ages (<30–>100 yr). Their ability to grow at old ages enables development of a range of silvicultural systems with a broad range of tree sizes and ages and with short or long rotations (Curtis 1992, Curtis and Marshall 1993). Their continued high rates of production at older ages is related to species characteristics such as a very long period of sustained height growth and to site productivity—deep soils, plentiful soil water, and mild weather throughout the year (Waring and Franklin 1979b).

Not surprisingly, methods of measuring forest growth and yield have traditionally focused on the fraction of NPP harvested for human use (merchantable volume generally includes the main stem above the stump to the top of the tree or to a specified top diameter).

In even-aged stands, each species has its own pattern of volume growth, which can vary with site characteristics (figure 5-7). Ponderosa pine grows fairly rapidly at early ages, but after about 20 yr it loses its momentum and thus does not have the potential to produce volume at the rate of Douglas-fir. This is probably because ponderosa pine occurs most frequently on dry sites, where the growing season is limited by lack of soil water in the spring and summer and cold temperatures during the rest of the year. The rapid early volume growth of redwood is also attributable to stump sprouts that avoid the period of relatively slow growth during seedling establishment as well as to the very productive sites typical of redwood forests.

Red fir and other true fir species have a long establishment period because of their slow juvenile growth as seedlings and saplings. However, because they are shade-tolerant and have large amounts of leaf area, they can grow at high densities (Schumacher 1928). Thus, they can attain large standing volumes at older ages, even though they grow on sites with short growing seasons and are limited by drought in the summer and by low soil and air temperatures from October through June.

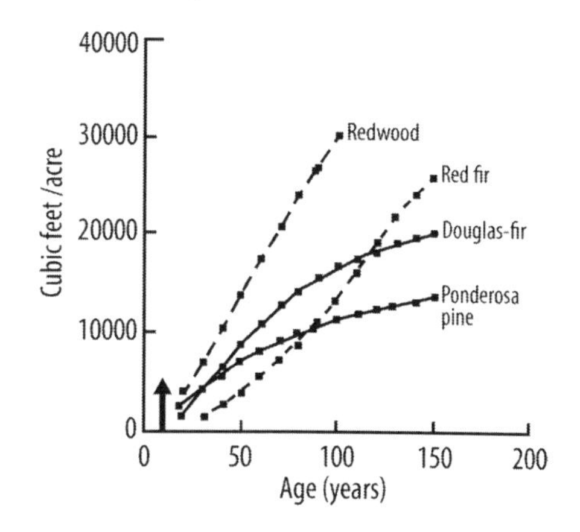

Figure 5-7. Volume expected on productive sites for Douglas-fir (SI 170 ft @ 100 yr, McArdle and Meyer 1961), ponderosa pine (SI 120 ft @ 100 yr, Meyer 1938), redwood (SI 180 ft @ 100 yr, Lindquist and Palley 1963), red fir (SI 120 ft @ 100 yr, Schumacher 1928); SI = site index, followed by base age. These values are for well-stocked, unmanaged natural stands and do not include volume of mortality from self-thinning. The black arrow indicates volume in a hybrid poplar plantation.

Figure 5-7 depicts the growth of natural stands with no management during the period of regeneration establishment and early growth and no commercial thinning. The early production of volume could be considerably higher than that shown in this figure if the stand had been established with site preparation, planting, and early control of competition and tree density. Also, the volumes do not account for trees that died from inter-tree competition nor for the effects of variables such as pathogens or insects. Thus, the potential volumes from 20–50 yr might be 20% greater or more than those shown. Stands with mixtures of shade-tolerant true fir and shade-intolerant ponderosa pine can yield more volume than ponderosa pine alone. However, if the initial density of the rapidly growing ponderosa was too high, it would suppress the volume production of shade-tolerant true fir because the latter has slow juvenile height growth (Garber and Maguire 2004).

VOLUME GROWTH AND STAND DENSITY

Stand volume growth is quite variable from stand to stand. When stands have very few trees and low basal area or density, volume growth is low because there are too few trees to use the site's capacity for volume or biomass production. The trees are "open grown," and although individual trees grow rapidly, stand growth is low. As stand density increases, stand volume growth also increases. This increase is linear until inter-tree competition occurs. After this point, growth increases with increasing density, but at a reduced rate. A point is reached where increasing density may not increase stand growth, and there is possibly a region where increasing density will result in a decrease

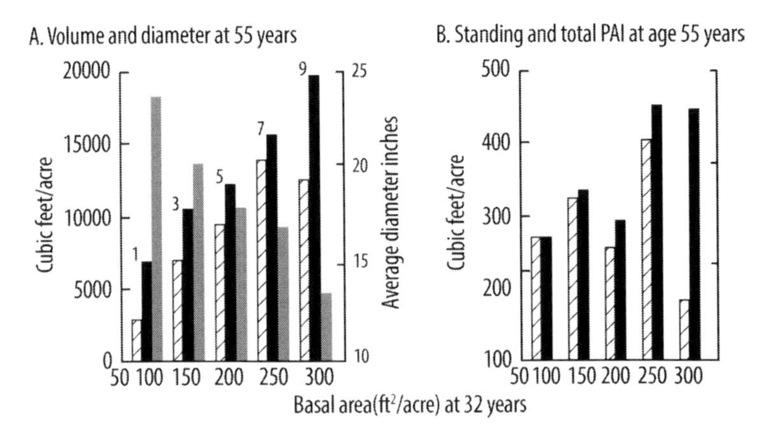

Figure 5-8. A: Standing cubic volume (hashed bars), net cubic volume (standing plus thinnings or mortality in the unthinned control –black bars), and diameter (shaded bars) at ages 55 yr that resulted from a range of basal area at 32 yr. Note that diameter decreases as basal area increases. Numbers above the bars represent different thinning treatments: heavy thinning (1), light thinning (7) and no thinning (9). B: periodic annual volume increment (PAI) at 55 yr. Standing (hatched bars) and total including mortality (black bars). Adapted from Marshall and Curtis (2002).

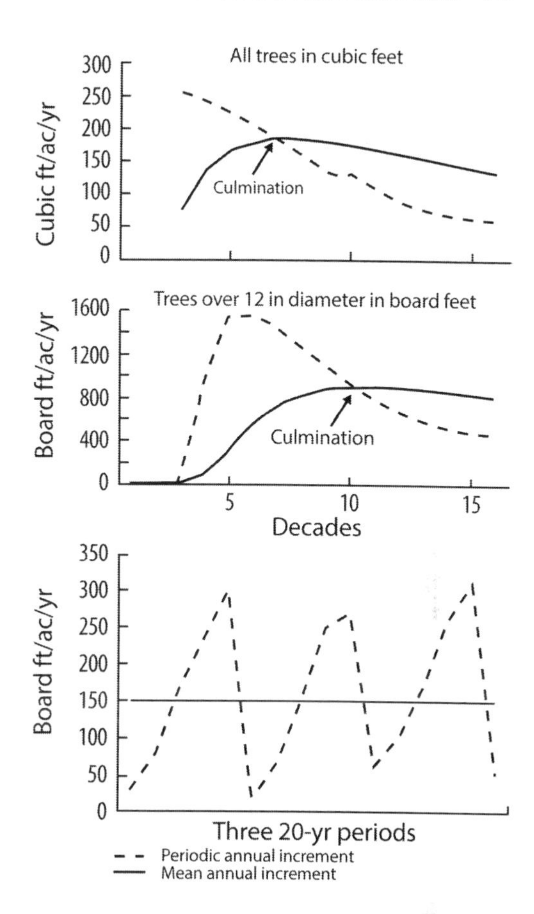

Figure 5-9. (A.) Mean and periodic annual increment for Douglas-fir site index 170 (McArdle and Meyer 1961) for all trees and for trees >12 in in diameter. Culmination of MAI occurs at about 60 yr for all trees and about 80 yr for trees >12 in in diameter.

(B.) Theoretical MAI and PAI for an uneven-aged stand over three cutting cycles. PAI is the volume growth in one cutting cycle. MAI is the average PAI over several cutting cycles. In theory, MAI in uneven-aged stands is the same as MAI at its culmination for even-aged stands of the same density and site class.

in stand growth. As stand density increases, tree diameter typically decreases, and the largest trees are found at the lowest densities.

The following example from a growth and yield experiment in a young Douglas-fir stand (figure 5-8) illustrates the nature of this relationship and how it relates to stand growth and thinning (Marshall and Curtis 2002; Marshall et al. 1992). Stand volume and volume growth at 55 yr are related to stand density at 32 yr. The range in basal area of 100–250 ft^2 reflects the intensity of thinning, with 100 ft^2 representing an intense thinning with many trees removed, and intensity decreasing through treatment at 200 ft^2; 250 ft^2 is an unthinned control. Leaf area also increased as residual basal area increased (Waring et al. 1981). The difference between standing and total volume for the control treatment (9) is mortality from self-thinning. For the rest of the treatments, the difference is due mainly to the volume removed in thinning and the decrease in volume growth from stand density reduction.

At 55 yr the net volume growth was less in the unthinned control (9) than in the most lightly thinned treatment (7), probably because so many trees were dead or dying from competition. At 55 yr, the unthinned stand had still produced the most total volume; however, the light thinning had the most standing volume and the highest volume-growth rate. In the past 10–15 yr the growth of the control slowed and at 55 yr it had the slowest growth of all treatments. The heaviest thinning treatments (1 and 3) were the only ones in which volume growth increased in the previous 10 to 15 yr. Possibly in the next 10–20 yr, growth will decrease in treatment 7 as it did in treatment 9. Then treatment 5 could have the most standing volume and highest growth rate.

We suggest two reasons for the greater volume production in high-density young stands: (a) the high-density trees in these stands more fully use the site resources; and (b) rapid height growth in 20–40-yr-old stands can be a major component of volume growth, especially when height growth is unaffected by stand density, as is the case with Douglas-fir on productive sites (Marshall 1990). Similar relationships have been reported for red pine (Buckman 1962; Evert 1964). A slowing of height growth plus the loss of trees from mortality may be the reasons that at older ages very dense stands may not produce as much volume as stands of moderate density.

Currently, there are no practical ways to predict within-stand growth variability not accounted for in measurements of density like basal area, relative density, or trees per acre (see chapter 6). The differences are probably caused by variables such as wind, soils, and possibly root pathogens, as well as subtle differences in stand structure. Volume growth is also variable from period to period (Curtis 1995), possibly due to weather fluctuations. Managed stands that cover a greater area would probably have even greater variability in stand growth than the small, precisely thinned research plots discussed above. There would be greater within-stand differences in site productivity, species composition, stand structure, and small-scale disturbance caused by factors such as root rot or wind damage. Thus, although there is a strong relationship between stand density and growth, the relationship is best considered a "band of response" and not a line (Oliver 1988).

On dry sites, a shrub understory may reduce the growth of the overstory trees even after they have grown well above them (Oliver 1990). Barrett (1982) found that 20-yr diameter and volume growth of ponderosa pine saplings was reduced by about 40% when an understory of manzanita and ceanothus was present. This trend continued another 15 yr, and 35-yr differences in volume reduction ranged 50–70%, with the greatest differences at the widest spacing (Cochran and Barrett 1999). The leaf area of the overstory trees was less when the shrubs were present (Oren et al. 1987), helping to explain the reduction in growth. Only at very high tree densities (>700 trees/ac) did the trees shade out the shrubs. In time, however, the lower densities (300–500 trees/ac) might also increase their growth rate and shade out the shrubs.

MEAN ANNUAL AND PERIODIC ANNUAL GROWTH OF EVEN-AGED STANDS

Stand volume growth is generally evaluated in terms of mean annual increment (MAI) and periodic annual increment (PAI). These values are used to assess the current productivity of a stand and to help determine rotation ages—the age to harvest an even-aged stand and start a new one, from a stand-growth efficiency perspective. MAI is the average annual growth of the stand from establishment to any given year, generally not including mortality. It is calculated as volume (V) divided by the age of the stand (A). PAI is the difference in volume at the beginning and end of a period (often 10 yr), divided by the number of years in the period. It is calculated as: V_2-V_1/A_2-A_1.

As discussed briefly above, the units by which growth and yield are measured reflect the output or end use for the material desired by the landowner. Here we are concerned primarily with cubic volume of stemwood (ft^3/ac). As would be anticipated from figures 5-6 and 5-7, stand growth in units of yield follows a peaking behavior generally proportional to NPP; hence, cumulative volume growth follows a classic sigmoid or "S-shaped" pattern (Husch et al. 2003). Sometime after PAI peaks, MAI also necessarily reaches a peak (figure 5-9a). Hence, repeated rotations of the age corresponding to the peak in MAI for a given species produce the greatest yields of volume (Husch et al. 2003). This age is most easily identified as the age at which the PAI curve crosses the MAI curve (figure 5-9a).

It is important to distinguish between gross and net growth and yield when evaluating forest production and assessing differences among silvicultural treatments. The potential for a site includes the standing volume of wood plus the volume removed in thinning or other treatments plus mortality and the in-growth of new trees in the stand (Avery and Burkhart 2002; Husch et al. 2003). The latter is especially important when estimating yield from uneven-aged or two-aged stands. Net growth or yield usually excludes volume of mortality that is not harvested.

Rotation age based on culmination of MAI can vary widely even within a species depending on site quality (table 5-2). For example, based on data from normal yield tables (McArdle et al. 1961), peak MAI for natural stands of Douglas-fir ranges 59–216 ft^3/ac/yr (4.1–15.1 m^3/ha/yr), for low- and high-quality sites, respectively. In fact, site quality is sometimes designated by yield classes indexed by maximum MAI in units of ft^3/ac/yr (table 5-2, Cannell 1982). Individual-plot PAIs have been measured as high as >530 ft^3/ac/yr (37 m^3/ha/yr) in intensively managed stands of Douglas-fir at 40 yr (Marshall et al. 1992). MAI for Douglas-fir grown under typical rotation lengths probably ranges up to 214 ft^3/ac/yr for unmanaged stands (McArdle et al. 1961) and up to 257 ft^3/ac/yr for managed stands. PAI and MAI for ponderosa pine on a few very productive sites may reach 286 and 228 ft^3/ac/yr (20 and 16 m^3/ha/yr) respectively (Oliver 1997).

Table 5-2. Forest site class at culmination of mean annual increment (MAI) and corresponding site indices from yield tables for selected species. Adapted from USDA Forest Service FSM 2409.6, Region 5, supplement 232, 1980, prepared by G. Davies.

Forest Survey Site Class	*1*	*2*	*3*	*4*	*5*	*6*	*7*
Maximum culmination at MAI (ft^3/acre/yr)	>225	224-165	164-120	119-85	84-50	49-20	<20
Mixed conifer[a] Site Index (50yr)	110-100	90-80	70-60	50	40-25		
Ponderosa pine[b] Site Index (100yr)	160-120	150-130	120-110	100-90	80-70	60-40	
Ponderosa pine[c] Site Index (50yr)	150	130	110	90	70		
Douglas-fir[d] Site Index (50yr)	145	125-105	85				
Douglas-fir[c] Site Index (50yr)	150	130	110	90	70		
Redwood[e] Site Index (100yr)	240-160	140	120	106			

a. Dunning and Reineke (1933); b. Meyer (1938); c. Hann and Scrivani (1987); d. Curtis et al. (1982); e. Lindquist and Palley (1963)

There are several important points to consider if stand rotation ages are to be based on culmination of MAI. First, MAI curves are relatively flat for several decades on either side of the culmination. MAI for Douglas-fir on site class II is estimated at 173 ft^3/ac/yr at 50 yr and 173 ft^3/ac/yr at 90 yr, with culmination of 188 ft^3/ac/yr at 60 yr and 70 yr (McArdle et al. 1961). Also, the shape of the MAI curve and its time of culmination for a species are determined by tree size goals, units of measure, and site productivity. If the goal is to produce volume in large trees, the rotation will be longer (Curtis 1992; Curtis and Marshall 1993). In well-stocked natural stands, MAI and PAI curves cross at 80 yr for trees with dbh>12 in and 20 yr earlier for the volume of all trees regardless of size (figure 5-9a; McArdle et al. 1961). Similarly, MAI culminates at about 100 yr using board foot measures, but at 60–70 yr with cubic measures simply because the ratio of board feet to cubic feet increases as the trees get larger. Curtis and Marshall (1993) also point out that culmination occurs sooner on more productive sites. Finally, the time of culmination of MAI is not well defined for Douglas-fir and probably not for other species either. Curtis (1995) examined the MAI (net, including thinnings) and PAI (net and gross, including thinnings and mortality) curves of thinned and unthinned stands ranging in age from 50 yr to over 117 yr and found that culmination had not occurred in any of them. PAI over time was erratic, and in some stands PAI net dipped below MAI for a growth period and then rose above it. However, MAI curves at older ages were relatively flat (Curtis 1995), indicating that there does not appear to be a critical time at which to harvest a stand to achieve large gains or avoid losses in yield. Comparing actual measurements of PAI and its intersection with MAI (Curtis 1995) to predictions from yield tables may cause this apparent discrepancy (McArdle et al. 1961). Other techniques for determining rotation, such as the present net worth or PAI as a percentage of standing volume, will generally result in shorter rotation ages (Davis et al. 2001). The fact that MAI does not have a sharp culmination and that PAI remains high beyond 100 yr provides flexibility in managing stands in an ecosystem context. Long rotations can be used to provide large trees and stands with characteristics of older forests, fire resistance, or large trees for habitat and aesthetics, without a sacrifice of yield.

MEAN ANNUAL AND PERIODIC INCREMENT OF UNEVEN-AGED STANDS

The relationships between NPP, growth, and yield in multi-cohort stands are consistent with the trends discussed for even-aged stands. In uneven-aged stands, average annual yield (AAY) serves as the analogue of MAI in even-aged stands (Guldin and Baker 1988). Rather than being tabulated by age and site quality, AAY for uneven-aged stands is typically summarized by initial volume or basal area, site quality, and length of cutting cycle. In managed uneven-aged stands, PAI is below AAY after cutting but rises above it at the end of the cutting cycle (figure 5-9b).

In general, stands with greater initial volume or basal area have greater PAI. However, the variety of stand structures that could aggregate into a given initial volume is infinite. This influences PAI the same way it does in even-aged stands. Specifically, stand structure will determine the proportion of GPP lost to maintenance respiration, the leaf area attained by the stand, growth efficiencies of individual trees and stands, and the proportion of growth that is usable vs non-usable. PAI in uneven-aged stands decreases after trees are cut and then rises (figure 5-9b) as the site is reoccupied by remaining trees and by seedlings and saplings growing into larger size classes (ingrowth). Over several cutting cycles average PAI converges on AAY. For a given site and species, MAI for even-aged stands is thought to be approximately the same as AAY for uneven-aged stands.

The relative productivity of even-aged vs uneven-aged stands is frequently debated. Clearly there are periods in the life of uneven-aged stands when PAI is much lower than it is in even-aged stands of similar species growing on similar sites. On the other hand, uneven-aged stands are always stocked with both large trees and seedlings or saplings—there is no regeneration period during which large areas are unoccupied and underused. Experimental assessment of which type of stand is more productive is difficult for several reasons: (1) uneven-aged stand increments depend strongly on stand structure; (2) there is little general agreement on optimal stand structures in both even- and uneven-aged stands, but a wider variety of options makes finding the

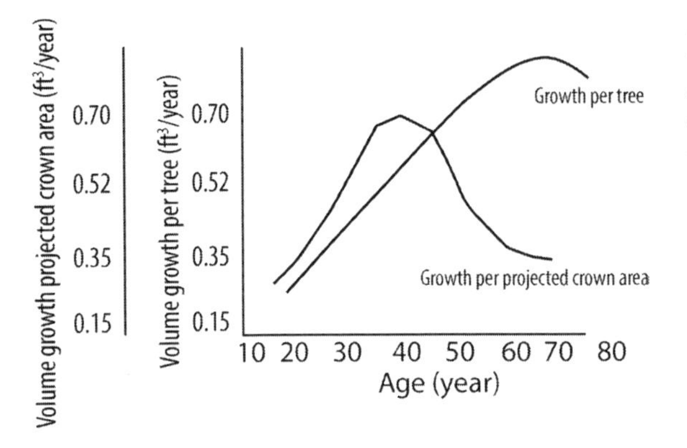

Figure 5-10. Volume growth of dominant and codominant trees per tree and per crown area projected on the surface of the ground. Adapted from Assmann (1970).

optimum stand structure particularly vexing in uneven-aged stands; and (3) implementation of silvicultural treatments in uneven-aged stands may be logistically more difficult. Mathematical optimization of both uneven- and even-aged stand structure relies completely on the accuracy of the growth models and usually assumes no site disturbance (for example, possible loss of production to skid trails); hence, considerably more field-testing is required to establish the relative productivity of these silvicultural systems. However, the debate is often moot when considerations other than timber productivity compel the application of uneven-aged silvicultural systems.

Simulation of hardwood stands in the northern United States suggested that, although uneven-aged management can produce greater merchantable cubic and dimension lumber volume, total cubic stemwood volume produced by even- and uneven-aged stands were virtually identical (Hasse and Ek 1981). In one long-term field study in mixed loblolly and shortleaf pine stands, total merchantable cubic foot yields were highest for even-aged systems, but board-foot sawtimber yields were greatest in uneven-aged stands (Guldin and Baker 1988). Also O'Hara and Nagel (2006) suggest that multi-aged ponderosa pine stands were as efficient at turning light energy into volume as even-aged stands of the same species. Because of this theoretical basis and the wide variation in structure and density that occurs in forest stands, they found no reason to think that productivity should be inherently different between these silvicultural systems.

It is important to recognize that growth and yield are strongly dependent on individual tree growth rate, tree size, and stand density; hence, yields are sensitive to silvicultural treatments, especially when solid wood products are the desired output.

MEAN ANNUAL AND PERIODIC INCREMENT OF DOMINANT AND CODOMINANT TREES

Individual trees can maintain high growth rates at very old ages. The culmination of MAI for stands of trees occurs at a much younger age than the culmination of MAI for individual trees. Assmann (1970) found that, out of a sample of 30 spruce trees, the

MAI (cubic volume) of only two had culminated (one at 103 yr), even though some of the trees were 300+ yr old (figure 5-10). As trees get older and larger, their rate of growth relative to the area of the stand they occupy (their crown area) decreases. They use their space in the stand less efficiently, even though their absolute growth rate remains high. A population of trees using growing space less efficiently causes the PAI and MAI for the stand to decrease. Assmann (1970) reported that the PAI of individual spruce trees peaked at 60–65 yr, but their volume growth in relation to their crown area (the area of the stand occupied) peaked at 35 yr (figure 5-10). Similarly, Douglas-fir, ponderosa pine, and sugar pine trees maintain rather constant basal area and, presumably, volume-growth rates beyond 300 yr (Latham and Tappeiner 2002; Poage and Tappeiner 2002). However, the culmination of cubic MAI of Douglas-fir and ponderosa pine stands occurs at less than 100 yr (McArdle et al. 1961, Meyer 1938).

ATTRIBUTES OF WOOD VOLUME

Ultimately, many other stand and tree attributes, in addition to volume per acre, may be of interest from a utilization standpoint. Tree and log size distributions are important in any effort to plan a harvesting system, to appraise timber accurately, or to perform an economic analysis. The quality of wood is reflected in relative wood density, uniformity, branch and knot structure, and micro-anatomical properties such as fiber length and microfibril angle (Haygreen and Bowyer 1982). Yield expressed as product grade distribution introduces another level of complexity beyond the variations in utilization standards imposed on trees and logs. Equations have been constructed to describe the grade distribution of recoverable products, including visually graded lumber, machine stress-rated lumber, and visually graded veneer (Fahey et al. 1991). Yields measured in terms of manufactured products vary tremendously by silvicultural regime (Briggs and Fight 1992), despite relatively small differences in cubic volume yield or NPP. Quality considerations have also been recognized in primary production studies (Lieth 1975), but have generally not been researched very much (Chung and Barnes 1977).

Thus, considerations of wood quality introduce a fourth layer of complexity to the comparative evaluation of forest production. These layers can be summarized as follows:

(1) Different stand ages or developmental stages are characterized by different capacities for total and above-ground NPP due to trends in LAI, respiratory load, and water relations.

(2) Different utilization standards and mensurational units emphasize different biomass components and physical dimensions.

(3) Different silvicultural regimes can produce stands of a given age or stage of development that may have similar total biomass but very different allocation among stems and tissue types.

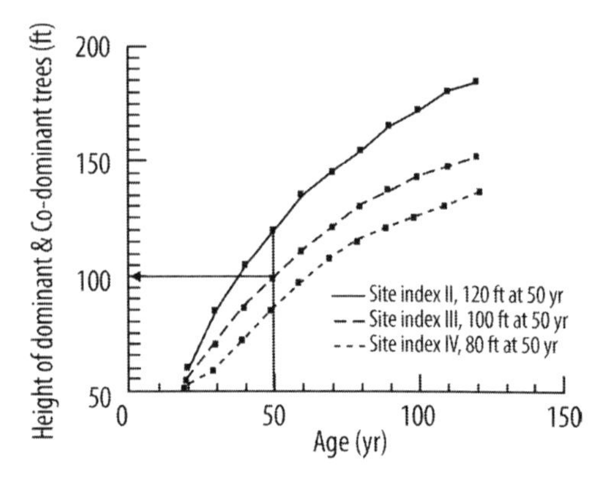

Figure 5-11. Average site index curves for Douglas-fir (site indexes 120, 100, and 80), adapted from King (1966). Base age is 50 yr. Ages are measured at a height of 4.5 ft (breast height) by counting rings.

(4) Different stands may produce similar total wood volume or biomass but very different wood quality and value in both a physical and economic sense.

Estimating Site Productivity

Several methods are used to estimate site productivity, including the height of trees at a certain age, height intercepts, plant community types, and volume-growth rate (MAI). The cumulative height growth of the dominant and codominant trees indicates the capacity of a site to produce wood (King 1966; figure 5-11). Dominant heights of 150 ft at 50 yr indicate the most productive site (site class I) for Douglas-fir, whereas lower heights at 50 yr indicate sites of lower productivity (site classes II, III, IV, etc.). As discussed above, trees can grow taller with more available water because it is easier to move water to greater heights against the force of gravity if soil water potential is high, and tree height is an important component of stand biomass and yield of wood. In general, sites with plentiful resources for tree growth can support higher basal areas, another important component of biomass and yield. However, soil water is only one of the growth-limiting resources of a forest ecosystem, and height growth can be affected by variables such as wind and variability in soil horizons as well as the ability of a soil to supply nutrients. Assmann (1970) reports that for well-stocked, even-aged stands, volume predicted from site index vs actual volumes varied by ±12% and MAI by ± 25% in European growth and yield studies. These values may represent a reasonable margin of error in the prediction of potential yield based upon site index curves.

Height growth varies with soil type (Carmean 1956) and elevation (Curtis et al. 1974). McArdle et al. (1961) report that 90% of the site class I areas in their studies were on north to northeast slopes. Other variables like wind may also affect height growth on certain sites. Thus it is important to verify that local site index curves adequately describe height growth. Trees on gravel soils grow more rapidly at early ages than trees on soils derived from basalt—a 30-ft height difference at 50 yr of age (Carmean 1956). This could lead to an overestimation of productivity, because after 50 yr, height growth slows for trees on gravel soils, and at age 100, dominant and codominant trees on the two soils were the same height. At 100 yr, height growth for trees on basalt was continuing; it had nearly stopped for those on gravel soils. Height growth of Douglas-fir at higher elevations (2000–5000 ft) in the Cascades is slower at younger ages than it is at lower elevations and in the Coast Ranges, but those trees grow more rapidly after 100 yr (Curtis et al. 1974). Differences in height-growth pattern likely indicate differences in volume-growth rates and the amount of volume produced in young stands (Curtis et al. 1974). Height-growth curves will likely become more localized or site-specific to improve estimates of productivity; however, when local site index curves are not available, it may be possible to calibrate regional curves to local conditions.

As discussed above, height growth of dominant and codominant trees on less productive sites may be reduced by high stand density (Alexander et al. 1967; Garber and Maguire 2004; Holmes and Tackle 1962). Alexander et al. (1967) developed a method using crown competition factor (CCF) to adjust the measurements of tree heights when determining site index. At 100 yr, Rocky Mountain lodgepole pine trees growing at low densities (CCF 125) were 11 and 25 ft taller than trees growing in high-density stands (CCF 300 and 500, respectively). Similarly, a site index for 43-yr-old Douglas-fir, based on the height of the dominant and codominant trees, was 82 ft at 100 yr at a 4-ft spacing and 120 ft/100 yr at a 12-ft spacing (Curtis and Reukema 1970). Like Alexander et al. (1967), these researchers suggest that measures of crown development along with relative position in the crown might be used to select trees for estimating site index in dense stands on low-quality sites. Dominant and codominant conifer trees with large crowns that had been growing at low densities would likely provide better estimates of productivity than trees in the same crown classes with small crowns that were growing under dense conditions.

It is important to ensure that the trees selected to evaluate site productivity truly express the height-growth potential. Vincent (1961) has pointed out that height growth can be suppressed for a time by insects, frost, and animal browsing. In addition, wind and snow can break treetops. As we discussed above, early growth can be suppressed by competition from shrubs and also by large trees in uneven-aged stands. Rapid early rates of height growth can be expected in intensively managed plantations. Using trees that have been damaged or suppressed will result in an underestimation of productivity (Hann and Hanus 2002; Hanus et al. 1999). Damage to Douglas-fir, white fir, and ponderosa pine by fire, mistletoe, animal damage—especially porcupine feeding on ponderosa pine—and broken tops reduced height growth 6–25% compared to undamaged trees. Species were affected differently; growth reductions on fire-scarred Douglas-fir and white fir were 14% and 28%, respectively, but no reduction was reported for ponderosa pine. Damaged trees may reach the same height as undamaged free-to-grow trees; however, it may take them considerably longer, resulting in an underestimate of site index. In uneven-aged stands, it may be difficult to find trees that have been free to grow and that adequately express the height-growth potential for the site. The number of years that a tree has been suppressed by competition (determined from increment cores) might be subtracted from total age and the result (effective age; Smith et al. 1997) used with current height to determine site index.

The height intercept method of estimating site productivity has been developed to address the problem of estimating site index in young stands. In young ponderosa pine stands, where there are no large trees suitable for estimating height, site indexes can be estimated by measuring the height growth from whorl locations on seedlings and saplings. Four-year height growth above 4.5 ft is correlated strongly with height at 50 yr, provided the trees are free from shrub competition (Oliver 1972). Similarly, a site index of western hemlock, which has no distinct whorls, can be estimated from age at breast height and average annual height growth of the entire stem above breast height (Nigh 1996).

Stockability (the potential of a site to support basal area) is also an issue in determining site productivity, and one that is not addressed by measuring the height and age of the taller trees in a stand. The height of the dominant and codominant trees may indicate a certain level of productivity. However, productivity may be limited by the basal area or density carrying capacity of a site rather than height growth potential, especially on dry sites (MacClean and Bolsinger 1973). Hall (1983) defined growth basal area (GBA) as the basal area at which 100-yr-old ponderosa pine can grow 1.0 in in stem diameter in 10 yr. In eastern Oregon, GBA varies considerably, and for a given site index, GBA may vary by more than 100 ft^2/ac. Hall (1983) developed a method to determine stockability or to estimate potential stand densities based on GBA.

Use of plant community classifications (Hall 1973) to predict site productivity appears to work only in very general terms, although plant communities may be a way to classify sites of low stockability (Cochran et al. 1984). Verdyla and Fischer (1989) found that on the Dixie Forest in Utah the most productive sites occurred in two habitat types; however, there was overlap of site quality among sites and a broad range of variability within types.

Where sites are well stocked with stands of the appropriate ages, actual volume production should probably be used to evaluate productivity. The USDA Forest Service has developed a site-classification scheme based on the MAI at culmination that reflects the relationship between site index and average volume production for different species. Table 5-2 summarizes this information for selected species. Estimates of stand volume and PAI at different ages derived from stand examination, inventory plots, and growth models can be used to estimate site-specific productivity and the yields that can be expected under different silvicultural systems or types of stand management. These estimates are probably much more reliable than estimates based on height growth alone.

The best estimates of productivity are those derived from measuring the volume and growth rates of well-stocked stands growing under the silvicultural system that the landowner would like to use. Indices based on height over age relationships or plant communities (plant associations) are not as accurate as direct measurements. However, they may be the only methods available for sites where stand stocking is too low, for very dense stands on sites of low productivity, or for very young stands or sites that are currently occupied by shrubs or other plant communities.

Use of Tree and Stand Growth Information

HEIGHT GROWTH

Long-term studies of Sierra Nevada ponderosa pine have shown that height growth is related to seed source and elevation (Kitzmiller 2005). Trees from higher elevations grew more slowly in height than those from lower elevations for nearly 30 yr, possibly as an adaptation to climate and snow damage. Thus seed source and climate need to be considered when establishing plantations in mountainous terrain where climate and site productivity are variable. Likewise, seed sources that begin height growth early in the spring may not be suitable for sites that are prone to late spring frost.

Understanding the potential rates of height growth is useful in planning for the development and future structure of mixed-species stands (Garber and Maguire 2004; Oliver 1980). Control at an early age of competitors such as red alder, bigleaf maple, Pacific madrone (Hughes et al. 1990), and other hardwoods with rapid juvenile height growth is often needed to ensure survival and early growth of slower-growing species. Height-growth rates of the conifers are greater than those of the hardwoods after 30–50 yr. In older stands, the conifers may need to be thinned to maintain vigorous hardwoods in mixed-species stands. At close spacing, slow height growth of shade-tolerant species can lead to stratification beneath the canopy of a faster-growing species (Garber and Maguire 2004). At wider spacing, however, both species may become part of the main canopy.

The effects of shrubs on the height growth of conifers may persist to older ages, even if the conifers have overtopped the shrubs. Eventually the conifers may shade out the shrubs and increase their height growth, but this may occur after several decades of reduced growth. Similarly, trees that have been growing in the understory of conifer stands can recover from very slow initial height growth.

On productive sites, the height growth of the dominant and codominant trees is generally not affected by stand density. On low-productivity sites, however, growth of these larger trees can be reduced if stand density is high. Thinning such stands may therefore increase both diameter and height growth, whereas thinning on productive sites will increase only diameter growth. Stand stagnation may occur in very dense stands of lodgepole or ponderosa pine, such that height growth may be only inches per year, and no self-thinning will occur. In such stands, thinning is needed for some trees to express dominance and for normal stand development to occur.

STEM AREA AND CROWN GROWTH

Stem diameter, basal area growth, and tree crown expansion are very much affected by stand density. Trees grown at low densities maintain larger crowns and stems. Thus one of the effects of thinning is to promote the growth of large trees. Stem growth is generally greatest within or at the base of the crown. As the crowns of trees in dense stands recede upward, the center of greatest stem growth shifts up the tree (figure 5-3). This tends to make these trees less tapered than trees with large crowns that maintain stem growth lower on the stem.

In practice, stem radial or cross-sectional area growth is usually measured from increment cores extracted at breast height. Typically, rings near the pith are wider, because they developed when that part of the stem was in the crown. Rings gradually narrow in stand-grown trees as the stem gets larger, and as inter-tree competition occurs and the crown recedes (figure 5-3). In general, growth at breast height is indicative of growth elsewhere in the stem. Trees with wide rings at breast height will have wider rings throughout the stem than trees with narrow rings at this height. However, silvicultural treatments that increase the force of wind on a tree, such as thinning or fertilization (which increases the weight of the foliage) may cause changes in stem diameter or cross-sectional area growth that are not detected at breast height. Garber and Maguire (2003) found that site-specific taper and volume equations were needed to evaluate the effects of thinning on the volume growth of lodgepole pine and white fir under different thinning trials. Regional volume equations worked well for ponderosa pine, suggesting the stem form of this species may be less sensitive to changes in stand density. Trees with large crowns and tapered stems are generally more resistant to wind than are trees with small crowns and less tapered stems.

LEAF AREA

Most silvicultural treatments directly affect leaf area. Relationships between leaf area, stand vigor, stem growth, and resistance to bark beetles have been established (Mitchell et al. 1983; Waring and Pitman 1980). Conifer seedling and sapling growth and survival were increased by reducing the leaf area of surrounding shrubs and hardwoods (Hughes et al. 1987; White and Newton 1989). Thinning and fertilization have been shown to increase the leaf area per tree and fertilization has been shown to accelerate recovery of leaf area in thinned stands and growth of both thinned and unthinned stands (Binkley and Reid 1984; Brix and Mitchell 1983; Waring et al. 1981). In the future, it may be possible to use sapwood area or leaf area as a measure of stand density and plan silvicultural treatments or evaluate site productivity based on leaf area for various combinations of sites and species.

Currently, it is difficult to measure leaf area accurately and to use it effectively in silviculture. Remote sensing appears to make it possible to estimate LAI by measuring the surface area of foliage by reflection of various spectra and calculating the proportion of the ground area occupied by the supporting trees, but this approach has limited use in silvicultural prescriptions. Once efficient methods for estimating leaf area are available, they can be used to allocate the most vigorous trees to growing stock (number of trees/area by size class) in order to optimize stand growth (O'Hara 1996). Sapwood cross-sectional area offers the greatest promise for consistent estimates of tree-level leaf area across a range of stand densities (Grier and Waring 1974; Maguire and Bennett 1996; Monserud and Marshall 1999).

An allometric relationship between an easily measured tree dimension and leaf area will probably be the only practical approach within a stand. In shrubs, stem diameter or basal area are effective measurements for estimating leaf area of open grown shrub populations (Hughes et al. 1987). It is reasonable to assume that compared to other stand density measures (see chapter 6), using sapwood basal area would be a better way to asses stand vigor and allocate growing stock to the most vigorous trees, particularly across stands of varying density. However, estimating sapwood area on standing trees still represents a significant statistical and operational challenge. Sources of error or bias include (1) uncertainty about the numbers of trees per size and per species required to achieve an accurate estimate of sapwood for a stand, (2) potential irregularities in the shape of the heartwood and sapwood, (3) difficulty in accurately calculating the relationship of sapwood taper from breast height to crown base (Maguire and Batista 1996; Waring 1983), and (4) variations in the ratio of leaf area to sapwood area that may result from future silvicultural treatments (Espinosa-Bancalari et al. 1987).

USE OF GROWTH AND DENSITY IN STAND MANAGEMENT

Stand volume growth (PAI—the annual growth of a stand, usually for the previous 5–10 yr) is related to stand density (basal area, trees per acre, and SDI or relative density—see chapter 6). As stand density increases, volume growth increases until severe inter-tree competition occurs. Then growth and yield may decrease from the loss of trees by self-thinning, poor tree vigor, and in some cases stand stagnation. The nature of this relationship can be variable, because PAI fluctuates over time (Curtis 1995) and throughout a stand on apparently uniform sites (figure 5-9).

PAI can be estimated from forest inventory or stand examination plots and growth models. This measure can be used to evaluate stands for treatments such as regeneration and thinning. Given similar stand age, species composition, and site class, it makes sense to regenerate stands that have low PAI or those whose PAI is close to MAI and leave the stands with high PAI—those that are likely to produce more volume in the future. MAI can be calculated from the current volume in the stand divided by stand age. Note that this method will underestimate potential growth if thinnings and mortality are not accounted for. If the PAI is greater than the MAI, then the optimal rotation for wood production has not occurred. However, it is not likely that silviculturists will often be able to demonstrate that the PAI and MAI curves have crossed, due to the variability of PAI through time and throughout a stand. Using stand examination data in a stand simulator will project the current trends in stand growth and enable silviculturists to estimate the relative productivity of stands. But it is unlikely that fluctuations in PAI such as those observed by Curtis (1995) can be predicted.

From a wood production standpoint, both MAI and PAI can be used to prioritize stands for regeneration. Culmination of MAI is a traditional way of deciding when to regenerate a new stand, and a stand whose MAI has culminated is a candidate for regeneration. When using this method, it is important to recognize that culmination of MAI is not a peak. Rather, there is a period of several decades during which MAI changes very little. Also, in young stands, heavy thinning, damage from wind, or other causes of low stocking can severely reduce PAI and therefore delay culmination for decades. For long-lived species, the PAI curves may approach the MAI curve but not cross it (Curtis 1995), and there is consequently no clearly defined culmination. We suggest that PAI (along with MAI) can be used as a way of prioritizing stands for regeneration. For example, for a species whose culmination of cubic foot growth is estimated to be about 70 yr (from growth model projections), stands over 70 yr can be ranked using PAI. Stands with the lowest PAI, and thus the lowest future growth potential, would be candidates for regeneration. From an economic standpoint, those with lowest percent growth (PAI/MAI, or PAI/current stand volume) would be candidates for harvest and regeneration.

Similarly, proposed thinning treatments can be evaluated using PAI. With data from the stands and stand simulators, the PAI after thinning can be estimated to (1) determine if stand growth after thinning is acceptable and to adjust it if it is not, and (2) determine the timing and feasibility of future thinnings. Here too, however, silviculturists must realize there could be fluctuations in volume growth that cannot be predicted, and using longer periods (>10 yr) to determine PAI may reduce its inherent variability.

It is important to recognize that forest management goals are not always met by optimizing stand growth. For example the size of trees that mills can use efficiently is smaller than it was several decades ago. This trend could lead to short rotations. In addition, from an ecosystem standpoint rotation ages may be extended to grow large trees that improve resistance to fire, wildlife habitat, and aesthetics, for example. There is even flexibility in using culmination of MAI as a criterion for when to harvest even-aged stands, since MAI appears to remain relatively flat over a wide range of years, and PAI of western conifer stands remains high beyond 100 yr of age.

REVIEW QUESTIONS

Height growth of trees

1. Compare height growth of redwood, Douglas-fir, noble fir, ponderosa pine, lodgepole pine, red alder—early vs later growth.
2. What crown classes of trees are used to determine growth potential and why?
3. Why is the height of the largest trees a reasonable predictor of productivity on productive sites, but not such a good indicator on low-productivity sites?
4. How does competition early in the life of a tree affect height growth?
5. How does stand density affect height growth of the bigger trees?

Diameter growth of trees

6. How do ring width or diameter growth and basal area growth vary within a stand-grown and an open-grown tree?
7. What is reaction wood?
8. What is juvenile wood and where is it produced in a tree? How is it important for the tree? For wood quality?
9. How is tree taper related to crown size? Explain.
10. What are the benefits and difficulties of measuring stem growth at breast height?

Tree crowns and leaves

11. In what way(s) is it difficult to determine tree crown size?
12. How do tree crowns change with tree size (age) and stand density?
13. What is leaf area? How is it expressed?
14. How does leaf area change over the life of a stand (very young to old)?
15. How do self-thinning and commercial thinning affect leaf area? Discuss.

Stand growth

16. What causes the difference between net primary production and stand or total biomass?
17. Draw the general age-volume relationship between stand age and volume for even-aged stands (Douglas-fir, a true fir, ponderosa pine, alder, redwood) and for a "typical" uneven-aged stand.
18. Explain and calculate the MAI and PAI of even-aged stands.
19. What is the difference in net and total (gross) volume and MAI, PAI?
20. What is the significance of culmination of MAI and how is it determined?
21. How do tree size and measurement units affect MAI and PAI? How does thinning affect MAI and PAI?
22. Compare AAY (MAI) and PAI of even- and uneven-aged stands.

23. Why is there a difference between MAI of trees and stands? What is the difference?

Relationship between tree and stand growth and stand density

24. Are stand density and net and gross or total MAI and PAI related? Explain.
25. What are the trade offs between volume growth and tree and crown size?
26. How is stand growth related to basal area? Is future volume more predictable than growth? Explain.
27. What are the effects of understory on dry sites on stand development?
28. Compare the average diameter vs diameter of the largest trees in dense and less dense stands.
29. Compare the average height vs height of the largest trees in dense and less dense stands and in even- and uneven-ages stands.
30. What is stand stagnation?

Height growth and estimation of site productivity

31. How may height growth vary by soil type? Discuss.
32. How does height growth vary by stand density and site productivity? Explain.
33. How do effects of wind and snow damage affect estimates of productivity?
34. Discuss height growth in relation to (a) uneven-aged stands and (b) shrub competition.
35. How is culmination of MAI used to classify site? What are the pros and cons of using culmination of MAI vs height over age to classify site productivity?

CHAPTER 6
Measures of Stand Density and Structure

Stand density describes the degree of occupancy of a given area by trees and hence measures the intensity of inter-tree competition for site resources. Stand density has major implications for individual tree dimensions (e.g. crown size), stand structure (including understory vegetation), and individual tree vigor through its effects on the availability of water, nutrients, and light. State and federal forest practice rules, silvicultural guidelines, and specific objectives of forest management are often stated in terms of stand density or its variation over time; as a result, stand density is fundamental to the practice of silviculture. In this chapter we: 1) summarize the different measures of stand density; 2) discuss their proper use; and 3) provide a mathematical derivation and comparison of density measures for those who want a more rigorous treatment of this topic.

Stand structure is a general term that refers to the collection and spatial arrangement of species and key biotic and abiotic features within a forest stand (see chapter 3). Measures of stand density quantify one aspect of stand structure by describing some combination of the average number of trees occupying a unit of forestland and their size. Stand density regime refers to the progression of stand density over time, with or without management. While many silvicultural treatments call for direct manipulation of stand density, the desired outcomes are typically other stand or tree responses such as diameter and volume growth, crown length, canopy depth, diameter distribution, or understory vegetation density and composition. Stand density also directly affects a myriad of other ecosystem properties and processes like interception of solar radiation, the amount of precipitation that gets through the canopy, the amount of snow that accumulates on the forest floor (and where), and forest floor temperature regimes and decomposition rate.

Absolute and Relative Measures of Stand Density

Stand density has been described in many ways, and considerable confusion arises about its meaning and the efficacy of alternative measures. For example, "stocking" and "stand density" are often used interchangeably. Stocking, however, is a measure of stand density relative to a specific optimum that is set in accordance with management objectives (Avery and Burkhart 2002). For example, a stand with 200 trees/ac and average diameter of 6 inches may be well stocked for one management objective but poorly stocked for another. In contrast, stand density is a more neutral term that refers to an ***absolute*** measure of tree occupancy per unit area. ***Relative*** stand density is this absolute density relative to the biological maximum for a given site and given species or group of species (i.e., a percentage). Unlike stocking, absolute and relative density both imply a biological limit rather than an optimum derived from some management objective. A central idea behind relative density is that stands with the same relative density but different average tree size are thought to experience approximately the same level of inter-tree competition.

Crown closure, or canopy cover, is somewhat different from stand density. Both relate only to the proportion of the ground area covered by the vertical projection of whole crowns or crown components (foliage, branches); tree size and numbers are not part of this measure. Total stand leaf area or leaf area index (LAI, projected or all-sided leaf area per unit land area) likewise provides little information about the number and size of trees, but can serve as the basis for allocating growing space among size classes or cohorts if the relationship between tree size and leaf area is known (O'Hara 1996). A functional balance between tree sapwood area (the cross-sectional area of nutrient- and water-conducting tissues) and tree leaf area (the photosynthesizing tissue) has long been recognized (Büsgen and Münch 1929; Shinozaki et al. 1964), and more recently has provided a practical way to estimate tree leaf area and LAI among stands of varying density (Waring and Schlesinger 1985). This same relationship facilitates allocation of growing space where the maximum or desired LAI is established (O'Hara 1996).

In theory, it is assumed that the resources available to support live, respiring tissues control maximum density on a site. In this respect, it is similar to the idea of carrying capacity in general ecology (Kimmins 2004). Trees present a special challenge in equating stand density with carrying capacity, because live cells constitute so little of the organic mass of a living tree, and because the mass of live cells is virtually impossible to measure separately from the rest of the tree mass. For example, heartwood contains no respiring cells, and the sapwood of conifers contains only 4–11% live cells by volume (Panshin and deZeeuw 1964). There are no measures of stand density that attempt to distinguish respiring mass from non-respiring mass; hence, the proportion of respiring mass is implied to be constant. However, the variation in maximum density among tree species is attributable, in part, to differences in photosynthesis and respiration rates and, in part, to differences in the amounts of photosynthesizing and respiring mass per unit of total tree mass. Stand density measurements likewise do not directly account for fine-root mass and its associated mycorrhizae (important for water and nutrient uptake) or for the mass of larger roots that support above-ground tree mass. These parts of a forest stand are certainly related to site occupancy, current stand growth and vigor, and total standing biomass, so it is logical to include them when characterizing overall stand density. However, there is currently no practical way to measure them, so they have not been incorporated into measures of stand density.

Most measures of relative density (table 6-1) are approximately equivalent, because they can be expressed as the *observed* number of trees per unit area (N_o) relative to an *expected* number per unit area (N_e) (Curtis 1970). The expected number is based on two different standards of comparison: (1) open-grown trees (no competition between trees), and (2) trees from stands at maximum carrying capacity (average maximum or normal stand density). Indices using the standard of open-grown trees are less common, but the most prevalent is crown competition factor (CCF; Krajicek et al. 1961). In CCF, the expected number of trees is the number that would just fully cover the area with their crown projections, assuming the trees are open grown and hence have maximum crown width or projection area for the observed diameter. When the basis of comparison is maximum carrying capacity, the expected number of trees per unit area represents the maximum number of trees possible for the observed average size. The measure of size can be quadratic mean diameter (Long et al. 1988; Reineke 1933), mean tree volume (Drew and Flewelling 1979), mean tree height (Wilson 1979), or occupied area (Chisman and Schumacher 1940; Gingrich 1967).

BASAL AREA AND VOLUME

Stand basal area (BA) and volume per acre are both often used to measure stand density and can be useful, but these measures cannot be directly expressed as a comparison of observed to expected numbers of trees. Instead, some measure of relative basal area (or volume) is required, for example:

Table 6-1. Common measures of stand density.

Measure	*Definition*	*Conditions of use and comments*
Absolute measures		
Trees/area	Numbers of individuals per unit area	Useful in stands of small trees (regeneration), for small numbers of large trees, or trees of a certain species or characteristics, etc.
Basal area/ac (BA)	Total cross-sectional area of stems (ft^2/ac, m^2/ha) at breast height (4.5 ft or 1.3 m)	Related to volume and volume growth but does not account for numbers of trees; older yield tables establish average maximums for a species, site, and age
Volume	Volume of wood in tree stems (from ground level or stump to a certain top diameter or tree tip) per unit area (ft^3/ac, board ft/ac, m^3/ha)	Describes merchantable wood, closely related to biomass; does not account for numbers of trees; several board-foot measures available; older yield tables establish average maximums for a species, site, and age
Cover	Percentage of ground surface area covered by foliage, crown projection areas, or foliage, branches, and stems together	Useful to describe shrub and understory vegetation, total tree cover or cover by layer; related to total leaf area or leaf area index (LAI); often estimated visually
Leaf area index (LAI)	Projected or all-sided surface area of foliage per unit ground area	Has been used for allocating growing space among size classes or cohorts; estimated from light transmission, allometric equations, foliage litterfall, or remote sensing; predictors for allometric equations include sapwood area at breast height or base of live crown (trees), basal stem diameter (shrubs), diameter at breast height, crown length, crown volume; related to stand-level photosynthetic capacity, less so to total respiring biomass

Table 6-1 cont.

Measure	*Definition*	*Conditions of use and comments*
Relative measures		
Reineke's stand density index (SDI)	Maximum number of trees per unit area with a given average diameter (10 inches or 25 cm)	Maximum varies with species; values can be calculated for combinations of numbers of trees and sizes; related to stand development and self-thinning; useful for describing and projecting stand development; provides little information on size-class distribution; maximum values depend on approach to estimation, for example, locating a boundary line graphically, regression through stands judged to be at maximum; regression estimate of asymptote based on observed time series, or stochastic frontier analysis
Curtis's relative density	Combination of any two of basal area per unit area (BA), quadratic mean diameter (Dq), and numbers of trees per unit area	Used mainly for Douglas-fir; closely related to Drew and Flewelling's relative density index (1979) and stand density index; flexible and readily computed from standard inventory data
Drew and Flewelling's relative density index	Ratio between observed number of trees per unit area to maximum number possible with the same average stem volume	Related to tree and stand volume, thus takes tree height into account as well as tree diameter; depicts stand development and volume production at different densities; basis for the density-management diagram for Douglas-fir (Drew and Flewelling 1979, Long et al. 1988)
Spacing and height; relative spacing (RS)	Ratio of average tree height to average tree spacing	Useful for specifying stand density regimes in young to mature stands
Crown competition factor	Percent of area that would be covered by crown projections if all trees had maximum crown widths for their diameter	Useful as a measure of crowding in mixed-species stands where potential crown widths vary widely by species (e.g., Bravo et al. 2005)

$$BA_o / BA_e = N_o k_1 D_o^2 / N_e k_1 D_o^2 = N_o / N_e$$

where BA_o and BA_e are the observed basal area and expected basal area, respectively, at carrying capacity; D_o is the observed quadratic mean diameter; and k_1 is $\pi/(4 \cdot m)$ with m=144 when D_o, BA_o, and N_o are measured in inches, ft^2/ac, and trees/ac, respectively, and m=103 when D_o, BA_o, and N_o are measured in cm, m^2/ha, and trees/ha, respectively. To create a relative stand density measure, BA_e would have to be established for a given site and age, or N_e would have to be established for the observed quadratic mean diameter, D_o. As shown below, many common measures of stand density can represent some variation on the ratio of observed number of trees per acre to expected number of trees per acre with the same quadratic mean diameter but at carrying capacity for the species.

STAND DENSITY INDEX

Reineke's (1933) stand density index (SDI) is a function of quadratic mean diameter and number of trees per unit area:

$$SDI = N_o(D_o/10)^{1.605}$$

where N_o is observed trees per unit area and D_o is the observed quadratic mean diameter. The equation for SDI takes any given combination of D_o and N_o and re-expresses stand density as the competitive equivalent of N trees with quadratic mean diameter of 10 in. Hence, SDI can be interpreted as the number of 10-in trees that would experience approximately the same level of inter-tree competition as the observed number of trees per ac (N_o) with the observed quadratic mean diameter (D_o).

A relative density measure is derived from SDI by expressing it as a proportion of maximum SDI (SDI_{max}):

$$SDI_{rel} = SDI/SDI_{max} = N_o(D_o/10)^{1.065} / N_e(D_o/10)^{1.605} = N_o/N_e$$

Relative SDI therefore represents the ratio N_o/N_e where N_e is the maximum capacity for number of trees per unit area with quadratic mean diameter D_o. As discussed in the section below on maximum SDI, SDI_{max} is estimated empirically by measuring a large number of stands at maximum stand density. Given an established SDI_{max}, the expected maximum number of trees can be expressed as a nonlinear function of the observed D_o, as shown by the following:

$$SDI_{max} = N_e(D_o/10)^{1.605}$$

or

$$N_e = SDI_{max}(D_o/10)^{-1.605}$$

so

$$SDI_{rel} = N_o/N_e = N_o/SDI_{max}(D_o/10)^{-1.605} = (N_o/SDI_{max})(D_o/10)^{1.605}$$

RELATIVE SPACING

Some indices of stand density have been based on observed number of trees per unit area (or observed spacing) relative to stand average height (Wilson 1979). Because the relationship between tree height and diameter is consistent among stands at maximum stand density, the spacing to height ratio, too, is interpretable as a form of N_o/N_e. Stand density indices based on average tree height are often referred to as relative spacing (RS) and are typically defined as:

$$RS = S/H$$

where S is the average spacing between trees, and H is average tree height. Because spacing and number of trees per unit area are analytically related ($S = [area/N_o]^{½}$), it follows that

$$RS = [area/N_o]^{½}/H$$

At maximum stand densities, Douglas-fir and other species exhibit the following approximate nonlinear relationship between individual-tree height and diameter (Curtis 1970):

$$H = k_2DBH^{0.8}$$

or

$$DBH = k_2^{-1.25}H^{1.25}$$

where H is total tree height, DBH is tree diameter (at breast height), and k_2 is an empirically determined constant. As shown above for SDI, the expected number of trees at maximum stand density can be expressed as a function of their quadratic mean diameter; that is,

$$N_e = k_3D_o^{-1.605}$$

where k_3 is $SDI_{max}10^{1.605}$. The relationship between individual-tree height and DBH identified above can be applied to a tree with DBH equal to D_o so that $[k_2^{-1.25}H^{1.25}]$ can be substituted for D_o, yielding:

$$N_e = k_4(H^{1.25})^{-1.605}$$
$$= k_4H^{-2}$$

or

$$H = (k_4/N_e)^{½}$$

where k_4 is $k_3k_2^2$. The expected number of trees at maximum stand density is therefore a function of H and vice versa, so H in the relative spacing index can be substituted with $(k_4/N_e)^{½}$ to give

$$RS = (area/N_o)^{½} / (k_4/N_e)^{½}$$

or

$$RS = k_5(N_o/N_e)^{½}$$

where k_5 is $[area/k_4]^{½}$. Therefore, relative spacing can also be regarded as a function of the ratio between observed number of trees per unit area and the maximum possible with average height H. Note that in this case, the relative spacing increases with declining number of observed trees (N_o).

RELATIVE DENSITY INDEX

Drew and Flewelling's (1977, 1979) relative density index (RDI) is similar to Reineke's (1933) SDI, but the former was derived from Yoda et al.'s (1963) –3/2 power law of self-thinning. The rationale for this "law" is based on two assumptions:

(1) The area occupied by a plant is proportional to the square of some linear dimension of the plant, L^2.

(2) The biomass or volume of the plant is proportional to the cube of some linear dimension, L^3.

The first assumption allows one to deduce the following relationship between the observed mean linear dimension of the tree and the maximum tree density:

$$N_e = k_6 L_o^{-2}$$

or

$$L_o = k_6^{-1/2} N_e^{-1/2}$$

where L_o is the observed linear dimension, N_e is the expected tree density at carrying capacity, and k_6 is a coefficient that is empirically determined from stands that are at maximum carry capacity (maximum tree density). The second assumption asserts that:

$$V_e = k_7 L_o^3$$

where V_e is the expected tree volume given the observed linear dimension L_o and k_7 is another coefficient that is determined from trees in stands that have reached maximum density. After substituting $k_6^{1/2} N_e^{-1/2}$ for L_o in the V_e equation, the relationship between average tree volume and the observed linear dimension becomes:

$$V_e = k_7 [k_6^{1/2} N_e^{-1/2}]^3$$

or

$$V_e = k_8 N_e^{-3/2}$$

where $k_8 = k_7 k_6^{3/2}$. The relationship is typically presented in this form or as a corresponding stand density management diagram with logarithm of mean tree volume on the Y-axis and logarithm of tree density on the X-axis. The exponent of -3/2 in the equation and slope of -3/2 in the diagram give the law its name. A relative density measure follows by expressing the expected tree density at carrying capacity, N_e, as a function of observed mean tree volume, V_o:

$$N_e = k_8^{2/3} V_o^{-2/3}$$

Drew and Flewelling's (1979) relative density index is the ratio of observed to maximum tree density, where the expected number of trees at maximum tree density is a function of average tree volume:

$$RDI = N_o / N_e$$

or

$$RDI = N_o V_o^{2/3} / [k_8^{2/3}]$$

CURTIS'S RELATIVE DENSITY

Recognizing the close relationship among the above stand density measures, Curtis (1982) developed a simpler relative density metric that could also be expressed in relative form, i.e., a form equivalent to the ratio N_o/N_e. Curtis's (1982) relative density, CRD, can be expressed as a function of any two of the following quantities: (1) basal area per unit area, BA; (2) quadratic mean diameter, D_q; and (3) number of trees per unit area, N_o. The standard form of the relative density measure is:

$$CRD = BA / \sqrt{D_q} = BA D_q^{-1/2}$$

and the two alternative forms are:

$$CRD = k_9^{-1/4} N_o^{1/4} BA^{3/4}$$

$$CRD = k_9 N_o D_q^{3/2}$$

where k_9 is $\pi/[4 \cdot 144]$ for imperial units and $\pi/[4 \cdot 10^4]$ for metric units.

Maximum CRD (CRD_{max}) must be empirically determined for a given species and perhaps for the same species on different sites. At maximum CRD the expected tree density, N_e, could be expressed as a function of observed quadratic mean diameter based on the last equation above:

$$CRD_{max} = k_9 N_e D_q^{3/2}$$

or

$$N_e = CRD_{max} / (k_9 D_q^{-3/2})$$

where CRD_{max} is maximum CRD that is empirically determined from stands at carrying capacity. The ratio of observed CRD to maximum CRD can therefore be interpreted as the ratio of observed tree density (N_o) to expected tree density at maximum CRD (N_e), as follows:

$$CRD_{rel} = CRD_o/CRD_{max} = k_9 N_o D_q^{3/2} / k_9 N_e D_q^{3/2} = N_o / N_e$$

Given CRD_{max}, CRD_{rel} can be computed as follows:

$$CRD_{rel} = CRD_o/CRD_{max} = k_9 N_o D_q^{3/2} / CRD_{max}$$

As an example of the range of values observed for CRD, stands of Douglas-fir that McArdle et al. (1961) identified as normal averaged 9.5 in metric units (BA in m^2/ha and D_q in cm) and 65 in imperial units (BA in ft^2/ac and D_q in in). The biological limits for managed stands of coastal Douglas-fir seem to be about 14 in metric units and 100 in imperial units (Curtis and Marshall 1986). Therefore, CRD expressed in imperial units can be interpreted approximately as a percentage of maximum possible CRD for coastal Douglas-fir.

ESTIMATING MAXIMUM SIZE-DENSITY RELATIONSHIPS

Reineke (1933) described a maximum density for an even-aged stand by placing a line along the outer boundary of a scatter plot of numbers of trees/ac and tree size on log/log scales. He used the intersection of the line at an average D_q of 10 in to describe his "maximum" SDI (i.e., about 600 10-in trees/ac). For example, an SDI of 600 is considered the maximum for Douglas-fir west of the Cascades. In reality, permanent plots containing coastal Douglas-fir (*Pseudotsuga menziesii* var. *glauca*) self-thin at a slope very close to Reineke's slope, but at widely varying SDIs (Hann et al. 2003). The maximum can also change among larger geographic regions that correspond to different subspecies; for example, maximum SDI for Douglas-fir (*Pseudotsuga menziesii* var. *glauca*) in the Rocky Mountains sites is about 380. The maximum also varies by species, with shade-tolerant conifers generally having higher SDIs than shade-intolerant conifers and hardwoods (Harper 1977); for example, maximum SDIs are approximately 1000 for redwood and red fir, 830 for white fir, and 800 for Sierra Nevada mixed conifer (table 6-2). Conversions to and from Curtis's relative density (CRD) can be derived from these maximum SDI values by algebraic manipulation (table 6-3).

In some cases the exponent in the SDI definition (1.605) has also been allowed to vary among species; for example, Cochran et al. (1994) found that estimates ranged from 1.51 to 1.77 for various conifer species in eastern Oregon. In general, the slopes of a maximum size-density line are expected to vary due to different factors related to stand and tree dimensions, including: 1) allometric relationships between a specific linear dimension and the area occupied by a tree (for example, DBH and crown projection area); 2) allometric relationships between a specific linear dimension and the volume or mass of a tree (for example, DBH and total stem volume); 3) the specific tree components included in total volume or mass (for example, total cubic stem volume vs total woody above-ground biomass vs total below- and above-ground biomass vs merchantable stem volume); and 4) ratio of live, respiring cell mass or volume to total mass or volume (Perry 1984).

In general, species with a lower rate of increase in growing-space occupancy per unit increase in DBH will have a steeper (more negative) slope. A simple look at a generalization of the assumptions behind the –3/2 law underscores the expectation of different slopes for the maximum size-density line among species. Recall that Yoda et al. (1963) assumed that:

$$N_e \propto 1/L^b$$

and:

$$V \propto L^c$$

where b=2 and c=3.

Algebraic manipulation results in the logarithmic version of the maximum size-density line with slope of −3/2. However, the parameters b and c will vary by species for the reasons listed above, so that, more generally, the maximum size-density line is:

$$\ln(V) = a - c/b \ln(N_e)$$

where *a* is a parameter estimated empirically from a sample of stands judged to be at maximum carrying capacity. The slope of this line is −*c*/*b*, so it is determined by the allometric relationships between the linear dimension of the tree (typically diameter), the area occupied by the tree (typically crown projection area), and the volume of the tree (typically the stem volume).

In mixed-species stands, several approaches have been recommended for establishing a maximum stand density based on maximum SDIs of the constituent species. Cochran et al. (1994) recommend selecting the SDI of the species with the lowest maximum value, but Hann and Wang (1990) calculate a weighted average SDI in which the weights are the basal areas of the respective species. Woodall et al. (2005) developed a method for calculating SDI of uneven-aged, mixed-species stands based on the specific gravities of component species. Cochran et al. (1994) produced SDI values for common forest types in eastern Oregon that account for "stockability" (see section below) and the effects of bark beetles (table 6-2).

Stands may reach but not maintain high densities, so the maximum SDI may sometimes need to be adjusted. Oliver (1995) suggests that ponderosa pine stands are "maintained" at an SDI of 365 by *Dendroctonus* spp. Although stands with greater SDI do occur, bark beetles soon reduce their density, and thus 365 is a more reasonable measure of maximum density. Peterson and Hibbs (1989) provide an example of adjusting the maximum SDI for lodgepole pine from 700 (McCarter and Long 1986) to 450 based on stockability in central Oregon. In practice, however, the former authors suggest that an SDI of 160–250 will reduce bark beetle risks and result in healthier stands, greater net wood production, and stable wildlife cover.

DENSITY MANAGEMENT DIAGRAMS

The concept of SDI and relative density has led to the development of density management diagrams that are useful for depicting and planning stand density-management and thinning regimes (figure 6-1). The diagrams can be used to depict development of even-aged stands and they indicate different levels of SDI at which crown closure and self-thinning will likely occur. As we will discuss in the next chapter, these diagrams are useful for planning thinning or planting densities and for anticipating when self-thinning or susceptibility to bark beetle infestation may occur. The rate of stand development can be estimated from growth models or from evaluating stand growth on similar sites. These diagrams probably work best in plantations or stands with uniform canopies, but even in these stands it is not clear if SDI consistently measures stand potential for growth or development. For example, if two stands reach the same SDI but with different histories of density, self-thinning or no thinning (figure 6-1), will they be equally productive or resistant to wind or insects? In stands where seedlings are established over several years or where there is vigorous advance regeneration, the relationship to diameter growth and stand density may be different than these diagrams indicate.

In uneven-aged mixed-conifer forests of the northern Rocky Mountains, Sterba and Monserud (1993) found that the slope of the maximum size-density line varied with differing shapes of the diameter distribution. This dependence on stand structure complicates interpretation and application of SDI and stand density-management diagrams in uneven-aged stands. However, approaches that assume a constant slope seem sufficient for managing uneven-aged stand density, particularly if SDI is calculated by the following summation method (Cochran 1992; Long 1996; Long and Daniel 1990; Puettmann et al. 1993a):

$$SDI = \Sigma(D_i/10)^{1.605}$$

where D_i is the diameter of the *i*th tree, or

$$SDI = \Sigma\, N_i(D_i/10)^{1.605}$$

where D_i is the average or midpoint diameter of the *i*th diameter class and N_i is the number of trees per unit area in the *i*th diameter class.

STOCKABILITY

The maximum stand density attainable for a given species is generally regarded as constant over a relatively large geographic area, if not the entire range of the species. With some coniferous species, particularly on dry sites, a relatively large range in maximum carrying capacity has been observed (Cochran et al. 1994; McCay 1985). This variation in the biological limit of stand density has given birth to the concept of stockability. Unfortunately, this term conflicts with the standard convention of using "stocking" to describe density relative to some optimum dictated by management objectives; however, the fact of differing maximum densities is widely recognized. Hall (1983) developed a measure called growth basal area (GBA) to characterize the biological carrying capacity of forest sites in eastern Oregon. GBA is the basal area at which dominant trees grow 1 in in diameter per decade at total age 100 years (see chapter 5).

It has been repeatedly demonstrated that stockability is not correlated with site index but rather indicates a different facet of the site's potential productivity. Early attempts to predict stockability were based on plant indicators (MacLean and Bolsinger 1973). This indicator approach, however, was superseded by direct measurement of

Table 6-2. Maximum recommended SDI (maximum number of 10-in trees/ac) for some common western trees. Note that SDI is lowest for the most shade-intolerant pine species, and that SDI for Douglas-fir on moist sites is greater than for dry sites. Adapted from Cochran et al. (1994) with addition of species from other sources.

Species	*SDI*	*Slope*	*Reference*
Ponderosa pine	365	1.77	DeMars and Barrett 1987
Ponderosa pine	450	1.60	Long and Shaw 2005
Lodgepole pine	277	1.74	Cochran et al. 1994
Western larch	410	1.73	Cochran 1985
Douglas-fir (dry)	380	1.51	Seidel and Cochran 1981
Douglas-fir (moist)	598	1.6	Long et al. 1988
White/grand fir	560	1.73	Cochran 1983
Western hemlock	850	1.605	Hyink et al. 1987
Subalpine fir	416	1.605	Cochran et al. 1994
Red fir	1000	1.605	Reineke 1933
Red alder	450	1.605	Puettmann et al. 1993a
Sitka spruce	810	1.605	Peterson et al. 1997
Coast redwood	1000	1.605	Reineke 1933
Western redcedar	720	1.605	Smith 1989

GBA, which requires coring trees to determine present growth rates and converting observed basal area and growth rate to an estimate of the basal area required to allow growth of 0.1 in/year on 100-yr-old dominants. In practice, GBA for a site is often inferred from the plant association, because the classification schemes on which plant associations are based are accompanied by tables of average growth basal area (for example, Hall 1973, 1983).

For example, Hall (1983) reports that for ponderosa pine growing on a site of a particular index, the GBA varied with plant association across the ponderosa pine types in Oregon, ranging from 35 to 135 ft^2/ac (8 to 31 m^2/ha; table 6-4). Cochran et al. (1994) used averaged GBAs and average site index for each plant association to determine its maximum SDI. The ratio of actual GBA for a given plant association to GBA predicted from its average site index was applied to adjust the maximum SDI assumed for that species to a specific plant association:

$$SDI_{plant} = SDI_{max}(GBA_{plant}/GBA_{site})$$

where SDI_{plant} is the maximum SDI for plant association *plant*, SDI_{max} is the maximum SDI for the species (table 6-2), GBA_{plant} is the growth basal area listed for plant association *plant*, and GBA_{site} is the predicted growth basal area for all plant associations with site index *site*.

In areas of the inland West where the carrying capacity of a site is not strongly correlated with site index, evaluation of local productivity and maximum stand density through plant associations may often be needed to establish stand density guidelines

(Cochran et al. 1994; Peterson and Hibbs 1989). Variation in carrying capacity has also been observed in moister forests west of the Cascades crest, but site characteristics that can account for this variation have not been identified (Hann et al. 2003).

Size-Class Structure of Even- and Uneven-Aged Stands

Two key points must be made when considering applicability of maximum size-density relationships to uneven-aged stands. First, identification of the maximum relative density may be complicated by the expectation that this maximum is a function of size-class distribution (numbers of trees by height or diameter class), of which there is almost an infinite variety in uneven-aged stands. Second, as mentioned above the slope of the size-density relationship also may depend on size-class distribution, particularly if the uneven-aged structure leads to slightly different average allometric relationships for a given species. Sterba and Monserud (1993) found that the slope of the maximum size-density line in uneven-aged, mixed-species stands became more negative with

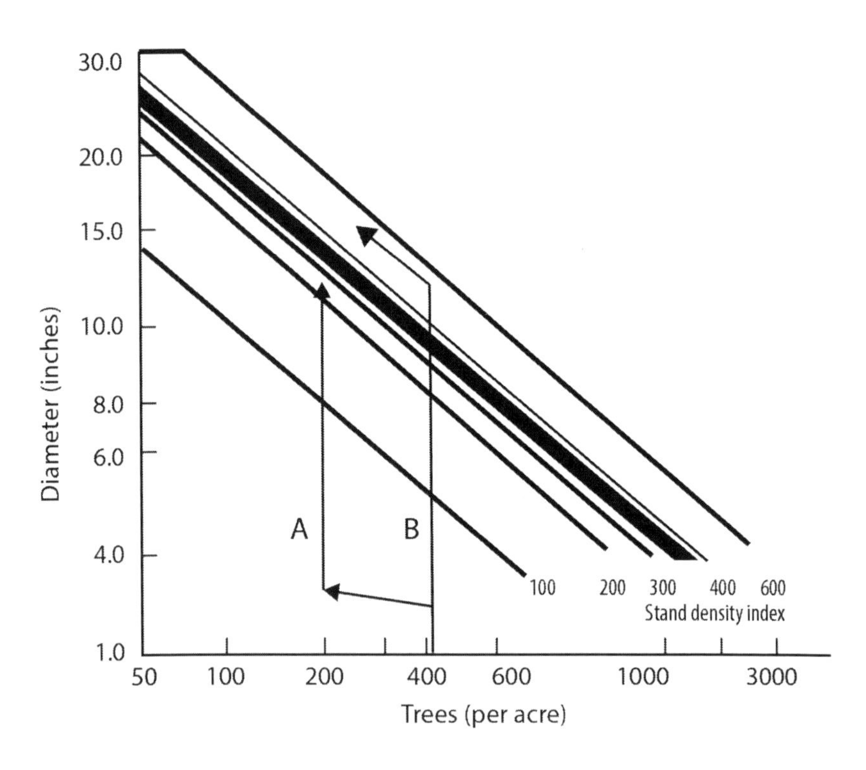

Figure 6-1. Density diagram for Douglas-fir (Long et al. [1988]). Black area indicates zone where self-thinning is likely to begin. Development of two theoretical stands is shown. Stand A started at 400 trees/ac, and was thinned to 200 trees/ac; grew to about 55% of maximum SDI and 14 in diameter. Self-thinning has not yet occurred. Stand B started at 400 trees/ac and grew to about 14 in diameter when it entered the zone of self-thinning; it grew to an average diameter of about 16 in.

Table 6-3. Formulas for computing SDI and CRD from BA, D_q, and N. Imperial units include in for D_q, ft²/ac for BA, and trees/ac for N, with constants $k = \pi/(4 \cdot 144) = 0.005454$ and D_S=10. Metric units include cm for D_q, m²/ha for BA, and trees/ha for N, with constants $k = \pi/(4 \cdot 10000) = 0.00007854$ and D_S=25.4.

Stand attribute	*Formula*
SDI	$= N \cdot (D_q / D_S)^{1.605}$
	$= (k \cdot D_S^{1.605})^{-1} \cdot BA \cdot D_q^{-0.395}$
	$= (k^{1/2} \cdot D_S)^{-1.605} N^{0.198} BA^{0.803}$
	$= (k \cdot D_S^{1.605})^{-1} \cdot D_q^{0.105} \cdot CRD$
CRD	$= BA/D_q^{1/2}$
	$= k \cdot N \cdot D_q^{3/2}$
	$= k^{1/4} \cdot N^{1/4} \cdot BA^{3/4}$
	$= k \cdot D_S^{1.605} \cdot D_q^{-0.105} \cdot SDI$

increasing skewedness of the dbh distribution (i.e., a slope with the same direction as in figure 6-1 but steeper).

Stand density and the stand-level attributes from which it is computed (basal area, quadratic mean diameter, stem density, etc.) are inadequate for understanding more detailed aspects of stand structure. Most silviculturists would hesitate to claim they knew anything about stand structure without, at minimum, knowing something about the size distribution of the trees. The diameter distribution (number of trees per diameter class) is by far the most commonly applied descriptor of stand structure, and it is related to the distribution of other variables such as total height, crown length, and stem volume or biomass.

The most readily interpretable form of diameter distribution is the histogram of absolute or relative stem frequency by diameter class and species (figure 3-2). In forests with relatively simple structure, such as many even-aged coniferous plantations, the diameter distributions are sufficiently regular that they can be described well by a smooth probability distribution. Several distributions have been applied in this context, but the most common has been the Weibull distribution because of its flexibility in accommodating not only symmetrical, bell-shaped diameter distributions, but also those with both a positive and negative skew, including negative exponential forms (Bailey and Dell 1973).

When fitting smooth functions to diameter distributions, considerable information on stand structure is stored in the parameters of the fitted curve. For example, the mean (location parameter) and the variance (scale parameter) of a normal distribution would contain all the information required for reconstructing the distribution of diameters from a stand conforming to a normal distribution. Similarly, a location, scale, and shape parameter would fully specify the distribution of diameters from a stand that conformed to a Weibull distribution.

Table 6-4. Growth basal area for ponderosa pine growing on four different plant associations.

Location	*Plant association*	*Growth basal area (ft^2/ac)*
Blue Mountains (site index 60 ft @ 100 yrs)	Wheat grass/savanna	35
	Elk sedge	90
Southern Oregon (site index 80 ft @ 100 yrs)	Bitterbrush/sage/fescue	50
	White fir/snowberry/starwort	135

The combined distribution of diameters and heights in a stand has received less attention than diameter distributions, but a two-dimensional histogram can depict the absolute or relative frequency of stems by diameter-height combinations. Due to the close correlation between height and diameter within an even-aged stand, the combination of a diameter distribution and an appropriate curve for estimating height from diameter (i.e., a height-diameter curve) is often more than adequate for describing stand structure. However, univariate probability functions have also been extended from their univariate application in describing diameter distributions to multivariate applications for describing the bivariate distribution of height and diameter (Hafley and Schreuder 1977).

TARGET SIZE-CLASS DISTRIBUTIONS FOR UNEVEN-AGED STANDS

The size-class distribution of uneven-aged stands has traditionally been described by a negative exponential curve. Alexander and Edminster (1977) described a simple method (BDQ approach) for specifying diameter distributions when the objectives for managing uneven-aged stands are best met by a progressive decline in the number of trees in larger-diameter classes (see chapter 10). The perceived adequacy of the negative exponential model for describing uneven-aged stands in a large number of different forest types, and the apparent conformity of old unmanaged stands to this structure, has led to widespread adoption of the BDQ approach for regulating uneven-aged stand structure (Bailey and Covington 2002, Nyland 2002). However, Nyland (2002) found that a reverse-J distribution that departed somewhat from the strict negative exponential curve was superior for regulating structure in northern hardwoods in the northeastern United States, primarily because strict use of a negative exponential distribution often demands unnecessarily large numbers of trees in the small-diameter classes. For shade-intolerant species in particular, the more small trees that are required by this negative exponential approach, the greater is the tendency to lower the density of larger trees and thereby forfeit total stand volume growth. The strict use of a reverse-J (negative exponential) distribution can lead to high density of small trees in many forest types, including ponderosa pine. The result would likely be stagnation or trees with unnecessarily small crowns and low vigor entering the larger size classes.

SDI can also be used to set goals for uneven-aged stands as an alternative to the BDQ approach. Long and Daniel (1990) noted that the summation method of determining SDI in uneven-aged stands lends itself well to allocating residual SDI to separate parts of the diameter distribution. Cochran (1992) took this approach in ponderosa pine by allocating SDI units to four or five merchantable diameter classes and one sub-merchantable class, maintaining only enough trees in the sub-merchantable class to replace the desired number in the smallest merchantable class (see chapter 10). As in the BDQ method, the manager decides on a maximum tree diameter, but then allocates SDI to the different size classes. Having too many trees in the small-size classes or too few in the large classes, for example, may be corrected by adjusting SDI among classes (see chapter 10 for an example of its use). The flexibility in the SDI approach is useful because it enables adjusting density of trees to reasonable levels within broad size classes (O'Hara 1996; O'Hara and Gersonde 2004).

Horizontal Structure

Forest stands can vary significantly in the spatial distribution of individual trees and understory plants. At a given stand density, individuals of one species may be clumped or evenly dispersed throughout a stand; clumps may vary in density and size. In the case of hardwood clumps, many stems may have sprouted from a single root system. The horizontal structure of a stand will have implications for how well specific objectives can be met. For timber-management objectives, relatively uniform spacing provides optimal growth of individual trees, the easiest thinning operations, and the greatest uniformity in log size and quality. Sometimes—to meet aesthetic or wildlife-habitat objectives, for instance—it may be desirable to have some heterogeneity in spatial distribution of stems. Departures from uniform spacing of forest plantations, however, have strong implications for individual tree development. In brief, if spacing is very irregular, individual tree development follows a course that would be expected under the higher mean stand densities commensurate with the local density in a clump of trees (Stiell 1982). However, edge effects on individual tree structure, such as asymmetrical crowns, dominate the character of the stand for a period of time determined by the rate of self-thinning and the initial inter-tree and inter-clump spacing.

Indices of spatial distribution are plentiful, particularly in the plant ecology literature. Cressie (1991) provides a relatively complete listing and description of these indices, including those based on plot-count data and those based on point-sample or nearest-neighbor data. One of the most common and potentially useful plot-count indices is the variance-to-mean ratio. This ratio should have a value of 1 if the individuals are randomly distributed. Values move closer to zero as the spatial distribution becomes more regular and will be greater than 1 if the spatial distribution becomes more clumped. The variance-to-mean ratio is directly connected to the expectation of equal mean and variance of the Poisson distribution. A slightly different index was developed by Morisita (1959) to detect differences in mean counts among patches, where the distribution of individuals within patches is assumed to be random. It is important to realize that the spatial scale at which horizontal structure is detected by any of the plot-based indices is limited to the plot size used. Varying plot sizes or scales of sampling can easily lead to detection of totally different spatial patterns (for example, see Maguire 1985). This limitation has led to a preference for multi-scale analyses such as K-function analysis (Moeur 1993, Cressie 1991). Although these indices and analyses of horizontal spatial structure have found numerous applications in forest research, they have not been applied in specifying silvicultural prescriptions. However, they do offer potential both for defining a target structure and for monitoring implementation and results of silvicultural prescriptions.

Frequency (number of occurrences/number of plots or points) is a simple, easily understood measure related to horizontal structure. Calculation of frequency might be used to determine the percentage of plots with shade-tolerant saplings, trees of a particular species, trees with root disease, or trees over a certain size, as well as the percentage of plots with a certain basal area, SDI, or shrub cover. Such frequency information could be used to determine whether species-composition goals are being met. Are particular species of trees or understory vegetation lacking in certain parts of a stand, for example? Here again, varying plot size and plot distribution, as discussed above, can make a big difference in the spatial patterns detected (Maguire 1985).

Vertical Structure

The vertical structure of forests has received considerably more attention in silviculture than horizontal structure, in part due to the strong influence it exerts on stand dynamics resulting from the effect of vertical foliage distribution on light absorption and gas exchange, as well as its long-recognized importance in characterizing wildlife habitat (e.g., MacArthur and MacArthur 1961). The varying growth potential associated with different competitive positions of trees within stands was recognized quite early in the development of forestry, as were the influences of those differences on crown development (Curtis and Reukema 1970), seed production, and stem form development (Larson 1963).

The most enduring descriptor of vertical structure has been the Kraft crown classes (Kraft 1884; Smith et al. 1997), which in most common applications include predominant, dominant, codominant, intermediate, and suppressed trees. These crown classes generally work well in even-aged stands. However, in uneven-aged stands the horizontal structure is often a variable mixture of tree species and sizes and these classes do not apply (see chapter 8 for further discussion).

A biologically more reasonable approach for uneven-aged stands combines features of two systems, the Kraft crown classes and the vertical layers designated by Richards (1964) for tropical rain forests (Oliver and Larson 1996). In this approach, the canopy emergent layer is labeled as the A-layer and successively lower layers are designated as the B-layer, C-layer, D-layer, and so forth. Within each of these layers, individual trees are assigned to a Kraft crown class. Basal area and volume per tree or per acre can be calculated for each of the crown classes to quantify that class's overall contribution to stand density and structure and to evaluate its growth potential. In temperate forests, however, this approach to uneven-aged stands has two shortcomings. The first is that in many uneven-aged coniferous stands it is difficult to identify distinct layers, even when it is clear that only three or four cohorts are present. The second is that the light conditions experienced by trees of a given crown class in a given layer may vary widely due to the effects of other layers.

With respect to the second shortcoming, the usefulness of any classification in predicting tree growth is somewhat limited. For example, consider an uneven-aged forest with some large trees in the A-layer that receive full light on most of their crowns. Dominant trees in the C-layer may be receiving considerable direct sunlight if growing in a canopy gap, or they might get virtually none if growing beneath a continuous canopy. These two hypothetical trees would be placed into the same classification, but their growth patterns would be very different because of smaller-scale site conditions that the classification does not consider. Perhaps one solution would be to further classify the trees according to their positions relative to the overlying layers, or quantifying competitive pressure with distance-dependent competition indices (e.g., Biging and Dobbertin 1995). However, even in mixed-conifer stands with complex structures, information on spatial location and local competitive environment has been shown to explain relatively little of the variation in individual tree growth (Mainwaring and Maguire 2004).

In even-aged stands, where trees are well-spaced, the relative height of individual trees may be sufficient or even superior for describing the social position of the tree and the various attributes associated with differing social position, such as relative foliage distribution by crown class (Gilmore and Seymour 1997) or by relative tree height (Maguire et al. 1998). In other stand structures, there will no doubt be significant value associated with assessing social position or crown class directly in the field rather than inferring it from height and crown measurements or assuming that it is correlated with relative height.

Vertical distribution of foliage has often been quantified by various probability distributions, including the beta distribution (Kellomäki 1986; Maguire and Bennett 1996), the Weibull distribution (Baldwin et al. 1997; Schreuder and Swank 1974), and the normal distribution (Meyer and Stevenson 1943). These descriptions have been developed at both the individual-tree level (Baldwin et al. 1997; Kellomäki 1986; Maguire and Bennett 1996) and the stand level (Schreuder and Swank 1974). In contrast to vertical foliage distribution, Dubrasich et al. (1997) developed an approach to characterizing vertical structure that may be more practical for defining silvicultural objectives and assessing their attainment. In their approach, tree crowns were viewed as solids, and crown area profiles were constructed to represent the total cross-sectional area of crowns at any given level within the stand, from ground level to the tip of the tallest tree (figure 6-2; Dubrasich et al. 1997). The potential utility of crown area profiles for characterizing wildlife habitat is worth further pursuit, because the open space between crowns is available for wildlife to move around in the canopy. This space can be measured as the vacant complement of the crown area profile, and has been referred to as the "porosity" of the stand.

Using Measures of Stand Density and Structure

The preceding section emphasizes that there are several ways of characterizing stand density and structure. The usefulness of each depends on the reason for describing the stand, resources available, and the user's preference. In some cases, state forest practice rules or owner policies specify stand-management goals in specific units, such as basal area or numbers of trees/ac, and so those measures are required. However, in many cases silviculturists will use a variety of measures depending on the stand structure, tree sizes, number of species, and the other components of stand structure that are to be characterized. Table 6-1 and the following discussion summarize the use of density measures.

Number of trees/ac is a simple measure of density that is very useful in stands of seedlings and small trees up to 4–6 in dbh. Using numbers of trees (usually summarized by species) is appropriate in stands of small trees because each tree contains little volume or basal area, and many small trees usually do not amount to high levels of SDI or relative density. Also, measuring individual trees is expensive. The density of young stands after thinning or planting is often stated as trees/ac, but the goal is to produce a stand with a desired D_q, basal area, volume, or relative density in the future. The actual goal is the future stand; the immediate one is a density expressed in trees/ac. Growth models or density-management diagrams can help determine the numbers of trees/ac needed to produce a stand of desired average diameter or stem volume in the future and the time required to produce it. Trees/ac may also be a useful measure where the goal is to manage stands for a few trees of a particular size or species, particularly if the value of the tree for meeting management objectives has little relation to the exact size beyond a specified minimum.

Basal area and volume/ac are strongly related to growing stock and wood production. Although total volume and volume growth are measures of production of wood, they are also closely related to overall biomass production (net primary production) and easier to measure. Basal area and basal-area growth are even more easily measured than volume and volume growth and are good predictors of growing stock and its rate of change, respectively. For trees of a particular size, stands with high basal area have greater volume, because the product of this cross-sectional area of wood, tree height, and form factor (a measure of stem form or taper) yields an estimate of total stem volume. Basal area and volume/ac are also both related to the amount of living or functioning tissue, although the only living cells are in the sapwood and cambium directly under the bark. Stands with high amounts of volume or basal area have high amounts of sapwood volume and sapwood cross-sectional, so the former are also somewhat related to the leaf area in the crown—generally the major photosynthesizing/transpiring part of the stand. Therefore, basal area and volume have both a mensurational and some physiological relationship to stand growth. However, sapwood basal area and volume of sapwood are much more strongly related to leaf area than are total basal area or volume. As stated above, maximum basal area and volume vary with stand age, site quality, and species composition, and therefore are not useful measures of relative density. Sapwood area and sapwood

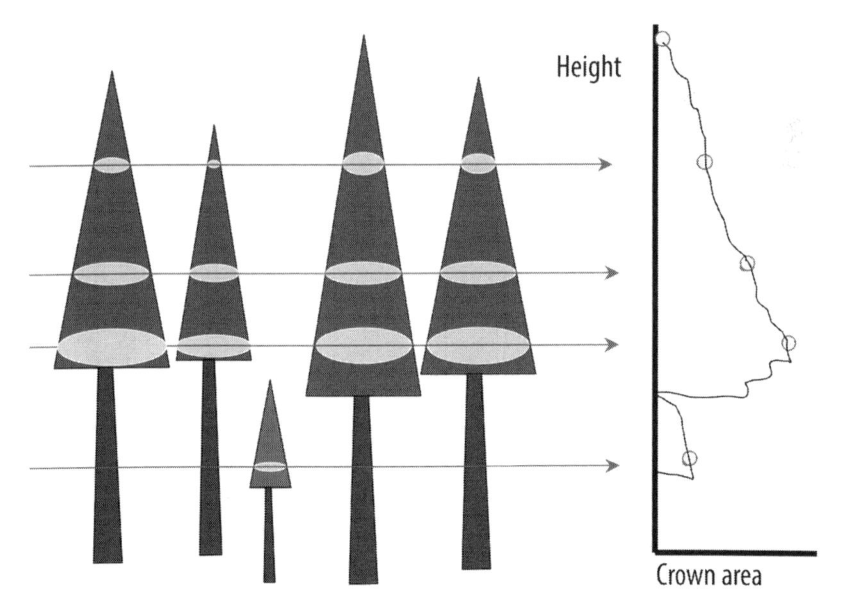

Figure 6-2. Crown profiles summed for a stand. The profile on the right represents the crown area by height for all trees in a stand (See Dubrasich et al. 1997). Crowns are assumed to be solids. This method of summing crown enables calculation of space among crowns from the top of the canopy to the forest floor, and it shows the vertical distribution of the crowns throughout a stand.

volume have not been estimated in a sufficient number of stands for us to know how well they might serve as an estimate of stand density.

Measures of relative density (such as SDI) probably best describe the level of inter-tree competition experienced by individual trees within a stand, because they take into account both tree size and numbers of trees/ac. As trees grow, the crowns eventually close and the maximum potential leaf area or biomass for a site is approached. For surviving trees to continue to grow, they have to gain leaf area at the expense of smaller trees as inter-tree competition increases. SDI reflects the degree to which the fixed amount of leaf area available in a stand is partitioned among individual trees (Long and Smith 1984). For a given tree size, a stand with many trees has a high SDI, high inter-tree competition, and a small amount of leaf area per tree. As tree numbers decline from self-thinning, the leaf area held by each tree increases, but total leaf area either remains constant or declines slightly.

Stand Density Index, as both a measure of stand density and growing stock, is related to self-thinning and the growth and yield potential of the stand (Marshall and Curtis 2002). Conversely, lower residual SDI after thinning implies greater growing space available for establishment of regeneration (Bailey and Tappeiner 1998). SDI also indirectly expresses tree crown size and crown cover in a manner analogous to leaf area, and indicates the total amount of stem surface area; thus, it provides a guide for managing stands for both hiding and thermal cover for elk (Long and Smith 1984; Smith and Long 1987). Maximum SDI is thought to be independent of stand age, site index, and possibly other measures of site quality. However, as described above, maximum SDI in some drier forest types appears related to aspects of site quality indicated by plant association (Cochran et al. 1994; Peterson and Hibbs 1989).

Neither SDI nor stand basal area measurements are sufficient to characterize stand structure accurately. Although tree size and numbers are both used to calculate these values, their ranges and relative frequencies are not represented explicitly. A stand of a given SDI, basal area, and volume with many small trees is likely to be quite different from one with the same measures but with fewer large trees, although they are probably more similar in regard to the level of inter-tree competition in the case of SDI. In stands of relatively uniform species composition and trees of about the same size, SDI and basal area seem to work equally well as measures of growing stock and growth potential (Marshall and Curtis 2002). Growth and yield and response to fertilization are best evaluated by volume or basal area because the goal of fertilization is normally an increase in yield. Also, fertilization may simply hasten self-thinning of small trees without a change in SDI. Measures of relative density seem to be useful for evaluating hiding cover, understory development, insect susceptibility, and potentially light and other aspects of microclimate. Numbers of trees/ac is an ideal stand-density measure for small trees or variables that occur in small numbers, such as snags or trees of a particular species or size.

In stands with variable structure, density, and species composition, probably no measure of overall stand density will sufficiently characterize the complexity of the stand. In such cases, it may be necessary to stratify the stand into layers or groups based on tree sizes, density, species composition, etc. This grouping is important when the criteria for meeting stand-management objectives, such as growth rates, presence of pathogens, and species diversity, vary by these layers. Different measures of density may be most efficient for characterizing different groups or layers—for example, numbers of trees for saplings vs basal area, volume, or SDI for larger trees.

Stand Growth Models

Stand growth models summarize stand density, stand structure, and growth rates, report their distribution across species and size classes, and project these attributes under different initial stand structures and silvicultural regimes. Forecasting stand development with simulation models helps to evaluate the likely efficacy of silviculture prescriptions. These models typically require descriptors of stand density, tree attributes, and stand structure from fixed or variable density plots (Avery and Burkhart 2002). The growth models most commonly used in the western United States operate on the level of individual trees, so require tree diameters, heights, diameter growth rates (from increment cores), live crown lengths, and expansion factors to calculate stand-level attributes (basal area, trees/ac, SDI, volume, and volume growth) by size classes and species and for the stand as a whole. This information is usually summarized in a stand table and a stock table (table 8-1). The models also simulate the future growth of the trees and stand in 5- or 10-yr increments and report stand tables and summaries for each growth period. Thus, silviculturists can assess the implications of hypothetical prescriptions on future stand structure and the degree to which a given silvicultural regime will meet management objectives (table 6-5).

Growth models are ideal for plantations and uniform stands, but they can also be used effectively in stands with more complex structures and variable species composition. Silviculturists can divide complex stands into subunits based on differences in density, age or size, species, or other variables such as pathogens. Most individual-tree growth models commonly used in the western United States do not account for the spatial location of trees during simulation; rather, they assume that the combination of individual-tree characteristics and stand-level structural attributes are sufficient for characterizing the local competitive environment (these models are referred to as distance-independent or spatially implicit). In reality, the stand may be a mosaic of groups that are distinctly different with respect to density, species composition, or some other

Table 6-5. Projection of no thinning (A) and thinning (B) of a 45-yr-old Douglas-fir stand over 40 yr. This example shows many but not all of the projections that can be made in stand growth models using data collected from specific stands (Hann 2005).

Age (yr)	*Species*	*Trees /ac*	*Height tallest 40 trees/ac*	*Diameter (in)*	*Basal area (ft²/ac)*	*SDI*	*Volume (ft³/ac)*	*Volume growth (ft³/ac/yr)*
A. No thinning								
45	Douglas-fir	112	95	12	90		2410	
	Grand fir	27		11	20		730	
	Maple	52		10	30		670	
	Total	191			140	270	3810	167
65	Douglas-fir	105	116	15	140		5090	
	Grand fir	26		17	40		1550	
	Maple	47		11	30		940	
	Total	178			210	400	7580	200
85	Douglas-fir	97	128	19	190		7900	
	Grand fir	26		21	60		2470	
	Maple	42		13	40		1180	
Total		**165**			**290**	**481**	**11550**	**185**

Age (yr)	*Species*	*Trees /ac*	*Height tallest 40 trees/ac*	*Diameter (in)*	*Basal area (ft²/ac)*	*SDI*	*Volume (ft³/ac)*	*Volume growth (ft³/ac/yr)*
B. Thinning								
45	Douglas-fir	70	95	13	73		2130	
	Grand fir	19		13	20		690	
	Maple	8		15	13		470	
	Total	97			106	192	3290	130
65	Douglas-fir	68	116	17	100		4590	
	Grand fir	19		17	28		1480	
	Maple	8		16	11		590	
	Total	95			139	227	6660	154
85	Douglas-fir	47	128	22	119		5430	
	Grand fir	14		21	34		1770	
	Maple	8		15	12		500	
Total		**69**			**165**	**273**	**7700**	**162**

attributes. For example, shade-tolerant conifers or hardwoods can appear as understory species in growth model projections, but they commonly occur as discrete aggregations of trees scattered throughout the stand. Likewise, when a few well-stocked plots of seedlings are averaged into the stand as a whole, one might conclude that there is a high density of seedlings in the understory when, in reality, seedlings may be distributed in clumps; therefore, users must consider variation in seedling numbers among plots and the frequency of areas with low or high seedling density. Also, hardwood density may be difficult to interpret because numbers of stems per tree may range from one to more than 10, particularly for species that form stump sprouts.

In short, a visual examination of the stand is often needed to interpret how the model characterizes stand structure with respect to within-stand variation in local structure. Comparison of the observed stand to model summaries may require the collection of data and their aggregation into tree lists in a manner that ensures the model projections represent likely stand development as closely as possible. A helpful feature of some models is the capacity for the user to enter tree characteristics such as crown classes, tree classes, or degree of mistletoe infection, and have the model summarize those characteristics (e.g., in terms of basal area, volume, and trees/ac).

REVIEW QUESTIONS

1. Which stand and tree attributes are related to stand density, and how?
2. What is a stand-density regime?
3. Distinguish among stocking, absolute stand density, and relative stand density.
4. How are crown closure, canopy density, and leaf area different from stand density?
5. What are two general measures of expected (maximum) numbers of trees per area?
6. What measures of average tree size can be used in relative density measures?
7. Sketch the maximum SDI or CRD relationship.
8. Explain how SDI measures vary with species and site. What parts of the SDI relationship vary?
9. Discuss the concept of growth basal area (stockability).
10. Draw a density diagram and explain how to use it.
11. Discuss the use of the various measures of stand density. When are they best used? What are the strengths and weaknesses of various measures?
12. What are growth models? What do they forecast and what are their strengths and weaknesses?

CHAPTER 7
Regeneration of Forests

Natural Regeneration
Advanced Regeneration
Artificial Regeneration
Site Preparation
Competing Vegetation and Establishment of Regeneration
Regeneration and Silvicultural Systems

Regeneration of forests is a major silvicultural activity in forest management. It is a process that includes ensuring a supply of seeds or seedlings, preparing the site, planting or establishing natural regeneration, evaluating stocking, and protecting seedlings by controlling competing vegetation and animal damage. Under most management objectives, forest stands will eventually need to be replaced or supplemented with new trees. The manner in which forest stands are regenerated has long-term implications for their subsequent management, habitats, and products. The rate of seedling establishment, initial stand density, tree size, species composition, economic value of the stand, and density and species composition of the understory vegetation are just some of the variables largely determined during regeneration of forest stands. In this chapter, we focus on seedling establishment; however, controlling stand density and species composition through precommercial thinning (chapter 8) might also be considered a step in the process of regeneration.

Regeneration is part of all silvicultural systems, and it is also used to establish new stands after severe fires or other disturbances and to afforest sites that previously had no forest. Silvicultural systems are closely linked to the regeneration method that establishes or maintains them (chapter 2; Smith et al. 1997). In even-age management, reforestation occurs at the time a stand is harvested and replaced by a new one at the end of a rotation. If a stand is clearcut, forest practice regulations for public forests and for private forests in several western states, as well as best management practices generally, mandate establishment of a minimum number of seedlings per acre. If a shelterwood method and natural regeneration are used, regeneration may occur over a longer period, possibly 10+ yr, or as long as shelterwood trees remain in the stand. In uneven-age systems, which are designed to maintain a relatively continuous forest cover, regeneration is ongoing to maintain a population of seedlings and saplings that can replace the overstory trees as they are cut or die.

Forest regeneration has been a major part of western forest management for several decades. By the early 1900s in the western United States, millions of forest acres had been burned by severe, stand-replacing fires, while others had been logged to provide wood for the development of the West and for export to the eastern states. Much of the early forestry research in the western United States therefore focused on forest regeneration (Hoffman 1924; Isaac 1938, 1940; Munger 1911; Pearson 1923). By 1940, state forest practice regulations required leaving seed trees for regeneration. But establishing regeneration remained a problem, especially on sites where hardwoods, such as red alder, shrubs and grasses, competed with seedlings and provided habitat for seed and seedling predators (Baker 1955).

Since those days, much research has been devoted to the science of reforestation (Cleary et al. 1978; Hobbs et al. 1992; Curtis et al. 1998). Today, forest practice regulations in several states specify when reforestation practices are to begin and the minimum number of "free-to-grow" seedlings that must be present after a certain period following timber harvest (Rose and Coate 2000). With existing technology, forest plantations can be quite productive and capable of producing substantial volumes of wood (Hermann and Lavender 1999). In addition, they can provide a range of forest resources (Muir et al. 2002).

Forest regeneration focuses on establishment of trees because they are the major vegetative components of forest stands; however, the concept is evolving to take into consideration a variety of species. Regeneration of shrubs, forbs (broadleaf herbs), and grasses from seeds or sprouting buds occurs simultaneously with that of tree seedlings, especially on disturbed sites. Many of these plants may interfere with or prevent regeneration of trees, whereas some may facilitate tree regeneration, and others may have little or no effect. Rapidly growing shrubs, sprouting hardwoods, and excessively dense conifer regeneration may shade out other desirable plants.

Because the variety of sprouting hardwoods, shrubs, and grasses that grow on recently disturbed sites can threaten the survival and growth of tree seedlings, intensive practices that require labor and considerable investments of money—such as site preparation, planting large seedlings, and releasing seedlings from competition—are used to control competing vegetation. These practices are common to Douglas-fir, ponderosa pine, and mixed-conifer forests. In other forest types, such as lodgepole pine in the interior west or western hemlock-Sitka spruce in southeastern Alaska, very little effort is required to establish conifer regeneration; in those forests, because of the abundance of seedlings that become established, control of seedling density is more of a concern than seedling establishment.

There are two basic regeneration strategies: natural regeneration and artificial regeneration. Natural regeneration consists of establishing new seedlings from the seed of overstory trees, release of advance regeneration (older seedlings already established), and stimulation of vegetative reproduction (sprouts from bud banks at the base of aerial stems and on roots, rhizomes, and burls). Artificial regeneration consists

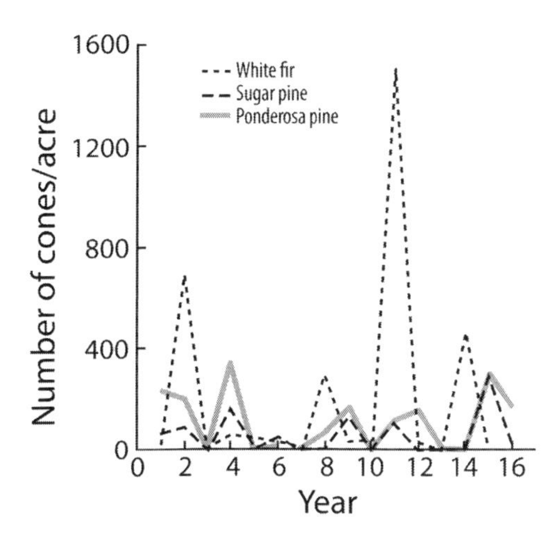

Figure 7-1. Number of cones/ac produced by three species in a mixed conifer forest during 17 yr between 1933 and 1953. Adapted from Fowells and Schubert (1956).

primarily of planting nursery-grown seedlings, but sometimes it includes sowing seed or planting cuttings.

The regeneration strategy chosen for a stand will depend on the characteristics of the site, such as topography, the need for site preparation, and the vegetation present. Will recolonization by shrubs and grasses require planting and competition control? Will there be sufficient natural regeneration? Will overstocking or poor distribution of seedlings be a problem? Are browsing animals likely to damage new seedlings? Other important factors include the landowner's objectives, government regulations, cost, and available resources (for example, personnel availability, ease of access to planting sites, quality of planting stock, etc.). On a particular site, a regeneration strategy will often use a combination of methods. For example, in a coastal redwood forest, reforestation could include sprouting of redwood trees, natural regeneration of western hemlock and red alder, and planting of Douglas-fir and redwood seedlings.

Natural Regeneration

Natural regeneration is the establishment of forest plants from the seed of overstory trees or sprouts from existing bud banks. It includes reforestation following timber harvest, fire, and other disturbances, and it also covers the establishment of advance regeneration. Classical regeneration methods such as shelterwood, seed tree, and group selection were originally designed to obtain natural regeneration. Early silvicultural research in western forests focused on natural regeneration of conifers on logged and burned forest sites (Dunning 1923; Franklin 1963; Haig et al. 1941; Hermann and Chilcote 1965; Isaac 1938; Lavender 1958; Pearson 1923). These researchers investigated the steps in natural regeneration from flowering and seed production through seedling establishment and the growth of seedlings, all of which must occur for seedlings to become established. It is important for practitioners to determine which steps they may need to manage by silvicultural practices to ensure that the process can be completed successfully—and which steps may fail on their sites (Minore and Laacke 1992; Zasada et al. 1986, 1992).

SEED AVAILABILITY

The availability of seed for seedling establishment depends on seed production and quality, seed dispersal, and seed predation. Seed production varies from year to year and is therefore unpredictable. Long-term records from mixed-conifer stands (Fowells and Schubert 1956) indicate that large cone crops may occur every 4–10 yr for white fir, ponderosa pine, and sugar pine. However, moderate crops occur more frequently, and some seed is produced every 1–2 yr (figure 7-1). Similarly, during a 24-yr study in a Sierra Nevada mixed-conifer forest, McDonald (1992) found seven medium to heavy cone crops for ponderosa pine, five for white fir, four for tanoak, and fewer than three for Douglas-fir, incense-cedar, and sugar pine, as well as fewer than three acorn and berry crops for California black oak and Pacific madrone, respectively. However, most species had several light crops during this period.

Older trees with large crowns are more likely than younger, smaller trees to produce large cone and seed crops (Tappeiner 1967). Past cone production (evidenced by cones on the forest floor beneath potential seed trees) is good evidence of a tree's ability to produce seed in the future (Sundahl 1971; table 7-1). Sundahl found that only 30–40% of the shelterwood trees in a ponderosa pine stand produced cones in a year of heavy seed production, and 7–35% produced cones in poor seed years. He suggested that these percentages could be increased if evidence of past cone crops was used as a criterion for leaving shelterwood trees. Seed production in the coming year can be estimated by observing the abundance of female flowers during the spring (Allen 1941).

Seed production can also be influenced by stand density. Reukema (1982) observed young Douglas-fir stands for 29 yr (39–68 yr of age) and found that thinning increased seed production per unit area, with more seed produced in heavily thinned stands

Table 7-1. Douglas-fir cones per tree in years of high (1962) and low (1965) cone production. Trees that had produced the most crops before 1962 produced the most cones in the next two crops. Evaluating the abundance of cones beneath trees is a viable way to judge if trees are likely to produce cones and seed in the future (Tappeiner 1967).

Cone crops per tree before 1962	*Cones per tree in 1962*	*Cones per tree in 1965*
3	600–3500	60–860
2	150–1200	10–460
1	5–2000	0–320
0	5–400	0–80

than in unthinned or lightly thinned stands. Additional thinning might be needed to maintain seed production because within the study period production in thinned stands returned to that of unthinned stands.

The number of seeds per cone is greater in years with good cone crops than in years with poor crops, and larger trees produce the most cones (table 7-2). Cone and seed insects (Keen 1956) can reduce seed quality, especially in poor seed years. A seed predator (*Megastigmus* spp.) infested only 1.6% of the seed per cone of the large seed crop in 1962 and 10–50% of the seed (1.0–13.0% of the seed per cone) of the light crop in 1965 (Tappeiner 1967). Thus seed quality (percentage of sound seed) is often best in years of large seed crops (Gashweiler 1969; Pearson 1923; Tappeiner 1967; table 7-2). All the reasons for this are not known, but in years with large seed crops, predation by insects and rodents is relatively low in relation to the number of cones and seeds produced; that is, production of seed exceeds the capacity of the predators to destroy it. The reason for more viable seed is that the high rates of pollen production in those years may lead to high rates of pollination and viable seed.

Weather and predation by birds, mammals, and cone and seed insects affect seed supply and quality in true fir, especially in years of low seed production. Owens and Morris (1998) estimated that only 18–22% of seed were sound for a Pacific silver fir seed crop on two sites in British Columbia. Pollination failure (26–31%), possibly caused by frost, and insect damage (32–39%), were the main causes of seed loss.

Little can be done to control cone and seed predation. However, Shearer and Schmidt (1970) found that it was important to protect seed trees from rodent predation of cones in order to obtain natural regeneration in western Montana. Collars or sheets of metal have been placed around tree trunks to prevent rodents from climbing trees and clipping cones in seed-production areas. Along with the collars, spacing trees at 30–45 ft will deter tree-to-tree movement of arboreal rodents. This type of protection is not likely to be practical except on special sites like commercial seed production areas, or where mass selection for blister rust resistance is being attempted (Hoff et al. 1976).

Seed availability is related to seed weight because weight, along with presence and

Table 7-2. Numbers of seeds per cone and percentage sound seed produced in two seasons by large, medium, and small Douglas-fir trees in the Sierra Nevada (Tappeiner 1967).

DBH (in)	*Cones/tree 1962*	*Cones/tree 1965*	*Seed/cone 1962*	*Seed/cone 1965*	*Sound seed (%) 1962*	*Sound seed (%) 1965*
30–53	1000–7000	100–1100	25–84	7–29	33–97	16–52
20–30	250–470	80–860	26–42	10–20	55–93	19–86
10–20	20–240	2–90	14–60	0–25	80–94	10–68

size of seed wings, primarily determines wind-dispersal distance (Fowells and Schubert 1956; Roy 1960; figure 7-2). Dispersal of small-seeded species such as western hemlock and red alder is much greater than that of large-seeded species such as black oak and sugar pine. Seed can be dispersed long distances (Isaac 1930), but it is often uncertain whether the amount of seed dispersed is sufficient to meet reforestation objectives. Combined seed fall from Douglas-fir, western redcedar, and western hemlock declined at a ratio of about 7 to 2 to 1 at 75, 225, and 375ft from the forest edge into clearcuts (Gashweiler 1969).

Animals also have considerable effects on seed dispersal. Seed dispersal of some pines (Lanner 1996), black oak, tanoak, and hazel probably depends on animals because the seeds are too large or produced too close to the ground for effective wind dispersal. Birds and small mammals disperse seeds of small shrubs such as salal, huckleberry, and bitterbrush. Groups of young (1–5-yr-old) ponderosa pine, Jeffrey pine, and bitterbrush seedlings—the result of rodent caches of seed—are common in eastside pine forests; however, there is no evidence that seedlings in caches have higher long term survival rates than those that are not cached (Keyes 2005, personal communication; Vander Wall 1992, 1994, 1995). Dispersal by animals is likely to be important in the long term for extending the range of a species within and among forest types. Animal dispersal will add seedlings to a population of advance regeneration, but it will probably not contribute to prompt, well-distributed regeneration following large stand-replacing fires or timber harvest.

SEED BANKS

Some forest plants store their seeds in the forest floor for periods ranging from a few months to hundreds of years, but in general, most western North American commercial tree species have little or no long-term seed storage in the soil seed bank. Knobcone, Monterey, and lodgepole pines store their seeds in serotinous cones (resinous cones retained on the tree that store seed and release it when they are opened by high air

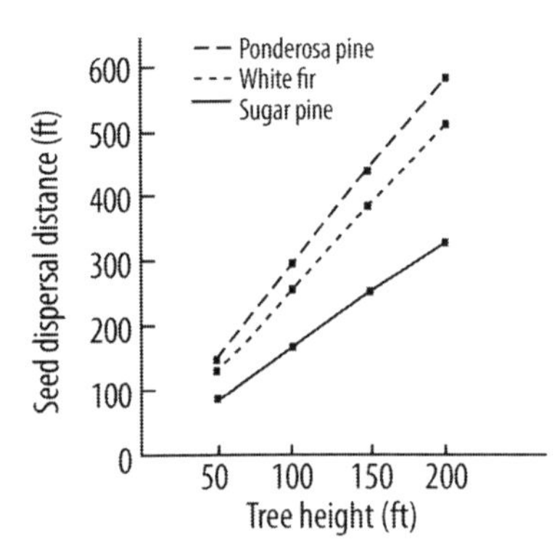

Figure 7-2. Theoretical seed dispersal of three conifer species by tree height and an assumed wind speed of 10 miles per hour. From Fowells and Schubert (1956).

temperatures) that provide a readily available source of seed for several years following disturbance, particularly from fire. Thus when high temperatures (35 to 55°C; Perry and Lotan 1977) occur during fire or at the surface of the soil after clearcutting, seed stored in the cones is released. Seed supply is often plentiful in such cases and dense stands frequently result after disturbance. Lotan (1967) reports that lodgepole pine cone serotiny varies among sites in the Rocky Mountains from about 90% of the trees on some sites to about 10% on others. Thus it is important to know the degree of serotiny in order to determine if seed supply is apt to limit regeneration (Lotan and Jensen 1970). Low serotiny may allow cones to open at normal (<25 to 30°C) summer temperatures (Perry and Lotan 1977).

Thousands of seedlings of forest shrubs such as ceanothus, Scotch broom (a non-native), and manzanita occur immediately after a fire, even though there may have been only a few plants on the site before burning (Cronemiller 1959; Hughes et al. 1987; Quick 1956 and 1959; Zavitkovski and Newton 1968). These seedlings germinate from seeds stored in the forest floor since the previous fire. The plants that produced the seeds died as conifers overtopped them, perhaps decades or even centuries earlier.

SEED TREES

The choice of seed trees to leave after harvest can have a major effect on seed supply. Large trees that have produced several past seed crops (evidenced by the cones on the forest floor beneath them) are likely to be the best seed producers (table 7-2; Sundahl 1971). Some good seed producers may have poor stem form, mistletoe, or other factors that would make them unsuitable seed trees, because their progeny might have poor form, and they might break from snow or wind, or spread mistletoe to the new stand. No practical, direct means of control over timing and amount of seed availability has been discovered; therefore, leaving large seed trees scattered throughout the stand will facilitate dispersal, especially of large-seeded species.

For many species, timing the preparation of the seedbed with the presence of good cone crops is likely to increase the number of viable seeds and the potential for germination and establishment of many seedlings. Schearer and Schmidt (1970) recommend delaying site preparation until a good cone crop is present as a way of increasing the likelihood of successful regeneration of ponderosa pine in Montana. Prescribed burning on the Fort Valley Experimental Forest in Arizona increased germination of ponderosa pine twenty-fold because it exposed mineral soil and increased soil moisture (Haase 1986). Variation in the coincidence of a good seed crop with a disturbance is probably a main reason for the mixed success of natural regeneration of Douglas-fir in the shelterwood method (Seidel 1983; Tesch and Helms 1992; Williamson 1973). After seed dispersal, a number of factors reduce the probability that seed will germinate and produce a seedling (Houle 1995).

OVERWINTERING AND GERMINATION OF SEEDS

Seeds of most western forest tree species are dispersed and fall to the forest floor from September through November. During the winter months, seeds overcome seed-coat dormancy through natural decomposition processes, and they imbibe water. Cold winter temperatures chill hydrated seeds in a process called stratification, and the seeds then break embryo dormancy, enabling them to germinate as temperatures warm. During the overwintering period, seeds are susceptible to fungi and predation by insects, birds, and small mammals. Lawrence and Rediske (1962) followed isotope-labeled seed over winter and found that those predators destroyed more than 80% of the seeds they placed in the field. In British Columbia, only 30% or fewer of sown conifer and hardwood seeds produced seedlings that emerged above the forest floor (Wright et al. 1998). Others (Haeussler et al. 1995; Tappeiner 1967) have reported similar losses of overwintering seed.

Rodent and bird predation (and the banning of seed-predator repellants) was a major reason that sowing seeds (direct seeding) on burns and clearcuts was abandoned as a regeneration practice in forests of the western United States. Site preparation does temporarily reduce the activity and population levels of rodents by modifying their habitat and probably making them more vulnerable to predation. Disturbance of the forest floor and litter layers provides a mineral-soil seedbed and reduces predation by invertebrates and pathogens. But although site preparation may reduce the effects of predators initially, the reduction is not sufficient to make sowing seed a reliable reforestation practice, particularly where rodents are the main predator.

Homma et al. (1999) found that predation of seed by rodents was directly related to snow depth in beech forests throughout Japan. Sites with deep snow had less predation and also more beech regeneration than sites with a low snowpack. Deep snow protected beech seed from rodent predation and increased the likelihood that the proportion of beech would remain high relative to other hardwood species.

Seeds germinate as solar radiation increases and soil temperatures warm in the late winter and spring. Germination occurs as the radicle or primary root grows through the seed coat. This is followed by development of the hypocotyl (young stem) and appearance of the cotyledons or seed leaves. The rate of germination *per se* (proportion of seed that germinates) does not appear to limit natural regeneration, but it is important to be aware that rates of germination will vary by year, site condition, and species. Haeussler et al. (1995) found higher rates of germination of red alder seed, and lower loss to pathogens and invertebrates, when they placed it on bare soil in clearcuts and in alder or conifer stands than when they placed it on undisturbed forest floor. Lower light levels and predation by insects and pathogens in the forest floor (Daniel and Schmidt 1971) were the primary reasons for reduced germination in the understory of mixed-conifer stands in the Rocky Mountains.

The common statement that "seed did not germinate here" is probably an inaccurate explanation for a lack of natural regeneration. It is more likely that the seed germinated but that most germinants did not survive. Germinating seed is easy to overlook during the brief period of germination. It is common to see small-seeded species such as red alder and redwood—and even species with medium-sized seed (table 7-3) such as ponderosa pine and Douglas-fir —germinating on the forest floor, where there is little likelihood that their roots will reach moist mineral soil. We have observed Douglas-fir seedlings germinating 6–8 in below the surface of a loose litter layer. These seedlings were etiolated, with very weak hypocotyls, and it was unlikely their roots would reach mineral soil or their tops would emerge above the forest floor. Thus it is a lack of survival—not a lack of germination—that leads to the failure of natural regeneration in these cases. There are often higher rates of seedling emergence (appearance of seedlings above the soil surface) on disturbed than on undisturbed sites (Haeussler et al. 1995; Zobel 1980). Soil moisture levels also affect emergence; however, for a given level of soil moisture, emergence was greatest on disturbed sites (Haeussler et al. 1995). Western hemlock, Sitka spruce, salal, and red huckleberry are species that can germinate on dead wood on the forest floor, but deep organic layers would likely inhibit their establishment. In practice, it is usually only emergence that is observed, so care must be exercised in assuming that emergence is a surrogate for germination.

The seeds of some species require more than one season to break dormancy and germinate, so not all the seed from a particular crop will germinate at the same time. Germination of vine maple seed (evidenced by emerging seedlings) that was protected

Table 7-3. Seed weight and presence or absence of wings for representative western forest trees (USDA Forest Service 1974).

Species	*Seed/lb*	*Wings*
Abies magnifica (red fir)	4,000–8,900	yes
Acer macrophyllum (bigleaf maple)	2,700–4,000	yes
Alnus rubra (red alder)	383,000–1,087,000	no
Larix occidentalis (western larch)	98,000-197,000	yes
Picea englemannii (Englemann spruce)	69,000-322,000	yes
Pinus edulis (pinyon pine)	1,500–2,500	no
Pinus lambertiana (sugar pine)	1,500–2,700	yes
Pinus ponderosa (ponderosa pine)	6,900–23,000	yes
Pseudotsuga menziessi (Douglas-fir)	15,400–53,000	yes
Quercus kellogii (black oak)	52–147	no
Sequoia sempervirens (coast redwood)	59,000-300,000	no
Thuja plicata (western redcedar)	200,000-592,000	yes
Tsuga heterophylla (western hemlock)	189,000–508,000	yes

from rodent and bird predation in the understory of conifer stands averaged 20–60% of the seed sown (Tappeiner and Zasada 1993). Most of the emergence (70–90%) occurred the second year after sowing, and the remaining—up to 5%—emerged in the third year (Tappeiner and Zasada 1993). They noted similar patterns of germination for hazel (*Corylus sp).*

EARLY GROWTH AND SURVIVAL

Early seedling establishment depends on providing safe sites for the newly germinated seedlings (Grubb 1977). Safe sites (or microsites) are places where the effects of the many interacting variables causing mortality are reduced to the point that at least some of the newly germinated seedlings survive. Even low rates of survival can ensure sufficient regeneration if large enough numbers of seedlings germinate.

Seedling establishment can be considered in two phases: early establishment (1–3 yr) and later seedling growth (3–10+ yr). The following discussion summarizes factors that enhance or hinder the survival and growth of established seedlings during both phases. Studies referenced are listed and summarized in table 7-4.

Newly germinated seedlings are quite small and can die from any of several causes. Early in the growing season, frost can kill the hypocotyl (Haeussler et al. 1995), and frost heaving, caused by formation of ice crystals in the soil can rip small seedlings from the frozen soil. Excessive heat also kills seedlings. Surface soil and litter temperatures that exceed about 130°F cause stem lesions in the fleshy hypocotyls of seedlings up to about 2 months old (Baker 1929; Helgerson 1990b; Maguire 1955). Seedlings at this stage are also susceptible to attack by root and stem pathogens and herbivory by rodents and invertebrates.

Newly germinated seedlings have small root systems (Hermann and Chilcote 1965; Muelder et al. 1963), and if their roots cannot remain in contact with sufficiently moist soil, they are susceptible to drought, especially in microsites where the air temperature or evaporative demand is high (Tappeiner and Helms 1971, See figures 3-6, 7). Brooks et al. (2002) report that during the summer the roots of large ponderosa pine and Douglas-fir diurnally lift water from lower soil horizons and deposit it in the upper horizons. This input of water might aid the early establishment of natural seedlings on dry sites, but not their subsequent growth: several studies have documented that the growth of conifer seedlings decreases with decreasing distance from large trees (McDonald 1976b; Zenner et al. 1998).

Partial shade generally favors seedling establishment for the first several years because it moderates soil surface temperature extremes, reduces the evaporative demand of the air, and helps keep the soil from freezing. However, leaves that shade the seedlings may cover and crush them as they become saturated with water or packed down by snow. Pathogens and insects in the forest floor cause mortality when they kill the tops of crushed seedlings during the winter and early spring (Koroleff 1954).

Seedling establishment is often related to the size of the seed. Seedlings of large-seeded species are often more vigorous than those of small-seeded species in the first several years after germination (Harrington and Bluhm 2006, Seiwa and Kikuzawa 1996). Tanoak (Tappeiner et al. 1986), California black oak, bigleaf maple (Fried et al. 1988), vine maple (Tappeiner and Zasada 1993), and Oregon grape (Huffman and Tappeiner 1997) produce large seedlings that soon develop relatively stiff stems with rudimentary bark. Seedlings of these species are less susceptible to heat, pathogens, and invertebrates, and have higher early survival rates than small-seeded species such as western hemlock, salal, Pacific madrone, redwood, and red alder (Haeussler et al. 1995; Seiwa and Kikuzawa 1996; Tappeiner et al. 1986; Tappeiner and Zasada 1993). Ponderosa and sugar pine are exceptions. Their moderately large seeds germinate into large seedlings, but the hypocotyls remain unsuberized for several weeks after germination, and they consequently have high rates of early mortality.

Small-seeded species may benefit from special sites, such as those containing an abundance of dead wood. Lusk (1995) found that decayed wood on the forest floor favored establishment of small-seeded species, whereas large-seeded species were readily established on the forest floor itself. Western hemlock (Christie and Mack 1984; Harmon and Franklin 1989), salal (Huffman et al. 1994), and *Vaccinium* spp. all germinate better on sites with dead wood. However, alder is a small-seeded species that does not establish well on organic seedbeds. Germination rates are higher on mineral soil, where the forest-floor-dwelling invertebrates and pathogens that attack seedlings are less likely to be present (Haeussler et al. 1995). Roots of alder seedlings develop N-fixing nodules within the first several weeks after germination, an indication that they are adapted to disturbed sites where N may be low in the upper soil layers.

Safe sites (microsites that favor seedling establishment) are often suitable for different species. More western hemlock and Douglas-fir seedlings became established on the shady south side of 0.4-ha gaps than on the sunny north side (Gray and Spies 1997). On the north side, survival of hemlock seedlings on logs was 60%, whereas survival of seedlings on the forest floor was 30% (Gray and Spies 1996). Even 30% early survival is probably more than adequate to provide enough saplings and trees in the long term for most forest management objectives. Similar results were reported for conifer and hardwood species in British Columbia (Wright et al. 1998). Establishment of both Douglas-fir and white fir on the same sites in the Sierra Nevada was better in partial shade where the evaporative demand and loss of soil moisture were significantly less than on more exposed sites (McDonald and Abbot 1994; Tappeiner and Helms

Table 7-4. Summary of results of representative studies on the early survival and establishment of seedlings from seed.

Study and location	*Conditions favoring regeneration*	*Causes of mortality*
1. Ponderosa pine, northern Arizona (Pearson 1923)	Gravel or rocky soil that retards evaporation. Shade from herbs and trees for 1–2 yr to protect seedlings from temperature extremes. Greater water availability on sandy soils.	Temperature extremes. Frost heaving. Drought. Herbivory.
2. Douglas-fir, coastal Oregon (Hermann and Chilcote 1965)	High rates of germination on charcoal and heavily burned seedbeds.	Herbivory. High soil temperature. Damping-off fungi.
3. Douglas-fir, coastal Oregon (Lavender 1958)	Greater germination and survival on north slopes.	Low soil moisture.
4. Douglas-fir, western hemlock, true fir, western Cascades (Minore 1986)	Shade from stumps. Protection from herbivory. Soil scarification.	N/A
5. Douglas-fir, western Washington (Isaac 1938)	Shade on disturbed soil. Increased soil water. Low soil surface temperatures.	Soil temperature. Drought. Insects.
6. Western white pine, larch, Douglas-fir, etc., Idaho (Haig et al. (1995)	Shade, i.e., protection from temperature extremes. North slopes. Proximity to coast (fog?). Compacted soil. Bare mineral soil.	Heat lesions. Drought. Frost in open sites. Pathogens. Invertebrates in the understory.
7. Red fir, Sierra Nevada (Selter et al. 1986)	Increased radiation from sun flecks in the understory of mature red fir stands.	Fungi. Herbivory. Drought.
8. Douglas-fir, Sierra Nevada (Tappeiner and Helms 1971)	Shade. Reduced evaporation of soil moisture. Reduced evaporation capacity of the air.	High soil temperature. Root and stem pathogens. Drought. Herbivory. Litterfall. Low light.
9. Douglas-fir, western hemlock, western redcedar, western Oregon Cascades (Gashweiler 1970)	N/A	Herbivory. High temperatures. Drought. Fungi.

Table 7-4 cont.

Study and location	*Conditions favoring regeneration*	*Causes of mortality*
10. Ponderosa pine, western Oregon (Wagg and Herman 1962)	Shade. Cool weather during germination.	Drought. Heat. Frost. Herbivory.
11. White fir, red fir, northern California (Gordon 1970)	Shade. Mineral soil seedbed.	Heat. Drought. Insects.
12. Red alder, Oregon Coast Range (Haeussler et al. 1995)	Shade — i.e., protection from temperature extremes. North slopes. Proximity to coast (fog?). Compacted soil. Bare mineral soil.	Heat lesions. Drought. Frost in open sites. Pathogens. Invertebrates in the understory.
13. Pacific madrone, southern Oregon (Tappeiner et al. 1986)	Shade. Soil moisture. Bare mineral soil.	Drought. Pathogens. High soil temperature. Litterfall.
14. Tanoak, southern Oregon (Tappeiner et. al 1986)	Protection from rodents. Light shade.	Rodent herbivory. Low light.
15. Bigleaf maple, Oregon Coast Range (Fried et al. 1988)	Protection from rodents. Light shade.	Rodent herbivory. Low light.
16. Lodgepole pine, birch, subalpine fir, western hemlock, western redcedar, British Columbia (Wright et al. 1998)	Shade on south edge of openings. Removal of moss cover from forest floor.	Low light. High temperatures. Probably pathogens and insects in the forest floor.
17. Western hemlock, Douglas-fir, Pacific silver fir, Oregon and Washington Cascades (Gray and Spies 1996)	Shade on south edge of openings. Decayed wood.	N/A

1971). The same was found for the establishment of seedlings of several conifers in the northern Rocky Mountains (Haig et al. 1941).

Safe-site sufficiency may vary from year to year. Cain (1994) found that reducing the density of mid-story hardwoods made a sufficient number of safe sites for establishment of loblolly and shortleaf pine in a good seed year (30,000 seed/ac), but the same conditions did not result in adequate stocking in a poor seed year (2,500 seed/ac).

Safe sites may be temporary—that is, they may be safe for only a limited part of the regeneration process. Sites that enable early seedling establishment may not support later growth. For example, some seedlings, especially those of shade-intolerant species, may die after several growing seasons because the same shade that favors early survival hinders growth. Seedlings of shade-tolerant species such as true firs and hemlocks may survive but grow very slowly in such environments. On sites where shade continues to hinder growth, seedling banks (Grime 1981) of advanced regeneration may be established and persist in the understory. Through time, older seedlings die and are replaced by new germinants. In the northern Rocky Mountains, northerly aspects provide shade suitable for seedling establishment that does not substantially reduce seedling growth (Ferguson et al. 1986).

SUCCESSFUL ESTABLISHMENT OF NATURAL REGENERATION

Successful establishment of natural regeneration generally requires lots of seed and many young seedlings to overcome the high rates of seed predation and seedling mortality. Several steps can be taken to help establish natural regeneration:

- Provide a seed source capable of disseminating seed throughout the area to be regenerated. This generally means leaving enough well-spaced, large, windfirm trees with full crowns, evidence of past cone crops, and no severe mistletoe infection. It is best if the trees are distributed so that they provide only partial shade, because overstory trees can substantially inhibit the growth of newly established seedlings.
- Provide partial shade, especially on exposed sites that lie on southeastern to western slopes. Shade reduces summer soil temperatures and the evaporative capacity of the air so that less soil water is lost to evaporation and transpiration in the top 6 in of soil. Shade also moderates winter temperature extremes and reduces the chance of frost damage.
- Ensure enough soil water. Shade plus a favorable soil-water regime will help seedlings minimize their water stress and survive summer drought. Reducing dense covers of grasses and shrubs will help increase the soil water available for seedlings. Partial shade of live vegetation with low rates of transpiration or dead shade is likely most beneficial.
- Reduce the density or depth and continuity of the litter layer by logging or by site preparation with fire or mechanical means to expose mineral soil and reduce predation

by fungi and insects—and probably by rodents—on newly germinated seedlings. If possible, schedule logging or site preparation and exposure of mineral soil to coincide with a good cone and seed crop.

- Protect advanced regeneration from damage, using reasonable care during logging and site preparation. This helps to ensure stocking of natural seedlings and reduces the need for planting.

Table 7-5 provides an example of how natural regeneration can be successful even with very high rates of early mortality. Natural regeneration can be considered a numbers game. The larger the seed supply and the higher the rate of early establishment, the greater the likelihood of success. If, for example, weather is favorable for a year or two, the effects of summer drought might be much less than anticipated and rates of seedling establishment might be quite high. If prompt regeneration is needed to avoid problems with competition, or to meet owner's objectives and comply with regulations, planting may be a better alternative. It is often better to err on the side of too much regeneration and to correct over-stocking by thinning. Failure of natural regeneration can lead to additional, expensive site preparation and planting. Natural regeneration becomes increasingly more useful on sites where it occurs promptly or where it can occur over several years and still meet management goals and regulations.

Advanced regeneration

Discussions such as the one above, emphasizing all that can go wrong in the process of natural regeneration, may lead one to conclude that it cannot be used in silviculture, and in many cases this is so. However, advanced regeneration—populations of natural seedlings and saplings that become established in the understory of a forest stand and survive from several years to many decades (Fried et al. 1988; Hett and Loucks 1971; McDonald 1976a; Tappeiner and McDonald 1984; figure 7-3)—is viable

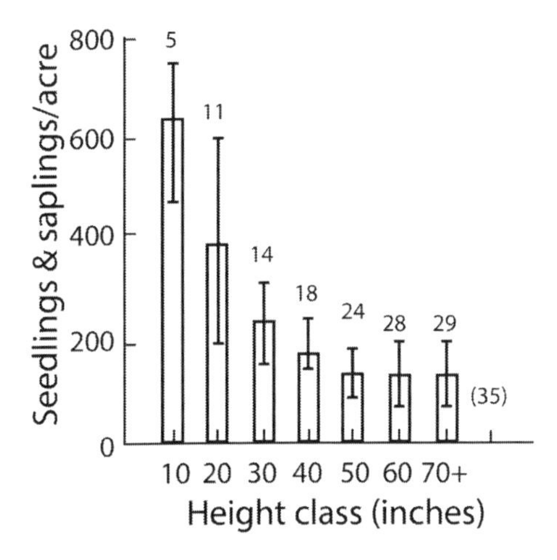

Figure 7-3. An example of a seedling bank: tanoak seedlings in the understory of a Douglas-fir stand in southwestern Oregon. Numbers above the bars are average seedling/sapling ages (yr) for each height class. Number in parentheses is age of oldest seedling found. From Tappeiner and McDonald (1984).

Table 7-5. Summary of studies on growth of advanced regeneration of western conifers following overstory removal.

Study and location	*Results*
1. White fir and Douglas-fir, southern Oregon (Tesch and Korpela 1993)	Height growth doubled 5 yr after release. Advance regeneration grew as well as planted seedlings for 20+ yr and had better growth than planted seedlings on low-productivity sites. Pre-release growth indicated rate of post-release growth.
2. White fir and Douglas-fir, southern Oregon (Tesch et al. 1993a)	Trees recovered from logging damage within 6 yr. 26% mortality from logging all in trees <30 in tall. Prerelease crown size best indicator of post-release growth.
3. White fir and Douglas-fir, southwestern Oregon (Korpela and Tesch 1992)	Advance regeneration likely to grow as well as free-to-grow planted trees and better than planted trees under competition. Advance regeneration grew better than planted trees on low-productivity sites.
4. White fir, red fir and Douglas-fir, northern California (Helms and Sandiford 1985)	Pre-release live crown ratios, rate of height growth, and increasing height growth rate all predictors of potential post-release height growth for 10 yr. Generally able to classify potential vigor correctly for about 70% of the trees. Live crown size most important on sites of low productivity.
5. White and red fir, northern California (Oliver 1986)	Pre-release height growth, crown diameter, and crown length predictors of post-release growth. Sunscald and snow damaged >35% of released trees. Potential for root disease in older trees to infect advance regeneration.
6. White and red fir, northern California (Gordon 1973)	Height growth rates increase 5 yr after overstory removal. Mortality greatest (44%) in trees <12 in tall. 7yr after release, 32% of damaged trees died; <10% of all trees died.
7. Lodgepole pine, western Montana (Perry and Lotan 1977)	Advance regeneration grew as well as new seedlings. Growth rates increased sharply when trees reached about 1m in height.
8. True fir, mountain hemlock, southern and central Oregon (Seidel and Head 1983)	Residual overstory, basal area, live crown ratio, and 5-yr pre-release height growth were related to growth after release; age was not a factor.
9. Pacific yew, Oregon Cascades (Bailey and Liegel 1997)	Average diameter growth of yew trees more than doubled after partial overstory removal. 10 of 23 trees responded positively; 11 trees did not respond; growth of 2 trees decreased.
10. Mixed conifers, northern Idaho (Bassman et al. 1992)	Shrubs also released and reduced conifer growth. Conifer growth increased with shrub control.

for many species and sites. It is suited to relatively open stands and species including: ponderosa pine, lodgepole pine, larch, western hemlock, grand and white fir, Douglas-fir, and incense-cedar (Lilieholm et al. 1990; McDonald 1976a; Olson and Helms 1996; Tappeiner and Helms 1971), as well as oaks (McDonald and Tappeiner 2002). Western hemlock, Sitka spruce, and bigleaf maple seedlings often become established in relatively dense Douglas-fir and Sitka spruce-western hemlock forests, especially after thinning (Shatford et al. 2009, Deal and Tappeiner 2002; Fried et al. 1988; Ruth and Harris 1979). Advanced regeneration is also common in true fir forests (Gordon 1973).

Unless the canopy is opened, older seedlings eventually die as light intensity drops below the light compensation point—the point below which light levels are too low for the seedling to produce enough photosynthate to maintain respiration or replace tissue lost to herbivory or pathogens (Lieffers et al. 1999). Tree growth in the understory may also be limited by water. Trenching to limit root competition has been shown to increase seedling growth even when light levels remained constant (Riegel et al. 1992). Browsing and mistletoe infection from overstory trees can also impact growth of advanced regeneration in some stands. Fried et al. (1988) found that continuous browsing along with low light severely limited bigleaf maple seedling growth in the understory of Douglas-fir forests.

Advanced regeneration is dynamic. Seedlings may die unless gaps in the canopy occur. It is not uncommon to find slowly growing seedlings and saplings 30+ yr old in the understory, with new seedlings becoming established among the older ones.

Messier et al. (1999) provide a useful conceptual model for evaluating the vigor and growth potential of regeneration in the understory of forest stands (figure 7-4). This model is supported by the work of Jain et al. (2004), who found that western white pine in the Rocky Mountains became established in the understory in canopy openings >23% of full sky, gained a competitive advantage over the more shade-tolerant western

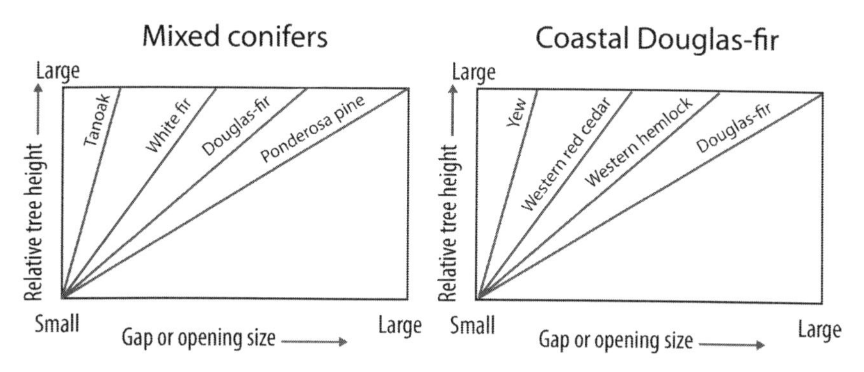

Figure 7-4. Relative heights that species in mixed conifer and coastal Douglas-fir forests would reach in various size gaps in the canopy. Lines represent theoretical relative heights at which light compensation points are reached. Adapted from Messier et al. (1999).

hemlock under canopies >50%, but was free to grow only under very low canopy densities < 5%.

The model emphasizes several important points. The first is that shade tolerance is relative. For example, Douglas-fir is shade-tolerant relative to ponderosa pine in mixed-conifer forests, but it is intolerant in Douglas-fir-hemlock forests that have high leaf-area indices (figure 7-4). If both shade-intolerant and shade-tolerant species are established in the understory, the shade-tolerant species can reach greater heights than intolerant species before they become stunted or die. Second, shade tolerance is related to the light compensation point. Third, although not illustrated in this diagram, it is probably not just light or canopy density that affects the ability of advanced regeneration to grow, but also soil water availability. Finally, variability in canopy density makes a difference in how well advanced regeneration survives and grows.

The suitability of advanced regeneration may vary considerably. In some stands it may be present but its species composition, vigor, density or distribution are not satisfactory. Also, advanced regeneration of trees is not always present. A dense, continuous cover of shrubs such as salal, bear clover, or salmonberry prevents the regular establishment of natural conifer and hardwood seedlings. Relying entirely on advanced regeneration in mixed conifer-hardwood forests may convert a mixed-conifer stand to a stand dominated by hemlock, true fir, or hardwoods (Lilieholm et al. 1990; McDonald 1978a; Tappeiner and McDonald 1984). These species survive in low light conditions that will not support the establishment or growth of shade-intolerant species such as ponderosa pine or Douglas-fir. Nevertheless, in many stands advanced regeneration can be relied on exclusively, or supplemented with planted seedlings.

Several studies have shown that advanced regeneration responds favorably to release or removal of overstory trees (table 7-5). Tesch and Korpela (1993) found that within 5 yr of clearcutting, height growth of more than 80% of the advanced regeneration on the site had increased, and the naturally regenerated seedlings grew as well as the planted seedlings for 20 yr. Although the response may not be immediate, growth of advanced regeneration has been found to compare favorably with that of planted seedlings (Korpela and Tesch 1992). Even seedlings that appeared severely damaged during logging recovered; their stems straightened and their rates of stem diameter and height growth increased (Oliver 1986; Tesch et al. 1993a). It appears that it is mainly the smaller seedlings (<30 in tall) that are killed or severely damaged by logging (Gordon 1973). Douglas-fir seedlings 20–40 in tall survived best during logging of overstory trees in southwestern Oregon because they were flexible enough and sufficiently established to survive some injury and changes in growing conditions.

The ability of a seedling to respond to release can be predicted with reasonable certainty. Morphological characteristics of seedlings and saplings are better indicators of potential growth than age. Crown size and rate of height growth before overstory removal are the best predictors of growth afterward (table 7-6).

Stem decay of true fir seedlings and saplings damaged by logging or fire must be considered. Seedlings may grow well after release, but pathogens and decaying fungi in the lower part of the stem may attack the wood and cause the tree to become unstable as it grows (Aho et al. 1983). It is also important to ensure that advanced regeneration is not seriously infected by mistletoe.

Often, poor distribution of advanced regeneration will require planting of seedlings or establishment of new natural regeneration. Furthermore, Bassman et al. (1992) point out that advanced regeneration density might be very high, or that shrubs may become established with conifer seedlings in the understory. In such cases, thinning of excess stems of advanced regeneration and of the associated shrubs would likely be needed to ensure the desired species composition and stand density.

An example of overabundant advanced regeneration comes from the hemlock-spruce forests of southeastern Alaska (Alaback 1982; Ruth and Harris 1979). Large clearcuts are naturally regenerated with hemlock and Sitka spruce seedlings and saplings that were growing in the understory of the harvested stand. More than 20,000 seedlings and saplings per acre are common in the understory of these forests (Ruth and Harris 1979). Regeneration results from both release of advanced regeneration and establishment of new seedlings; because the clearcuts are quite large and seed trees are not left, however, it is unclear what proportion of the regeneration is from seedlings established after harvest. Here, severe inter-tree competition will ensue if stands are not thinned at a young age.

Artificial Regeneration

Artificial regeneration includes planting nursery-grown seedlings or cuttings as well as sowing seed on areas to be reforested. Methods of artificial regeneration, especially planting, have been a major emphasis in silvicultural practice and research during the past several decades. Planting has become the primary method of reforestation in many parts of the world, because it is more predictable and reliable than natural regeneration in many forests. Artificial regeneration by sowing seed is similar to natural regeneration from seed, discussed above. Sowing seed has been used on some sites in the Douglas-fir region —especially for reforestation after large fires— but it has not been consistently successful. The principal difficulties are the need for large quantities of seed, primarily because of high seed predation by rodents, and the normally high mortality of newly germinated seedlings. With the development of forest nurseries and the capacity to produce large quantities of seedlings for planting, along with the ability to control stand density by planting, aerial sowing of conifer seed has been generally

Table 7-6. Comparison of the natural and artificial regeneration processes. Steps 1 through 4 in the natural regeneration process occur in the forest. In the artificial regeneration process these steps occur in seed orchards or seed-collection programs whereby seed is provided to a nursery where seedlings are grown for planting. All the causes of mortality and loss of seed and seedlings in these steps are controlled by nursery practices. After natural seedlings are established, the two processes are similar (also see Shearer and Schmitt 1970).

Steps in the process	*Causes of mortality*	*Expected losses * Seed/seedlings remaining/ 100,000 seeds/ac*	*Implications for silvicultural treatments*
Natural Regeneration from Seed			
1. Seed production	Frost and cone- and seed-eating insects, birds, and rodents	Initially 100,000 seeds/ac. 50% mortality. 50,000 seed remain	Recognize variability in size and time of seed crops and species differences. Leave seed-producing trees, evaluate past cone production. Consider seed tree density and species.
2. Seed dispersal	Seed size (weight and wings), wind	50,000 seeds/ac. No expected mortality and animal dispersal	Seed tree density and position
3.Overwintering	Fungi and insects in the forest floor, rodents, and birds	Assume 50%+ mortality. 25,000 seeds/ac	Disturbance to the forest floor to reduce litter depth by fire or site preparation to reduce predator habitat.
4. Germination and early survival	Losses from fungi, insects, high soil temperatures, drought, frost, litterfall, and combinations of those factors	Assume 90% mortality. 2500 seedlings/ac	Reduce litter depth. Reduce cover to lessen competition for water and light. Reduce litterfall and insect and pathogen habitat. Leave some cover to reduce temperature extremes.
5. Later survival and growth (3–4 yr) **	Drought, competition from grasses, shrubs, pathogens, and insects, low light and temperatures	Assume 90% mortality. 225 seedlings/ac	Control competing vegetation

Table 7-6 cont.

Steps in the process	*Causes of mortality*	*Expected losses * Seed/seedlings remaining/ 100,000 seeds/ac*	*Implications for silvicultural treatments*
6. Advance regeneration **	Fewer causes of mortality: low light, browsing	Numbers may fluctuate as older seedlings die and as new seedlings become established	Control competing vegetation. Reduce overstory density to increase light and water and increase seedling/ sapling growth in uneven-age or two story systems. Remove overstory trees in the shelterwood or seed tree methods.
Artificial Regeneration			
Planting Seedlings			
1. Early planting and survival (1–3 yr) **	Drought, competition for water, browsing, poor planting or seedling handling practices, and combinations of those factors	Survival is generally high (80+%) with good planting stock, planting practices, site preparation, and vegetation control	Control competing vegetation (mainly grasses and herbs) to ensure survival. Protect seedlings from browsing on some sites.
2. Growth (3–5yr)	Drought, low light, and browsing	Same as above	Control competing vegetation (mainly shrubs or hardwood sprouts) to ensure growth and survival. Protect seedlings from browsing on some sites. Reduce overstory density if seedlings are planted in an understory.

*Expected losses are estimates based on the literature rates. Causes will vary considerably among species and sites.

** These steps are similar because they are concerned with the survival and growth of large seedlings.

abandoned. Sowing seed on scarified spots and protecting it from rodents could be a successful regeneration practice on some sites (Franklin and Hoffman 1968), but we know of no widespread use of this technique.

Planting seedlings has been found to be a reliable way to regenerate productive and high-yielding forest stands (Cafferata 1986; Hermann and Lavender 1999). There are several compelling reasons for regenerating forest stands by planting:

1. Planting avoids the steps in the process of natural regeneration from seed production through early survival (table 7-6) and helps ensure prompt regeneration as required by state and federal forest practice regulations (Rose and Coate 2000).
2. There may be few suitable seed sources or seed-producing trees of the proper species present after a large disturbance, such as a wildfire. Given annual variation in seed crops and the dispersal limitations of seed, natural regeneration could thus be delayed for many decades. Furthermore, on many sites it is difficult to achieve the coincidence of a good seed crop with a suitable seedbed just after a fire or site preparation, and so a cover of herbs, shrubs, and sprouting hardwoods will often develop that inhibits or prevents the establishment of tree seedlings. This is often the case in ponderosa pine, Douglas-fir, and mixed-conifer forests, where dense communities of manzanita and ceanothus become established from seed banks after fire (figure 7-5).
3. Planting helps ensure that the future stand will have the desired mix of species, and it allows for potential genetic improvement. Planting will also ensure the presence of species whose natural regeneration is difficult to achieve.
4. The desired density and distribution of seedlings is better achieved through planting. The uneven distribution of seed sources, safe sites, and patches of existing advanced regeneration are all apt to lead to uneven spatial distributions of seedlings; planting affords more control over this distribution. Planting can also help achieve a prescribed heterogeneity in seedling distribution and species composition when uniform spacing and a single species will not meet management objectives.
5. Planted seedlings are more likely to become established on sites that are difficult to regenerate. Vigorous, large, well-planted seedlings are more apt to survive and grow well on sites with high soil-surface temperatures and droughty soils.
6. Planting may be necessary to meet management policy or governmental regulations. State and federal forest practice rules often require establishment of a minimum number of well-distributed, free-to-grow seedlings within a specified number of years following harvesting. Because of the uncertainties of natural regeneration, it is difficult on many sites to meet requirements such as these without planting.
7. On sites where competition from invading, rapidly growing vegetation is likely, it is most important to reforest the site quickly and avoid the expense and possible

Figure 7-5. A shrub community of manzanita (A) became established after a fire 35–40 yr previously. There was no site preparation or planting, and no natural regeneration has occurred. On the same site (B) is a ponderosa pine stand established following shrub control and planting. Photo by Peter Tappeiner.

disturbance of repeated site preparation if initial reforestation were to fail. It is often prudent to take advantage of periods when there is reduced vegetative cover and competition following fire, logging, or site preparation. Planting large seedlings minimizes the period of susceptibility to competition and animal damage.

8. Planting may help to establish a variety of species and genotypes that may mitigate potential negative effects of climate change.

GENETIC DIVERSITY AND FOREST PLANTATIONS

The forests of western North America have a tremendous legacy of genetic diversity. In general, most of the forest area that existed before European settlement is still forested. Exceptions occur at low elevations, where forests merge with agricultural land and urban development, and in areas where forests have been cleared for extractive uses. With all the differences in sites caused by variation in elevation, slope and aspect, soils, wind patterns, air temperatures, etc., there has been ample opportunity for local species adaptations.

Reforestation practices likely have the capacity to maintain or even augment this genetic diversity. Incorporating genetic diversity into planted forests directly affects the forests grown today, and it may greatly enhance the adaptability of these forests in response to future changes in climate and other elements of the biophysical environment such as fire regimes, and to human use and impacts (Ledig and Kitzmiller 1992). Preserving genetic diversity is vital to ensure the growth and survival of forests not only today, but also into the future.

Forest genetics is a major consideration in the regeneration and management of young stands (Adams et al. 1992). Selecting the appropriate seed sources for artificial and natural regeneration ensures that trees grown for wood products have desirable qualities, including small branch sizes and angles, good stem form, high specific gravity of wood (a measure of wood strength), resistance to pathogens, rapid growth rate, and strong seed production potential. Each of these characteristics is subject to a high degree of genetic control (Daniel et al. 1979). Traits such as rates of height growth, timing of bud burst, bud set, and seed set have also been shown to be related to seed source (Conckle 1973; Hermann and Lavender 1968; Rehfeldt 1974, 1989).

The major reason to pay attention to seed sources in the context of regeneration is that, especially because of their longevity, trees must be adapted to fluctuating conditions under which they will grow. Trees of the same species from different latitudes and elevations grow quite differently, and when they are grown together in a common environment their performance will be highly variable. Elevation and latitude govern such traits as rate of height and diameter growth, and timing of bud burst, bud set, and seed set (Conckle 1973; Hermann and Lavender 1968; Rehfeldt 1974). This work suggests that trees from local seed sources usually grow better than those moved from other elevations or latitudes to the planting site. However, there may be exceptions to the rule that "local is safer" (Kitzmiller 2005) and some genotypes may be widely adapted and grow well in a large range of environments. In general if non-local sources are to be planted it would be wise to plant local sources as well. How seed source selection will be done in light of the uncertainties of climate change is an unanswered and important question.

SEED COLLECTION ZONES

An example of the genetic control of tree growth comes from a study conducted on the west slope of the Sierra Nevada (Conckle 1973). In 29-yr-old ponderosa pine plantations, trees from seed collected at mid-elevation sites (approximately 3,000 to 4000 ft) grew taller than trees from lower and higher elevation sites. At high-elevation sites, the trees from high-elevation seed sources grew best, because snow damaged the trees from the lower elevation sources. An important finding was that, for trees planted at the full range of elevations, those from high-elevation seed sources consistently had greater diameter growth than those from low- or mid-elevation sources, and consequently had a lower height/diameter ratio, giving them more taper. (Kitzmiller 2005 found similar relationships for Sierra Nevada ponderosa pine on other sites). This suggests that trees from high-elevation seed sources may be better adapted to snow and ice than those from low- or mid-elevation seed sources.

Trees from mid-elevation seed sources grew well at all elevations, even growing better than those from local seed sources at the lower elevations. This suggests that in the Sierra Nevada, at least, trees from mid-elevation sources can be used at a range of elevations. It may be that the adage "Local sources are safest but not always best" applies in this case. It may be possible to improve yields from plantations by mixing rapidly growing phenotypes from mid-elevation seed sources with those from local low-elevation seed sources. However, it may not be prudent to use rapidly growing mid-elevation trees at high elevations, where they could be susceptible to snow breakage. Also, considering the short duration of the study (29 yr), longer-term variability in weather and climate (Ledig and Kitzmiller 1991), and the unpredictable effects of periodic drought and bark-beetle outbreaks, it would be risky to rely solely on mid-elevation sources at either low or high elevations.

On the west slope of the Sierra Nevada, where the site quality changes gradually with elevation, it may be possible to plant trees from a wide variety of seed sources across a wide range of sites. This is one example of how the genetic potential varies across sites, but it should not be generalized too far. In the Douglas-fir region, productive sites often occur at low elevations. Furthermore, Hermann and Lavender (1968) (also see Campbell 1979 and 1986) observed that seedlings from north-aspect seed sources grew taller than those from south-aspect sources and suggested that there may be genetically determined "aspect races" in Douglas-fir. In areas with sharp differences in environment resulting from variation in slope aspect, cold air pockets, and elevation, adaptation might be somewhat more site-specific. Rehfeldt (1974) found considerable variability in Rocky Mountain Douglas-fir half-sibling families from north and south aspects, but there was no predictable relationship between tree growth and aspect or elevation. Likewise, Campbell and Franklin (1981) found that elevation of habitat alone accounted for 56% of the variability in seedling growth in half-sibling families from 190 Douglas-fir seed sources in western Oregon. Habitat type or community classification explained 36% of the variation. Thus it appears that some of the considerable genetic variability in forest trees is related to site conditions, expressed through indices such as aspect, elevation, latitude, and distance from the coast (Campbell 1986), but that it is not readily predictable. Because predicting genetic suitability is difficult, it is important to have many local seed sources represented in plantations.

To ensure a broad range of genetic diversity in planted forest seedlings, seed is collected from many seed trees of apparently good quality. Seed trees are selected at low densities (<3/1000 ac) within seed-collection zones and within 500-ft elevation bands to ensure a broad range of genetic variability in the seed collected (Kitzmiller 1990; Silen and Wheat 1979). Initially seed comes from carefully selected, large, well-formed trees in the forest. The genetic quality of these trees is then tested by planting them in test sites that are similar to future planting sites (progeny test sites), and trees that produce poor progeny (e.g., poor stem form, low growth rates, etc.) are eliminated as seed sources. Seed or grafts from the best-performing trees may be used to establish seed orchards. Genetic diversity is enhanced in seed orchards (and in natural stands as well) by the large amount of wind-dispersed pollen. Cones are pollinated by nearby trees and also by distant trees as the pollen is dispersed and mixed by wind.

Collecting seeds during good seed years further ensures a broad range of variability in the collections. Because pollen and cone production are plentiful in good seed years, out-crossing is high and many parent trees contribute to the seed source (Adams 1992). Natural regeneration, either newly established seedlings or advanced regeneration, further adds to genetic variability. Adams et al. (1998) evaluated genetic variability in a Douglas-fir stand regenerated by the shelterwood method. Seedlings produced through natural regeneration mirrored the genetic variability of the parent stand. Genetic diversity of planted seedlings was greater than that of natural seedlings because the seed came from many more trees and sites than the shelterwood trees, the parents of the natural seedlings. These researchers recommend retaining some smaller trees in the shelterwood to increase the diversity of natural regeneration. Techniques such as these may be important to ensure a broad genetic base as a hedge against long-term climate change and other future shifts from current ecosystems.

Comparisons of progeny from seed orchards and controlled crosses from selected trees with open-pollinated "woods-run seed" have shown that after planting, progeny from selected trees grew 3 to 6% more in height and diameter than trees from open-pollinated parents. The increase in these two stem dimensions amounted to an increase in individual tree stem volume of over 25% (St. Clair et al. 2004). Stand volume increase of 3 to 6% at 40 yr and 2 to 3% at 60 yr have been estimated from growth model projections with the greatest gains on the more productive sites (Gould and Marshal 2010).

In forests where severe fires or other agents have killed trees over large areas, seed for natural regeneration may be lacking. In addition to establishing a forest, planting seedlings can be a way to restore genetic diversity if there are only a few seed trees or none, or if seed production occurs several years after the fire and the presence of shrubs prohibits natural regeneration. As George and Bazzaz (1991) point out, shrub cover mostly filters out shade-intolerant species such as ponderosa pine, Douglas-fir, and larch. Thus, restoring genetic diversity on severely disturbed sites may require planting shade-intolerant tree species and controlling shrubs to ensure their establishment.

SEEDLING PRODUCTION

Once the decision has been made to reforest by planting, several factors are important to consider. Planting, like natural regeneration, begins with seed production and ends with a stand of young seedlings or saplings. With planting, however, seed is collected and stored, and seedlings are raised in a nursery. In carefully monitored nursery environments—greenhouses or nursery field beds—seedlings are protected from disease, invertebrates, and predation by birds and rodents. They are watered and shaded to avoid moisture stress and extreme temperatures, and they may be fertilized. Roots are pruned or wrenched to develop dense, compact, fibrous root systems that can readily produce new roots after planting.

Several guides and manuals outline the methods for production and handling of seedlings for planting (Duryea and Landis 1984; Jenkinson et al. 1993; Rose et al. 1991).

They cover all phases of seedling production, from seed collection, cleaning, storing, and sowing through procedures for growing, lifting, and storing seedlings. It is beyond the scope of this section to provide detailed information on nursery practices. When silviculturists are developing methods for reforestation by planting, it is important that they work with seedling growers to obtain seedlings suited to the environment in which they will be planted.

Seed source and the size and vigor of the seedlings are two major considerations in selecting seedlings for planting. The seedling's size (height, stem diameter, and root volume) is important for its ability to withstand high soil temperatures, litterfall, animal damage, and intense competition from herbaceous and shrubby vegetation. Size is also an index of stored photosynthate and the potential to produce new foliage. The physiological condition of the seedlings—especially their ability to regenerate new roots after planting (root growth potential)—is important to their survival (Jenkinson 1980; Jenkinson et al. 1993, Rose 1991). Physiological condition is a critical factor in seedling establishment in regions with prolonged summer drought because water stored in the soil is the primary source of moisture for new roots. Vigorous seedlings aid the establishment of plantations and shorten the length of time they are susceptible to competition and animal damage.

Each nursery has its own methods for producing seedlings, adapted to the tree species the nursery grows, nursery soil, local microclimate, and customer requests. Nursery practices are being refined so that a vigorous or target seedling (Rose et al. 1991) can be produced to help ensure seedling survival and growth after planting. Characteristics of a target seedling include height, stem diameter, root growth potential, root volume, and lack of elevated plant moisture stress. Optimal characteristics are species- and site-specific.

Nurseries can grow seedlings of different sizes and physiological conditions, subject to the constraints of the biology of the species and the environment and operational procedures of the nursery. For example, seedlings to be planted on dry sites with competing vegetation might be grown in the nursery to have very dense and fibrous root systems, large stem diameters, and a low shoot–root biomass ratio. These characteristics might require growing seedlings initially in containers and then transplanting them to nursery beds for one growing season, where the roots can be pruned by undercutting to produce a dense root system while maintaining a low shoot-root ratio.

SEEDLING SIZE AND CONDITION

The size of seedlings produced by a nursery depends upon the species, the seed source, and the conditions at the planting site. For the same species, some seed sources (for example, those from higher elevations or more eastern parts of the range) might require 2+0 or 1+1 seedlings (where the first number refers to years in the seedbed and the second number to years in the transplant bed). For planting on a site with the potential for dense shrub cover and severe browsing, silviculturists might request a seedling 18–24-in tall with a 10-12-in root system, and a stem diameter (caliper) of 0.25 in, for example. Such a seedling would be more expensive and more difficult to plant than smaller seedlings, but it would also be more likely to survive and grow on such sites. To produce this seedling, growers might (a) reduce seedling density in the seedbed and grow seedlings for 2 yr (2+0), (b) transplant the seedling after 1 yr to another seedbed at a lower seedling density (1+1), or (c) grow the seedling for 1 yr in a container in the greenhouse and 1 yr in a nursery bed in the open (plug-1). To grow large seedlings with a root system short enough to be planted and with enough growing tips to produce many new roots after planting, it might be necessary to undercut or root-prune the seedlings in the nursery bed during the summer. Nursery practices can affect the success of plantation establishment, and with experience silviculturists can learn the type of seedlings that are best for their sites.

Container-grown ponderosa pine had higher rates of growth and survival than bare-root seedlings on dry sites in the Rocky Mountains, but both stock types did well on productive sites (Sloan and others 1987). However, 5 yr after planting, many of the roots of the container-grown seedlings were still somewhat restricted to the volume of the container. Possibly growing the seedlings in larger containers or another year in the nursery bed (plug-1 seedlings) would have enabled better root system development.

Seedlings that are lifted too soon in the fall may not be dormant. Those lifted too late, in late winter or in early spring, may have begun top and/or root growth. In both cases, the seedlings would not have the stored carbohydrate needed to initiate new roots. In addition, the volume of fine roots is an important part of root growth capacity and is strongly associated with field performance (Rose et al. 1997). Root pruning (undercutting with a sharp blade) is often done in the nursery bed to promote development of compact root systems with a high level of fibrosity (number of fine root tips). The ends of fine roots are the sites at which most new roots are initiated. Thus, there is a physiological and a morphological component to root growth potential (Jenkinson 1980; Rose et al. 1990).

Careful handling and storage of seedlings from the nursery until they are planted is crucial to maintaining the vigor of the seedlings and their ability to grow new roots and foliage after planting. After seedlings are lifted from nursery beds or growing containers, they are usually sorted and graded, packed, stored in coolers, and shipped—all procedures that can reduce seedling vigor if not done carefully. Seedlings must be kept in a cool (±33-34°F, sometimes 28-30°F), moist environment after lifting until they are planted, in order to prevent initiation of growth (tops or roots), loss of energy from high rates of respiration, drying out of tissues, and fungal damage to foliage and roots.

Although planting seedlings immediately after lifting is the ideal approach for realizing all the gains of nursery technology, in many cases it is necessary to store seedlings. Nurseries often cannot fill all the orders to coincide with particular planting times. Nurseries are often at low elevations, and seedlings may begin to grow in the nursery bed long before planting sites at higher elevations are free enough from snow to be accessible and before soil temperatures are warm enough for planting. Wet soils at the nursery often restrict the number of days when seedlings can be lifted with minimum damage to seedling roots and the soil. Nurseries must often take advantage of a narrow lifting window and store seedlings before they are sent to the planting site. A common nursery practice is to lift seedlings early in the winter, then pack and store them at cold temperatures for several months before planting.

It is important for the silviculturist who is planning reforestation to know how seedlings are grown, lifted, stored, and shipped. Nursery visits during the growing and lifting periods to evaluate seedling quality and ensure careful seedling storage and handling can contribute to optimal seedling quality at time of planting.

McDonald and others (1983) provide an example of the importance of nursery practices on seedling establishment in the northern Rockies. Lifting Engelmann spruce, larch, and lodgepole pine seedlings from the nursery beds in the fall and storing them at freezing or near-freezing temperatures enabled them to be successfully planted into mid-June. The 3rd yr height of larch seedlings planted on June 20 differed significantly by lifting date. Seedlings lifted in November and stored at 28°F were 33 in tall; seedlings lifted March 18 and stored at 34 to 36°F were 24 in; those lifted the day before planting were 22 in tall. Weeding and irrigation in the nursery so that soil water did not fall below 30% of field capacity was important for seedling survival and growth. Seedlings that were grown at low soil water stress and were lifted and stored as outlined above had 85 to 90% survival rates when planted from April 4 to June 27, compared to 8 to 60% survival when grown under moderate stress.

Seedlings can also be grown in containers for planting in the forest. Seedling size and potential root growth will vary with the size of the container, species, and nursery practices. In greenhouse conditions it may be possible to grow large seedlings suitable for planting in a shorter period of time than outside in a nursery bed. Since the root system of a container seedling is not disturbed, it should have a high capacity to grow new roots after planting. Handling and transporting container seedlings in the field may be more difficult, especially for seedlings grown in large containers, because the root system and the soil around it remain intact. It is not clear that there is an advantage in planting container seedlings, but this will depend on species, planting sites, and nursery practices (Owston et al. 1992).

SEEDLING HANDLING AND PLANTING

Planting is a straightforward task, but even apparently minor mishandling or less than optimal techniques can result in high rates of seedling mortality and plantation failures. Low levels of soil moisture (Rose and Haase 1993) and other factors that limit seedling access to water have been shown to reduce seedling top growth and foliage expansion (reduced needle growth or "bottle-brushing"). Although planted seedlings may not die immediately from limited top growth (planting shock), their later growth is delayed and they are more susceptible to damage or mortality from browsing and competition. Following are some important guidelines (Rose 1992) to help ensure seedling establishment:

1. The seedling's top and roots should not be allowed to become warm and dry as they are being transported from cold storage to the planting site, temporarily stored on site, or carried by tree planters in planting bags. Planting on warm (above 70°F), windy days risks seedling desiccation.
2. Fine roots and shoot buds are particularly susceptible to damage during handling and should be given special care. They are the points of initiation of new roots and foliage when planted seedlings begin to grow (Hermann 1967).
3. Plant seedlings with top and roots vertically aligned. Bending the roots during planting ("J" rooting) may cause trees to produce a weak root system that is unable to supply sufficient water for the seedling or to support a large tree. This may be most important for deep-rooted species like ponderosa pine. Although there is little support in the literature for this statement, our limited observations on dry sites suggest that "J" rooting definitely contributes to mortality of ponderosa pine, but root deformity by planting apparently has little effect on the survival and growth of Douglas-fir (Haase et al. 1993) if the roots have good contact with moist soil. However, root deformity by planting has been shown to reduce growth of loblolly pine (Harrington and Howell 1998), and we suspect that it affects survival of ponderosa pine.
4. Seedling roots should be placed firmly in moist soil, with no air spaces, organic matter or dry soil around seedlings' roots that could cause them to desiccate.
5. Seedlings planted in the shade of large logs and stumps may benefit from reduced evapotranspiration in the shaded microsite that these objects provide. Likewise, shading seedling stems from direct radiation (e.g., with shade cards) also can encourage successful establishment on exposed sites (Helgerson 1990a, b).
6. Seedlings planted in cold, wet soil can become desiccated if root growth is delayed by cold soil (<38°F) or anaerobic conditions. If the soil is too cold or too wet, seedlings cannot take up water and meet evapotranspiration demands when the above-ground environment is dry (Bassman 1989; Running and Reed 1980; Teskey et al. 1984). Planting seedlings in cold soils also risks desiccation because the roots

are unable to make close contact with the soil until it thaws.

7. Planting procedures should be adapted to the characteristics of the site. Often "micro-site planting" on the shady side of logs or stumps, and in moist sites or micro-topography, and avoiding rock outcrops and difficult planting spots, will favor plantation establishment.

Guidelines such as these should be used to ensure consistent survival of planted seedlings. In years when the weather is favorable (during mild, wet springs, for example) it may not be necessary to rigorously follow guidelines. On severe sites or in years when dry, cold weather follows planting, and in cases when seedlings may not be vigorous, it becomes quite important to follow strictly guidelines like these.

The addition of forest soil and consequently mycorrhizal fungi from forested sites to the planting hole might help reforest certain sites. However, information is lacking about sites where soil transfer may be beneficial. For example, Amaranthus and Perry (1987) found that soil transferred from a vigorous conifer plantation to a previously burned and clearcut site enhanced mycorrhizae formation on planted conifer seedlings. It also enhanced seedling growth and survival on a dry and cold clearcut site. The effects were quite variable depending on the site from which the soil was transferred and the planting site. It appeared that there was little effect on mycorrhizae formation and seedling growth and survival at the lowest-elevation site.

MONITORING PLANTED SEEDLINGS

Monitoring begins during the planting operation, primarily to ensure that seedlings are carefully handled and planted. A sample of newly planted seedlings may be dug up to ensure that roots are straight and in firm contact with moist soil. If needed, planters can be instructed to improve procedures as planting proceeds. A survey about 30 days after planting to check for dead foliage and root growth will determine if the seedlings were in good condition at the time of planting.

Seedlings are usually monitored during the first several years after planting (Stein 1992). Plantations are surveyed at the end of the first year to evaluate initial survival and, if possible, determine causes of mortality. Some of the important variables to evaluate are:

1. The number of seedlings surviving per acre.
2. The causes of mortality. This is determined by digging up dead seedlings to look for rodent damage to roots, lack of root production (poor physiological condition), "J" rooting (poor planting techniques), and other factors. Digging up live seedlings may also be necessary to verify that factors like "J" rooting are the actual cause of mortality or slow growth.
3. Actual or potential impacts of rodents, deer, livestock, or other animals. Is protection of seedlings or control of animals needed?

4. Areas needing replanting if mortality is too great to ensure a satisfactorily stocked plantation. Mortality is frequently not distributed uniformly, so it is necessary to locate areas where attention is needed. Where non-stocked areas are small, the cost of planting may not be warranted, depending upon regulations, cost of replanting, and forest owners' policies.
5. Development of sprouting shrubs and hardwoods, a dense cover of shrubs, and other forms of competing vegetation. These may require some control to ensure seedling survival and growth.
6. Areas of seedling overstocking where thinning may be required.

Two monitoring methods are generally used in plantation surveys (Stein 1992). Plantation-wide grids of plots are frequently established to evaluate variables such as those outlined above. This grid is especially important for evaluating damage from rodents, deer, and weed competition; for identifying areas that need replanting; and for evaluating the overall condition of the plantation. A combination of areas of overstocking or lack of seedlings may be expected in areas of natural regeneration.

Transects of staked seedlings established at the time of planting may be used to quantify percent mortality and identify its causes (numbers 1 and 2 above). Dead seedlings are difficult to locate. Locating a sample of seedlings to examine their vigor or reasons for mortality is most effectively accomplished using staked-seedlings throughout the plantation. Information from staked-seedling transects is used to correct mistakes in planting or handling of seedlings and to assess animal damage. Samples of seedlings from various seed lots might be planted and staked by those in charge of reforestation to ensure that seedling quality or handling and planting are not the reason for poor seedling establishment, should this occur.

One important task during early surveys is to determine if control of competing vegetation is needed. Estimates of cover and level of overtopping by competing vegetation can be used on moist sites (Chan and Walstad 1987; Comeau et al. 1993; Howard and Newton 1984; Wagner and Radosevich 1991a, b). Ruth (1956) found that growth was reduced when competing vegetation shaded the lower half or more of the crowns of Douglas-fir saplings, but saplings did not die until they were entirely overtopped. On drier sites, other measures of cover of competing vegetation are more useful, because grasses and low-growing vegetation are also potential competitors for limited soil moisture (Crouch 1979).

Site Preparation

Site preparation facilitates both planting and natural regeneration (Helgerson et al. 1991; Stein 1995). Site preparation often begins with wildfire or logging. These disturbances may leave the site suitable for regeneration with no further treatment. Often, however, a

Figure 7-6. Slash following logging (photo A). Slash was burned (photo B) to prepare the site for planting and to help control shrubs. Photo B shows the site 1-2 weeks after the burn. The fire consumed most of the smaller pieces of slash and made the site accessible for planting. To avoid costs of burning and problems of smoke dispersal, sites like this may be planted, at a higher cost, without slash control. Photo courtesy John Zasada.

site might be left with enough slash to be a fire hazard or a hindrance to planting, or there may be the potential for vigorous sprouting of hardwoods and shrubs. Such a site may need some application of fire (figure 7-6), herbicides, mechanical scarification, piling of slash, or a combination of these techniques to prepare it for planting. With the shelterwood or seed tree method, site preparation is done throughout the stand (i.e., a broadcast treatment) and care is taken to protect the seed and shelterwood trees from prescribed fire or mechanical damage. With uneven-age methods, site preparation may occur on small areas (i.e. ± 0.5 acres) for regeneration of groups of trees, or it may be done as a broadcast treatment when seedlings or saplings are lacking in a single-tree selection system.

Often the distribution of slash can be managed during logging, minimizing the need for later treatment. For example, trees can be topped and limbed along skidding routes or at the landing, thereby concentrating slash for burning during site preparation. On steep slopes, concentrations of slash can be piled during logging or hand-piled and burned afterwards.

There are several reasons for site preparation:

1. Successful planting requires areas that are safe and easily accessible. Thick slash makes it difficult to walk on—and often impossible to reach—the soil surface for planting (figure 7-6). It is generally the smaller pieces of slash (approximately 4 in or less in diameter) that are a problem. Fortunately, broadcast burning can effectively reduce the cover of wood pieces of this size, or it can be piled and redistributed. The larger pieces (12 in or more in diameter), which are more useful than the smaller ones as habitat for various plants or animals, generally do not substantially hinder planting, and they may provide shade that favors seedling establishment for the first several years. If reforestation is from advanced regeneration, then site preparation and disturbance should be minimized. It should be used mainly on those areas where planting will supplement poor seedling distribution or add tree species to the stand.
2. Successful seeding and planting generally require some control of competing vegetation. Planted seedlings would be unlikely to survive in dense, well-established populations of manzanita or salmonberry, and planting in this dense vegetation is very difficult indeed. Applying fire or herbicides, or both, just after harvesting reduces the sprouting vigor of species such as vine maple, tanoak, etc. On sites with very dense shrub and hardwood cover, herbicides may desiccate woody vegetation and prepare the site for burning. The timing of site preparation and planting is important. Sites should be planted as soon after site preparation as is practical. Delays can jeopardize plantation success as shrubs and grasses invade, compete with the trees for water, and provide habitat for animals that damage seedlings.
3. Site preparation may be done to prepare a mineral-soil seedbed for natural regeneration. As discussed above, rates of germination of seeds and seedling survival are highest on mineral soil for many species. Disturbance of the forest floor reduces high surface temperatures and disturbs the habitat of pathogens and invertebrates that can kill young seedlings (Daniel and Schmidt 1971; Haeussler et al. 1995; Hermann and Chilcote 1965). Some scarification may be needed when sufficient areas of bare mineral soil are not exposed by wildfire or logging. Where dense covers of shrubs or grasses are likely to develop, it is desirable to schedule exposure of bare mineral soil with occurrence of a seed crop of the desired tree species. Synchronizing site preparation with a seed crop may require recognizing a good cone crop in April or May and conducting site preparation before seedfall in September. However, this may not always be practical. There may be 5 or more years between good seed crops for many western tree species, and inhibiting covers of shrubs or grasses could easily become established in this time.
4. Site preparation can be used to regulate species composition. Burning and scarification greatly increased the proportion of larch and reduced the proportion of true fir

while having little effect on Douglas-fir and Engelmann spruce regeneration (Boyd and Deitschman 1969).

5. Site preparation can help reduce fire potential. Because of the small size of the trees and the high density of shrubs and sprouting hardwoods, young stands are very susceptible to fire. Concentrations of slash from logging or site preparation may act as ignition points. Stands that were broadcast burned at the time of site preparation are less likely to burn during a major fire (Weatherspoon and Skinner 1995). Realistically, the goal of fuel management in young stands should be to reduce the potential for very intense fire. Slash and litter that accumulate after logging or wildfire can be broadcast burned. However, current management techniques are relying less on broadcast burning and more on burning piles of slash. Reasons for this trend include smoke management regulations, cost, and potential impacts to site productivity from severe fire.
6. Bark beetles (*Dendroctonus* spp. and *Ips* spp.) often breed in slash left after logging or in trees that have been killed by fire or windthrow. Burning slash or dispersing it on the site so that it will dry are ways to reduce the potential habitat for insects and to kill insects already present in the wood. These treatments should take place soon after the disturbance. Beetles can inhabit recently killed trees within a few days or weeks; usually only a few months pass before insects colonize the slash and emerge from it to inhabit live trees. Thus slash treatments have to be planned well in advanced and implemented promptly. In general, green slash should not be piled against living trees because of the increased risk of bark beetle attack.

SITE PREPARATION AND SOILS

The effects of site preparation on soils depend upon its potential for erosion, susceptibility to compaction, and depth and fertility (Gomez et al. 2002; Powers et al. 2005; chapter 3). Piling, erosion, and severe compaction of the topsoil, as well as severe fire, should be avoided, especially on shallow soils, where most of the organic matter and nutrients are near the soil surface. Raking or piling shrubs and slash to expose mineral-soil seedbeds are sound practices on gentle slopes and stable, non-erodible soils. On steep slopes, however, herbicides or prescribed fire are used, separately or in combination, to expose mineral soil with a minimum of disturbance. It is important to distinguish between a light disturbance of the litter layer that exposes mineral soil and benefits natural seedling establishment, and severe soil degradation that occurs when the topsoil is carried off the site into streams or elsewhere. In the northern Rocky Mountains, scalping the top 6 in of soil caused poor seedling growth, while mounding the top 6 in of soil increased water and nutrient availability to planted Douglas-fir seedlings (Graham et al. 1989). Best growth occurred with the combination of weed control and mounding treatments. The effects of removing or adding the top soil indicate that severe site-preparation treatments could reduce productivity on these sites. A promising method of site preparation on sites having pinegrass or other sod-producing species is to use a tractor-mounted plow shear blade ("salmon blade") that disrupts the roots of the grasses and exposes mineral soil without significant soil displacement (Graham et al. 2005).

Large pieces of dead wood add organic matter to the soil and provide shady microsites for planting, as well as habitat for amphibians, rodents, insects, and fungi in the next stand. Some state forest practice regulations require leaving large pieces of dead wood for habitat.

Competing Vegetation and Establishment of Regeneration

EFFECTS ON TREE SEEDLINGS AND SAPLINGS

Control of competing vegetation may be needed to ensure the establishment of natural and planted seedlings, especially to meet state and federal reforestation regulations. Control of shrubs may be particularly important to ensure the establishment of ponderosa pine or other shade-intolerant species, because only shade-tolerant species such as true firs can become established beneath them (Conard and Radosevich 1982a, b). Nursery-grown seedlings also require a favorable environment for early establishment. When they are lifted from a bare-root nursery, they lose a large portion of their root systems. Consequently, they require sufficient soil moisture through their first growing season(s) to develop fine roots and rebuild a root system.

In many forests, a cover of grasses, forbs, and shrubs is quickly established after disturbance (Tappeiner et al. 1992). Plants invade a site by means of wind-blown seed, seed stored on-site in soil seed banks or in cones on trees, or by sprouting (see chapter 4). Collectively, these plants may interfere with the planted and natural tree seedlings.

On the other hand, herbs, shrubs, and hardwoods are an important part of the forest ecosystem. They reduce soil erosion and retain and recycle nutrients that might otherwise be leached from the soil when the overstory is removed. They all add organic matter to the soil, and some fix nitrogen. They may also provide sources of beneficial mycorrhizal fungi, cover and browse for wild and domestic animals, and other benefits. Therefore, the goal of managing vegetation during reforestation is usually to establish vigorous tree seedlings with minimal long-term impact to the associated vegetation.

Site preparation affects the potential vegetation that will invade a site, and to a large extent these responses are predictable. For example, exposing bare mineral soil for natural tree regeneration also prepares the site for invasion by grasses, light-seeded herbs, and shrubs and hardwoods such as red alder. Broadcast burning will stimulate sprouting shrubs and hardwoods and germination of buried seed. Fire intense enough to cause high temperatures below ground may damage bud banks and reduce the sprouting of shrubs and hardwoods, but it may also destroy soil organic layers, soil micro-organisms, and soil structure.

Herbicides are an important site-preparation tool for controlling woody and herbaceous plants directly or by reducing germination of their seed. If a disturbed site is not immediately regenerated, it often contains a dense cover of shrubs, grasses or forbs which must be controlled before establishing artificial or natural regeneration, but there may be too little slash to produce a fire intense enough to consume their aboveground parts. Therefore, spraying herbicides has the secondary effect of preparing the site for broadcast burning, with the killed tops of plants serving as dry fuel. However, as mentioned previously, the trend in site-preparation techniques is toward less reliance on broadcast burning.

As seedlings grow into saplings and small trees, they can generally grow well with considerable shrub cover beneath them if they are on productive sites. On dry sites, however, a shrub understory can strongly reduce the growth of overstory trees (Barrett 1982; Gordon 1962; Oliver 1984). When managing early-successional vegetation, it is

Figure 7-7. Growth of planted Douglas-fir seedlings after two treatments to control tanoak sprouts at the time of planting. Treatments achieved 100% (A), 50% (B), and no control (C). After Harrington and Tappeiner (1997).

Table 7-7. Typical results of shrub–conifer competition studies in young stands.

Study	*Site productivity*	*Duration*	*Competitive status and results*
1. Douglas-fir, manzanita, canyon live oak (Hobbs and Wearstler 1985)	Low	3 yr	Shrubs overtopped or same height as conifers. Relative growth rate increased 48%; xylem sap tension of Douglas-fir increased 50% when shrubs were controlled. More than 320,000 shrub stems/ac 1 yr after cutting
2. Douglas-fir, tanoak, Pacific madrone, manzanita (Jaramillo 1988)	Moderate	3 yr	Hardwood overtopped or same height as conifers. Douglas-fir seedling growth increased with control of shrubs and hardwoods up to a 12-ft radius around seedlings.
3. Douglas-fir, red alder, grass (Cole and Newton 1987)	Moderate -high	5 yr	Red alder overtopped conifers. Douglas--fir growth least with grass on dry sites; least with alder on moist sites. Alder overtopped Douglas-fir within 3 yr.
4. Douglas-fir, ponderosa pine, western hemlock, noble fir, *Ceanothus velutinus* (Zavitkovski and Newton 1968)	Moderate	14 yr	Shrubs overtopped or same height as conifers. Height growth and survival of all species and both natural and planted seedlings increased with *Ceanothus* control. Natural seedlings established with the *Ceanothus* seedlings grew better than natural and planted seedlings established after a cover of *Ceanothus* had developed.
5. Douglas-fir, *C. Velutinus* (Petersen et al. 1988)	Moderate	5 and 10 yr	Conifers same height as or overtopped shrubs. Shrub control reduced water stress in Douglas-fir seedlings and increased water potential in the soil. Stem diameter in complete and partial control averaged 1.9 and 1.5 times greater, respectively, after 5 yr and 1.4 and 1.2 times greater after 10 yr.
6. Douglas-fir, greenleaf manzanita, canyon live oak (Tesch and Hobbs 1989)	Low-moderate	3 yr	Shrubs same height as or overtopped hardwoods. Douglas-fir survival did not differ among three levels of shrub control. Root and shoot biomass increased 25 and 103 times, respectively, under least competition (12% cover); both increased <5% with 25–60%+ shrub cover.
7. Ponderosa pine, sugar pine, bitterbrush, grass (Gordon 1962)	Low	5 yr	Pines overtopped low shrubs. Control of shrubs and grasses beneath trees 20+ yr increased basal area growth.

Table 7-7 cont.

Study	*Site productivity*	*Duration*	*Competitive status and results*
8. Ponderosa pine, bearmat (Tappeiner and Radosevich 1982)	Moderate	19 yr	Pines overtopped low, dense shrub cover. Survival averaged 9-, 66-, and 90% after no, moderate, and complete control of bearmat; after 19 yr, tree heights were 1.2 and 3.6 times greater in the moderate and complete control treatments, respectively, than those in control.
9. Ponderosa pine, juniper, mountain mahogany (Ross et al. 1986)	Low	8 yr	Pines taller than competing trees and shrubs. Site preparation treatments that reduced shrub density to very low levels increased ponderosa pine height 3 times compared to no treatment. Other treatments were intermediate and related to the amount of shrub cover.
10. Ponderosa pine, manzanita (Powers and Jackson 1978, Powers and Ferrell 1996)	High and low	1 to 10yr	Pines overtopped shrubs. Shrub removal from beneath pine and fertilization doubled pine height growth on poor sites. No effect of shrub removal on tree growth on productive sites.
11. White fir, manzanita (Conard and Radosevich 1982b)	Low	4 yr	Shrubs overtopped white fir. Control of shrubs did not increase seedling growth without shade. Shade with reduction of competition from water increased fir growth 2.3 times compared to other treatments.
12. Douglas fir, ponderosa pine, whiteleaf manzanita (White and Newton 1989)	Low	3 yr	Conifer seedlings overtopped shrub seedlings. Conifer seedling size at leaf area index (LAI) of 0 and 0.6 m^2/m^2 was 3.0 and 1.5 times greater, respectively, than at LAI of 1.2 m^2/m^2. Control of herbaceous vegetation increased the survival and growth of both manzanita and conifers.
13. Ponderosa pine, snowbrush, manzanita, bitterbrush (Barrett 1982)	Low	20 yr	Pine much taller than shrubs. At 100 trees/ac, conifer diameter was 1.3 times larger when understory shrubs controlled; at 250 trees/ac it was 1.1 times greater with shrub control. Overstory tree density had little effect on understory density.
14. Ponderosa pine, manzanita (Oliver 1984)	Low	15 yr	Conifers overtopped shrubs. No effect of tree density on tree diameter growth when shrub cover exceeded 30%.

Table 7-7 cont.

Study	*Site productivity*	*Duration*	*Competitive status and results*
15. Ponderosa pine, grass (Crouch 1979)	Low	10 yr	Pine overtopped grass. Control of grass increased pine height growth 1.5 times and survival from 25 to 55%. A diminished population of pocket gophers from reduction in grass cover was primary reason for higher survival.
16. Ponderosa pine, *Calamagrostis rubescens* (Petersen 1988)	Low	4 yr	Pine overtopped grass. Pine foliage, wood tissue above ground, and root weights were 430, 450, and 460% greater where grass was controlled than where it was not.
17. Ponderosa pine, grass (Baron 1962)	Low	3 yr	Pine overtopped grass. Pine mortality was 80% when planted in year-old grass; 30% with no grass.
18. Douglas-fir, Pacific madrone (Hughes et al. 1990)	Low	5 yr	Madrone overtopped conifers. Shrub and herbaceous cover were 45% less under dense cover of sprouting madrone, 8–30% less under moderate cover, than in open. Douglas-fir seedling diameter averaged 1.8 times greater in the open than under dense madrone, 1.3 times greater than under moderate madrone cover. Douglas-fir height–diameter ratios increased from about 50 in the open to 80 in dense madrone. Shrubs, grasses, and herb density increased as madrone density decreased.
19. Douglas-fir, tanoak (Harrington and Tappeiner 1997)	Moderate	11 yr	Conifers and hardwoods about the same height. Shrub and herb cover decreased from 75% to 0% as tanoak cover increased from 0 to 75%+. Douglas-fir diameter was 2.2 to 1.2 times greater in the treated plots than in the controls. Suppressing herbs did not increase Douglas-fir growth except in the absence of tanoak.
20. Ponderosa pine, sugar pine, white fir (Lanini and Radosevich 1986)	Moderate	3 yr	Trees and shrubs about the same height. Ponderosa pine was 90 and 50% taller and had 164% and 75% larger stems, respectively, when growing with low and moderate shrub canopy volumes than when growing under high canopy volumes. For sugar pine, height was 75 and 45% and diameter 130 and 55% greater, respectively, in low and moderate shrub volumes. White fir was only 20% taller and had 50% greater diameter with low shrub volume compared to high shrub volume.

most important to distinguish between environments sufficient for seedling establishment (i.e., management for survival) and environments that enable seedling growth after establishment. Many studies document the effects of shrubs and hardwood sprouts on the growth of conifer seedlings (table 7-7). They generally started within 1–2 yr after seedlings were planted, with replicated levels of competition produced by controlling shrubs and herbs to a range of densities, from no control (dense cover) to complete control (very little cover) (figure 7-7). They were conducted according to operational planting and site-preparation procedures and indicate that early survival is generally of little concern on mesic, productive sites if healthy seedlings are carefully planted within 1–2 yr after sites are logged or burned (studies 1, 2, 3, 6, 18, 19, and 20 in table 7-7). On dry sites, however, control of even sparse covers of grasses or small shrubs may be needed to ensure adequate survival (studies 8, 12, 15, and 17).

These studies have documented the effects of competition on stem diameter, height, basal area, and crown size. In general, stem diameter is more responsive to control of competing vegetation than is height (studies 3, 12, 18,19). Root growth is also affected, and there is likely a strong, direct relationship between stem growth and root growth (Newton and Cole 1991).

It is important to distinguish between effects on early seedling survival and on later growth. Studies on the effects of competing vegetation suggest that without managing dense covers of associated herbs, shrubs, and sprouting hardwoods, the following scenarios might occur:

1. Dense, tall covers of shrubs and hardwoods will overtop planted conifers. Conifers will eventually die, or at best, conifer seedling growth will be very slow. Examples are Douglas-fir and hemlock competing with red alder, sprouting bigleaf maple, and salmonberry on productive sites (study 3 in table 7-7).
2. Sprouting hardwoods and shrubs (for example, Pacific madrone, tanoak, chinkapin, and manzanita) or shrubs reproducing from seed (for example, manzanita and ceanothus) will grow at the same rate as conifers. Conifer growth will be reduced; seedling mortality will occur. Semi-shade-tolerant species such as Douglas-fir and tolerant species such as western hemlock and white fir, depending upon site quality, are more apt to emerge from this type of competition than are shade-intolerant species such as ponderosa pine (studies 2, 4, 5, 18, and 19; figure 7-5). However, conifer trees may be susceptible to breakage from ice and snow for several decades because of reduced diameter growth relative to height growth. Examples are Douglas-fir or ponderosa pine growing with tanoak, black oak, Pacific madrone, and snowbrush on productive sites in the Coast Range and western Cascades, and the same species growing in mixed-conifer forests in the Sierra Nevada.
3. Conifers overtop low-growing shrubs, grasses, and forbs on dry sites. However, many trees may die from drought, become infected with reproduction weevil, be

killed by gophers, or grow very slowly for 20+ yr (studies 7, 8, 9, 13, 14, and 15). Examples are ponderosa pine overtopping a dense cover of whiteleaf manzanita, bear clover, bitterbrush, greenleaf manzanita, and grasses on dry sites.

These studies also show that seedling and sapling growth is greatest for the first 10+ yr on sites where competing species have been intensively controlled and offer only minor competition to planted seedlings (figure 7-7). For example, Douglas-fir saplings growing in treatments where there was intensive vegetation control (<5% hardwood cover) averaged 4–6 in in stem diameter 11 yr after treatment, whereas those in moderate cover (50% of non-treated stands) averaged 6-8 in in diameter (study 19 in table 7-7). Reducing hardwood (tanoak and Pacific madrone) cover to about 25% at 1 yr increased seedling diameter at 11 yr by 150–200% compared to no control of hardwoods (Harrington and Tappeiner 1997). Seedling diameters in the 25% treatment were also 18–24% smaller than those of trees in the most intensive treatment, where hardwoods were virtually eliminated. Without hardwood control, understory shrub and herb cover was less than 5%; after hardwood cover was reduced by 75%, shrub and herb cover was 35–40%, and with complete removal of hardwoods, understory cover reached 60–75% after 11 yr.

The conventional belief has been that woody species are the competitors of primary concern when establishing a new stand of conifers. However, forest scientists have confirmed that herbaceous vegetation (grasses and broadleaf forbs) can be highly competitive with conifer seedlings, especially during the first 3 yr of stand development (Newton and Preest 1988). Rose and Ketchum (2002) present strong evidence for the effects of herbaceous/grass species on growth of planted conifers. On four of five sites the stem volume of planted ponderosa pine, redwood, and Douglas-fir increased as the area of control of grasses and herbs around them increased, and the magnitude of the differences increased for 3–4 yr. Several studies conducted on moist sites in the Coast Ranges of Oregon and Washington have shown that herbaceous vegetation competes strongly with Douglas-fir (Harrington et al. 1995) and with ponderosa pine on drier sites (Baron 1962; Crouch 1979; Gordon 1962; Petersen 1988).

Nursery practices and the quality of the seedlings planted can have a very important effect on seedling growth and the degree to which it is affected by competition. Planted Douglas-fir, western redcedar, western hemlock and grand fir seedling volume after 8 yr increased with seedling size at the time of planting and intensity of weed control (Rosner and Rose 2006). Furthermore the effects of seedling size and weed control were synergistic, and volume growth from increasing weed control was greatest for the largest seedlings. For example, the estimated volume of seedlings 4 yr after planting and with thorough weed control vs minimal weed control was 3.0 dm^3 (6.0 vs 3.0 dm^3) greater for seedlings that were 9 mm in diameter when planted. With the same weed control the difference was 2.0 dm^3 (4.0 vs 2.0 dm^3) for seedlings 5 mm in diameter at planting.

Most importantly, the studies summarized in table 7-7 quantify the trade-offs among treatments. Knowing these trade-offs is essential to making informed silvicultural decisions to meet specific objectives. For example, the results of study 19 in table 7-7 suggest that controlling large shrubs and sprouting hardwoods to promote conifer growth may increase the abundance of herbaceous cover. In practice, it is typically neither practical nor cost effective to achieve near-complete control except on the most productive sites and on sites where production of wood volume is the primary management goal. Moreover, it is usually desirable to maintain some cover of shrubs and hardwoods for wildlife habitat and other values. Of particular significance is the fact that moderate to high levels of vegetation control can greatly accelerate tree growth while retaining many, if not all, the plant species found in a non-treated stand.

CONTROL OF COMPETING VEGETATION TO PROTECT SEEDLINGS

Vegetation control begins with site preparation. As already mentioned, soil disturbance, especially from fire, may encourage establishment and growth of shrubs, herbaceous species, and grasses, as well as conifers. However, it may also reduce sprouting from shrubs and hardwood trees if their buds are removed from the soil or killed by fire (Stein 1995, 1997).

With experience and local knowledge of plant succession, silviculturists can anticipate the control that may be needed before the regeneration process begins. However, they should monitor vegetation development during first several years of regeneration establishment to determine whether the anticipated vegetation treatments are actually needed, or whether additional treatments might be needed to control unexpected establishment of vegetation from seed or sprouts.

Usually, more than one type of treatment can be used to control a given species (Walstad and Kuch 1987). These include manual or herbicide and mechanical treatments and browsing by sheep, cattle, or goats. The cost and effectiveness of treatments vary considerably. It is beyond the scope of this section to recommend specific treatments for specific cases. Techniques for controlling vegetation change as new mechanical methods or herbicides are developed and as regulations change. Worker safety is a concern in using both manual and herbicide methods. Therefore, it is important to get current information on the effectiveness, safety, and application regulations of any method before beginning a treatment. See Harrington and Parendes (1993), McDonald and Fiddler (1993a and b), Newton 2006, Walstad and Dost (1984), and Walstad and Kuch (1987) for more information on controlling vegetation in reforestation.

Some additional discussion of important considerations for treating different species of vegetation and regeneration development follows.

Preemergent herbicides, foliar herbicide applications, mulching, and hoeing around seedlings can control forbs and grasses. Early control of these species is important to ensure seedling survival and early growth. These plants often form dense covers within 2–3 yr that can deplete soil water while seedlings are developing root systems. They also provide habitat for rodents that browse seedlings or eat seed (Crouch 1971, 1979). Control by herbicides is effective in reducing the cover of forbs and grasses for several years. One recent technological development is fall (preemergent) application of soil-active herbicides. Applied during site preparation and before planting, these herbicides prevent germination of many herbaceous species and eliminate most competition for the new conifer seedlings (Ketchum et al. 1999; Newton 2006). Plastic or fiber mats (usually 3x3 ft or larger) placed around the seedlings also control forbs and grasses (Harrington and Tappeiner 1997; Helgerson et al. 1992; Hughes et al. 1990; McDonald and Helgerson 1991). The mats prevent the emergence of herbs and grasses, reduce evaporation from the soil, and, depending on the durability of the mulch material, kill woody vegetation growing underneath. Hoeing around seedlings is also effective but it is labor intensive (McDonald and Fiddler 2001). Hoeing may be needed for two or more years because grasses and herbs will invade the scalped area. Timing is important—competing plants should be removed as they germinate in the spring before their transpiration has depleted much of the soil water.

Shrubs and hardwoods, both those sprouting from bud banks and those regenerating from seed, can be effectively controlled with herbicides and manual treatments. It is important to distinguish between treatments applied during site preparation, before conifer seedlings are present, and release treatments, which are applied after conifers are present. Treatments applied during site preparation are often done in the fall, when shrubs are still growing and therefore susceptible to herbicides, while the conifers are dormant and thus relatively tolerant of herbicides. Release treatments are typically applied in the spring as the conifers are beginning to grow, and herbicides must be applied carefully so they do not damage them. A few herbicides are reasonably safe to apply directly over the tops of newly planted conifer seedlings for spring control of herbaceous vegetation (Newton 2006). Some soil-active herbicides will limit seedling development of germinating shrubs such as Scotch broom, blackberry, and ceanothus (Ketchum and Rose 2003; Rose and Ketchum 2002; Harrington 2014).

Injecting herbicides into their stems or spraying the stem base or surface of cut stumps with herbicide can effectively control hardwoods. These treatments can be applied at different times of the year to achieve different levels of control (Tappeiner et al. 1987a, Wagner and Rogozynski 1994). Cutting sprouts of hardwoods and shrubs to release conifer seedlings can be reasonably effective, although expensive. Timing of cutting is important in several ways. Cutting early in the spring or in the fall generally provides less control than cutting in the late spring or early summer, after leaves have reached their full size and nonstructural levels of carbohydrates are low (Zasada et al.

1994). Cutting at this time reduced the number and vigor of salmonberry resprouts, although the plants still sprouted prolifically (Zasada et al. 1994). Two cuttings may be required for satisfactory control of salmonberry even if cutting is done at the optimal time. Cutting of red alder at 6–7 yr was more effective in reducing sprouting than cutting of younger trees (DeBell and Turpin 1989; Harrington 1984). As with salmonberry, cutting trees at this age, after full leaf development (late spring to early summer), resulted in a further reduction in sprout number and growth, with many alder actually dying (DeBell and Turpin 1989). Belz (2003) refined this approach by identifying that peak mortality (>80% of trees) occurred when the trees were cut 13–15 weeks after bud break. However, waiting until alders are 6–7 yr old may slow the growth of the conifers, and it may also make them susceptible to sunscald or thinning shock as well as exposing them to damage from felling of the alders.

Repeated cutting of sprouting tanoak and Pacific madrone in radii of 4–12 ft from individual Douglas-fir increased Douglas-fir seedling size compared to untreated controls (Jaramillo 1988). Clearing to 8 ft at least once a year doubled Douglas-fir stem diameter in three growing seasons. Waiting to control hardwoods will result in conifers with small crowns and stems and very high height/diameter ratios, which will render them very susceptible to damage from snow or wind when the hardwoods are removed. Douglas-fir will likely die under dense covers of red alder and bigleaf maple that grow rapidly in height. Injecting hardwoods with a chemical could be a good treatment because they would not damage the seedlings as they gradually decayed and broke apart.

Shrubs can often be effectively controlled at the seedling stage. Timing of the treatment is crucial. Seedlings of ceanothus, manzanita, Scotch broom, and gooseberry often germinate from seed banks 2–3 yr after site preparation. These seedlings can be pulled, or grubbed, from around conifers when they are small (1–2 yr) to prevent establishment of a dense shrub cover. Grubbing is most effective when shrub seedlings are small (probably <1 ft tall). Pulling larger seedlings is expensive and difficult—even a year's growth can make seedlings much more difficult to pull. Because seed-bank germination may continue over several years, it is important to examine the site 1–2 yr after pulling to see if additional germination has occurred and more control is needed. Species such as snowbrush ceanothus can grow quite tall (about 12+ ft in 12 yr). Thus, wide circles may need to be weeded around seedlings for effective control (Rose et al. 1999).

Deer and elk browse has controlled *Ceanothus* and *Prunus* spp. on some sites, with no subsequent vegetation control required. However, this type of "natural control" is not dependable.

When stems of bigleaf maple, tanoak, chinkapin, madrone, and California black oak are killed by fire or cutting, the stem base produces vigorous sprouts within several months. The size of the resulting sprout clumps is strongly related to the diameter of the parent tree. Thus, the cover of hardwood can be predicted up to 15+ yr following fire or clearcutting from estimates of tree density and size prior to cutting (Harrington et al. 1984, 1991, 1992). When the goal is to regenerate well-stocked, productive stands of conifers on sites with dense hardwood cover, these large hardwoods are often removed. When the goal is to produce mixed hardwood-conifer stands, the hardwood sprout clumps may be allowed to grow among the conifers; however, suppressed growth and mortality is likely for conifers growing under the hardwoods.

Hardwood sprouting can be controlled by injecting stems of uncut trees with herbicides, spraying a thin line of herbicides around the stem, or treating freshly cut stump surfaces with herbicides (Tappeiner et al. 1987a). Stem injection and cut-stump treatments of tanoak were more effective when applied in the fall, winter, and early spring than when applied in the late spring and summer (Tappeiner et al. 1987a). Cutting low stumps (<6 in from the ground) reduced bigleaf maple sprout clump size somewhat, as did cutting several years before the overstory conifers were clearcut (Tappeiner et al. 1996). However, these treatments were not nearly as effective as herbicides (Newton 2006).

Treatments to control competition prior to harvesting are not used frequently, but they may be effective. The understory of well-stocked conifer stands is less vigorous than the understory in open areas, and it has a smaller bud bank (Tappeiner et al. 1979, 2001) composed of clone fragments (Huffman et al. 1994; Tappeiner et al. 1991) and seedlings of shrubs and hardwoods (Fried et al. 1988; Tappeiner and McDonald 1984, 2002; Tappeiner and Zasada 1993). These small individuals are much more easily controlled by low-intensity fire or light application of herbicide (Tappeiner 1979) than when they are large or growing rapidly in the open. In addition, shrubs and hardwoods damaged by machines during logging and site preparation are less susceptible to herbicides than undamaged plants because of their damaged vascular system. Prescribed fire might stimulate germination of shrub seeds, and subsequent burning would reduce the seed bank and the sprouting potential of small shrubs (Tappeiner 1979). Thinning treatments in the previous stand that increase the vigor of shrubs and hardwoods will make these species more difficult to control when the stand is being regenerated.

Grazing by cattle, sheep, and goats has been used to control vegetation in plantations (Doescher et al. 1987, 1989; McDonald and Fiddler 1993a, b; Sharrow et al. 1989, 1992). This practice requires careful control of animals and knowledge of their behavior and preferred browse species. For effective control, grazing needs to be concentrated to have sufficient effect on the competitors, and then the animals must be moved before they browse the tree regeneration too heavily. An important requirement is that target species be palatable (Leininger and Sharrow 1987, 1989), and grazing is therefore not successful in controlling salal, tanoak, or manzanita, but it is effective on most *Ceanothus*, *Prunus*, and *Rubus* spp., and on vine maple at low densities. It is important to schedule grazing to coincide with vegetation development, to ensure that animals browse competing vegetation when it is most palatable and avoid the less palatable new conifer shoots. Once the target species have gone unbrowsed for several growing seasons after the initial disturbance, the plants may be too big, vigorous, or unpalatable for browsing. Also, animals may not be able to control large areas of very rapidly growing plants such as sprouting vine maple, bigleaf maple, Pacific madrone, or red alder seedlings. Some minimal damage to trees is to be expected throughout a plantation even with careful control of grazing animals. However, grazing does not necessarily result in poor plantation establishment. Browsing caused 25% mortality in ponderosa pine plantations in Idaho. Gophers, deer and elk caused 21% of mortality, cattle only 4% (Kingery and Graham 1991).

Damage is likely to be high where animals concentrate near water or salt licks or on bedding grounds. Grazing has promise in a managed agroforestry context, but uncontrolled grazing is likely to cause unacceptable damage to young, small trees.

PROTECTION OF SEEDLINGS FROM ANIMAL DAMAGE AND HEAT

In some cases protection from animals and heat may be needed during the first few years of seedling establishment. Plastic mesh tubes placed over the seedling prevent the leader from being browsed by deer or elk; however, such barrier treatments require maintenance to prevent them from damaging the seedlings themselves (Schaap and DeYoe 1986). Mountain beavers in coastal Douglas-fir forests can climb saplings and girdle them up to 12+ ft high on the trunk. They may be trapped to control their populations.

In northern Idaho, the first 4 yr after planting, gophers caused over 80% mortality of planted Douglas-fir and larch, but only 42 to 44% mortality of Engelmann spruce and lodgepole pine and 12% mortality of western white pine (Ferguson 1999). The research suggests that western white pine and Engelmann spruce are the most suitable species for planting in this habitat because the pine suffers relatively little gopher damage while the spruce has greater resistance to snow damage.

Shade cards should be used only if there is concern for the vigor of seedlings and their ability to tolerate high temperatures. They are placed on the south side of newly planted seedlings to protect them from high soil temperatures, which can girdle their stems and desiccate their foliage (Helgerson et al. 1990a, b). Shade cards are most likely to be needed on dry, hot south- to southwest-facing slopes and when newly planted seedlings are small and most susceptible to extreme temperatures.

The following important points summarize the effects of controlling vegetation during regeneration:

1. Vegetation control is often needed to ensure adequate seedling survival and sufficient growth to protect seedlings from intense competition, herbivory and other factors that cause mortality.

2. Control of vegetation (shrubs, grasses, and sprouting hardwoods as well as conifers) will increase the rate of tree growth and stand development and lead to earlier commercial entry into the stand or earlier cover for wildlife, aesthetics, etc. Rapid development is well documented (table 7-7).

3. Planting vigorous seedlings and facilitating rapid growth by controlling competing vegetation enables seedlings to more quickly outgrow their susceptibility to browsing (Black 1992; Gourley et al. 1987) and extreme soil temperatures.

4. Intensive control may not only increase the rate of volume production, but may also increase the total volume production of a site. Long-term evaluation of studies such as those in table 7-7 and careful evaluation of operational plantations are needed to discover to what extent this potential for an increase in yield can be achieved by vegetation control.

With regard to point 4, studies on the effects of tanoak competition on Douglas-fir over a 23 yr period have shown that although the average volume growth per area was greatest when all the tanoak was removed, the result was not significantly different from that achieved with a treatment that left 25% tanoak cover at planting (Harrington and Tappeiner 2009). This suggests that the most intensive treatment might not be cost-effective.

USING HERBICIDES TO CONTROL COMPETING VEGETATION

Since their development in the 1940s, herbicides have had widespread use in agriculture, and more recently in forestry. Relatively few compounds are used today to control competing vegetation in western forests, and they vary widely in their mode of activity and species controlled (table 7-8). Hence it is a common practice to combine different herbicide products and formulations to manipulate the density of a plant community and achieve silvicultural objectives. Herbicide treatments are used in forest vegetation management for site preparation, herbaceous weed control, competition release, and timber stand improvement. Products vary from those that are water or oil soluble to ones that are suspended in a liquid carrier or applied as a dry formulation. These product types affect whether plant uptake of the herbicide occurs through the foliage, stem, or roots.

In general, herbicide "selectivity" (i.e., differential control of species) is achieved by differences in plant physiology, morphology, or phenology, and by differences in herbicide placement. For example, site preparation treatments avoid herbicide contact with planted seedlings altogether, whereas herbaceous weed control and competition release treatments are typically applied when planted seedlings are dormant. Competing vegetation is most susceptible to herbicide treatments when it has a fully developed canopy, an active metabolism, and is uninjured.

Because herbicide use is regulated by federal, state, and local laws, the label represents a legal document that must be carefully followed by the user to ensure safe practices. Most forestry herbicides have a low toxicity to animals, comparable to that of table salt (Newton and Knight 1981). Thus they provide a safe, reliable, and cost-effective method for controlling competing vegetation when they are used properly.

Regeneration and Silvicultural Systems

REGENERATION OF EVEN-AGED STANDS

As discussed in chapter 2, even-aged stands are regenerated by the clearcut, shelterwood, and seed tree methods. Using these methods, regeneration is established within a relatively short time. The result is often a stand of trees of uniform age, spacing, and size. The choice of whether to use natural or artificial regeneration depends largely on site and species characteristics and management objectives. On many ownerships in the Pacific Northwest, clearcutting followed by artificial regeneration by planting has become the established method for regenerating Douglas-fir on the west slope of the Cascades and in coastal forests. To a limited extent, planting is also being used for western hemlock, western redcedar, and red alder; however, these species may reproduce naturally among the planted seedlings where a seed source is present. In mixed-conifer and ponderosa pine forests, ponderosa pine and sugar pine may be planted, and additional seedlings of Douglas-fir, true fir, and incense-cedar may also be planted or may seed in naturally among the planted seedlings. Red alder (Hibbs et al. 1984) and hybrid poplar are planted to ensure the spacing and density needed to produce high yields of valuable straight-stemmed trees free of branches on the lower 20–30 ft.

CLEARCUTS

Planting of clearcuts and burned areas is supported by considerable research, by an infrastructure of nurseries that produce high-quality planting stock, and by woods workers and contractors who plant trees, control vegetation, and prepare sites for planting. Planting has been adopted largely because: (a) seed production is relatively infrequent and unpredictable, (b) well-developed canopies of shrubs often become quickly established after fire and timber harvest and prevent successful natural regeneration, and (c) most commercial tree species grow well in the open on disturbed sites.

Studies evaluating the results of extensive planting programs generally indicate that most stands are stocked after planting. For example, in the Blue Mountains of eastern Oregon, clearcuts were well stocked with planted seedlings and natural regeneration (Seidel 1979). Establishment of ponderosa pine varied with elevation; below about 5300 ft this species grew well and was the dominant tree species in the stand, but survival

Table 7-8. Characteristics of herbicides commonly used in forest vegetation management in the western United States (Newton and Knight 1981; Wehtje et al. 1987; Tu et al. 2001; Kelpsas et al. 2012).*

Herbicide	*Characteristic*	*Forestry application*
2,4-D (Weedone®, Esteron®)	Labeled uses	Site preparation and release (herbaceous and woody plant control); ester and amine formulations are used for foliar and stem injection treatments, respectively.
	Susceptible species	Broadleaf herbaceous species, blackberry, alder, aspen, manzanita, *Ceanothus* spp., tanoak, canyon live oak, and madrone.
	Tolerant species	Grasses, bracken fern, rhododendron, salal, cherry, and bigleaf maple.
	Activity	Foliar uptake (mimics auxin).
	Timing	Wide range of timings, but early spring is often best for foliar treatments; rising sap in early spring can limit efficacy of injection treatments.
	Weather	No special considerations.
	Mixtures	With triclopyr for broad spectrum control of woody species.
	Other information	Relatively inexpensive; foliage curling is a typical symptom of damage.
Atrazine (Aatrex®)	Labeled uses	Herbaceous weed control in conifer plantations.
	Susceptible species	Annual grasses and some germinating broadleaf species.
	Tolerant species	Conifers (can be applied over the top of seedlings), corn family.
	Activity	Primarily root uptake (inhibits Hill reaction in photosynthesis).
	Timing	Pre- or post-emergent, but typically applied in late winter or early spring.
	Weather	Rainfall activated.
	Mixtures	With glyphosate , 2,4-D, or clopyralid for broad-spectrum herbaceous weed control.
	Other information	Relatively inexpensive; rate depends on soil texture; restricted use pesticide (i.e., a certified applicator's license is required for its purchase and use) because of its high mobility in coarse-textured soils and potential groundwater contamination.
Clopyralid (Transline®)	Labeled uses	Herbaceous weed control in conifer and hardwood (*Populus* spp.) plantations.
	Susceptible species	Many annual and perennial forbs (including thistles and composites), elderberry, emerging Scotch broom seedlings, and legume family.
	Tolerant species	Grasses and other monocots, mustard family.
	Activity	Foliar with some root uptake (mimics auxin).
	Timing	Pre- or post-emergent.
	Weather	No special considerations.
	Mixtures	With glyphosate for broad-spectrum herbaceous weed control.
	Other information	Can be safely sprayed over the top of conifer seedlings.

Table 7-8 cont.

Herbicide	*Characteristic*	*Forestry application*
Glyphosate (Roundup®, Accord®)	Labeled uses	Site preparation, herbaceous weed control, and individual stem treatment; formulations without surfactant are safest for conifer release in the fall.
	Susceptible species	Broad range of deciduous woody and actively-growing herbaceous species.
	Tolerant species	*Ceanothus* spp., chinquapin, madrone, manzanita.
	Activity	Foliar (inhibits synthesis of aromatic amino acids); adsorbed tightly to soil, preventing root uptake.
	Timing	Spring (herbaceous weed control) or early-late summer (site preparation); weeds must be physiologically active.
	Weather	No rain through 6+ hours after application.
	Mixtures	With imazapyr (site preparation) or with sulfometuron, or atrazine (herbaceous weed control).
	Other information	Pre-plant or directed sprays for control of herbicide-tolerant weeds.
Hexazinone (Velpar®, Pronone®	Labeled uses	Site preparation, herbaceous weed control, conifer release.
	Susceptible species	Many herbaceous species, greenleaf manzanita, deerbrush, and snowbrush.
	Tolerant species	Conifers, *Vaccinium* spp., cascara, cherry, elderberry, hazel, maple, madrone, manzanita, poison oak.
	Activity	Primarily soil active with some foliar activity of herb species (inhibits photosynthesis and destroys lipids and proteins); moderate persistence in the soil.
	Timing	Pre-emergent to early post-emergent.
	Weather	Rainfall activated.
	Mixtures	With sulfometuron or atrazine (herbaceous weed control).
	Other information	Rate depends on soil texture and percentage organic matter; eye irritant.
Imazapyr (Arsenal®, Chopper®)	Labeled uses	Site preparation, conifer release, herbaceous weed control (rate dependent), and individual stem treatment (i.e., basal spray and injection).
	Susceptible species	Very wide range of woody and herbaceous species; Douglas-fir and other conifers can be injured.
	Tolerant species	Blackberry and legume species.
	Activity	Soil and foliar (inhibits synthesis of branched chain amino acids); moderate persistence in the soil.
	Timing	After foliage is mature; do not use for stem injections when sap is flowing.
	Weather	Rain soon after application will reduce foliar uptake
	Mixtures	With glyphosate (site preparation, especially for deciduous species), triclopyr (site preparation, especially for evergreen species), or sulfometuron (herbaceous weed control; applied during site preparation).
	Other information	Low pH soils (<6) can reduce solution concentration.

Table 7-8 cont.

Herbicide	*Characteristic*	*Forestry application*
Metsulfuron (Escort®)	Labeled uses	Primarily for site preparation or directed spraying.
	Susceptible species	Many broadleaf herbaceous species, blackberry, snowberry, and ferns.
	Tolerant species	Alder, hazel.
	Activity	Primarily soil active with some foliar activity (inhibits amino acid synthesis); moderate persistence in the soil.
	Timing	Early post-emergent most effective (<2″ ht).
	Weather	Warm moist conditions best.
	Mixtures	With imazapyr, glyphosate or sulfometuron for site preparation.
	Other information	Rate depends on soil texture and percentage organic matter.
Sulfometuron (Oust®)	Labeled uses	Herbaceous weed control and release in conifers and some hardwoods.
	Susceptible species	Annual and perennial grasses and forbs, ferns, and Douglas-fir (can cause first-year growth reduction).
	Tolerant species	Many woody species, including alder; thistles.
	Activity	Primarily soil active with some foliar activity (inhibits amino acid synthesis; inhibits root growth).
	Timing	Pre-emergent or early post-emergent.
	Weather	Rainfall activated.
	Mixtures	With hexazinone, imazapyr, or glyphosate (herbaceous weed control).
	Other information	Rate depends on soil texture and percentage organic matter; low pH soils (<6) can reduce solution concentration; fall applications provide herbaceous weed control throughout the following year.
Triclopyr (Garlon®)	Labeled uses	Site preparation, individual stem treatments, and Douglas-fir release.
	Susceptible species	Broad spectrum of woody and forb species, including Scotch broom and gorse; ponderosa pine.
	Tolerant species	Grasses and other monocots, ferns.
	Activity	Foliar (mimics auxin).
	Timing	Growing season (site preparation) or dormant season (individual stem treatments).
	Weather	Rainfast in 4 hours.
	Mixtures	With imazapyr (site preparation of evergreen species).
	Other information	Potential volatilization of ester formulation (Garlon® 4) above 90°F; amine salt formulation (Garlon® 3A) can cause eye injury.

*This table presents descriptive information about forestry herbicides commonly used in western US forests. Please note that all uses of herbicides must be registered by appropriate state and federal agencies, or both, before they can be recommended. Applicator licenses may be required.

was lower at higher elevations, where true fir and hemlock were the primary species. On 61 clearcut sites in northwestern California, stocking after about 10 yr averaged around 400 trees/ac (range 80–1250), and northerly exposures were better stocked than southerly exposures (Strothman and Roy 1984). Stocking increased with the age of the plantation, probably because natural seedlings became established among the planted ones. Larger clearcuts had fewer seedlings than smaller ones, probably because of limited seed available for natural regeneration.

Cold temperatures, competition from grasses and shrubs, and gopher damage were the primary factors limiting the establishment of Douglas-fir, ponderosa pine, and white fir in the southern Oregon Cascades (Minore 1978; Williamson and Minore 1978). However, natural and planted lodgepole pine survived on these sites, and as the trees grew they reduced the temperature extremes and provided cover for the establishment of Douglas-fir and white fir regeneration (see chapter 12, case 3).

Lodgepole pine is naturally regenerated following clearcutting and fire. Regeneration is not generally limited by lack of seed, and competition from shrubs is generally not a major concern in the plant communities it dominates. The western hemlock-Sitka spruce forests of southeastern Alaska also regenerate naturally after clearcutting. Ruth and Harris (1979) found stands in this forest to be overstocked (>20,000 seedlings/ac) from a combination of new seedlings and probably some advanced regeneration. The self-regenerating nature of lodgepole pine and hemlock-spruce forests means that regeneration costs are nil, but stands may need to be thinned when the seedlings are young. Ponderosa pine natural regeneration may be successful in clearcuts on some sites (McDonald 1983); it is not likely to succeed with intense shrub and grass competition. In the central Rocky Mountains, natural regeneration of clearcut openings 300–400 ft wide on north-facing slopes and with mineral soil exposed regenerated within 5 yr (Alexander 1986). Seed was plentiful or adequate on these sites for 9 of 15 yr; however regeneration did not occur on south exposures. In the grand fir-cedar-hemlock forests of the northern Rocky Mountains, density of natural regeneration of Douglas-fir and grand fir was plentiful on north slopes, and increased with the basal area of overstory trees on south aspects (Ferguson et al. 1986). Also natural regeneration did not always occur rapidly and increased with time since harvesting (up to 20 yr). Soil type and distance to a seed source did not improve predictions of seedling density. Density of ponderosa pine, cedar, and other conifers were not predictable on these sites.

SHELTERWOOD

The shelterwood method has been used in a number of forest types. Tesch and Mann (1991) compared the shelterwood method with the clearcut method and outlined the pros and cons of each. Considerations are ecological (availability of seed, presence of competing vegetation, goals for wildlife habitat, and effects of insects and pathogens), managerial (difficulty of site preparation and release of seedlings, removal of the shelterwood trees [Laacke and Fiddler 1986]), and aesthetic. Trees may be left to protect young seedlings and saplings from extreme temperatures on frost-prone sites or sites with high summer temperatures (Childs and Flint 1987); site preparation, vegetation control, and planting of seedlings occur beneath them. Conifers can be regenerated under a shelterwood, but it is not always necessary to use this method if regeneration is the only goal. Tesch and Mann (1991) cite examples of successful regeneration of ponderosa pine and Douglas-fir on exposed dry sites both in clearcuts and under shelterwoods. Both methods require vegetation control and proper planting techniques for successful seedling establishment. If there is uncertainty about the timing of seed production by shelterwood trees, seedlings may be planted under them. This is especially important on sites where grasses and shrubs are likely to invade quickly, because they would provide habitat for herbivores and compete for soil water.

Shelterwoods have been successful in regeneration of Douglas-fir forests (Williamson 1973), mixed-conifer forests (Dunlap and Helms 1983; McDonald 1976b) and true fir forests (Gordon 1979; Laacke and Tomascheski 1986). The shelterwood method is particularly well suited to spruce and true fir forests because seed production is fairly regular. McCaughey and others (1991) found successful regeneration of spruce-fir forests on several sites in the intermountain west using the shelterwood method and clearcutting. Seed production appeared to be plentiful and competition was not particularly severe, and seedling establishment occurred for at least 20 yr on their sites.

Natural regeneration may be a way to develop blister rust–resistant western white pine—and possibly sugar pine as well—in the interior west. About 19% of western white pine trees naturally regenerated by the shelterwood method were found to be resistant to white pine blister rust (Hoff et al. 1976). The shelterwood method of regeneration might be used to regenerate plentiful resistant seedlings and develop resistance to this disease (Graham et al. 2005). It is likely important to search for rust-resistant seed trees, though it may be difficult to identify rust-resistant trees at an early age during thinning treatments.

The overstory trees reduce temperature extremes, and they can slow the invasion of sedges and other vegetation. Advanced regeneration is also plentiful in true fir forests, and it can be supplemented with natural seedlings from the shelterwood (Gordon 1973; Oliver 1986). Another advantage of the shelterwood method in these forests is that they occur at high elevations, where snow accumulation often makes planting difficult.

It has become a more common practice to leave trees, either as scattered individuals or in groups, in what would otherwise be a clearcut. This strategy is called green tree retention (Franklin et al. 1997). Large trees or snags are left for their scenic and aesthetic value, as well as to provide wildlife habitat and sites from which lichens, bryophytes, and other organisms can occupy the new stand.

Depending on their density, trees retained in shelterwoods may delay the growth of understory tree seedlings and, if left long enough, may affect the growth and yield of the developing stand. The height growth of ponderosa pine seedlings was greater as their distance from the shelterwood trees increased from 10 to 40 ft (McDonald 1976b). At these distances the height growth of seedlings was greater in stands with 4 trees/ac than in stands with 8–12 trees/ac. Zenner et al. (1998) report that, after 50+ yr, the growth rate of trees that had grown under an overstory of 5 trees/ac was reduced by 19%, whereas the growth rate of trees that had grown under an overstory of 50 trees/ac was reduced by 47%. The reduction in growth rate was greater on north slopes than on south slopes. The mean annual increment of the understory at 97 yr was reduced by about 26% for every 40 ft^2/ac of overstory trees (Acker et al. 1998).

When choosing shelterwood trees, and especially trees that will remain in the overstory for a number of years, it is important to be sure that they are nearly free of dwarf mistletoe, which can infect the younger trees beneath them, and that they are stocky (H:D<70), to improve wind resistance.

REGENERATION OF UNEVEN-AGED AND TWO-STORY STANDS

(Also see chapter 10.)

Natural regeneration and planting are used in uneven-aged systems to produce multistory stands. In mixed-conifer forests, shade-tolerant species such as true firs, incense-cedar, and black oak frequently occur in the understory as advanced regeneration, as does ponderosa pine on drier sites. Uneven-aged true fir forests can be managed using these sources of regeneration.

In uneven-age management, selective harvesting provides the soil disturbance and increased light necessary for the establishment of natural regeneration. Ideally, a seedling bank of advanced regeneration of desirable tree species accumulates. These seedlings grow into saplings and poles, and eventually into overstory trees, as thinning reduces stand density around them. Thus, recruitment of new seedlings and facilitating their growth into the upper canopy is an ongoing process. On productive sites, this process might also result in the regeneration of shrubs and an overstocking of shade-tolerant sprouting species such as salal, tanoak, or bigleaf maple. Competition from these shrubs and hardwoods is of particular concern in uneven-aged management of Douglas-fir on productive sites in the Coast Range and Cascades. They may preclude the planting of conifer seedlings unless they are controlled (Brandeis 1999; Brandeis et al. 2001; Coates 2000).

Heavy commercial thinning of young, dense Douglas-fir stands often leads to establishment of natural regeneration of western hemlock, Douglas-fir and bigleaf maple (figure 10-3; Levy 2010, Shatford et al. 2009, Bailey and Tappeiner 1998; Fried et al. 1988; Miller and Emmingham 2001). Similarly, partial cutting of western hemlock-Sitka spruce stands in southeast Alaska leads to abundant regeneration of both species (Levy 2010, Deal and Tappeiner 2002). Although establishment of regeneration may not have been intended in these examples, seedlings would likely grow into a second story of trees with future partial harvesting of the overstory (another thinning or actual multi-aged entry). In this way, a "two-aged" silvicultural system could emerge when regeneration follows thinning. Smith et al. (1997) call this the "telescoping" of one rotation into the next, and it avoids some of the unproductive regeneration period typical of even-aged systems. Some level of vegetation control and planting may be needed to ensure a vigorous seedling bank, even in the mixed-conifer, ponderosa pine, and true fir types more suited to multi-aged management.

When planting shade-intolerant species in uneven-aged stands, it is necessary to ensure sufficient light and soil moisture. In mixed-conifer forests, seedling height growth increased by 34% as opening size increased from about 0.25 to 2.5 ac (York et al. 2003, 2008). Seedlings were generally taller on the north side of the openings than on the south. Height growth of Douglas-fir and giant sequoia was related to increased light and soil moisture, whereas for ponderosa pine it was related only to light. On similar sites, natural regeneration was plentiful—well over 1000 seedlings/ac—in 48 openings ranging 30–90 ft in diameter (McDonald 1976a; McDonald and Abbot 1994). However, after 10 yr, seedlings were only 1.0–1.5 ft tall. It is questionable whether the ponderosa pine in this instance would be vigorous enough to respond to thinning, but the Douglas-fir and white fir would likely respond. White fir may respond poorly to understory vegetation control without reduction in overstory density (Oliver and Uzoh 2002). Small openings may favor natural seedling establishment, but not subsequent growth. Gray and Spies (1996), in a study in Douglas-fir stands where gaps ranged from 400 to 20,000 ft^2, found that establishment of western hemlock seedlings was greatest in the smaller gaps and was aided by shade in larger gaps. However, seedling size was greatest in large gaps, and western hemlock seedlings grew more on the forest floor than on dead wood. Seedlings are not expected to grow as rapidly in uneven-age systems as they are in even-age or two-story systems because they are growing among larger trees. They must be able to respond to release, however.

The competition from both overstory trees and shrubs can affect the establishment of trees in two-age or uneven-aged stands. Survival of western hemlock and grand fir decreased as overstory density increased from about 50 to 120 ft^2/ac; survival of western hemlock was about 55% on plots where vegetation was controlled and 30% where there was no control (Brandeis et al. 2001). Survival of grand fir ranged from about 50% on plots with high overstory density to more than 90% on plots with low overstory density. Survival of western redcedar ranged from about 75% to 95% at all densities, however, and was unaffected by understory vegetation control. Vegetation control improved the growth of grand fir and hemlock but not western redcedar. With the same basal area distributed unevenly, seedlings generally had greater rates of growth and survival in the gaps or least-dense parts of the stand (Brandeis et al. 2001). Miller and Emmingham (2001) estimate that it will be necessary to keep the canopy at low densities (40–50% canopy cover) for successful Douglas-fir regeneration. Competitive effects of overstory trees and understory vegetation are generally additive. In a five-year study in western Washington, Harrington (2006) found that stem volume of planted Douglas-fir, western redcedar, and western hemlock seedlings in clearcuts averaged four to eight times that which was observed in stands containing a moderate density of overstory trees (basal area=112 ft^2/ac), and with vegetation control it averaged two to four times more than what was observed without it.

In addition to over- and understory density, variation in microclimate—as affected by aspect, slope, and other variables—may be important for regeneration in the understory. Rates of survival and growth may be greater on southerly exposures than on shadier slopes. In young stands, overstory closure occurs quickly, especially after light thinning, and frequent thinning may be needed to ensure the growth of established seedlings. Also, the results will likely vary by species and type of regeneration—whether seedlings were planted or naturally regenerated, and if it was the latter, whether from seed or from release of advanced regeneration. For example, we have observed seedlings and saplings of western hemlock advanced regeneration triple their height growth within 4yr after 60–70-yr-old Douglas-fir stands were thinned to 40–50 trees/ac (Shatford et al. 2009). Growth rates of Douglas-fir were less than half that of western hemlock. The effects of deer and elk browsing vary from site to site with very heavy browsing of some species, especially redcedar on some sites but not others. Similarly to the results discussed above for conifer stands, Emmingham et al. (1989) found that control of mountain beaver, as well as overstory and shrubs, was needed to regenerate western hemlock in red alder stands.

The information to date clearly shows that both under- and overstory density affect the survival and growth of seedlings. Until more is known about regeneration and the growth of trees in the understory of conifer stands, seedling survival and growth, herbivory, changes in understory and overstory density, and other ecosystem variables will need to be evaluated to obtain satisfactory development of conifers in the understory.

REVIEW QUESTIONS

1. What is the process of natural regeneration? Discuss the steps in the process.
2. What are the factors that affect seed/seedling survival at each step in the process?
3. How would you judge a good seed tree—what are its characteristics?
4. Why is "bare mineral soil" important for early seedling establishment? Describe the microsite that it provides.
5. How can litterfall affect seedling survival?
6. What are seedling banks? How can they aid in reforestation?
7. What is the light compensation point and how does it affect the growth and survival of advanced regeneration?
8. What realistic steps can silviculturists take to favor natural regeneration?
9. Discuss and explain the differences between natural and artificial regeneration.
10. What nursery practices can be used to produce vigorous planting stock? Explain.
11. How can seedling handling (lifting from the nursery bed, storage and transportation) and planting affect seedling chances for survival? Explain the major factors that can reduce seedling vigor in this process.
12. What are two major soil conditions that favor survival of planted seedlings? Explain.
13. Why does site preparation favor establishment of natural and planted seedlings?
14. What other forest management/ecosystem variables can site preparation affect? Discuss.
15. Explain how grasses, shrubs, and hardwoods can affect establishment of artificial and natural regeneration. What are several outcomes (+) of the interaction between these species and planted and natural seedlings?
16. What methods are used to control competing vegetation when establishing tree regeneration? What are the important pros and cons of the various methods?
17. Discuss the effects of animal (rodents and ungulates) damage on regeneration and how it can be controlled.
18. Discuss the use of regeneration in the various silvicultural systems. What are some of the important factors that affect the choice of the regeneration method and the decision to use natural or artificial regeneration in different systems?

CHAPTER 8
Management of Even-Aged Stands

Crown and Tree Classes
Thinning Methods and Severity
Effects of Thinning on the Growth of Individual Trees
Effects of Thinning on the Development of Stands
Measures of Stand Density and Thinning
Tree Spacing
Thinning Early in Stand Development
Thinning Late in Stand Development
Height/Diameter Ratios: Maintaining Tree Stability
Thinning Shock
Effects of Thinning on Wood Quality
Effects of Density Management on Forest Stands as Ecosystems

In this chapter we discuss the management of even-aged stands, methods of managing stand density, and the effects of stand density on tree and stand growth, structure and development. There has been considerable long-term research throughout the world on the management of even-aged stands. Stands of different species have been thinned to various densities, at different ages and frequencies, and fertilizers (chapter 9) have been applied to stands of varying ages and densities. The effects of these treatments on the growth and development of individual trees and of entire stands have been documented.

New information is emerging on the effects of regulating stand density on such variables as understory shrubs, microclimate, streamside habitat, bird habitat, fuels and fire potential, insects, etc. This is valuable information, but it is not yet well integrated into stand management. Moreover, the responses of these variables may be site-specific. For example, the response of understory plants to stand density may vary as understory species composition and density change within a stand. Nevertheless these "ecosystem variables" are becoming increasingly important in contemporary forest management.

Initial stand density and tree and shrub species composition are determined by regeneration practices (planting and natural seeding) and by vegetation control during seeding establishment. Thinning is the major way to manage stand density and species composition after a new stand has been established. It is carried out for reasons related to wood production, aesthetics, and wildlife habitat, to maintain tree vigor and resistance to insects and diseases, to minimize the effects of fire, and for other purposes. Thinning may be done before the trees are of a marketable size (precommercial thinning) or it may be done in older stands to yield commercial wood. In stands managed for commercial wood production, thinning not only yields wood but also ensures the species composition, volumes, tree vigor and sizes that will support future economic returns to the landowner.

Site-specific analysis of the stand is usually needed to ensure that thinning methods are compatible with the variability of a stand's density and species composition, the presence and vigor of advance regeneration, and other important characteristics of the overstory and understory. Such analysis will often reveal how slight modifications to standard thinning practices can benefit several resources simultaneously.

Fertilization (chapter 9) is also an important aspect of stand management. It affects the growth rate, yield, and value of the wood produced. It also affects shrub density in young stands of seedlings and saplings, as well as the understory in older stands. The effects of fertilization depend on thinning and stand density.

Crown and Tree Classes

Trees assume different relative positions in the canopy as stands grow and develop. Crown classes are used to designate trees for thinning in even-aged stands. As noted in chapter 6, Kraft (1884) stratified these relative positions into crown classes and gave them the labels predominant, dominant, codominant, intermediate, and suppressed. Crown classes are commonly used to describe stand structure and tree growth potential (Kraft 1884; Oliver and Larson 1996; Smith et al. 1997).

Competition and self-thinning are a natural part of forest stand development. Intertree competition occurs as trees grow and their crowns and roots begin to interfere with one another. The larger trees become dominants or codominants with superior crown position; their crowns form the uppermost part of the canopy. The smaller trees that experience the most competition become intermediate and suppressed (trees in the 8–12 in diameter classes in figure 8-1 and table 8-1). Competition from the dominants and larger codominants increases as the stand grows. Slower-growing codominant trees may lose their position in the canopy and, depending on stand density, become intermediate or suppressed, and may die. The following general definitions are often used to identify a tree's crown class (Smith et al. 1997):

Predominant trees—those that extend above the main canopy—occur at very low densities if at all. They may have been established in the previous stand, or at least before the main cohort of the current stand. The term "wolf tree" is often used to describe predominant trees, probably established early in the life of a stand, that have large branches extending along most of the stem.

Dominant—crown extending above the general level of the canopy; receiving sunlight from above and partly from the side.

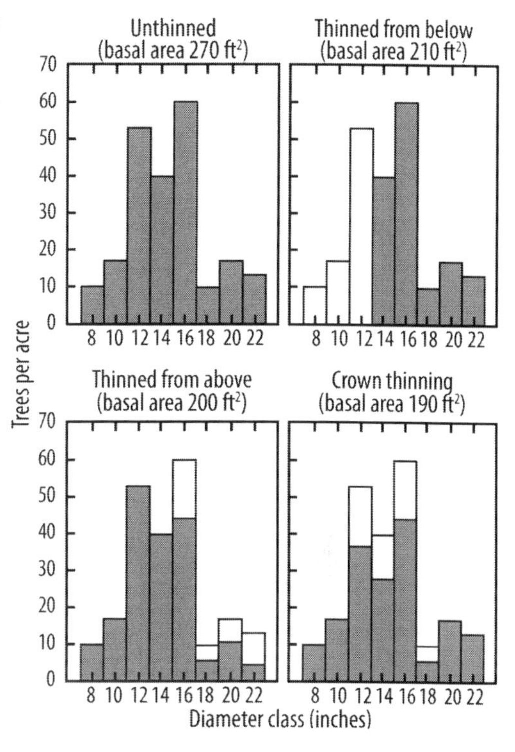

Figure 8-1. Effects of different types of thinning on stand structure, D/d ratio (diameter after thinning/diameter of the cut trees), and basal area. The data is from the 55 yr old Douglas-fir stand described in Table 7-1. White bars show tree removed. Basal area after thinning.

Codominant—crown forming the general level of the canopy; receiving sunlight from above but little from the side.

Intermediate—crown extending into the general level of canopy; receiving some sunlight from above and none from the side (i.e., half-in and half-out of the upper canopy layer).

Suppressed—crown entirely below general level of canopy; no direct light from above or from side.

Dominant and large codominant trees are the major components of the overstory. They are the most vigorous trees and account for most of the biomass and stand density. Most leaf area and growth of merchantable wood occurs on codominant trees (table 8-1). Generally a stand has very few dominant trees; the majority are codominant. Intermediate and suppressed trees contain relatively little of the total stand volume; they have the smallest crowns and are generally slower-growing than the dominant and codominant trees. They are in inferior positions in relation to the dominant trees and are likely to produce little growth and die from suppression, insects, snow and ice damage, or other causes (table 8-1).

Crown classification is most easily used in thinning single-story, single-species stands where inter-tree competition and differentiation into crown classes has begun.

Table 8-1. Stand table for a 55-yr-old Douglas-fir stand on low site class II.

Diameter class (in)	*Crown class*	*No. trees/ac*	*Basal area (ft^2/ac)*	*Volume (ft^3/ac)*	*Volume growth ($ft^3/ac/yr$)*	*% stand volume growth*	*Cumulative growth ($ft^3/ac/yr$)*	*% cumulative growth*	*Average tree growth (ft^3/yr)*
8	Suppressed	10	4	130	7	2.3	7	2.3	0.7
10	Intermediate	17	9	360	15	5.1	22	7.4	0.9
12	Int./codom.	53	42	1710	59	19.7	81	27.1	1.1
14	Codominant	40	43	1790	52	17.4	133	44.5	1.3
16	Codominant	60	84	3560	90	30.1	223	74.6	1.5
18	Codominant	10	18	760	17	5.7	240	80.3	1.7
20	Codominant	17	36	1580	31	10.4	271	90.7	1.8
22	Dominant	13	35	1540	28	9.3	299	100	2.2
	Totals	220	271	11430	299	100			

% volume growth for the stand is (299ft3/ac/yr)/(11430ft3/ac) = 2.6%/yr

Crown classes are a convenient way to assess tree vigor and the potential of individual trees to grow vigorously or die from competition and suppression. Dominant and co-dominant trees are likely to maintain their position in the stand and resist insects and pathogens. Intermediate and suppressed overtopped trees are the least vigorous and are likely to grow slowly or die. As we will discuss, however, hierarchy does not always ensure growth and survival; pathogens and insects can kill large dominant and codominant trees.

In mixed-species stands, crown classes may not be as useful for predicting future growth responses as they are in single-species stands and tree classes may be more useful (see chapter 10). For example, in Douglas-fir-western hemlock stands (Wierman and Oliver 1979), and in mixed pine-true fir stands, early height growth of the hemlock and fir is generally slower than that of the pine and Douglas-fir. Thus the fir and hemlock might occur in the smaller codominant or even the intermediate crown classes. Nevertheless, because they are shade-tolerant, hemlock and true fir retain relatively large crowns and may, after thinning, increase growth much more rapidly than Douglas-fir or pine of the same size.

Thinning Methods and Severity

Thinning can be described in terms of both method and grade (also known as thinning severity). Grade (Nyland 2002, Helms 1998, and Smith et al. 1997) refers to reduction of stand density—that is, the reduction in volume, basal area, trees per acre, SDI, etc.—and the effect on stand density. This may range from grade A (high removal of volume or reduction of density) to grade D (low removal). Intensity is the severity and frequency of thinning combined. The method or type of thinning refers to the crown classes of the trees removed. Thinning from above, or "high" thinning, removes trees from the larger crown classes (larger trees), whereas thinning from below, or "low" thinning, removes trees from the smaller crown classes (smaller trees). The proportion of large to small trees removed can be expressed mathematically by the D/d ratio, where D is the average stem diameter of the stand after thinning and d is the diameter of the cut trees. A ratio of >1.0 indicates that smaller trees were removed (thinning from below), and a ratio of <1.0 indicates that larger trees were removed (thinning from above). A ratio of 1.0 occurs when trees from both larger and smaller crown classes are removed (figure 8-1). The more severe the thinning from below or above, the more the ratio deviates from 1.0.

THINNING FROM BELOW

Thinning from below leaves the dominant trees and generally the large codominant trees while removing intermediate and some small codominant trees, the trees most likely to die from competition or suppression. Severe thinning may remove considerable numbers of codominant trees. For example, a very light thinning from below might remove only small trees (in the 8-, 10-, and 12-in diameter classes; table 8-1 and figure 8-1); a severe thinning from below might remove these trees and some from the 14- and 16-in classes as well. Thus the D/d ratio increases with an increase in the severity of thinning. Non-merchantable trees in the small size classes may be left because of the cost and limited benefit associated with their removal. They may also provide cover for vertebrates; on the other hand, they may be fuel ladders that predispose a stand to crown fire.

Low-severity thinning from below is the most conservative type of thinning and is likely to have the least impact on individual tree growth and overall stand growth. Effects of wind are minimized because the canopy generally remains intact and the large, windfirm trees are left in the stand. The yields of wood from such a thinning may be low; however, overall volume production is generally reduced the least by this method.

CROWN THINNING

Crown thinning is the removal of trees from the main canopy to provide growing space for vigorous trees with desirable stem characteristics. It can be considered a crop-tree thinning in that selected trees are deliberately provided with growing space by removing trees from around their crowns in one or more thinnings. Crown thinning may have major effects on the growth of individual trees and the entire stand. The trees removed are generally some of the larger ones in the stand—for example, trees from the 14-, 16-, and 18-in diameter classes (figure 8-1 and table 8-1). Their removal reduces the density of the overstory canopy and may reduce total stand volume production, but it will favor the growth of the remaining large trees much more than thinning from below. Often, dominants with poor crown or stem form are removed in a crown thinning to favor growth of other dominant or codominant trees with better form. Crown thinning might lead to windthrow if a severe (grade A or B) thinning is applied to dense stands, because it leaves gaps in the canopy and increases wind turbulence and depth of wind penetration into the canopy.

If there are trees in the smaller codominant, intermediate, and suppressed crown classes, such as those in the 8- and 10-in diameter classes in table 8-1, a two-aged stand structure may develop after crown thinning (Bailey and Tappeiner 1998). Opening the canopy around selected codominants may provide enough light for the trees in smaller crown classes to persist and grow in the lower canopy or the understory, especially if they are shade-tolerant species such as western hemlock or true firs.

THINNING FROM ABOVE

Thinning from above, also called selection thinning, is the removal of trees from the dominant and codominant classes. This type of thinning decreases the D/d ratio.

Thinning from above may be done to remove large trees with poorly formed stems or to harvest valuable, straight-stemmed trees for poles, pilings, or sawlogs. Large trees (some trees from the 18–22-in diameter classes in table 8-1) may be removed deliberately and carefully without reduction of potential growth or damage to the remaining stand, provided that a sufficient density of vigorous codominant trees is retained (Buckman et al. 2006, Buckman 1962; O'Hara 1988). Large, dominant or codominant trees with big limbs and long crowns (wolf trees) produce low-quality wood with large knots, but they may be potential nest or den trees or have other value for wildlife habitat.

Careful removal of selected large trees is quite different from simply removing all the large or commercially valuable trees from the stand without regard for the density, vigor, and potential growth of the remaining trees (high grading). Codominant trees are vigorous, and most of the stand's volume growth occurs in these crown classes (Buckman et al. 2006; O'Hara 1988). Too-frequent or too-intense thinning from above could leave stands at low density and stocked with trees of low vigor, susceptible to breakage from heavy snow, windthrow, and other stresses. Selection thinning can also shift the species composition of the stand toward shade-tolerant species such as western and mountain hemlock, true firs, and incense-cedar.

FREE THINNING

Free thinning, as the term suggests, favors desired trees by combining various crown-class and spacing criteria. The D/d ratio may vary from <1.0 to >1.0 throughout the stand. Free thinning may be a practical way to thin stands for a range of forest management objectives, and it may be particularly useful when thinning mixed-species stands with variable density, spacing, or species composition. For example, thinning to maintain a species that occurs infrequently might call for releasing small conifers or hardwoods by removing dominant and codominant trees in selected places in a stand. If this strategy were used throughout the stand, it would be called a thinning from above or a crown thinning. Stands with heterogeneous size classes, density, and species distribution may be so variable that they would require a mix of strategies over the whole stand. In stands with trees damaged by fire or infected with mistletoe, for example, it may be best to disregard crown classes altogether and simply use stand-specific criteria for tree removal or retention. Free thinning can also be used to stimulate greater uniformity within a stand that has considerable variability in its structure.

With a stand table (table 8-1), silviculturists can evaluate the growth potential of trees in different crown or diameter classes, and manage a population of trees. Carefully removing some trees in the codominant crown classes, with little taper, that are valuable for poles, can be profitable (they often have three times the value of other trees). Careful removal includes causing minimum damage to surrounding trees and leaving trees that are resistant to wind and ice/snow damage and also possess sufficient leaf area to respond to thinning by maintaining stand and tree growth.

Free thinning has been termed "variable density thinning" (Carey and Curtis 1996). This method was developed to create spatial heterogeneity in composition and structure similar to that found in old-growth forests, and therefore, it is assumed, habitat for old-growth-associated species such as the spotted owl (Carey et al. 1999). The approach retains and at times enlarges existing clumps and openings within the stand to increase spatial heterogeneity. Typically, trees from the codominant and intermediate classes are removed to provide more space for overtopped trees and shrubs of other species. Large standing dead trees are retained to add to the "decadent" structure typical of an old-growth forest. Trees with the potential to provide long-term wildlife habitat, such as those having cavities or stem rot, are usually retained, as are some hardwoods and mast producers.

STRIP OR ROW THINNING

Strip thinning and mechanical thinning (also called geometric thinning, Smith et al. 1997) leaves trees at predetermined spacing without regard to crown class and tree vigor, species composition, or other stand attributes. Thus, these methods typically remove trees from all diameter classes. The D/d ratio will remain close to 1.0, indicating very little change in the diameter distribution. A common form of strip thinning is one in which strips are cut or plowed through the stand, releasing trees along the strip edge and aiding self-thinning and differentiation into crown classes. Similarly, rows of trees might be removed from a uniform plantation. Only trees immediately adjacent to the cut strip (or row) may respond, however, unless thinning is also done between the strips. Depending on the density of the strips, there may not be much effect throughout the stand (McCreary and Perry 1983). Strip thinning might be used in stands of small, low-value trees that are so dense that access is difficult and there is no practical way to select individual trees and thin around them. Strip thinning can be followed by another method of thinning, such as low thinning, to provide both a reduction in density and an attempt to improve stand structure to favor the larger, more productive trees.

Effects of Thinning on the Growth of Individual Trees

Thinning has a major effect on the growth of individual trees. The effect may be positive as trees are provided more growing space, water, and light and increase their rate of annual increment. There may also be negative effects, at least initially, from sunscald, logging damage, and wind (Harrington and Reukema 1983).

Thinning affects basic tree vigor by increasing the availability of soil water. As canopy area is decreased, soil water increases, probably from a combination of increased precipitation reaching the forest floor (precipitation throughfall) and decreased transpiration (Brix 1986; Donner and Running 1986). Consequently, after thinning, leaf water potential increases in trees in thinned stands, enabling them to increase rates of photosynthesis (Helms 1964; Sucoff and Hong 1974; see table 3-7). For example, Donner and Running (1986) found that soil water depletion in thinned lodgepole pine stands with 50–72% of their basal areas removed was much less than it was in the unthinned controls. They estimated that plant water potential was an average of 0.3 MPa greater in the thinned stands and that seasonal photosynthesis was 21% greater after thinning. This increased photosynthesis supported the processes that increased stem growth.

Trees respond to thinning by expanding their crowns and maintaining or increasing the density of their foliage (table 3-7). Thinning provides space for crown expansion (see figure 5-4). Curtis and Reukema (1970) report that for 43-yr-old Douglas-fir stands, crown width and length increased as spacing increased. Increase in crown length can result from three processes: upward expansion of the crown from height growth, downward expansion of the crown from epicormic branching, and increased longevity of the lower branches. For example, at 12-ft spacing, crown length averaged 40, 28, and 22 ft for dominant, codominant, and intermediate trees, respectively, whereas at 8-ft spacing, crown length averaged 25, 18, and 14-ft. These researchers report similar trends for crown width, demonstrating that total crown volume and surface area increased as spacing increased.

Crown expansion is likely accompanied by larger and perhaps more vigorous foliage. Ponderosa pine fascicle length and weight increased by a factor of 1.2–1.5, increasing more as thinning severity increased (Wollum and Schubert 1975), with the greatest increase associated with the most severe thinning. Also, the density of interwhorl buds of young Douglas-fir saplings and seedlings increased within 1–2 yr after reducing the density of surrounding trees and shrubs. Increases in the number of interwhorl buds preceded increases in diameter growth, probably because they were the precursors of a dense crown (Harrington and Tappeiner 1990; Maguire 1983; Tappeiner et al. 1987b). Brix (1981) found that 7 yr after a 24-yr-old Douglas-fir stand was thinned, needle mass per tree had increased. Thus, both increased foliage mass and reduced water stress increase the potential for photosynthesis.

The rate of crown expansion can be very rapid after thinning. Chan (personal communication) reports that 38% sky was visible through the canopy of a 37-yr-old Douglas-fir stand 1 yr after thinning to 60 trees/ac, only 18% sky was visible 8 yr after thinning. Reduction in visible sky was the result of crown expansion and an increase in foliage density.

The crown may expand downward, below the crown base, and density may increase within the crown from epicormic branching. When the stems of species such as bigleaf maple, redwood, Sitka spruce, Douglas-fir, and true firs are suddenly exposed by windthrow, severe thinning, or fire, new branches may sprout from buds under the bark. A very narrow new crown of dense, short branches is often produced below the original crown. These branches may persist on Douglas-fir and form clusters distinct from the large primary branches initially produced from the terminal meristem. Epicormic branches that occur throughout the crown are a characteristic of old-growth trees (Waring and Franklin 1979b), where they are probably the result of fires or windthrow that killed surrounding trees.

Thinning increases the growth of trees in the larger size classes as well as increasing average stand diameter (table 8-2). Cochran and Barrett (1993) found that the largest 30 ponderosa pine trees/ac in stands thinned to 147 trees/ac averaged 15.3 in. in diameter, whereas the 30 largest trees in stands with 577 trees/ac averaged only 12.1 in. in diameter.

Height growth of dominant and codominant conifer trees is apparently unaffected by thinning on productive sites (Marshall and Curtis 2002; figure 5-2). However, evidence suggests that in dense stands of shade-intolerant species such as lodgepole pine (Holmes and Tackle 1962) and ponderosa pine (Barrett and Roth 1985), and even Douglas-fir on poor sites (Curtis and Reukema 1970), height growth of the dominant and codominant trees is inversely related to stand density. Thus, thinning stands of these species could increase height growth if they are overstocked or growing on sites of low productivity.

Thinning can reduce height growth, at least temporarily. Thinning decreased height growth up to about 10% for the first 10 yr for Douglas-fir averaging about 30 yr of age (Hann et al. 2003). A similar response has been reported for loblolly pine; Ginn et al. (1991) noted reduced height growth of 8-yr-old loblolly pine following thinning. Possibly this reduction is caused by "thinning shock" and could be related to sudden exposure of trees growing in dense stands (Harrington and Reukema 1983). Severe thinning in pure hardwood stands may decrease height growth, at least in young stands. The crowns of hardwoods tend to expand horizontally, producing wider crowns at the expense of height growth following thinning. Hibbs et al. (1989) reported a 56% decrease in height growth in recently thinned 14-yr-old red alder stands, but 5 yr later found that this trend reversed (Hibbs et al. 1995b). For this reason, when the goal is to produce large, knot-free stems, it may be desirable to keep hardwood stands at a relatively high density until a sufficient length of branch-free stem is produced. Then the stand could be thinned to enable development of large crowns and large-diameter stems on selected trees. However, epicormic sprouting may at least partly reduce the area of knot-free stem.

Thinning of mixed hardwood and conifer stands might increase both the height and diameter growth of relatively shade-tolerant hardwoods such as bigleaf maple and California black oak. Sprouts of these hardwoods generally grow in height faster

Table 8-2. Characteristics of the largest 40 Douglas-fir trees/ac at 55 yr, following different intensities of thinning 15–30 yr earlier (Marshall and Curtis 2002).

Treatment number [total trees/ac at 55yr]	*Diameter (in)*	*Crown ratio (%)*	*Height to diameter ratio*
9 No thinning [307]	19.0	24	82
7 Light thinning [202]	20.8	30	77
5 Moderate thinning [145]	21.5	35	75
3 Moderate-heavy thinning [98]	22.4	35	71
1 Heavy thinning [52]	24.4	45	66

than conifers for several decades. Then their height growth slows while that of conifers continues, until at 30–50 yr the conifers are taller than the hardwoods. Thus, thinning conifers from around hardwoods (a crown thinning) might delay the time of conifer overtopping or enable overtopped hardwoods to resume height growth. Similarly, thinning of the taller Douglas-fir from around Oregon white oak, a shade-intolerant species, stimulates diameter growth, epicormic branching, and acorn production (Devine and Harrington 2004).

Thinning has been shown to stimulate seed production in larger Douglas-fir trees (Reukema 1982). Seed production was greatest following severe thinning and decreased as stand density increased. It is likely that this would be the case for other species, too. Exposure to sunlight along with increases in soil water may stimulate cone and seed production.

Effects of Thinning on the Development of Stands

The potential biomass production of forest stands is determined by the quality of the site (soil, water, climate, etc.) and the tree species present. However, for a given site quality, wood volume production is related to the density of the stand (figure 8-2). Planting density and precommercial and commercial thinning are ways of regulating stand density and thereby volume production.

In general, the denser a stand, the greater the volume of wood it produces, at least in the short run. For example, in an experiment in a young Douglas-fir stand on site class II land (Marshall and Curtis 2002), cubic volume growth ranged from nearly 300 to 500 ft^3/ac/yr as basal area ranged from 136-286 ft^2/ac. In the high-density stands, competition and self-thinning caused volume loss from mortality and decreased standing volume (figure 8-2), and volume yield dropped below its potential or total volume (surviving trees plus mortality). Stands that underwent intense thinning had few trees, so their total volume was less than those at higher densities. In another study in ponderosa pine stands at 45 yr (Oliver 1997; figure 8-2), growth also increased with stand density, but not as dramatically as with

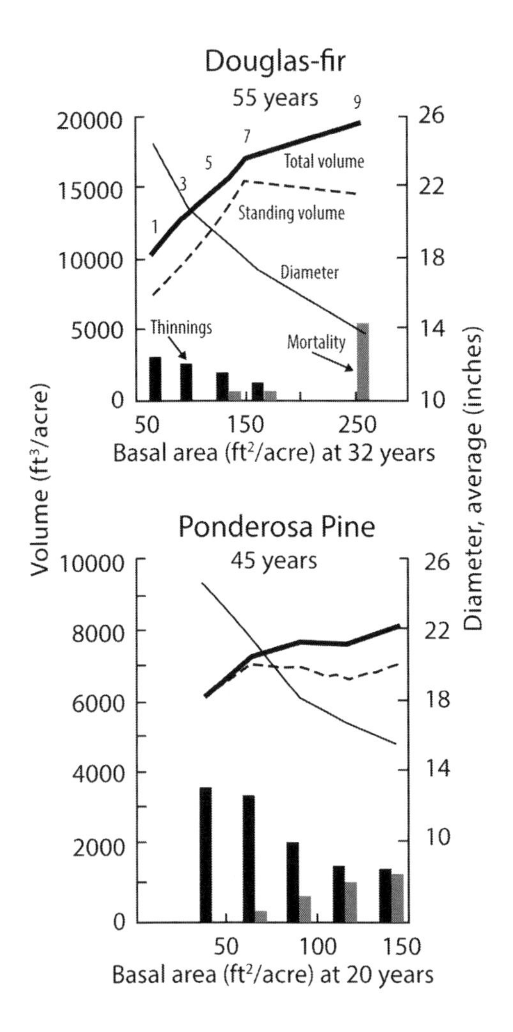

Figure 8-2. Total volume produced (standing plus thinned volume—standing plus mortality for the unthinned Douglas-fir stand), standing volume, thinning yield, mortality and average diameter of ponderosa pine (Oliver 1999). Douglas-fir (Marshall and Curtis 2002) stands at age 55 and 40 yr in relation to basal area after thinning at ages 32 and 20 yr. For Douglas-fir, numbers above the total volume line refer to the stands in Figure 8-6.

Douglas-fir. In both the pine and Douglas-fir stands, mortality was greatest in the unthinned or lightly thinned treatments. In the Douglas-fir stands, mortality was due to self-thinning (Marshall and Curtis 2002), whereas in the pine stands it was due to bark beetles and snow breakage. It is known that high-density ponderosa pine stands are susceptible to mortality from insects (Cochran and Barrett 1993; Oliver 1997), although the timing and extent of mortality are currently unpredictable. These results are corroborated by a thinning study in 45–50-yr-old redwood stands. Oliver et al. (1994) reported no significant difference in net volume growth between stands with 200 ft^2/ac basal area and those with 400 ft^2/ac basal area. However, there was a significant decrease in volume growth in stands with less than 200 ft^2/ac basal area; these were the stands in which severe thinning had removed all but 25% of the previous basal area. As expected, diameter growth decreased as stand density increased across all densities.

When evaluating the effects of thinning on stand growth, it is important to consider not only the severity of thinning but also the units of volume or value used to evaluate stand growth. If a stand is carefully and frequently thinned from below, the trees that would die from suppression, self-thinning, or some other cause would be harvested and the stand density would remain high. Under this type of thinning regime, the cubic volume of gross wood production would be close to the potential for the site. However, if the effects of thinning are evaluated using tree size or board feet, rather than cubic feet, the optimum production would likely occur at a somewhat lower density because the number of board feet per cubic foot increases with the increasing diameter of a tree or log (Avery and Burkhart 2002). Therefore, it may be desirable to reduce stand density and cubic volume growth and grow stands of larger trees to optimize log size and merchantable volume growth.

Thinning to achieve multiple resource values will also have effects on the development of a stand. For example, thinning stands to grow trees with low height–diameter ratios, which are resistant to wind and ice damage, or thinning to create stands that are resistant to insect attack during periods of drought, may require reducing stand density below the level required for maximum wood production. This is not to say that financial, biological, or multi-resource objectives can be met only at the considerable expense of wood volume production. However, all these considerations generally suggest maintaining stands at lower rather than higher densities. It is important to recognize that thinning, whether done primarily for commercial production of timber or for other purposes, generally reduces stand growth somewhat. Growth of large, vigorous trees usually occurs at the expense of overall stand growth.

There are exceptions to this generalization, however. Growth of shade-intolerant species such as lodgepole pine, ponderosa pine, and sometimes Douglas-fir may be curtailed in very dense stands on poor sites. In stands such as those with many trees (>500/ac) of a shade-intolerant species, height growth is so slow that self-thinning cannot occur, and both height and diameter growth essentially stop. In those situations, reducing stand density stimulates both diameter and height growth and, consequently, stand volume growth.

Measures of Stand Density and Thinning

Several measures of stand density can be used in prescribing thinning. Each measure has its uses, depending on the age of the stand, the size of the trees, the purpose of thinning, and personal preference (see chapter 6 for further discussion of measures of stand density).

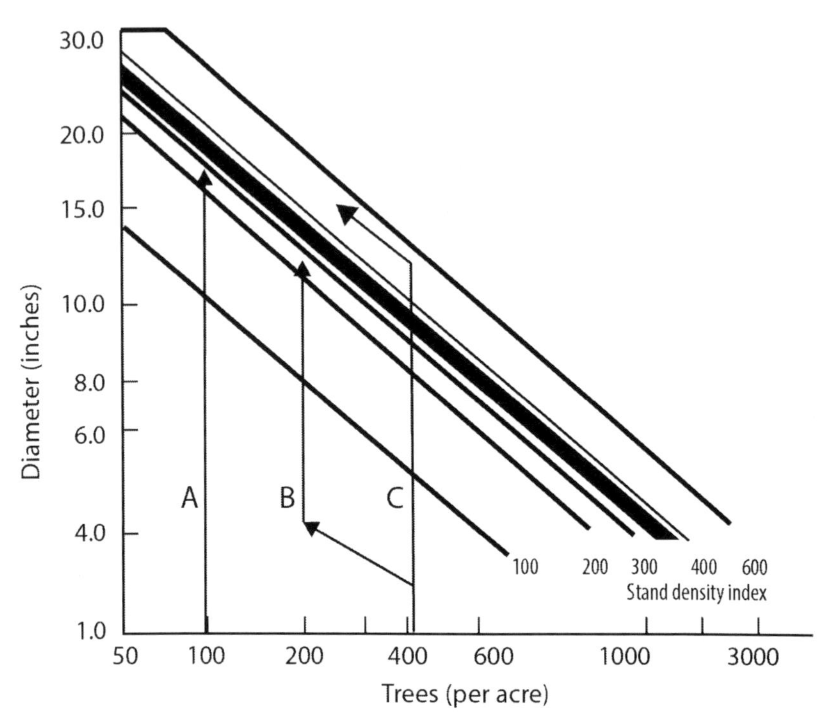

Figure 8-3. Density diagram (Long et al. 1988) showing the effects of stand density in three theoretical young stands. A starts at 100 trees/ac and reaches the zone of self-thinning (black area) at an average diameter of about 18-20 in. B starts at 400 trees per acre, and it is thinned to200 trees/ac and an average diameter 4 in. It reaches the zone of self-thinning at an average diameter of about 14 in. C starts at 500 trees/ac reaches the self-thinning zone at about 9-10 in average diameter

TREES PER ACRE

Trees/ac is a useful measure in young stands where small sizes (<4 in dbh) and high densities make it difficult to measure and interpret basal area, volume, or relative density. Because the trees are small, these stands have very little volume and basal area, so other density measures are too low to be meaningful. Also, when the goal of thinning is to grow stands of a few crop trees (<40–50 ac) or stands with old-growth characteristics, for example (Newton and Cole 1987; Tappeiner et al. 1997), trees/ac may be easier to measure and interpret than SDI or basal area.

BASAL AREA

Stand basal area is related to volume because it is one of the major dimensions (along with height and tree form) of the tree bole. Volume growth can be predicted from equations that use basal area as a dependent variable (Buckman 1962). However, basal area does not account for the number of trees/ac nor for their average size. For example, a stand of 150 ft^2 of basal area might have 150 trees averaging 13.5 in. in diameter, or it might have 250 trees averaging 10 in. in diameter. Clearly the potential of these stands for individual tree growth, self-thinning, susceptibility to insects, and damage from snow, ice, and wind is different, but these differences are not reflected in a basal-area measurement. Another disadvantage of basal area is that full stocking or "normal" basal area (McArdle et al. 1961) for a particular species varies with stand age and site quality.

RELATIVE DENSITY INDICES

Relative density indices are based on the maximum number of trees of a particular size that can be attained for a species (Curtis 1982; Drew and Flewelling 1979; Long et al. 1988; Reineke 1933; see chapter 6). Stand density index (SDI) is expressed as a competitive equivalent number of 10-in trees/ac. For example, an SDI of 500 means that a species can attain a maximum of 500 10-in trees/ac, or the competitive equivalent number of trees of other stem diameters (for example about 150 20-in trees/ac; Reineke 1933). Current estimates of maximum SDI are known for a number of conifer species (Cochran et al. 1994; Long 1985; Reineke 1933). All relative density indices are expressed as proportions of the maximum stand density for a species.

For example, the maximum SDI for Douglas-fir is nearly 600 (Reineke 1933), whereas for red alder it is about 450 (Puettmann et al. 1993b). Shade-tolerant species such as red fir have higher SDI values (1000). When Douglas-fir stands grow to a relative density of about 55% of maximum, mortality of the smaller trees is likely to occur from inter-tree competition (Drew and Flewelling 1979). However, as with basal area, it is important to know the sizes and numbers of trees/ac as well as the SDI. A stand at an SDI of 200 with 66 20-in trees/ac is quite different with regard to commercial value, fire resistance, appearance, etc., than a stand at the same SDI with 200 10 in trees/ac.

Density-management diagrams developed from these indices (figure 8-3) are quite useful in comparing stand development and in planning for density management in young stands. For example, three stands growing at different densities will have quite different tree sizes when reaching the zone in which self-thinning is expected (55% of maximum; Drew and Flewelling 1979; Long et al. 1988). Thus, thinning can be planned to produce trees of different sizes, and the severity of thinning needed to grow them can be estimated. Rates of growth depend on site quality, so the time it takes for the trees of a given stand to grow to a certain size can be estimated from growth models or from observing radial growth on similar sites. These diagrams have the potential for relating density management to a range of outcomes other than tree size and stand density. Susceptibility to insects (Anhold et al. 1996; Oliver 1997), cover for elk (Smith and Long 1987), and amount of understory conifer regeneration (Bailey and Tappeiner 1998) have all been summarized in relation to stand density index on density-management diagrams.

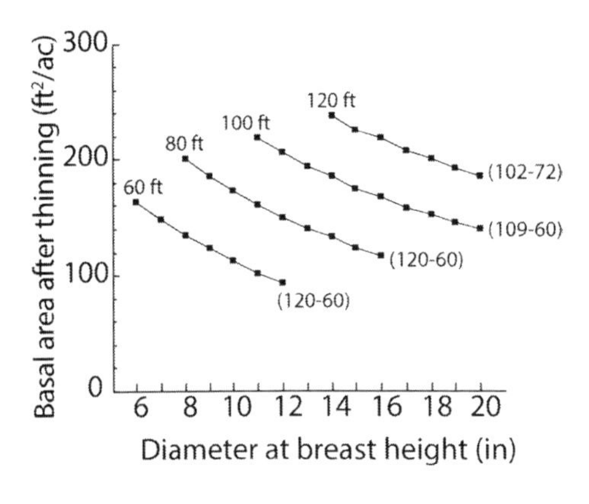

Figure 8-4. Thinning guidelines developed by Brieglib (1952) that take into account past stand development. For each height stands with small diameter, that developed at high density, more basal area is retained after thinning than for stands with larger diameters that developed at lower densities. These guidelines reflect the amount of canopy needed to provide substantial volume growth, and they also account for possible wind or snow damage. For any height trees in low density stands have lower height:diameter ratios (H:D) and are likely to be more stable than are trees in high density stands. Numbers in parentheses, at the bottom of each curve are the H:D ratios for the smallest and largest average stand diameters for each height. Ratios >70 indicate possible susceptibility to wind and snow damage.

STAND HISTORY

None of the density measures described above account for stand history, or the condition of the stand before thinning. Judgment is often needed in applying density measures, especially in stands with variable structures or previously unthinned stands. For example, three stands all starting at different numbers of trees/ac grow to SDI 330 (55% of maximum SDI for Douglas-fir) (figure 8-3). Stand A started at 50 trees/ac and was not thinned; stand B started at about 200 trees/ac and was thinned to 80 trees/ac. Stand C was not thinned until its SDI was >55% of maximum and then more than 150 trees/ac were removed. If all three stands are thinned from below to 30% of maximum SDI, will they grow in the same way? This could be a relatively low-severity thinning for stand A and possibly a moderate-severity thinning for B, but a very severe, possibly risky thinning for C. Would C grow in the same way as A and B because it had reached a higher density? The trees left after thinning would probably be less vigorous and windfirm than those in A or B.

This difficulty also occurs when applying measures of normal or maximum basal area. If normal basal area for a particular species, age, and site is determined to be 300 ft^2/ac, thinning to 50% of maximum might be very conservative for a stand with 170 ft^2/ac. But it might be too risky a thinning for a stand with 250 ft^2/ac, depending on the number of trees and the tree sizes or H:D ratios in each stand.

We know of only one thinning guideline that adjusts density goals depending on current stand attributes or takes into account the history of the stand in evaluating its current condition and potential response to thinning: Briegleb's (1952) methods suggest that the density of Douglas-fir to leave after thinning should depend on the height and diameter of the current stand (figure 8-4). For example, a stand with trees 100 ft tall with an average diameter of 18 in should be thinned to 120 ft ^{2}/ac. According to Briegleb's guides, however, a stand with an average diameter of 15 in at the same height should be left at a higher density of 180 ft^2/ac. The stand with the larger-diameter trees undoubtedly grew at a lower density than the stand with smaller trees. Therefore, the crowns of these trees are bigger, and fewer of them will be needed to provide foliage or leaf area for optimal stand volume growth. A greater number of smaller trees would be needed to provide the same leaf area. We calculated the H:D, (the height–diameter ratio) and found that Briegleb's guides also provide for tree stability. For the stand with large trees H:D was 66, which is considerably lower than the ratio of 81 for the stand with small trees. The stand with small trees is likely to be more susceptible to wind, and so it is reasonable to leave more trees for greater stand stability after thinning. We know of no formal testing of the validity of these guidelines (Briegleb 1952). However, we believe the principles they illustrate regarding past stand history should be considered routinely when thinning.

Current thinning guidelines are soundly based on measures of stand density and principles of stand density, growth, and yield. But they do not account for variables such as soil depth and rooting depth, root and stem pathogens, fire potential, wind, and so on. All of these may affect the outcome of thinning. These variables require thoughtful, qualitative consideration and local experience to estimate their possible effects and suggest ways to minimize undesirable effects.

Tree Spacing

Density measures are often converted to spacing between trees when they are used to select trees to cut or retain during thinning. For example, leaving trees at 16x16-ft spacing will result in 170 trees/ac: 16x16 = 256 ft^2/tree. 43,560 ft^2 per ac/256 ft^2 per tree = 170 trees per ac. If the average diameter of the stand is 15 in and the basal area to be left after thinning is 100 ft^2/ac, then 81 trees/ac are to be left, and average spacing is about 23 ft: a tree 15 in dbh has a basal area of 1.23 ft^2. Thus 100 ft^2/1.23 is the equivalent of about 81 trees/ac, and 43,560/81 = 537 ft^2/tree, $537^{1/2}$ = 23 ft spacing. This calculation assumes a square spacing and that the diameter distribution will be the same after thinning as before. If thinning is to be done from below, then the estimated average diameter of the larger trees left after thinning might be used in the calculation, not the average stand diameter. Larger trees would have more basal area per tree, and fewer trees would be left.

However, very rigorous use of spacing guidelines and strip or geometrical thinning is likely to remove some of the more vigorous or otherwise desirable trees. Dominant and codominant trees almost never occur at uniform spacing, and vigorous trees may occur in groups or at irregular spacing. For example, Marshall et al. (1992) found that when the growth of 80 evenly spaced crop trees/ac was compared with the largest 80 trees/ac regardless of spacing, the cubic volume growth of the 80 largest trees was 20% greater than the growth of the selected 80 crop trees. Spacing is important, however, especially in young uniform stands. In a 15-yr experiment (ages 13 to 28 yr) in a red pine plantation, Stiell (1982) found that trees left in clumps were smaller in diameter and had smaller crowns than more evenly spaced trees. The stands with clumped trees had less basal area and volume than those with spaced trees, even though the trees/ac were the same in both stands. However, young red pine plantations are often quite uniform, with little differentiation into crown classes. Therefore spacing will likely depend on previous stand development and variability in tree sizes or crown classes.

Guidelines for designating trees for thinning are often stated as average spacing between trees to achieve a certain density after thinning. Flexibility in spacing will usually be needed to leave the largest, most vigorous trees/ac, meet guidelines for species composition, and leave trees with other important characteristics. For example, species that occur with relatively low frequency, such as western redcedar in Douglas-fir-western hemlock stands, or sugar pine in Sierra Nevada mixed-conifer forests, do not occur at a uniform spacing. Therefore, it is often necessary to vary spacing guidelines considerably to develop mixed-species stands, especially to favor the growth of shade-tolerant conifers and hardwoods into the overstory.

Thinning Early in Stand Development

There are a considerable number of silvicultural advantages to thinning early in stand development when trees are small. Young stands in this case may include those that have been recently established (10-15 yr old), as well as stands of relatively old (30+ yr) small poles and saplings that have been released from an overstory in a multi-layered stand or have emerged from a dense community of shrubs. These trees are at the beginning of a period of rapid diameter and height growth, and they have the potential to

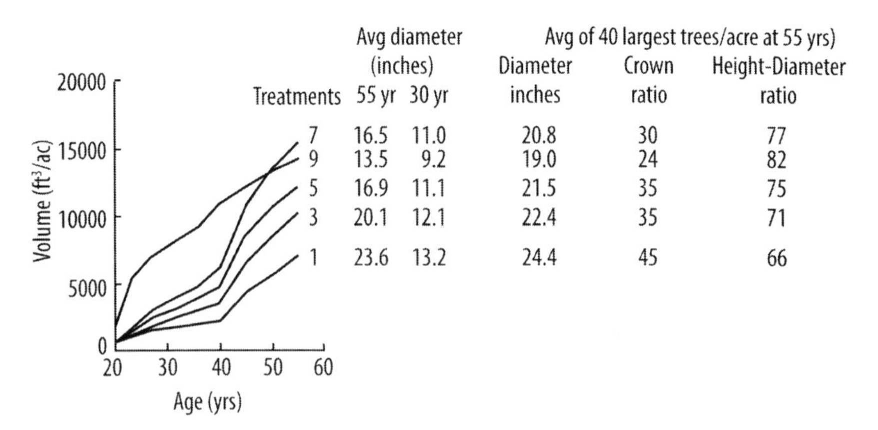

Treatments	Avg diameter (inches) 55 yr	30 yr	Avg of 40 largest trees/acre at 55 yrs) Diameter inches	Crown ratio	Height-Diameter ratio
7	16.5	11.0	20.8	30	77
9	13.5	9.2	19.0	24	82
5	16.9	11.1	21.5	35	75
3	20.1	12.1	22.4	35	71
1	23.6	13.2	24.4	45	66

Figure 8-5. Standing volume from ages 20–55 yr; average diameter of all trees at ages 30 and 55 yr. Diameter, live crown ratio and height/diameter ratio of the 40 largest trees/ac in Douglas-fir stands thinned heavily (1), moderately (3 and 5), and lightly (7) from ages 20–40 yr and unthinned (9). Adapted from Marshall and Curtis (2002).

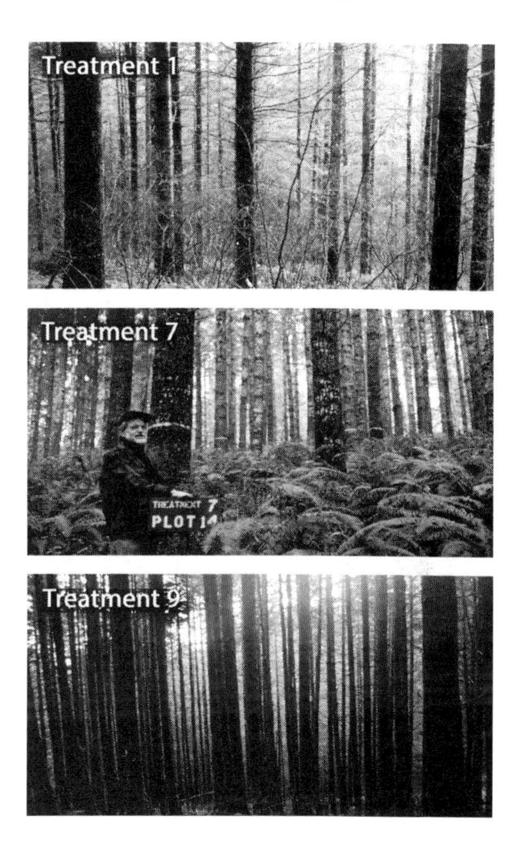

Figure 8-6. Douglas-fir stands 55 yr old with 52 (treatment 1), 202 (treatment 7), and 307 (treatment 9, an unthinned control) trees/ac. Numbers 1 and 7 were thinned to their current densities between 20 and about 35 yr of age. Understory consists of well-developed hazel and oceanspray in #1, sword fern in #7, and mainly herbs and a sparse cover of sword fern in #9. See Figure 8-5 for stand statistics. Adapted from Marshall and Curtis (2002).

develop large, live crowns (crown ratios of 40–50% or more). At this stage of stand development, trees respond rapidly to thinning. If thinning is delayed, trees may be slower to respond because their crowns have receded, and they have to reverse a trend of declining diameter growth, and possibly declining height growth as well.

Density management at planting or at an early age can affect the amount and timing of the yield of wood as well as the structure of the stand. When managing the density of young stands, it is important to set goals for tree size and volume at the first commercial thinning. Density-management diagrams are useful in determining average stand diameter when the stands become dense enough to begin self-thinning (figure 8-3). For example, in figure 8-3, stand A grew at a low density and its average diameter was more than 20 in dbh when it reached the self-thinning zone, whereas stand C had an average diameter of only 12 in at the self-thinning zone. Heavier thinning early in the life of the stand (as with stand B) might result in a more valuable commercial thinning than in stand C because of the larger diameters in B. Stand A grew large trees at the expense of decreased volume production, while C grew large volumes in small trees.

An example of the effects of early thinning of a stand on a productive site (site class II) is presented in figure 8-5, showing the wood volume produced and the tree diameter and crown characteristics. Stand structures for three of the treatments are illustrated in figure 8-6. Thinning in the first 20–35 yr of stand development decreased the volume production over that of the unthinned stand (stand 9); the heavier the thinning, the greater the reduction. However, unless the greater production of wood can be used, there is little advantage to keeping density high. The net and standing volume in the unthinned stand declined, and at age 55 the greatest volumes were in stand 7, the lightly thinned stand. The thinned stands had larger diameters at age 55 (figure 8-6) and at younger ages, too (Marshall and Curtis 2002). Even though there was less volume in the thinned stands (figure 8-5), the volume was in the larger trees, and so the wood yield of that stand might be more valuable. This effect also occurred in Rocky Mountain western larch, where the control had >2000 trees/ac (Schmidt and Seidel 1988). Control stands produced the most volume until about age 25 yr, when self-thinning began, but then greater, and possibly more valuable, volume production occurred at lower densities.

The live crown ratios and the height:diameter ratios (H:D) were affected by the early thinning. Crown size increased and H:D decreased as density decreased (table 8-2). Thus, trees became more vigorous and stable with decreasing density. Furthermore, thinning not only increased the average diameter but also the diameter of the larger trees.

On a less productive site (class IV), Reukema (1979) found that, after 50 yr, stands spaced at 10x10 and 12x12 ft produced more merchantable and total volume than stands at 8x8 ft or less. The stands with the closer spacing had greater mortality and slower height and diameter growth, which reduced both volume and volume growth. Similarly, Graham (1988) reports that Rocky Mountain mixed-conifer stands at 10x10-ft and 7x7-ft spacing produced more standing volume at age 75 than did either more dense stands (at 5x5-ft spacing) or less dense stands (at 14x15- or 20x20-ft spacing). The 5-ft spacing probably had considerable mortality, and the wider spacing did not fully use the site. In other forest ecosystems, pathogens, insects, and fire risks would be increased in stands growing for so long at high densities.

In general, stands of higher initial density have the most rapid rates of volume growth at young ages, but later their growth is surpassed by stands of lower initial density—a process known as the crossover effect (figure 8-5, Oliver and Larson 1996; Peet and Christensen 1987). The time at which crossover may occur varies with density, site productivity, and species.

In summary, thinning early in stand development has many advantages. Crown maintenance and development are enhanced because thinning delays or prevents crown recession and also because trees have a considerable period of height growth remaining during which crown ratios can expand. For example, at 30 yr, Douglas-fir has grown to about 35% of its maximum height and is in a period of rapid height growth (figure 8-7). At 80 yr, height growth is nearly 80% complete and the rate of height

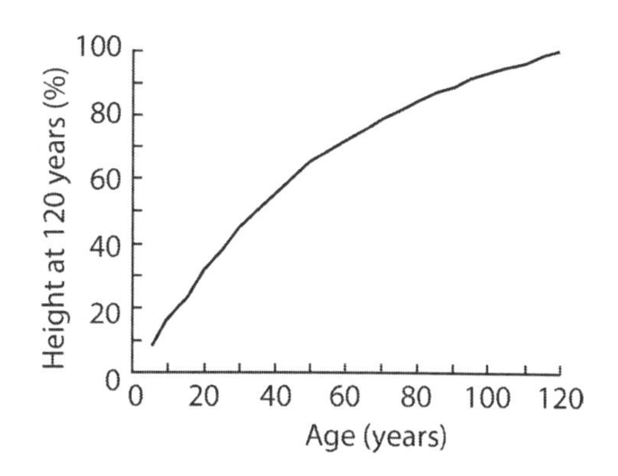

Figure 8-7. Percent of Douglas-fir total height at 120 yr from 10 to 120 yr. At ages 40 and 80 the dominant and codominant trees have completed about 55 and83% of their height growth respectively. Adapted from King's (1966) site index curves for site class II Douglas-fir.

growth is declining. At this age, if crowding has shortened crowns, the potential for developing large crowns still exists but the rate of development will be slower than in younger stands. Similarly, Hibbs et al. (1984) recommend thinning red alder at young ages when height-growth potential is still high. Trees thinned at age 20, when height growth was about 66% complete, had a much slower response to thinning than those thinned at age 12, when height growth was less than 50% complete.

Diameter growth and basal-area growth are rapid at young ages. Large trees can be grown more readily by maintaining rapid growth rates rather than waiting for diameter-growth rates to recover from the effects of high stand density (figures 8-5 and 8-6).

Species composition is often established early in the life of the stand. Species such as western redcedar and white fir, which have relatively slow early-growth rates, must be given space to grow if they are to become part of the main canopy, otherwise they will become suppressed in the understory or die during stem exclusion. Development of trees selected for wood production and wildlife habitat can be managed effectively by thinning at young ages. Even sprouting hardwoods need space if they are to maintain large crowns as conifers overtop them at 40 + yr.

Early thinning that increases tree vigor may enhance insect and disease resistance. In mixed-species stands, early thinning may increase the proportion or dominance as well as the vigor of resistant species.

Shrubs and herbs may be maintained in the understory because early thinning reduces the effects of stem exclusion; stands thinned at the right stage may even skip this stage of stand development altogether. If stem exclusion has occurred, thinning can accelerate the understory reinitiation stage.

There may be some important disadvantages to early thinning as well:

- The cost of thinning—if small trees cannot be sold—is often the major drawback to thinning young stands in some regions. Fluctuations in demand for logs may cause commercial thinning opportunities to become noncommercial within a few months. In times of low demand for small wood, thinning would be counted as a cost and an investment in the future value of the stand. Also, logging costs may be high relative to production in stands of very small trees.

- Snow and ice damage may occur after thinning of young, dense stands, especially when the height:diameter ratio is high. Megahan and Steele (1987) found that stands of ponderosa pine were susceptible to snow and ice damage for about 13 yr if they were thinned when average dbh was 5–6 in. Thinning stands at this stage of development could lead to stand damage and susceptibility to bark beetles (*Ips* spp.).
- Slash and its effects on fire potential and insect populations are major concerns in young stands, especially if the wood cannot be harvested. Green slash generated in precommercial thinning can provide habitat for bark beetles, which invade the slash initially and produce other generations of beetles that invade healthy trees. This problem is particularly acute when green slash is piled against living trees; once slash is dry, it is no longer habitat for bark beetles. Increased potential for a serious infestation may require disposal of slash or careful timing of thinning. Slash can also be a major fire hazard, because it increases fuel loads and makes it harder to control wildfire (see chapter 11). It is difficult to control slash in young stands, which are often too small to withstand the flame heights or fire intensity of burning slash. Mechanical treatment of thinning slash may be expensive and may damage trees.

Staebler (1960) developed a thinning schedule designed to maintain stands at a density that will yield a high volume of wood while growing large trees (table 8-3). This schedule calls for frequent thinning when the trees are young and growing rapidly in height and diameter. Under Staebler's schedule, trees will theoretically grow at 5 rings/in or 2 in in diameter every 5 yr. Also, there is very little mortality with this scheme because trees are thinned before severe competition occurs. Although probably impractical to implement over large areas because it is so intensive, Staebler's schedule illustrates that frequent thinning at young ages, when height growth and crown development are rapid, maintains a constant diameter-growth rate. Note that a constant rate of diameter growth indicates an accelerating rate of stem basal area and volume growth. Intensity of thinning decreases with age as rate of height growth decreases. Like Curtis (1992) and Curtis and Marshall (1993), Staebler's (1960) schedule also suggests that Douglas-fir grow well at older ages and that there is no clearly defined culmination of MAI. Considerations other than stand development are also important in the timing of thinning. Where habitat use, aesthetics, and the cost of thinning are major concerns, severe, less frequent thinning is likely to be more realistic than frequent low-severity thinning.

Thinning Late in Stand Development

Older stands of large trees have a history of development that may affect the type of thinning that is most appropriate for them. As mentioned above, stands that have grown for long periods at high densities with small diameters and short crowns should be thinned with care. Their ability to withstand wind, snow, and ice damage is questionable. Also, the small crowns suggest that the stand's response to thinning might better be delayed until crown size and density increase. Generally, these stands should be thinned lightly and left at high densities (figure 8-4). In old stands, where height-growth potential is low, crown development may be quite slow, and if crown ratios are low (probably less than 25% of total tree height), trees may not grow very much after thinning. Stands with trees that have large-diameter stems and high live-crown ratios (40% or more) probably developed under low densities and consequently are more resistant to the mechanical forces of wind and ice. Stands of these trees can be thinned to low densities (Briegleb 1952).

Old stands and old trees do respond to thinning, however. For example, Oliver (1988) found that diameter growth of 100-yr-old white and red fir tripled when the stand was thinned from 360 to 140 ft^2 basal area/ac. Net volume growth ranged from 321 ft^3/ac in the controls (which were left at 360 ft^2/ac) to 195 ft^3/ac in stands thinned to 140 ft^2/ac. Douglas-fir trees 110 yr old also responded to reduction in density, with trees in heavily thinned stands increasing diameter growth 1.3 times and those in lightly thinned stands 1.08 times before thinning (Williamson 1982). Similar responses to density reduction in old-growth Douglas fir, ponderosa pine, and sugar pine stands

Table 8-3. A schedule for intensive thinning of young Douglas-fir designed to produce large trees with a relatively constant diameter growth rate. Adapted from Staebler (1960) and summarized by decades.

Age (yr)	*dbh (in)*	*Height (ft)*	*Trees (no/ac)*	*Total yield (ft^3/ac)*	*Volume thinned (ft^3/ac)*	*Volume after thinning (ft^3/ac)*	*PAI (ft^3/ac/yr)*	*MAI (ft^3/ac/yr)*
20	8.0	44	212	1,550	200	1,350	256	77
30	12.4	75	101	4,110	1,400	2,500	339	137
40	17.0	100	87	7,500	1,020	4,405	400	188
50	21.0	119	61	11,500	1,210	6,800	200	230
60	25.0	130	56	13,500	1,220	8,400	200	225
70	29.0	144	52	15,500	580	10,000	260	221
80	32.0	154	47	18,100	506	12,000	300	226
90	35.0	164	45	21,000	488	14,000	350	233
110	39.0	177	45	24,500	187	17,000	130	223
120	41.0	181	45	25,800		19,100		215

occurred in southwestern Oregon and northern California. Latham and Tappeiner (2002) found that trees 200–400+ yr old increased basal-area growth after thinning from 10% to more than 100%. The increase in diameter or basal area growth did not always occur immediately; often ring width did not increase for 10+ yr. However the increased growth lasted for 20–30+ yr after thinning. Similar results were reported for 70–150-yr Douglas-fir stands (McDowell et al. 2003; Williamson and Price 1971), where thinning increased stem growth and reduced mortality from bark beetles. Newton and Cole (1987) have proposed a regime of thinning in old stands to promote the development of old-growth characteristics.

Height/Diameter Ratios: Maintaining Tree Stability

Trees with large crowns tend to have larger diameters in relation to their height. Greater diameter growth relative to height growth results in tapered stems and therefore trees that are better able to resist breakage from wind or heavy, wet snow. Height:diameter (H:D ratios, see chapter 5) are used as a measure of the ability of a tree to resist damage from these factors. Ratios below 70–80 generally indicate stability (Coutts and Grace 1995; Wilson and Oliver 2000). However, trees with such ratios growing on shallow soils or windy sites or with diseased roots or stems may not be strong enough to resist wind, snow, or ice damage. Knowing H:D ratios before thinning and leaving trees with low ratios may help predict potential damage to the stand after thinning and thereby prevent it. Wonn and O'Hara (2001) found that 80–100% of the lodgepole and ponderosa pine they studied suffered damage from snow, ice, or wind if their H:D exceeded 80. Only 11–45% of the undamaged trees had an H:D > 80 (see also Lanner 1985).

Because height growth is relatively unaffected by thinning, the main effect of thinning on H:D is to increase diameter relative to the height. However, because height growth is generally most rapid in the early years of tree or stand growth, thinning during periods of rapid height growth may not lower (improve) H:D. Nonetheless, it may stabilize the ratio or keep it from increasing (Wilson and Oliver 2000). The greatest improvement in the H:D ratio occurs through moderately severe thinning in young stands to enable diameter growth to keep pace with height growth (Wilson and Oliver 2000). Space for crown development is needed during periods of rapid height growth for the diameter-growth rate to be sufficient to maintain H:D ratios below about 70. Early, moderately severe thinning in Douglas-fir resulted in the 40 largest trees per acre having an H:D of <70 at age 55. This result was for the most intense thinning treatment in the trial; for the other treatments, the H:D of the 40 largest trees was over 70 at age 55 (Marshall and Curtis 2002). Thus many trees in young stands have high H:D ratios.

Old-growth Douglas-fir and ponderosa pine trees with H:D of <40 have remained standing as isolated trees for many years (Latham and Tappeiner 2002; Poage and Tappeiner 2002; Sensenig et al. 2013,). However, dense, even-aged stands of these same species often develop small live crowns and high H:D ratios (figure 8-4). Height:diameter ratios are based only on stem characteristics; however, large-stemmed trees are also likely have large roots at the base of the stem, which would aid stability. These ratios do not take into account variables such as rooting depth, stem and root diseases, or fluctuating water tables that damage root systems, all of which could decrease stability.

For lodgepole pine, H:D as an index of tree stability may be intimately linked to stand productivity in two ways. First, trees with high H:D values experience sway during strong winds that can result in crown abrasion among trees, loss of leaf area, and ultimately reduced stand growth (Rudnicki et al. 2003, 2004). Second, intense swaying of trees with high H:D values reduces their xylem permeability because of associated narrow growth rings and damage to the vascular tissue (Liu et al. 2003). Reductions in the permeability of vascular tissue limit conductance of the water needed to support gas exchange and photosynthesis. For these same reasons (reduced crown size, leaf area, and permeability of vascular tissue), trees with high H:D values are slow to respond to release from thinning.

Thinning Shock

The symptoms of thinning shock are that trees appear less vigorous, grow poorly, or blow over following thinning. Loss of vigor or poor growth may be either temporary or long-lasting effects. Harrington and Reukema (1983) report that the immediate effects of thinning a dense 27-yr-old Douglas-fir stand on a low-productivity site were detrimental. Trees suffered sunscald, foliage became chlorotic, and snow and ice broke tops and branches and caused some trees to lean. Damage was generally greatest on plots that were thinned most heavily.

Although height growth was reduced for the first 10 yr, it recovered 25 yr after thinning and was greatest in the heavily thinned areas (28x28-ft spacing). Basal area and growth per tree were also greatest at the widest spacing, as expected (Harrington and Reukema 1983). This stand was very dense before thinning, having grown at about 8x8-ft spacing. A less dense stand might not have been affected adversely. However, the stand was young, and the trees could increase height growth (and therefore crown size) and recover from the initial thinning effects. Harrington and Reukema (1983) concluded that the thinning was beneficial despite the initial shock.

Dense stands of older trees that receive intense thinning might not recover as rapidly. Sun scald of the stem and exposure of shade needles to temperature extremes may have contributed to thinning shock. However, DeBell and others (2002) found only minor shock (a 1-yr reduction in height growth) after thinning 10-yr-old Douglas-fir thinned from 8.5x8.5 ft to about 15x15 ft. They attributed the lack of shock to the local seed source and the fact that the stand had not grown for very long at high density.

Effects of Thinning on Wood Quality

Density management affects the value of wood by increasing tree size. High volumes of small trees (figure 8-6) may be inefficient to harvest and process, especially if the desired yield is dimension lumber. There is currently a trend away from paying premium prices for very large logs, and trees that meet size requirements for commercial logs can be grown with little sacrifice in volume growth. For example, in the study discussed earlier (figure 8-2), the light thinning (treatment #7) produced a good balance of high volume and large enough trees, making it the most economically efficient treatment even though treatments 1, 3, and 5 produced larger trees, and treatment 9 produced greater volume up to age 40 (Tappeiner et al. 1982). Severe thinning in young stands might increase tree taper as diameter growth increases relative to height growth, decreasing the H:D. Increase in stem taper could reduce the yield of poles and pilings, which are high-value products. Standards for these products are quite exacting with regard to length, diameter, straightness, and taper.

Another major effect of density management is the effect on limb size and the size and kinds of knots that occur in the major part of the stem (Grah 1961). Western conifers retain their branches at high stand densities; that is, they do not readily self-prune. Dead branches below the crown produce loose knots. Because the cambium in the branch is dead, the cambium in the stem of the tree grows around the dead branch. Live branches, in contrast, are fused with the xylem in the stem, and thus knots from live limbs are tight knots in sawn lumber. Large knots, especially loose ones, reduce the strength and overall quality of wood (Megraw 1986). They are also undesirable in pulp and remanufactured wood products.

The effect of branch or knot size on wood quality depends on the standards used to evaluate it. For example, DeBell et al. (1994a) studied the effect of early spacing on branch size in western hemlock stands 29 and 38 yr old. Stands were thinned at 11 and 12 yr from 8x8- to 22x22-ft spacing. Average branch size in the bottom 16 ft of the stem ranged from about 0.5 to 1.2 in. in diameter, and all branches were dead. By current log-grading standards, knot size had no effect on log grade, but knots more than 2.5 in. in diameter would result in a lower grade. Thus, managers could vary the spacing or density within the range of 8x8 to 22x22 ft to achieve high yields of wood, large log size, or other desired values, while keeping knots to a size that did not lower the value of the wood.

Pruning trees will eliminate branches and knots in the pruned part of the stem. Stand density and pruning are considered together. Pruning trees in dense stands will likely result in little knot-free wood, because diameter growth may be too slow for much wood to cover the pruned stem. Pruning for wood quality should probably follow a thinning and be limited to dominant or codominant trees (crop trees) after evaluation of rates of diameter growth, yields of knot-free wood, and the cost of thinning (Staebler

1963). Removal of a large proportion of the crown may cause a reduction in height and diameter growth. Stein (1955) found that removing 25% of the live crown in a 28-yr-old Douglas-fir stand had little effect on height and diameter growth for 13 yr. Removing 50% of the live crown reduced diameter growth about 0.04 in/yr and height growth 0.35 ft/yr (a 10% reduction in height growth). Removal of 75% of the crown reduced diameter growth 0.08 in and height growth 0.5 ft/yr (a 34% reduction in height growth). Therefore, in stands managed intensively for high-quality wood production, pruning might be done in stages, or "lifts," with sufficient time between the prunings that the crown could expand as trees grew taller and be maintained at 40–50% of total tree height. If pruning is managed in this way, the pruned trees will not lose their position in the canopy or their ability to grow in height and diameter. Also, the height of pruning should be conducted according to specific log lengths—half (16 ft) or full (32 ft).

Natural pruning may occur in mixed-species even-aged or two-storied stands. Thinning that encourages the growth of shade-tolerant understory trees may result in pruning of intolerant dominant and codominant trees. Shade-tolerant trees with dense foliage may grow into the lower crown of shade-intolerant trees and cause their lower branches to die. As mentioned above, intense thinning will keep lower branches growing and may result in epicormic branching in species such as coast redwood, Sitka spruce, white fir, and Douglas-fir.

Cahill et al. (1986) recommend pruning when trees are small and growing rapidly to keep the knotty core small and allow rapid growth of knot-free layers. They report an increased value of 10% for sawlogs and 20% for veneer logs 34 yr after pruning a 38-yr-old Douglas-fir stand. The knot-free radii range from about 1.8 in in 10-in logs to 4.5 in in logs more than 20 in at the small end.

Pruning should avoid scarring the bole, and care should be taken to prevent bark from being stripped down the bole as the pruned branch falls. Felling trees in older stands may prune some dead branches from the lower boles of the remaining trees. This incidental pruning is not thorough and often occurs only on part of the boles; nevertheless, some increase in log quality may result.

Because density management affects the diameter growth of forest trees, it may consequently affect wood density and strength. The major effect is on the core of wood produced from the pith up to 15–20 rings outward (Maguire et al. 1991). This core of "within-crown" or "juvenile" wood increases as stand density decreases. Juvenile wood has low specific gravity and other properties that are undesirable from a structural standpoint. The increased volume of juvenile wood must be evaluated in light of the increased volume of denser, older wood that will be produced outside the juvenile wood by the rapidly growing trees. Briggs and Smith (1986) caution that, although ring width has been used to screen juvenile wood from mature wood, there is not in fact a strong relationship between them. The wood's age and position in the tree also need to be considered. Trees growing rapidly at low densities will produce more juvenile wood than will slower-growing trees in dense stands, but they will also produce larger volumes of mature wood.

The effect of thinning on mature wood probably varies among species and sites. Megraw (1986) reports that the proportion of late wood produced in a xylem ring of Douglas-fir varies more from year to year than does early-wood production, and there is high variation among trees. He also suggests that trees in dense stands may produce a smaller amount of late wood than trees in less dense stands, because late-season competition for soil moisture slows xylem growth. Likewise, Smith (1980) reports that there is less late wood in trees growing at wider spacing in 20-yr-old stands of Douglas-fir, western hemlock, and western redcedar. However, he concludes that the larger volume of wood per tree offsets the lower density of the wood and that a reduction in wood quality from lower wood density is unlikely.

In evaluating Douglas-fir wood for density and strength, it appears that the overriding factor is whether it is juvenile or mature wood. Thinning that increases the production of mature wood will probably have a positive effect because there is not likely to be a major decrease in wood density with an increase in ring width. Treatments that encourage a large core of juvenile wood, however, will increase the proportion of low-density wood. DeBell et al. (1995b) found that, for a specific number of rings from the pith in western hemlock trees at a height of 4.5 ft, wood density decreased with an increase in growth rate because the proportion of early wood was greatest in the wider rings. But, depending upon the use of the wood, the lower wood density may not cause a commercially important decrease in wood quality. Reader and Kurmes (1996) found that thinning initially increased the amount of compression wood in ponderosa pine but that those amounts decreased after about 10 yr. Thinning severity had no effect on the amount of compression wood, which was greatest in the unthinned control (Reader and Kurmes 1996). The decrease in compression wood might be attributable to greater diameter growth and larger stems following thinning (See chapter 5 for discussion of reaction wood). Fluting and form of the lower stem of western redcedar and western hemlock are affected by stand density. Flute numbers and depth increased with spacing in western redcedar; however, the majority of the trees had very few flutes even at the widest spacing, about 16x16 ft (DeBell and Gartner 1997).

Effects of Density Management on Forest Stands as Ecosystems

Tree removal or density management of even-age and uneven-age stands affects a number of forest ecosystem attributes. Stands may appear quite open immediately after thinning, but this usually changes as tree crowns expand and become denser, and understory vegetation develops. The full effects of thinning may not be evident for 10–20 yr or more.

EFFECTS ON MICROCLIMATE

Thinning affects several microclimate variables simultaneously (Riegel et al. 1995). Both incoming and outgoing radiation increase as canopy density decreases; therefore, temperature extremes increase. Temperature extremes appear to be maximized when the ratio of canopy gap diameter (D) to height (H) of the surrounding canopy (D/H) is about 1.5 (Smith et al. 1997). Light intensity in gaps reaches a maximum when D/H is 2 or greater (Runkle 1985). Less dense canopies intercept less precipitation, and so less of the precipitation will evaporate from the canopy as throughfall increases. Thus snow accumulation on the forest floor increases as canopy density decreases. Sunlight reaching below the overstory canopy and above the understory will decrease as crowns close, but will remain higher than before thinning.

Changes in microclimate after thinning are most predictable at and below the main tree canopy. However, effects at the forest floor are often influenced by the response of understory vegetation to thinning. Increased light is accompanied by an increase in soil water (Donner and Running 1986), at least for a short time. Therefore, the understory is likely to grow and develop.

The increase in water is due to decreased transpiration of overstory trees and the increased amount of snow and rain reaching the forest floor (Teklehaimanot et al. 1991). For the same reason, thinning may also affect the amount and timing of water yield from snow (Troendle and King 1987). The rate and timing of snowmelt depend on canopy density (Golding and Swanson 1978; McNay et al. 1988; Meiman 1987). However, the rate of melting and evaporation of the snow that accumulates under thinned stands will be considerably less than it will be in the open. As a result, when practiced on large areas, thinning may affect not only the amount of snow that accumulates and its melt rate, but also the volume and timing of stream flow.

Wind speed in the stand increases as canopy and stem densities decrease (Green et al. 1995). Wind flow or turbulence is also affected by thinning. The canopy opening and gaps that occur from thinning result in an increase in canopy roughness. This may cause the wind to eddy in the canopy and increase the stress or mechanical forces on tree stems. As wind velocity increases in the stand, laminar airflow may become turbulent as it travels across a canopy that is irregular following thinning of codominant or large trees. Strong turbulent wind may break stems and uproot trees. Also, the trees remaining after thinning have to bear a greater snow load because adjoining trees have been removed (Megahan and Steele 1987). Damage from snow following thinning and fertilization in Sweden amounted to 9–20% of the basal area increment; the greatest damage occurred on plots that were thinned from above or unthinned (Valinger and Petterson 1996).

EFFECTS ON UNDERSTORY TREE AND SHRUB DEVELOPMENT

In uneven-aged stands, thinning is often the principal way to regulate the species composition and vigor of the smaller trees in the stand. In Sierra Nevada mixed-conifer stands, regeneration of white fir and incense-cedar—shade-tolerant species—accounts for 65% of the saplings and smaller trees; Douglas-fir accounts for 17%, ponderosa pine 10%, and sugar pine 8% (Lilieholm et al. 1990). Thinning is thus a way to encourage seedling establishment and the growth of the less shade-tolerant pines and Douglas-fir, and to reduce the proportion of white fir and cedar where needed. Without attention to the species composition and growth of smaller trees, mixed-conifer stands could become nearly pure white fir stands.

Redwood is unique because it stump sprouts after thinning. Oliver et al. (1994) report that 69% of the stumps in thinned stands sprouted, and that sprouting was not related to stand density, although sprout-growth was greater at lower density. Their predictions of height growth 15 yr after thinning ranged from 31 ft in plots with 75% of basal area removed to 11 ft in plots with 25% of the basal area removed.

Thinning often creates an environment for the establishment and growth of understory plants (Riegel et al. 1995, see chapter 4). After thinning, new trees and shrubs are established in the understory, and seedlings and saplings already established are released. Bailey and Tappeiner (1998) found that seedlings and saplings were common in the understory of Douglas-fir stands following commercial thinning (figure 8-8). The highest seedling frequency and density occurred on sites with the lowest SDI after thinning (figure 8-8), probably because of increased seed production (Reukema 1982) and exposure of some mineral soil during tree removal.

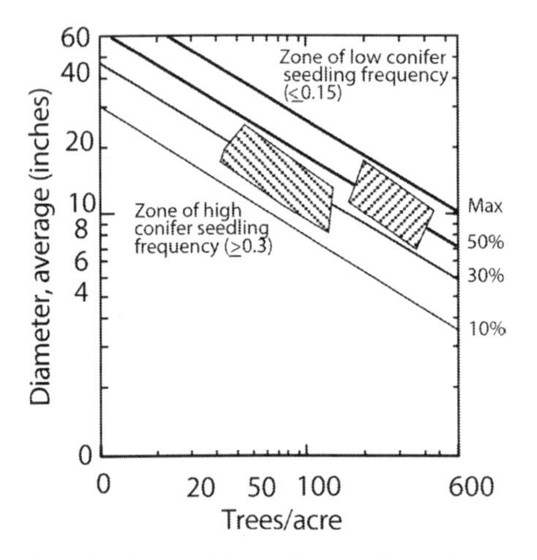

Figure 8-8. Zones of high and low conifer seedling frequency following thinning in relation to stand density index (Bailey and Tappeiner 1998) Seedling are established in the zone of low numbers of trees and large diameter, probably because of increased light, soil disturbance from logging, and because of increased seed production following thinning.

Thinning also stimulates the growth of shade-tolerant species established in the stand before thinning, such as western hemlock (Alaback and Herman 1988; Alaback and Tappeiner 1991), bigleaf maple (Fried et al. 1988), white fir, and tanoak (Tappeiner and McDonald 1984) and other hardwoods (McDonald and Tappeiner 2002). Seedling growth may be restricted as the canopy closes, and additional thinning, carefully done to reduce damage to the seedlings, encourages the development of multi-storied stands.

Shrubs respond to increased light and soil water by establishing new seedlings and growing their below-ground and above-ground parts (see chapter 4). Light levels in dense, unthinned stands are often insufficient to enable the germination and survival of shrub and tree seedlings; rates of emergence, survival, and growth are higher following thinning (Huffman et al. 1994; Tappeiner and Alaback 1989; Tappeiner and Zasada 1993). The disturbance of the organic layers in the forest floor that occurs during thinning stimulates germination and aids seedling survival of seed bank species such as ceanothus and manzanita.

Shade-tolerant clonal shrubs frequently persist as clone fragments in the understory of forest stands for many years and then respond to reduction of overstory density. For example, Huffman et al. (1994) report clonal fragments of salal occupying an area of up to 20 in in diameter in unthinned stands. Rhizome growth of these fragments more than doubled in thinned stands, and clone fragments occupied areas >30 ft in diameter.

When clones expand and seedlings become established and grow, they form dense patches of shrubs. Huffman et al. (1994) report that salal biomass in thinned stands (about 15,000+ lbs/ac) was nearly double the biomass in unthinned stands (about 8,000 lbs/ac). In addition, the greater values of total rhizome length in (13 ft/ft^2 of land surface area) stands than in unthinned stands (5–7 ft/ft^2) suggest that thinning also favored underground development and spread of a bud bank (below-ground buds on rhizomes that can give rise to new aerial stems and rhizomes). The cover of salal and bracken fern in thinned stands was nearly double that in adjoining unthinned stands (Bailey and Tappeiner 1998). Reducing basal area from 154 to 30 ft^2/ac increased browse production of grasses and forbs but not of shrubs in 65-yr-old lodgepole pine stands (Crouch 1986).

In ponderosa pine stands, understory shrubs have been shown to reduce volume production of the pine for 35 yr, especially in stands grown at wider spacing (>13x13 ft; Cochran and Barrett 1999). However, during this time the presence of understory has added organic matter to the soil and potentially increased the productivity of the site (Busse et al. 1996, See case 5 chapter 12).

Thinning may initially increase shrub biomass and cover, but without additional thinning, they may subsequently decrease (Alaback and Hermann 1988; Messier et al. 1989). Long and Turner (1975), working in stands of different ages and density, found that total shrub biomass decreased from about 7,000 lbs/ac to 250 lbs/ac as stand density increased, age increased from 22 to 73 yr, and basal area increased from about 70 to 200 ft^2/ac. Thinning would likely have maintained shrub biomass at a higher level. The cover and leaf area index of shrubs and the density of shrub stems in western Oregon forests were all greater in thinned stands than in unthinned stands, and they were very close to the levels found in old growth (Bailey and Tappeiner 1998; Schoonmaker and McKee 1988).

Thinning can have effects not only on understory density, but also on species composition; other than a general increase in cover, however, the responses are neither predictable nor well understood (Bailey et al. 1998, Halpern et al. 1999, Sean et al 1999). Immediately after thinning, there is the potential for the establishment of a variety of understory plants. However, on many sites—and within a few years on productive sites—the cover and leaf area of understory shrubs increase markedly, and the possibility for establishment of new plants diminishes until there is a disturbance, such as fire or logging, or a silviculture treatment that reduces understory density. On dry sites with more open overstory and understory canopies, ongoing recruitment of trees and shrubs is more likely. In a study throughout western Oregon, Bailey et al. (1998) found that the number of native plant species was greater in thinned than in unthinned and old-growth stands; however, their abundance was greatest in the old stands. There were increases in certain species groups following thinning, such as vine-like species, nitrogen fixers, grasses, and exotics. However, species composition was more related to region than to the effects of thinning. Similarly, Deal (2001) found that partial cutting in hemlock-spruce forests in southeast Alaska (Deal and Tappeiner 2002) did not affect the understory plant community structure, provided that no more than 50% of the basal area was removed. Also, this level of thinning did not change the abundance of important deer-forage species. A heavier thinning would likely have enabled the development of a dense, shrubby understory that would have shaded out the low shrubs and herbs that are preferred deer forage. In these cases, there were populations of native plants within or adjoining the thinned stands. Where there are very dense stands over large areas, the potential for establishment of native plants by thinning may be decreased (Halpern et al. 1999).

Thinning has the potential to increase exotic species. Although the cover of non-native species in thinned stands was greater than in unthinned stands (Bailey et al. 1998), it was still quite low (<0.1% cover). On sites where exotic species such as Himalayan blackberry, false brome, and Scotch broom are present in the understory, the cover of these species is likely to increase after thinning. Seed of non-native species may enter thinned stands either through wind dispersal or from machinery and foot traffic. Seedling establishment frequently occurs along roads and trails.

EFFECTS ON WILDLIFE

The development of herbs, shrubs, and tree communities and the growth of shrub and tree crowns after thinning provide cover and food for a variety of wildlife species. Bird populations in particular appear to respond favorably to thinning (Hayes et al. 1997).

The development of shrubs and trees provides cover for nesting and a substrate for insects that are prey for insectivorous birds such as the Swainson's thrush (Hayes et al. 1997). Species that feed and nest in the canopy, such as flycatchers and bats, are favored by large crowns with spaces among them, which enable these species to forage for insects.

Thinning may also increase the abundance of herbs and shrubs available for browse. All shrubs are not equally palatable. Some—salal, Oregon grape, and tanoak, for example—contain tannins and other defensive chemicals that make them less palatable to browsing herbivores. Others—seedlings and sprouts of ceanothus, bitterbrush, trailing blackberry, and bigleaf and vine maple—are preferred browse species. Shrubs growing in the understory of forest stands contain fewer defensive chemicals, higher nutrient concentrations, and lesser amounts of lignin and structural carbohydrates than those growing in the open. Thus, understory shrubs in general may be more palatable browse than those growing in the open (Happe et al. 1990; Lange 1998). The abundance of palatable grasses and forbs is usually greater in open sites, however. By stimulating the sprouting of existing woody plants, thinning can increase the amount of palatable browse (via growth of succulent, nutrient-rich vegetation) and reduce the height of browse, making it more available to ungulates and other herbivores.

In ponderosa pine forests, the collective yield of grasses, sedges, and forbs increased from 117 lbs/ac in dense 36-yr-old stands to 550 lbs/ac in stands thinned to 125 trees/ac (Sassaman et al. 1977). In addition, preferred browse species that were not in the stand before thinning began to become established after thinning. In addition to increased understory density, fruit production by shrubs also increases. Unthinned stands had well-developed salal, Oregon grape, ocean spray, and red huckleberry in the understory that produced little seed or fruit. There was a dramatic increase in fruit production in adjoining thinned stands, however (Wender et al. 2004).

Effects of thinning vary with forest type. Precommercial thinning increased the herbaceous community and thus deer browsing in southeastern Alaska. However, dense western hemlock regeneration can become established after thinning in these forests, and so the increase in forage may be short term, as the new hemlock forms a dense second story and the herbaceous plants are shaded out (Doer and Sandberg 1986, Alaback 1982; Alaback and Tappeiner 1991). Lower-severity thinning or pruning treatments may enable establishment of preferred browse species without stimulating the development of a new cohort of conifers. Precommercial thinning to favor Sitka spruce in southeastern Alaska forests appeared to increase porcupine damage to the spruce. Porcupines damaged 52% of the spruce and 26% of the western hemlock, possibly shifting future species dominance back to western hemlock (Eglitis and Hennon 1997).

To the extent that thinning reduces mortality from self-thinning, it will reduce habitat for wildlife species that forage on insects that live in small dead trees. In the long term, however, large trees in thinned stands are likely to become large, long-lasting snags and logs that can be used by foraging and cavity-nesting wildlife.

Young stands growing at low density maintain long, wide crowns because the lower branches remain alive. These crowns are large and close enough to the ground to provide hiding cover for elk and other large mammals (Smith and Long 1987; figure 8-9). As the crowns recede with increasing stand density, the quality of hiding cover may decrease. Young stands with closed canopies may provide better thermal cover and reduce temperature extremes.

EFFECTS ON INSECTS AND PATHOGENS

Insects. Thinning can either increase or decrease the relative susceptibility of a stand to insect damage. Likewise, the stand structures and species composition produced by thinning can have implications for insect population dynamics. Species composition may be the most important single factor in susceptibility; for example, if mixed-species forests contain a major proportion of true fir or Douglas-fir, they are at greater risk of major damage from tussock moth and budworm. For other insects, the vigor of individual trees, as controlled by stand density, may also be important (Safranyik et al. 1998).

There is considerable evidence that less dense stands of white fir and lodgepole and ponderosa pine are less likely to experience mortality from bark beetles than higher-density stands (Cochran 1998b; Sartwell and Stevens 1975). For example, Oliver (1997) found that stands of ponderosa pine that were thinned periodically to low SDI (71–229) and trees/ac (21–151) from ages 20 to 45 yr lost <7 trees/ac and 320 ft^3/ac volume to bark beetles and snow damage during this period. Stands growing at 108–507 trees/ac (SDI 227–348) lost 180 trees/ac and 2270 ft^3/ac. This general relationship was found in several other studies as well: Oliver (1997) and Cochran and Barrett (1993). Fiddler et al. (1989) report an 86–100% reduction in tree mortality from bark beetles and root diseases for 8 yr following thinning of a 70-yr-old ponderosa pine stand.

The mechanism by which stand density is related to damage from mountain pine beetles is still under debate. Two theories seem plausible, and they are not necessarily mutually exclusive. One proposes reducing the number of trees that will attract beetles to the stand—that is, trees of poor vigor and trees susceptible to damage by wind and snow (Berryman 1982). Thus, to make a stand less susceptible to insects, thinning should remove the less vigorous trees in the smaller-crown classes that may be breeding habitat not only for mountain pine beetles but also for other bark beetles (*Dendroctonus* spp.) that emerge to attack other trees. Similarly large, old trees may provide bark beetle habitat (Keen 1936; Wickman and Eaton 1962).

The other theory proposes that lower stand densities result in increased growth and vigor in individual trees. Thinning provides more soil water as a result of less interception of precipitation as well as reduced transpiration (Donner and Running 1986). This increases tree vigor (water potential and resin pressure in the xylem), making trees more resistant to attack, especially during periods of drought (Mitchell et al. 1983). Trees growing at low density are often vigorous enough to "pitch out" the beetles they attract. Vigorous trees are presumably better able to produce carbohydrates and thus synthesize protective secondary compounds such as resins, which form pitch tubes that trap the bark beetles. Pitch prevents insects from successfully boring into the phloem and depositing their eggs. The secondary compounds associated with pitch may also have insecticidal properties. In years of drought or when insect populations are high, even large, normally vigorous trees can be susceptible to invasion by beetles.

Mitchell et al. (1983) present evidence for the effects of tree vigor. They conclude that if stand density is kept sufficiently low, vigor (growth per unit of leaf area) remains above a threshold value (100 g annual stem growth/m^2 leaf area), and mortality from beetle attacks will be negligible. Waring and Pitman (1980) found that the intensity of insect attacks required to kill a tree became greater with increasing growth efficiency, and that no mortality was observed in trees with growth efficiencies greater than >100g annual stem growth/m^2 leaf area/yr. At the other extreme of growth efficiency, trees producing <20 g m^{-2} leaf area/yr failed to exude pitch after beetle attacks. Waring and Pitman concluded that managing under low stand densities has strong potential for reducing susceptibility to mountain pine beetle during epidemics, and that stand-density manipulation is generally preferable to fertilization due to the relatively short duration of fertilization responses (3–4 yr).

Although there is general agreement about the benefits of maintaining low stand densities where mountain pine beetle epidemics are a potential problem, alternatives to the growth-vigor mechanism have been proposed. Amman et al. (1988) propose that thinned stands are resistant to mountain pine beetle because temperatures in these stands rise above the level for successful insect reproduction. Larger trees are often killed preferentially during beetle attacks. In thinned stands of lodgepole pine 76–102 yr old, the mean dbh of trees killed by mountain pine beetle was significantly greater than the mean dbh of survivors, whereas growth efficiency was not significantly different. Growth efficiency of most trees in that stand was generally quite low, rarely exceeding the 100g× m^{-2} leaf area/yr threshold; however, Amman et al. (1988) noted that three of the seven trees with growth efficiencies exceeding this value were killed. It was also significant that, despite the relatively low growth efficiencies of even the thinned stands, mortality from mountain pine beetle attack was much greater in unthinned control stands (73–94%) than in the thinned stands (4–39%). Stuart (1984) similarly found growth efficiency to be a poor predictor of risk to mountain pine beetle infestation in stands of south central Oregon lodgepole pine; however, the growth-efficiency index was developed to differentiate trees at risk within a stand, not for entire stands.

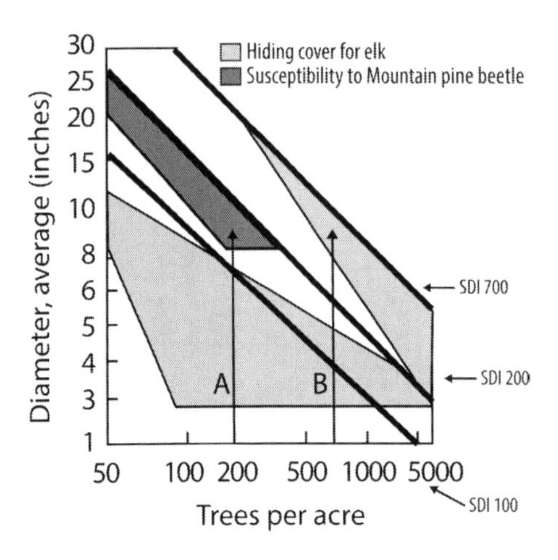

Figure 8-9. Density diagram for lodgepole pine indicating densities likely to provide hiding cover for elk and at which stands are susceptible to mountain beetles. Stand A growing at 200 trees/ac provides hiding cover from its crowns and stems while it is 3–8 in in diameter. Its crowns recede when it reaches about 8 in, and it no longer provides hiding cover. A becomes susceptible to beetles at about 9 in when it has stems large enough to provide beetle habitat. Stand B, growing at 700 trees/ac, provides hiding cover from its crowns and high stem density until it is about 5 in diameter, and its crowns recede, and again at 9 in+ as stem size increases. B is not likely to soon become susceptible to mountain pine beetles because it has too few large trees. Adapted from Smith and Long (1987) and Anhold et al. (1996).

Susceptibility of lodgepole pine trees to mountain pine beetle was related to both tree size and stand density (Anhold et al. 1996). These researchers found low numbers of trees attacked (<30%) in stands with SDI <125 and >250. They suggest that dense stands of small trees provide poor habitat for the beetles because the phloem tissue in the small trees is inadequate to support the eggs and larvae. As tree size increased and stand density remained high, beetle habitat improved and stands became more susceptible to beetle damage. The greatest mortality rates from mountain pine beetle occurred at relatively low SDIs of 125–150, and this suggests that mortality from bark beetles corresponds to the onset of stress from crown closure and inter-tree competition. To reduce effects of bark beetles, they suggest that lodgepole pine stands be grown at densities of less than 100 or greater than 400 trees per acre (Anhold et al. 1996; figure 8-9).

Factors such as root disease may predispose stands to bark beetle attack. Analyses of primary-focus trees for mountain pine beetle attacks in ponderosa pine (Eckberg et al. 1994) support the idea that predisposing factors can override the apparent benefits of lower stand density and greater growth efficiency. Primary-focus trees were those affected by lightning, wind breakage of the crown, *Armillaria* infections, and previous pine beetle attacks, all of which cause attractant pheromone compounds to be released from the trees. In their analysis, Eckberg et al. (1994) hypothesized that endemic levels of pine beetle are maintained in trees infected with *Armillaria* or attacked by *Ips*, but that the insect populations may be unable to increase in these trees due to poor tree growth and relatively thin phloem. These primary-focus trees were found to encompass a wide range of tree sizes, suggesting that a predisposing factor is more important than merely size. However, at least a minimum tree size is necessary for a successful insect attack. For example, Mitchell and Preisler (1991) found that infestations were highly correlated with the location of large-diameter trees and that the epidemic lost momentum when most of the large trees were gone.

The hazard-rating system presented by Schmid et al. (1994) for ponderosa pine accounts for both tree size and stand density. For stands with a given mean dbh, the hazard rating declines with increasing average spacing, and for a given spacing, hazard increases with increasing mean dbh. These researchers also noted that stands with basal areas >140 ft^2/ac represent a very high hazard based on levels of observed pine beetle mortality. In short, the higher the relative density of a stand (higher SDI), the greater the probability of significant damage from mountain pine beetle. This finding is consistent with that reported by Oliver (1997), cited above.

Variation in density within stands that makes them susceptible to bark beetles is not reflected by stand-density averages. Detailed studies of the interactions between mountain pine beetle epidemics, stand structure, and local variations in stand density emphasize the complexity of factors influencing beetle outbreaks and help explain the apparently inconsistent results of previous studies. Olsen et al. (1996) found that local centers of high ponderosa pine stand density (>200 trees/ac) corresponded to pockets of beetle attack.

The net implication for management is that the increased susceptibility of some trees is caused by increased competition in heavily stocked areas, particularly if the weakened trees are larger than the 8-in threshold. Altering local variations in stand density during silvicultural treatments can therefore increase resistance of both even- and uneven-aged stands. Repeated thinning may help maintain resistance.

However, in these forests, thinning may attract insects and promote population buildup to the point where serious damage to live trees becomes a major concern. Bark beetles may invade logging slash and then live trees. *Annosus* root diseases, which enter cut stumps after thinning, can make trees and stands susceptible to bark beetles in ponderosa pine, mixed-conifer, and Douglas-fir forests in northern California. Similarly, precommercial thinning, if done in wet periods of the year, may make Douglas-fir susceptible to root colonizing insects (Witcosky et al. 1986). Tree responses to wounds caused by pruning can attract pitch moths to pruned Douglas-fir. Controlling slash buildup and damage to trees and treating cut stumps in ponderosa pine stands to control *Annosus* root disease are ways to keep insects at low levels following thinning.

Thinning is not likely to directly reduce a stand's susceptibility to defoliators such as budworms, tussock moths, or sawflies by improving tree vigor. However, changing the species composition to favor resistant species, or changing the stand structure, or both, may make stands less susceptible to defoliators. For example, reducing the density of true fir and Douglas-fir stands in the understory of mixed pine and fir stands reduces the density of susceptible species, making it less likely that dispersing larvae will find a host before they land on the forest floor and are killed by predators. In addition, providing habitat for insect-feeding birds and protecting ant nests are ways of maintaining natural insect predators in forest stands. Wickman (1988) found that although thinning did not reduce tussock moth defoliation, trees in thinned stands recovered from defoliation more rapidly than trees in unthinned stands. Thinning that encourages a mixture of species may help reduce the effects of defoliators.

Pathogens. Stem and root pathogens are important considerations when thinning forest stands. Damage to stems from felling and skidding logs should be minimized to reduce stem decay diseases. For example, Aho et al. (1983) found that, about 13 yr after logging, 8–15% of red and white fir trees were so badly decayed that they were no longer suitable as crop trees. Most damage occurred at the base of the tree from the skidding of logs. Wounds higher in the tree from felling were less damaging. Western hemlock is also susceptible to stem fungi, but ponderosa pine and Douglas-fir are much less affected by logging damage. Hanus et al. (1999) and Hann and Hanus (2002) found that no rot was associated with logging scars less than 21 yr old for Douglas-fir. They found that if 20% of the trees grown for 50 yr after thinning were damaged, only 2% of the volume would be lost. However, given their assumptions for the value of logs, they estimated that it would pay to use careful logging practices to reduce the percentage of damaged trees from 20% to 5%. They also report that mechanical damage can reduce height growth.

It is difficult to predict the occurrence of root diseases from crown symptoms such as sparse foliage, and so thinning of stands to control root diseases may be difficult to prescribe. Thies and Nelson (1997) report that laminated root rot appears to occur in pockets early in the life of Douglas-fir stands but that a high proportion (66%) of the trees around these pockets are not infected. In older stands, the occurrence of laminated root rot is more diffuse and centered in individual trees.

Filip (1986) surveyed mixed-conifer stands in southern Washington and found that more trees had root disease than symptoms indicated. Of 119 trees with no symptoms, 78 had root disease. In these trees, only about 17% of the root system was decayed. Of the 23 trees with symptoms, 22 had the disease and 64% of their root systems were decayed. Thus, thinning large trees specifically to control these diseases does not seem practical (Roth et al. 2000). However, precommercial thinning of ponderosa pine decreased mortality of crop trees at 20 and 30 yr after thinning. Crop tree mortality in thinned stands was 8.6%, compared to 16.4% in unthinned stands (Filip et al. 1989a, 1999). The increased tree vigor in the thinned stands may have enabled the pine to overcome the pathogens in its roots.

It is important to note that thinning did not increase the incidence of this disease. The spread of *Armillaria* and *Phellinus* does not appear to be affected by thinning. The trees that die from these diseases are often larger than they would have been if the stand had not been thinned. Thus, the dead trees in thinned stands could be valuable enough to be salvaged, or they could be left to provide large snags/logs for wildlife habitat.

Annosus root diseases can enter healthy trees as spores, colonizing freshly cut ponderosa pine stumps and then invading the roots of adjoining trees through grafted roots. The effects of the disease can be minimized by growing mixed-species stands and controlling stand density when trees are small and stump infection less likely. In true fir stands, wounding by logging did not appear to increase the incidence of *Annosus*, but the amount of stem decay in wounded trees was more than twice that in undamaged trees (Sullivan et al. 2001). The spread of this disease from cut stumps to adjacent small seedlings does not appear to be a problem in true fir stands (Filip et al. 1992, 2000b). Slaughter et al. (1991) and Slaughter and Parmeter (1998) report that the mortality of trees that had been growing next to infected white fir stumps for 11–25 yr was less than 5%. Infected true fir trees are more prone to wind and snow breakage than uninfected trees.

Annosus disease is also of concern in ponderosa pine stands. It infects groups of trees that are then invaded by bark beetles. *H. annosum* is likely the primary cause of mortality (Bega 1978). Ponderosa pine is susceptible to air pollution, principally ozone. This photochemical oxidant increases ponderosa pine's susceptibility to *H. annosum* (James et al. 1980).

The parasite dwarf mistletoe is a major consideration when thinning in both even-aged and uneven-aged stands. Thinning increases the vigor and aids the dispersal of dwarf mistletoe into the canopies, and the increased light in the canopy increases the growth of established mistletoe plants. A less dense canopy enables mistletoe seed to disperse horizontally, because there are fewer adjoining trees to block dispersal. Severe mistletoe infections can reduce tree vigor and potential for growth and also make trees more susceptible to mortality from drought or bark beetles. Both height and diameter growth of Douglas-fir over a 20-yr period were reduced when over one-half of a tree's crown volume was composed of mistletoe brooms. Trees with infections in only the lower one-third of the crown grew well, especially in thinned stands (Knutson and Tinnin 1999; Tinnin et al. 1999). Mistletoe-infected western larch responded to thinning, but grew less in volume than did uninfected trees in the same stands (Filip et al. 1989b). The reduction in volume growth was proportional to the severity of the infection.

Mistletoe is most important in uneven-aged stands because infection of the understory is very likely if the overstory is infected (Roth 2001). After removing infected overstory trees, 22-yr tree mortality of the infected understory was 73% in unthinned stands and 38% and 24% in stands thinned to 500 and 250 trees/ac respectively (Roth 2001). A second thinning to 125 trees /ac, combined with pruning of infected branches, was suggested to obtain thorough control of the disease, but this level of control may be warranted only on very productive sites or in otherwise valuable stands. The removal of the infected overstory can be delayed until the understory trees are about 1.5 m tall

(Mathiasen 1998). Smaller conifer seedlings are apparently too small for establishment of mistletoe seedlings on their branches and stems. Similar results are reported for red fir. Therefore reducing potential dwarf mistletoe effects by removing heavily infected trees could be a goal for thinning. If trees are of low commercial value, or if snags are needed for habitat, heavily infected trees might be killed to make snags.

Many of the diseases and insects described above and in tables 3-1 and 3-2 are host specific. Thus, one of the most effective ways that thinning can be used to lessen their effects is by encouraging mixed-species stands. To do this, it will probably be necessary to deviate from rigorous application of spacing rules and adopt flexibility inusing crown classes when designating trees to thin or leave, especially in young stands.

REVIEW QUESTIONS

1. Explain the difference between tree classes and crown classes.
2. Discuss the development of crown classes in even-aged stands or groups of trees.
3. Why are crown classes not always useable in uneven-aged stands?
4. How are stand tables used in thinning? In an even-aged stand, in what crown classes does most of the volume growth occur?
5. What are the differences in the five types of thinning (thinning from below etc) discussed, and how do they affect stand structure? When is it appropriate to use each type of thinning?
6. How does thinning affect individual trees from a morphological/physiological point of view?
7. Explain how thinning affects: a) tree crowns, b) seed production, c) diameter growth, d) height growth. (This question has several different aspects to it depending upon species, site etc.). Be able to explain and discuss.
8. Discuss the effects of thinning on stand growth from the following points of view: average diameter, volume growth (MAI and PAI), the trade-off between volume growth and diameter growth, mortality, and wood yield.
9. In what types of stands will thinning increase volume growth? Explain. 10. Explain how the response to thinning may be affected by the density of the stand at the time of thinning.
11. Why is strict adherence to spacing guidelines likely to reduce the volume growth in a stand?
12. Explain the advantages and disadvantages of thinning early in the life of a stand.
13. What are the advantages of waiting to thin rather than thinning early in the life of a stand?

CHAPTER 9

Forest Stand Nutrition and Fertilization

Nutrient Demand
Nutrient Responses to Silvicultural Treatments
Responses of Tree Physiology and Leaf and Crown Morphology
Shifts in Biomass Allocation Induced by Fertilization
Forest Floor and Soil Properties
Direct and Indirect Effects
Tree and Stand Growth
Response of Young Stands and Understory Vegetation to Fertilization
Wood Quality and Economics
Past Disturbance and Response to Fertilization
Prescribing Fertilizer Treatments
Predicting Response to Fertilization
Fertilization Effects on Disease, Insects, and Animal Damage

All plants require nutrient elements in order to grow, function, and reproduce successfully (Marschner 1995). Soil fertility, nutrient availability, and nutrient uptake by trees have therefore long been important considerations in the practice of silviculture (Binkley 1986; Fisher and Binkley 2000). The basic building blocks of trees and other plants are organic molecules composed predominantly of carbon (C), hydrogen (H), and oxygen (O). Several other elements are also considered macronutrients because they are needed in relatively large quantities: nitrogen (N), calcium (Ca), potassium (K), magnesium (Mg), phosphorus (P), and sulphur (S). Micronutrients are needed in relatively small quantities, but are nevertheless essential for the proper physiological functioning of the tree: chlorine (Cl), iron (Fe), manganese (Mn), boron (B), zinc (Zn), copper (Cu), molybdenum (Mo), and nickel (Ni).

Carbon is fixed by the plant through carboxylation of atmospheric CO_2 during photosynthesis, and H and O are obtained by hydrolysis of water (H_20) molecules from the transpirational stream. As nutrients, therefore, C, H, and O are not typically limiting to the growth of trees. The primary source of N is biological fixation of atmospheric N_2 by a variety of microorganisms, although wet and dry atmospheric deposition can also be a significant source of N in forest ecosystems (or can even be problematic in some areas due to industrial and automobile emissions). Nitrogen is by far the most common limiting nutrient to forest productivity because: (1) the only significant source of N is the atmosphere (Schlesinger 1991), (2) N is easily volatized from organic matter during prescribed and wild fires, (3) large quantities of N are typically bound in forest floor and soil organic matter, and (4) significant amounts of N can be removed from forested sites by timber harvesting or severe site preparation. The vast majority of all other macro- and micro-nutrients are derived from weathering of parent material (Schlesinger 1991). Their availability therefore depends on the mineralogy of the parent material, weathering rates, and their relative mobility in the soil. The latter determines both availability to plants and potential leaching loss through the soil profile and to groundwater. Some of these nutrients (e.g., P in the form of PO_4^{3-}) are strongly adsorbed to soil particles at relatively low pHs typical of many coniferous forest soils, but are also retained and concentrated in organic matter; hence, the rate of recycling within the organic pool can control to a great degree the availability of nutrients (Fisher and Binkley 2000).

Because some forest management activities like timber harvesting remove organic matter containing macro- and micro-nutrients, forest scientists have long been interested in quantifying nutrient removals, turnover in organic matter, and replenishment by natural processes (Kimmins 1990, Kimmins et al. 1999). The objective is to ensure that forests are managed sustainably, or in a manner that does not erode long-term site productivity (e.g., Cole and Gessel 1988, Perry et al. 1989). A network of studies has been established in North America to test the effects of different experimentally imposed levels of organic matter removal by harvesting tree biomass and by artificial removal of the forest floor, in addition to the effects of soil compaction and competing vegetation control (Powers et al. 2005, Powers 2006). This network includes sites in western U.S. forests (for example, Ares et al. 2007a, b; Harrington and Schoenholtz 2010). Although this research is still relatively young, no evidence has accrued to suggest that any of the treatments that are generally considered severe, compared to modern harvesting practices, have had a negative influence on subsequent tree growth. Some studies have found positive effects from moderate retention of logging slash, probably as a result of amelioration of the microclimate and suppression of competing vegetation rather than from nutrient retention (Harrington and Schoenholtz 2010, Slesak et al. 2010, Harrington et al. 2013).

As discussed in chapter 3, the condition of forest soils—indeed, of any soil—is the result of centuries of soil formation overlaid with human use. This legacy of soil properties (fertility, organic matter content, porosity, water-holding content, etc.) is the basis of soil fertility. It is more effective to preserve soil productivity by careful implementation of silvicultural practices than to have to try restoring soils after they have been damaged by poor silvicultural practices or severe fires. Bengston (1981) argues that replacement of nutrients by fertilization is likely to be only about 50% effective, because nutrients from fertilizers are not completely taken up by crop trees. The main reason for fertilization should therefore be to correct nutrient deficiencies in certain soils or meet other silvicultural objectives, not to correct management mistakes.

Fertilization has been employed to: (a) increase the growth and yield of current forest stands (e.g., Gessel et al. 1973); (b) maintain or improve inherent soil properties such as organic matter and ability to supply nutrients (e.g., Schoenholtz et al. 2000); (c) ameliorate nutrient deficiencies that adversely affect tree growth and morphology (e.g., top dieback in lodgepole pine; Brockley 1996), and (d) ameliorate nutrient deficiencies that predispose trees to disease (e.g., *Dothistroma* in radiata pine; Lambert and Turner 1977; Lambert 1986; Turner and Lambert 1986). Walker and Gessel (1991) provided a comprehensive overview of the assessment and amelioration of macro- and micronutrient deficiencies in Douglas-fir and several other Pacific Northwest coniferous species, noting that successive harvests from a given site raise the potential for the emergence of deficiencies in nutrients other than nitrogen.

Fertilizers have been most commonly applied to forests in western North America to increase yield of merchantable wood. Bengston (1979) reported that more than 1 million ac in the Pacific Northwest had been fertilized by the mid-1970s, and fertilization continues to be a common practice; an average of 2% of all coastal Douglas-fir forests that are intensively managed for timber production were fertilized in the 1990s (Briggs and Trobaugh 2001).

Management of forest nutrition through fertilization involves several issues: 1) identification of stands within a density range that minimizes suppression mortality (self-thinning) from accelerated stand development promoted by fertilization; 2) identification of age and density thresholds that avoid stimulation of competing vegetation; 3) identification of sites that will respond to the nutrient amendment to a degree that ensures a favorable economic and environmental outcome; 4) determination of application rates that are optimal from economic and environmental perspectives (e.g., avoid damage to water quality); 5) estimation of the duration of direct and indirect effects of fertilization and the desirability of additional applications; and 6) minimizing the risk of inducing deficiency in nutrients other than the applied nutrient.

Nutrient Demand

Fertilization with N, usually after thinning, is the most common forest-fertilization treatment in western forests. Nitrogen can be lost from systems by removal during timber harvesting or volatilization during fire. As stands grow, N also tends to accumulate in organic matter in the forest floor or mineral soil horizons, but in a form that is unavailable for plant growth. This accumulation may limit rapidly growing young stands where the canopy is closing and there is a high demand for N (Miller 1981).

As stands grow and accumulate biomass, the amount of nutrients taken up can be estimated from the biomass and average nutrient concentrations in various biomass components (table 9-1; figure 9-1). Foliage contains the largest amounts of nutrients associated with the photosynthetic apparatus (N, P, K, Mg), and the stem wood (heartwood and sapwood) contain proportionally more elements associated with the structural support provided by cell walls (e.g., Ca). Because nutrient concentrations vary by biomass component, and because the magnitude of the different biomass components varies tremendously (e.g., stem-wood vs foliage; figure 9-1a), the distribution of total nutrient content among biomass components also varies (figure 9-1b, c; table 9-2). The pattern in cumulative nutrient uptake depends on the rate of biomass accumulation in various stand components (figure 9-2a) and on nutrient concentrations in those components (tables 9-1 and 9-2; figure 9-2b). Early uptake in given tree species depends on initial stand density, but at later ages total nutrient content seems to converge (figure 9-2c, d). Not surprisingly, the peak in demand for most nutrients occurs during the period of rapid growth occurring at and just following crown closure and up to the time of peak biomass increment (figure 9-3). Nutrient demands continue, however, especially for sapwood, which accounts for a large proportion of biomass accumulation throughout a stand's life (figure 9-2b). Biomass and associated nutrient accumulation are driven by overall site quality, that is, the combined effect of water availability, nutrient limitations, growing season length, and other factors controlling primary productivity. Annual uptake or demand provides a foundation for maximum potential uptake and insight into the stages of stand development during which nutrients are most likely to become limiting; for example, this period ranges from about age 5 to 20 yrs in coastal Douglas-fir planted between 500 and 3000 trees per hectare (figure 9-3).

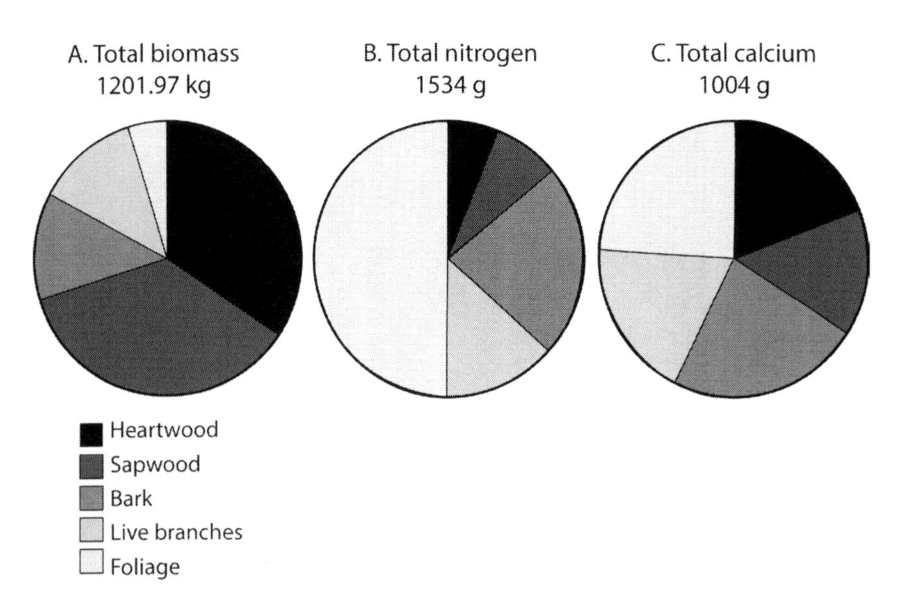

Figure 9-1. Distribution of above-ground biomass, nitrogen, and calcium in a 38-yr-old Douglas-fir tree (dbh=45.6 cm, ht=33.5 m, cl=19.94 m): (a) biomass; (b) nitrogen; and (c) calcium.

Nutrient Responses to Silvicultural Treatments

Forest fertilization is often combined with other silvicultural treatments. To understand responses and mechanisms specific to fertilization, it is therefore important to identify the effects of these other treatments on nutrient availability and uptake, in particular stocking (figure 9-2) and competing vegetation control (figure 9-3), thinning (Brix 1993), and prescribed fire (Brockley et al. 1992). The effects of prior or ongoing natural disturbances, e.g., wildfire, can create an important backdrop as well (DeBell et al. 2002).

Table 9-1. Nutrient concentrations (mg/kg) of primary biomass components in a 38-yr-old Douglas-fir tree (dbh=45.6 cm, height=33.5 m, crown length=19.9 m).

Quantity	*Heartwood*	*Sapwood*	*Bark*	*Branchwood*	*Foliage*
Nutrient concentrations:					
N	220	280	2320	1410	13734
Ca	469	340	1571	1236	4844
K	512	228	94	440	359
Mg	176	412	206	116	449
P	569	439	362	120	393
S	289	195	96	385	441
Fe	38	540	170	497	400
Mn	17	9	46	38	260
B	70	77	398	92	397
Zn	9	587	230	425	263
Cu	540	540	166	105	152

Table 9-2. Biomass and nutrient content of primary biomass components of a 38-yr-old Douglas-fir tree (dbh=45.6 cm, ht=33.5 m, Crown length = 19.9 m).

Quantity	*Units*	*Heartwood*	*Sapwood*	*Bark*	*Branchwood*	*Foliage*	*Total*
Biomass	kg	337.89	334.35	87.77	30.00	41.97	831.98
Nutrient contents (g):							
N	g	74.3	93.6	203.6	42.3	576.5	990.4
Ca	g	158.4	113.5	137.8	37.1	203.3	650.1
K	g	173.0	76.2	8.3	13.2	15.1	285.8
Mg	g	59.5	137.8	18.1	3.5	18.9	237.6
P	g	192.3	146.8	31.8	3.6	16.5	390.9
S	g	97.7	65.2	8.4	11.6	18.5	201.3
Fe	g	12.8	180.5	14.9	14.9	16.8	240.0
Mn	g	5.9	3.0	4.0	1.1	10.9	25.0
B	g	23.7	25.7	34.9	2.8	16.7	103.8
Zn	g	3.0	196.3	20.2	12.8	11.0	243.3
Cu	g	182.5	180.5	14.6	3.2	6.4	387.1

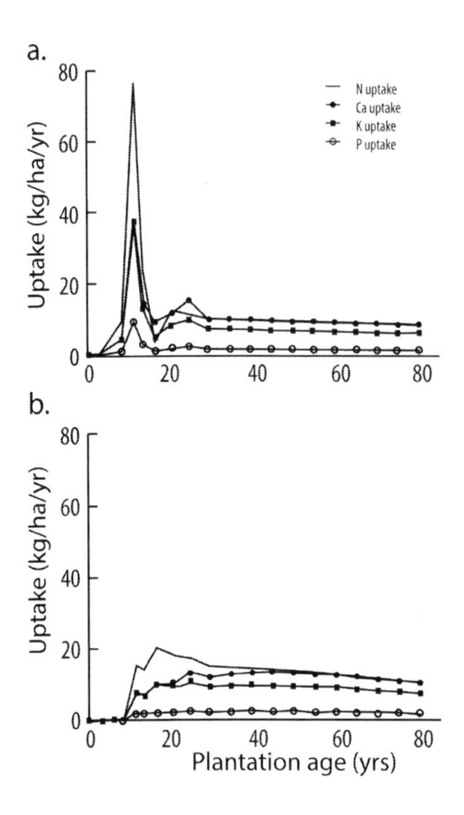

Fig. 9-2. Periodic annual uptake or demand for N, Ca, K, and P over the course of even-aged stand development in Douglas-fir planted at (a) 2x2 m spacing and (b) 5 x 5 m spacing. Based on tree measurements to plantation age 24 years and subsequent projection with the growth model ORGANON through 81 years.

COMPETING VEGETATION CONTROL

Control of competing vegetation has in some cases been shown to increase nutrient concentrations in the crop trees. For example, in the Sierra Nevada and southern Cascades, control of shrubs in ponderosa pine plantations increased nutrient concentrations in the needles (Powers and Ferrell 1996). Four-year old loblolly pine that had received two years of spring release from competing vegetation had significantly greater foliar potassium concentrations, but also significantly lower concentrations of calcium and magnesium (Sword et al. 1998). The latter reductions were attributed to the dilution effect of growth acceleration stimulated by control of the competing vegetation. Other work in loblolly pine presents a rather complicated response of nutrient concentrations to competing vegetation control, with surprisingly few significant differences 2-15 years after treatment among plots receiving herbaceous control only, hardwood control only, or both herbaceous and hardwood control (Miller et al. 2006a). Again, responses are complicated by positive growth responses and dilution of nutrient concentrations by stimulated production of foliage and other biomass components. By year 15, no treatment effects could be detected on foliar P, Ca, or Mg; however, foliar K was significantly lower under all three competing vegetation treatments, and foliar

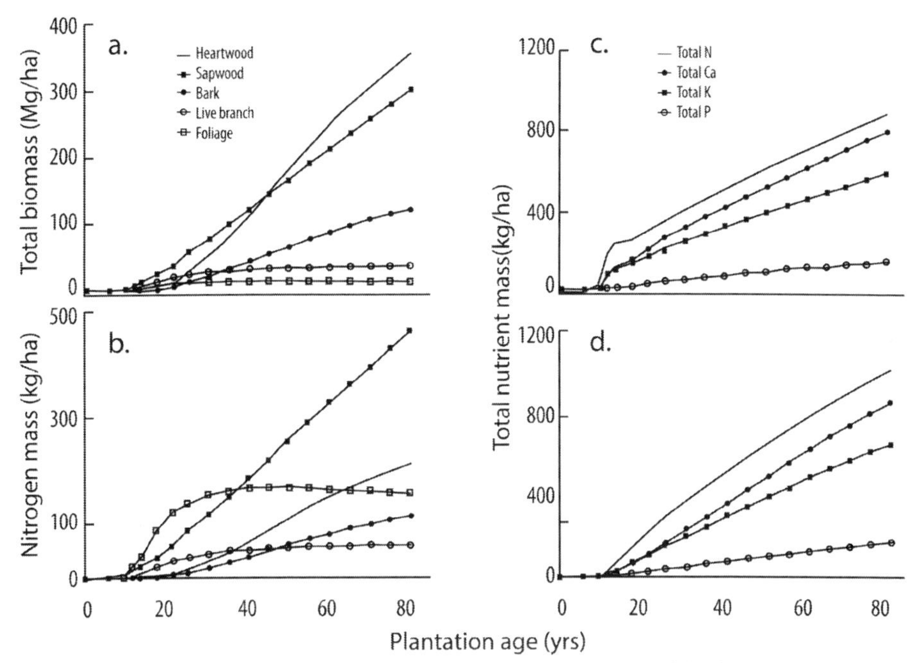

Fig. 9-3. Biomass and nutrient accumulation over the course of even-aged stand development: (a) cumulative biomass by above-ground component in a Douglas-fir stand planted at 5 x 5 m spacing; (b) cumulative N content by above-ground component in Douglas-fir stand planted at 5 x 5 m spacing; (c) cumulative N, Ca, K, and P content in above-ground biomass of Douglas-fir stands at 2 x 2 m spacing; and (d) at 5 x 5 m spacing. Based on tree measurements to plantation age 24 years and subsequent projection with the growth model ORGANON through 81 years.

N was significantly lower under herbaceous control (Miller et al. 2006a). Two-year results from the same study were somewhat site-specific, but herbaceous competing vegetation control increased N concentration at about half of the sites, decreased P concentration at about half the sites, and reduced Ca and Mg at all sites (Zutter et al. 1999). The dynamics of nutrient concentrations have to be interpreted in the context of availability of soil nutrients, translocation within the crown, and increasing demand and uptake rates in response to stimulation of tree growth and recovery of competing vegetation (Zutter et al. 1999, Miller et al. 2006a). Zhang and Allen (1996) speculated that loblolly pine stand growth would slow during mid-rotation to maintain foliar N above critical levels (1.0-1.2%). In an effort to predict response of loblolly pine to elimination of hardwoods at mid-rotation, Albaugh et al. (2013) found encouraging correspondence between estimates of N-uptake by hardwood foliage and loblolly pine growth response to this same amount of N if made available by hardwood control.

Foliar nutrient concentrations in *Pinus sylvestris* five years after low, medium, and high intensity control of competing vegetation were in general not significantly different from those that received no control (Hytönen and Jylhä 2011), except for a significant decrease in foliar Mg under the most intense removal of vegetation. However, the highest intensity of vegetation control produced needles with the largest mass, implying a dilution effect on all nutrients except Mg.

In coastal Douglas-fir, foliar N concentrations did not differ significantly among the following three treatments after four growing seasons: 1) complete control of competing vegetation; 2) seeded grass competition; and 3) planted red alder competition (Cole and Newton 1987). In contrast, foliar P was significantly higher in the grass-seeded treatments. Peterson et al. (2008) and Devine et al. (2011) also detected no significant differences in foliar macronutrient concentrations between Douglas-fir that received no competing vegetation control and those with control for five consecutive years. These results all suggest that growth responses of crop trees to competing vegetation control is quite rapid, and is probably attributable to the greater availability of soil water (Newton and Preest 1988, Dinger and Rose 2009). This growth response accelerates demand for nutrients at an apparently greater rate than nutrients are mineralized from the decaying competing vegetation and other organic matter in the forest floor and mineral soil. As a result, biomass dilution of nutrients in foliage and other tree components leads to either reductions in nutrient concentrations or at best constant levels if uptake rates can keep pace with biomass productivity. The extent to which competing vegetation control releases nutrients and raises the potential productivity of young trees beyond that stimulated by increased water availability is not clear, but total biomass and hence nutrient content of the crop trees increases relative to plots where competing vegetation is not controlled. The tripling in 6-yr growth after competing vegetation control in the Sierra Nevada (Powers and Ferrell 1996) was probably attributable to increased availability of water, suggesting a shift from water constraints on productivity to nutrient constraints, given the lack of change in foliar nutrient concentration in this and many other studies that assessed response of crop tree nutrient concentrations to competing vegetation control. Ultimately, the effects of enhanced water and nutrient availability can be difficult to separate.

DENSITY MANAGEMENT

As discussed in chapters 5 and 8, meeting stand management objectives often requires achieving a specific range in stand density. Thinning is probably the most common mid-rotation silvicultural treatment for managing even-aged stands. Thinning produces its own suite of tree, stand, and ecosystem responses (chapters 3 and 8), including changes in nutrient dynamics (figure 9-3). Thinning reduces total leaf area but increases growth per unit leaf area (Brix 1981a, Waring et al. 1981, Binkley and Reid 1984). Because fertilization is sometimes combined with thinning, it is important to understand nutrient responses to thinning to distinguish them from responses to fertilization alone.

Stands are often thinned prior to fertilization to ensure that growth responses are not lost to increased self-thinning mortality, and the fertilization accelerates site re-occupancy and minimizes the temporary loss in stand productivity caused by reduced growing stock or leaf area. Brix (1991, 1993) noted that thinning had no significant effects on foliar N concentrations at the Shawnigan Lake field trials in British Columbia during the first four years after thinning. Total leaf area was significantly reduced by thinning (Brix 1981a), so soil and shoot water potentials were increased, as expected (Brix and Mitchell 1986). In an analysis of the 9-year response at Shawnigan Lake, Pang et al. (1987) found a significant increase in total tree N concentration and a significant reduction in foliar Mg, with no difference in foliar P, K, or Ca concentration. In balsam fir, Piene (1978) documented an increase in N-mineralization rate due to greater throughfall precipitation and penetration of solar radiation, improving moisture and temperature conditions for decomposition of the forest floor. However, in contrast to tree responses to competing vegetation control as described above, the higher inorganic N availability led to higher foliar N levels after thinning. Carlyle (1995) also reported responses from thinning in radiata pine that were consistent with these other studies; i.e., elevated rates of N mineralization, higher foliar N, and greater individual tree growth three years after thinning. Other thinning studies, however, have detected no change in mineralization rate or foliar N concentrations in residual trees (Son et al. 1999).

Miesel (2012) assessed foliar nutrient concentrations following thinning and prescribed fire in ponderosa pine and white fir in northern California. None of the treatment combinations resulted in any significant differences in foliar N or P concentration for either species. After thinning to two different residual stand densities in 32-37-yr-old Scots pine, Blanco et al. (2009) found no significant thinning effect on foliar concentrations of N, P, K, or Ca, but did find a significant reduction in Mg concentration. Although they did not measure foliar N, Tan et al. (2008) report continued increased mineralization rate and soil N availability 24 years after thinning 22-yr-old lodgepole pine, relative to unthinned plots. Thinned trees were growing significantly faster than the largest trees in unthinned plots, but given that this is a typical thinning response, it is unclear how much of the response was attributable to greater N availability and associated increase in foliar N concentration, and how much was attributable to greater light and water availability to each residual tree in the thinned plots.

Direct and Indirect Effects of Fertilization

Insights into the use of fertilization treatments are facilitated by a basic knowledge of mechanisms of stand and tree responses. These mechanisms are closely linked to the distinction between direct and indirect fertilization effects. Direct effects are growth responses to fertilizer that exceed those of unfertilized stands with the same initial structural conditions. For individual trees this direct effect is typically the stem

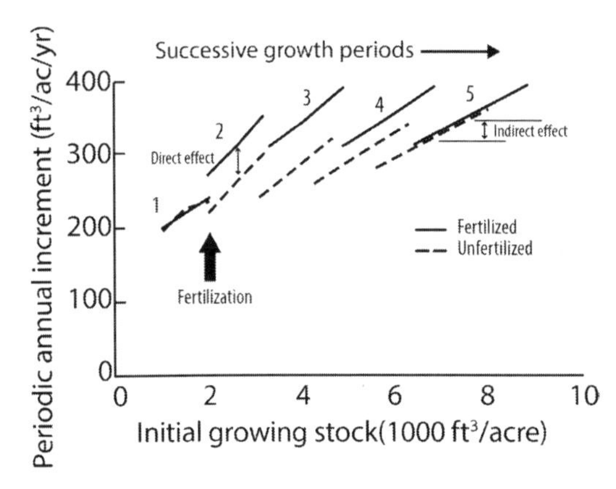

Figure 9-4. Periodic annual increment for fertilized and unfertilized Douglas-fir stands as a function of initial growing stock over five successive 5-yr growth periods, with the first occurring before fertilization and the next four after fertilization.

(biomass) growth based on initial dbh, total height, live crown length, stand structure (especially stand density and tree crown class), and pre-fertilization site conditions. For whole stands, direct growth effects are conditional on growing stock (stand density), size class structure, and inherent site quality. In general, the clearest silvicultural insights into direct growth effects for a given species are gained by expressing both growth and initial condition in the same units, e.g., stem cubic volume increment per unit initial stem cubic volume or top height increment for a given initial top height. Because growth of responding trees and stands accelerates, initial conditions for successive growth periods change with years since fertilization. Fertilization therefore also causes indirect responses due to changes in stand density, total leaf area, average tree size, or other stand attributes that change growth potential relative to the initial unfertilized stand. Before fertilization, cubic volume growth for a given initial cubic volume does not differ between fertilized and unfertilized stands (approximately equivalent lines in the first growth period in figure 9-4. However, after fertilization, growth of responding stands is greater for a given initial cubic volume (elevation of line for the fertilized stand in the second growth period in figure 9-4), for reasons that will be discussed in the next section. As growth responses to fertilization lead to more rapid accumulation of growing stock (initial cubic volume), the average growth expected for the fertilized stand increases due to both the direct effect of improved nutrition and the indirect effect of increased growing stock (upward AND rightward shift of line for the fertilized stand in the third growth period in figure 9-4). With increasing time since fertilization, the direct effect diminishes and the difference in growth between the fertilized and the unfertilized plots is driven to a greater degree by initial conditions at the start of successively later growth periods (lesser upward shift and greater rightward shift of line for the fertilized stand in the fourth growth period in figure 9-4). Eventually, the direct effect of the fertilization disappears and only an indirect effect is left (upward extension of the common line for the fertilized and unfertilized stands in the fifth growth period in figure 9-4).

Responses of Tree Physiology and Leaf and Crown Morphology to Fertilization

FOLIAR NUTRIENT CONCENTRATION, PHOTOSYNTHETIC RATES, AND GROWTH EFFICIENCY

The interaction between nutrient *concentration* and *content* is fundamental to understanding and managing forest nutrition. Any silvicultural treatments or other site factors that accelerate net primary production and biomass accumulation rates may cause the biomass dilution of nutrients and reduce nutrient concentrations, if uptake does not keep pace with biomass accumulation. In fact, productivity may slow to maintain the minimum nutrient concentrations required for proper balance in physiological functions (Zhang and Allen 1996). Laboratory experiments have demonstrated that seedlings generally maintain a narrow range in nutrient ratios for optimal growth (Ingestad 1979, 1982, Ingestad and Ågren 1992), suggesting a feedback between biomass accumulation and nutrient ratios. As will be discussed below, initial increases in the concentration of added nutrients return relatively quickly to original levels by biomass dilution even as the total nutrient *content* of trees and stands increase. This dilution effect and natural tendency to return to more or less constant ratios and concentrations probably explains the limited utility of foliar nutrient concentration as a sole diagnostic for fertilization.

The most immediate effect of N fertilization is a significant increase in foliar N concentration immediately after fertilization. This response has been observed in both coastal and Rocky Mountain Douglas-fir (Mika et al. 1992; Brix 1971, 1981b, 1993; Brockley 2006), western hemlock (Harris and Farr 1979, Radwan et al. 1991), sitka spruce (Harris and Farr 1979), ponderosa pine (Cochran et al. 1979), red fir (Powers 1979), and lodgepole pine (Weetman and Fournier 1982, Brockley 1991, 1996, Pinno et al. 2012). Similar elevation in foliar phosphorus concentration was achieved after P-fertilization of western hemlock (Radwan et al. 1991). Because an increase in N concentration promotes higher concentrations of photosynthetic enzymes like Rubisco (Brix 1993, Mitchell and Hinckley 1993) and higher concentrations of chlorophyll (Brix 1971), maximum photosynthetic rate and growth per unit leaf area also increase (Brix 1971, Brix 1981b, Binkley and Reid 1984). However, others have detected no difference in photosynthetic rate between fertilized and unfertilized Douglas-fir (Helms 1964). Regardless, the direct effect of fertilization, i.e., greater stem volume growth or total biomass increment for a given initial stand density or set of initial tree dimensions, is likely attributable to an increase in photosynthetic efficiency at some scale (figure 9-5). Early changes in leaf area density within the crown may also explain part of the direct effect of fertilization on stand growth. Several attributes associated with leaf area density have been shown to increase rapidly after fertilization of Douglas-fir with N: for example, increased needle width and length, number of needles per

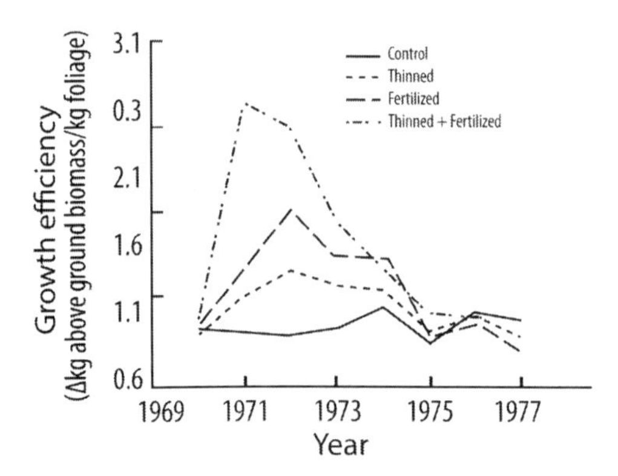

Fig. 9-5. Growth efficiency (total above-ground biomass increment per unit initial foliage biomass) during 7-yr period after fertilizing and thinning 24-yr-old Douglas-fir at Shawnigan Lake, BC. Based on data from Brix (1981a).

shoot, and number of shoots (figure 9-6a; Brix and Ebell 1969, Brix 1981a). Similar results have been observed for needle length in ponderosa pine (Gower et al. 1993) and lodgepole pine (Yang 1998). Brix (1991) reported that the increase in N concentration declined rapidly with years since fertilization, with fertilized treatments approaching unfertilized controls by year four. The corresponding trend in growth per unit leaf area in Douglas-fir lasted only about 2-6 years before the increase in foliage mass diluted the N back to pre-treatment levels (figure 9-5).

The duration of elevated foliar nutrient concentrations after fertilization seems to vary by species and site conditions and their effects on the rate of the dilution from increased foliar biomass. For example, by the tenth year after fertilizing red fir, significant differences were lacking in foliar N concentrations between fertilized and unfertilized plots, despite significant growth response to fertilization (Powers 1992). More surprising was the lack of significant differences in foliar N concentration two and three years after fertilizing lodgepole pine (Brockley 1996 and Pinno et al. 2012, respectively).

Fertilization with N or any other elements can change not only the concentration of the applied nutrient, but often the concentration of other nutrients as well. Pang et al. (1987) detected significantly lower average concentrations of N, P and K in Douglas-fir biomass nine years after fertilization with urea. Fertilization with N was found to induce S and perhaps B deficiency in lodgepole pine (Brockley 1991); it was therefore recommended that S and B be blended with N if foliar sulphate S levels are <80 ppm and foliar B levels are <12 ppm.

FOLIAGE BIOMASS AND CANOPY NUTRIENT CONTENT

After the short-term increase in foliar concentrations of nutrients applied in fertilizer, and associated increase or decrease in other nutrients, the long-term response has typically been an increase in total foliage biomass (Brix 1981a, 1993; Binkley and Reid 1984, Samuelson 1998). Although foliar N concentrations in Douglas-fir fertilized with N returned rapidly to initial concentrations (Brix 1991), total leaf area and total canopy N apparently remained elevated for a longer period (figure 9-6a, b; Brix

1981a, 1993). Binkley and Reid (1984) studied 53-yr-old Douglas-fir stands thinned and fertilized 30 yrs previously and confirmed the expectation that thinning reduced total stand leaf area but increased the rate of stem growth per unit leaf area. In contrast, fertilization increased both stand LAI and rate of stand growth per unit of leaf area. These latter responses increased with increasing rates of N-fertilization up to 470 kg/ha. At Shawnigan Lake, it has also been shown that the proportion of above-ground biomass allocated to foliage and branches generally increased (Barclay et al. 1986).

When N is added, trees use not only more N, but greater amounts of other nutrients as well. Turner et al. (1977) found that response to N was related to adequate S (>80 ppm) in the current year's needles of Douglas-fir. In an analysis of nutrient content of all above-ground biomass, Pang et al. (1987) found that the total amount of N, K, Ca, and Mg (but not P) increased after N fertilization at Shawnigan Lake, generally in proportion to the increase in biomass resulting from growth responses to the fertilization. These concurrent increases in uptake underscore the potential for other nutrients to become limiting after fertilizing with N alone.

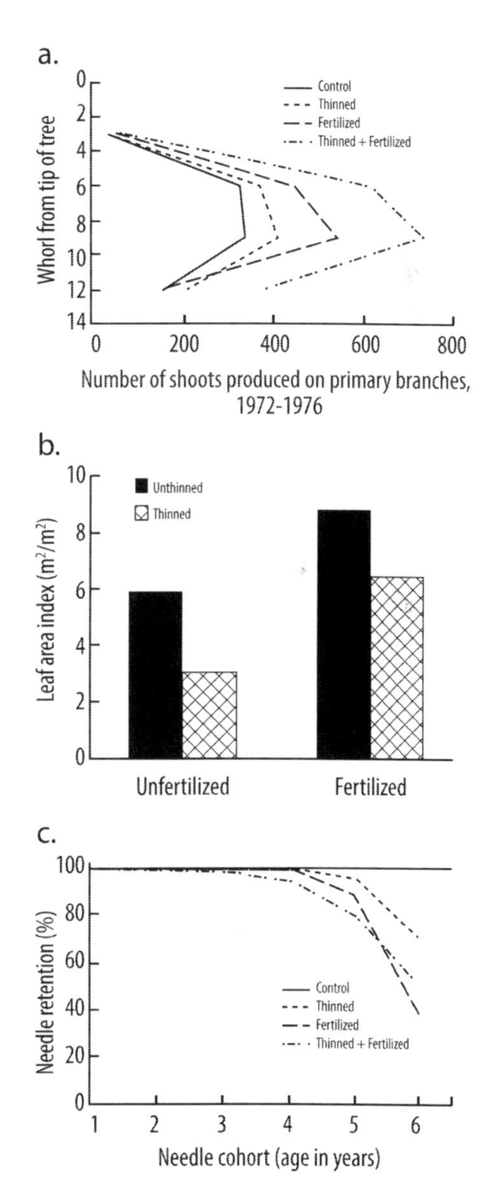

Figure 9-6. Canopy, crown, and foliage responses to fertilization and thinning of 24-yr-old Douglas-fir at Shawnigan Lake, British Columbia (based on data from Brix 1981a): (a) number of shoots produced on primary branches located at different depths into crown (whorl number from tree tip) in the four-year period following treatment; (b) leaf area index (LAI) seven years after treatment; and (c) Percent needle retention by needle cohort seven years after treatment.

NEEDLE LONGEVITY

Coniferous needle life span varies with a number of different site factors, and has been shown to increase as site quality declines (Reich et al. 1992). This effect has been proposed as a nutrient conservation strategy on nutrient poor sites. But others have proposed that greater needle longevity is caused by the slower height and branch growth on poor sites and, hence, the larger number of shorter annual segments required to produce a fixed foliated length on the main stem and on branches (Weidman 1939, Schoettle 1990, Schoettle and Smith 1991; figure 9-6a). In other words, at a fixed foliated length, a larger number of needle age classes would be needed to accumulate sufficient foliage and reduce interior light intensity below the light compensation point. This mechanism would help explain the decline in needle longevity observed after N fertilization in grand fir (Balster and Marshall 2000), Douglas-fir (figure 9-6c; Brix 1981a, Gower et al. 1992), and ponderosa pine (Gower et al. 1993). Gower et al., however, found that fertilization did not increase total leaf area in ponderosa pine, perhaps due to drought and water stress during the period of observation (Gower et al. 1993). However, they did find an increase in the amount and proportion of foliage in new needles after fertilization, creating the potential for a direct growth response if accompanied by the usual greater photosynthetic efficiency of younger foliage (Freeland 1952; Brix 1971). Results from other species suggests that the relationship between tree nutrition and foliage longevity may become more complex if other nutrients are considered; for example, Laclau et al. (2009) found an increase in eucalypt leaf longevity with K fertilization but not with N fertilization.

Shifts in Biomass Allocation Induced by Fertilization

Brix (1991) noted that the increase in photosynthetic efficiency of Douglas-fir at Shawnigan Lake after N fertilization, along with the delayed and more gradual increase in total leaf area, was insufficient to account for the magnitude of response to N fertilization. Because fertilization also increased respiration rate slightly (Brix 1981a), the most plausible explanation for the remaining increase in above-ground productivity was a shift in photosynthate allocation from below- to above-ground biomass components. The challenge of measuring below-ground production, including fine root turnover, precluded a direct estimate of below-ground allocation at Shawnigan Lake, but the hypothesized shift is consistent with observations on total production in Douglas-fir and its differential allocation on sites of inherently different quality (Keyes and Grier 1981). This shift in allocation is also consistent with the results of fertilization experiments in other species such as Scots pine (Linder and Axelsson 1982, Linder and Rook 1984).

Other allometric shifts occur after fertilization among above-ground components, and these shifts differ between sites for a given species (e.g., Albaugh et al. 2009). Nitrogen fertilization of Douglas-fir at Shawnigan Lake caused an increase in relative biomass allocation to branches, but had no effect on the relative proportion allocated to stem vs foliage (Barclay et al. 1986). Gower et al. (1993) found that two years after fertilization with a balanced blend of N, P, K, Ca, and Mg, no significant shift in the allometric relationship between tree dbh and either stem mass or branch mass could be detected in ponderosa pine and red pine. However, fertilized trees of both species had a greater amount of new foliage and new twigs for a given dbh, and only red pine had a greater amount of total foliage.

Response of Forest Floor and Soil Properties to Fertilization

Given the response of foliar nutrient concentrations to fertilization, parallel effects would be expected on needle litterfall and perhaps on fine woody litterfall. Foliar litter does tend to have a higher N concentration after N fertilization, and probably reflects higher N concentrations and a reduced need to translocate N within the tree on sites where the N is taken up in large quantities (Trofymow et al. 1991). Total mass of other nutrients tends to change little, because their concentrations are diluted by growth response, while total litterfall increases (Trofymow et al. 1991).

The fate of N applied as fertilizer is poorly understood, so it has been investigated through the use of ^{15}N-labelled fertilizer (e.g., Mead et al. 2008). Four potential sinks of applied fertilizer include: a) uptake by trees and other vegetation, b) immobilization in the forest floor, c) adsorption by the mineral soil, and d) leaching losses. Fertilization of Douglas-fir with urea or ammonium nitrate increases the quantity of NH_4^+, NO_3^-, Ca^{2+}, Mg^{2+} and K^+ leachates in the forest floor immediately after fertilization, but this response has been documented to last only for 5-10 months (Pang and McCullough 1982). The level of leachates at 10-30 cm depth into mineral soil fluctuated, but higher concentrations of NO_3^-, Ca^{2+}, Mg^{2+} and K^+ were detected after application of ammonium nitrate than after application of urea. Thinning combined with fertilization promoted deeper movement of these nutrients into the mineral soil, probably due to the reduced density of tree roots to take up the pulse of nutrients. Because urea is hydrolyzed into ammonium, NH_4^+ it can volatize to ammonia, NH_3, if conditions are not sufficiently wet after application (Nason et al. 1988). As result, urea is typically applied late in the fall to increase the probability of exposure to rainfall soon after application.

As mentioned above, on low-productivity sites N may be immobilized and slowly released. After N fertilization, Heilman and Gessel (1963) measured an increase of 200–700 lb/ac on a site with total N pool of 2050–3200 lb/ac to a depth of 24–30 in. Vegetation on the fertilized sites contained 2–10 times more N, and litterfall contained about twice as much N as the unfertilized sites. These increases suggest that fertilization may improve site productivity of low-quality sites for long periods of time. Footen et al. (2009) found that carryover effects from fertilization of low productivity sites in the previous rotation resulted in significant increases in stem diameter (29%) and height (15%) of 15-22-yr-old trees in the subsequent rotation.

If not volatilized or immobilized, fertilizer N moves into the soil solution or occupies exchange sites in the mineral soil, where it is available for plant uptake. Nutrients in fertilizer that are not taken up by plants can be immobilized in soil organic matter by microbes (Nason and Myrold 1992) or lost to leaching (Mead et al. 2008). Because of the typically large amounts of organic matter in the forest floor and surface soil horizons, much of the N fertilizer applied in forests is immobilized by soil organisms and converted from inorganic to organic forms as their populations grow in response to the increased N availability (Chappell et al. 1999). The rate of immobilization depends greatly on the form that is applied. A large portion of N fertilizers, including urea and ammonium sulfate, tends to be immobilized, averaging nearly 70% six months after fertilizer application (Nason and Myrold 1992). For nitrate-based fertilizers such as $Ca[NO_3]_2$, immobilization tends to be about 20% (Nason and Myrold 1992). Over time, immobilized N will be mineralized and released as bacterial populations die, but rates of mineralization can be very site-specific (Binkley 1986).

Tree and Stand Growth

Growth responses to fertilization represent the integrated effect of cumulative soil nutrient cycling processes and physiological and morphological responses of trees to the addition of nutrient amendments. Ultimately, these growth responses are the objective that motivates fertilization, although as mentioned above nutrient amendments

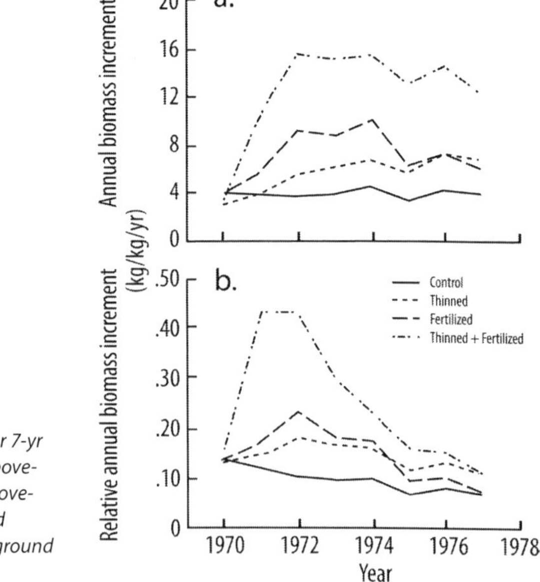

Fig. 9-7. Annual responses of individual, codominant, 24-yr-old Douglas-fir trees for 7-yr period after fertilizing and thinning: (a) Above-ground biomass increment (b) Relative above-ground biomass increment (above-ground biomass increment per unit initial above-ground biomass). Based on data from Brix 1981a.

sometimes address other management concerns such as growth deformities or the emergence or intensification of disease.

Enhancement of tree growth is often a long-term response to fertilization, particularly on sites that are severely limited with respect to the added nutrient. Responses are typically quantified by comparing the net or gross stem volume growth of fertilized plots or trees to the growth of unfertilized control plots or trees, so these long-term responses represent a combination of direct and indirect effects (difference between fertilized and control absolute growth in figure 9-7a). As alluded to above, however, the sequence of responses to fertilization is best understood as an initial direct effect (difference between fertilized and control treatments relative growth in figure 9-7b) that reflects immediate physiological responses of foliage to N fertilization, perhaps in combination with the initial phase of a shift in carbon allocation from below-ground tree components to above-ground components (Gower et al. 1992, Brix 1993). These direct effects result in accumulation of basal area, stem volume, and leaf area that contribute to a longer-term indirect growth response (Brix 1983). The magnitude of direct effects can be assessed by: 1) comparing absolute growth rates between fertilized and unfertilized treatments after correction for initial stand conditions, or 2) comparing relative growth rates—for example, volume growth per unit initial volume (figure 9-7b)—between fertilized and unfertilized treatments. The physiological driver of this direct effect is probably best regarded as the change in growth efficiency of foliage, typically expressed as stem volume growth or above-ground biomass growth per unit initial foliage mass or area, imposed by the fertilization treatments. Whether direct effects are assessed by considering covariates such as initial stem volume, or by relative growth rates, or by direct estimates of foliage growth efficiency, these effects typically last for approximately 4-5 years after fertilization (figure 9-7b). Brix (1983) estimated that the increase in growth efficiency accounted for 37% of the growth response during the first seven years after N fertilization, with the increase in foliage quantity accounting for the rest of the response. The former direct effect was initially larger (figure 9-7b; Brix 1983), and largely disappeared within 5-7 yrs after fertilization. Indirect effects of the increase in basal area, stem volume, and foliage amount accounted for the continued long-term response to fertilization (figure 9-8; Brix 1983). Gardner (1990) assessed the 15-year growth responses at Shawnigan Lake not only to the two initial levels of N-fertilization (224 and 448 kg n/ha), but also to the repeated treatments nine years later, superimposed on three thinning levels (no thinning, removal of 1/3 of the initial basal area, and removal of 2/3 of the initial basal area). After adjusting for thinning intensity, the combined direct and indirect responses to fertilization were significant at year nine, but were declining (figure 9-8a, b). Specifically, volume growth increased and remained high for several years relative to unfertilized treatments, but steadily declined after about 4 yr (figure 9-8b). Re-fertilization at nine years after the initial fertilization

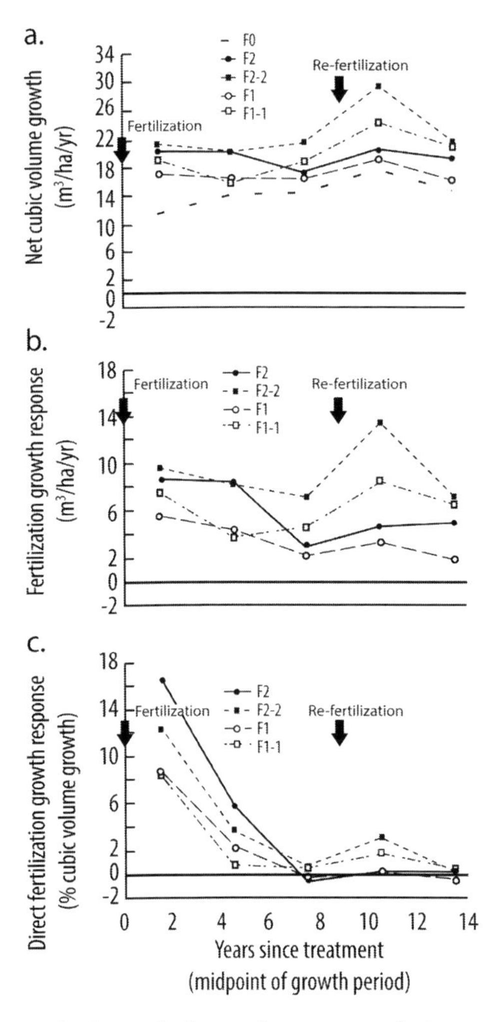

Figure 9-8. Growth response of Douglas-fir to fertilization during 15-yr period after treatment at Shawnigan Lake, BC (based on data from Gardner 1990): (a) average periodic annual increment of control (F0) and fertilization treatments (F1=224 kg N/ha at year 0, F1-1=224 kg N/ha at years 0 and 9, F2=448 kg N/ha at year 0, and F2-2=448 kg N/ha at years 0 and 9); (b) Combined direct and indirect growth response to fertilization (difference between average growth of fertilized plots and average growth of unfertilized control plots); and (c) Direct growth response to fertilization (difference between relative volume growth of fertilized plots and relative volume growth of unfertilized control plots).

evoked a marked growth response relative to controls (figure 9-8b). After adjusting for indirect effects by expressing volume growth as a percentage of initial volume for a given growth period, the direct effect of the initial fertilization treatments disappeared by year seven, and the second fertilization evoked a much smaller direct response than the initial fertilization (figure 9-8c).

In a classic fertilization trial in loblolly pine, growth response in the first two years after N fertilization was attributed to increased LAI (Vose and Allen 1988). In fact, growth efficiency declined with the increased LAI associated with fertilization, due to the lower average light intensity experienced per unit of additional foliage. The polycyclic growth habit of loblolly pine (multiple flushes per growing season; Harrington 1991) and its generally lower LAI (Vose and Allen 1988) may account in part for the apparently greater rate of foliage increase after fertilizing loblolly pine relative to fertilizing Douglas-fir.

Fertilization is mostly done in older stands (40+yr) to increase the yield of merchantable wood (Miller et al. 1986). Studies throughout the Pacific Northwest have shown that adding N in the form of urea or ammonium nitrate increased the growth of Douglas-fir (40+ yr) from 27 ft^3/ac/yr on site class I to 91 ft^3/ac/yr on site class IV over a period of 5+ yr (Miller and Fight 1979). Increased yield was achieved with rates of fertilizer ranging from 140 to 420 lb/ac of N, with an optimum yield occurring with about 280 lb/ac.

The rate and duration of response to N fertilization varies with site productivity, and it is important to recognize that not all stands respond to fertilization (Miller et al. 1986). Miller and Fight (1979) reported that, of 87 fertilization trials across site class I to site class V Douglas-fir stands, 87% showed at least a 10% volume growth increase in response to fertilizing with 200 lb of N/ac. The highest numbers of stands increasing their volume growth by ≥10% were on site class III and IV (84 and 80%, respectively); by contrast, stands on site classes I, II, and V (64–67%) averaged a 10% increase, a result corroborated by Hann et al. (2003) and Peterson and Hazard (1990). Absolute increase in growth was also related to site quality. On the more productive sites (I and II), N is probably not limiting, whereas on the least productive sites (class V), there may be too little water, or a lack of other nutrients, for stands to respond to application of N fertilizer.

Douglas-fir stands fertilized with 200 and 400 lb of N/ac were reported to increase volume growth by 60–80 ft^3/ac/yr for the first four years after fertilization, and by 30–50 ft^3/ac/yr for up to another eight years (Miller and Fight 1979). An additional 200 lb/ac of N eight years after the initial fertilization increased growth by another 40–60 ft^3/ac/yr for the next four years (Miller and Fight 1979). Most of this response appears to have been an indirect effect of greater growing stock in fertilized stands, similar to the response reported at Shawnigan Lake above (figure 9-8a,b; Gardner 1990). On these sites, N from the original fertilization seems to have been incorporated into live biomass and soil/forest floor organic matter, becoming generally less available after 8–10 years. However, fertilization nearly doubled the N capital on the poor site reported on by Binkley and Reid (1985). Consequently, N concentrations in Douglas-fir foliage on fertilized plots remained higher than those in unfertilized plots for 18 years (Binkley and Reid 1985), and volume growth increased 37–107% for 12–18 years after the initial fertilization (Miller and Tarrant 1983, Strader and Binkley 1989).

Responses to fertilization are similar in the more arid interior sites of western North America, but N often has to be supplemented with other nutrients to be effective. Brockley (1991) found a generally good but variable response of lodgepole to N fertilization, especially if S or B were added. Another study in young lodgepole pine showed increased volume growth following applications of 100 and 150 lb/ac of N along with a mixture of P, K, S, Ca, Mg, and micronutrients (Kishchuk et al. 2002). Adding N alone or the mixture alone, reduced volume growth. Foliar S was the nutrient in the mix that appeared to contribute most to tree response to N (Kishchuk et al. 2002). Similarly, Cochran (1989) observed up to an 8-yr direct response of lodgepole pine to heavy applications of N, P, and S, with indirect responses lasting longer. Stem volume growth of white fir also responded positively to N+S fertilization in central Oregon, although a subsequent budworm infestation caused considerable variability in height growth (Cochran 1991). In northern Idaho, grand fir and Douglas-fir volume growth responded significantly to fertilization with 220 kg N/ha (Shafii et al. 1989). The combination of direct and indirect responses lasted at least 14 years.

STAND DENSITY, THINNING, AND FERTILIZATION

Stands are often fertilized after thinning, both because the reduced stand density ensures that growth response does not accelerate mortality and because fertilization accelerates foliage increase and site re-occupancy. The latter effect minimizes the duration and hence amount of growth loss attributable to the reduced growing stock and under-utilization of site resources imposed by the thinning (Binkley and Reid 1984, Brix 1993, Brockley 2005, Omule et al. 2011). Stegemoeller and Chappell (1990, 1991) and Peterson and Hazard (1990) reported that fertilization of unthinned Douglas-fir stands accelerated mortality from inter-tree competition, but this result pertained primarily to stands that were at or near their maximum size-density limit at time of fertilization. Thinning results in at least a temporary delay in crown recession and re-accumulation of tree and stand leaf area as a result of increasing crown length and width promoted by improved light conditions (Marshall and Curtis 2002). Fertilization accelerates leaf area accumulation by promoting crown expansion and increasing crown density (Brix 1981, 1993).

The reduction in site occupancy and stand growth induced by thinning may be exacerbated by some level of "shock " attributable to physiological adjustments to light, water, and nutrient availability (DeBell et al. 2002). However, this shock is largely manifest in height growth, so may largely represent a shift in allocation of photosynthates to diameter growth (Gardner 1990; Marshall and Curtis 2002), crown lengthening and expansion (Gardner 1990, Marshall and Curtis 2002), or crown densification (Brix 1981a, Maguire 1983). Fertilization may ameliorate "thinning shock" by increasing tree-level net photosynthesis and overall growth rate (Stegemoeller and Chappell 1991, Brix 1993, Barclay and Brix 1985). In young stands of pre-commercial size, "thinning shock" may be eliminated entirely by fertilization (Debell et al. 2002, Stegemoeller and Chappell 1991).

Peterson and Hazard (1990), in an analysis of data from fertilizer trials throughout the Pacific Northwest, found that combining thinning and fertilization increased growth of young-growth Douglas-fir trees more than did either treatment alone. This finding was confirmed by Hann et al. (2003), who reported that a combination of thinning and fertilization increased diameter growth more than either treatment alone. Both diameter growth and height growth of Douglas-fir responded to N fertilization, with the response increasing with the rate of application up to 500 lb/ac of N. Diameter growth was greatest for the first 5 yrs and then decreased to pre-treatment levels after about 10 yrs. Proportionate growth increases were greatest on less productive sites. Thinning from below plus fertilization may increase the economic yield of fertilization because it adds wood mainly to high-quality trees while earning revenue from cut trees.

RESPONSE OF YOUNG STANDS AND UNDERSTORY VEGETATION TO FERTILIZATION

Fertilization in young stands presents the challenge of getting as much of the nutrient to the crop trees and as little to competing vegetation as possible. Shrubs, forbs, and graminoids are often the major cover in young stands of seedlings and saplings. Because they are so dense and because they often grow earlier in the season than the crop trees, they may preempt uptake of the applied nutrients by crop trees, and by doing so also compete more intensely than they would in the absence of fertilization. On sites with a pronounced summer drought, control of competition has been shown to be essential for young tree response to fertilization (Powers and Ferrell 1996). In the Sierra Nevada, fertilization of young stands without controlling understory vegetation increased shrub growth, which blocked the beneficial effects on the pine (Powers and Ferrell 1996; Powers and Jackson 1978). These authors concluded that, although fertilization increased tree growth, weed control plus fertilization may not offer much advantage over weed control alone in young stands on droughty sites. Roth and Newton (1996b) found that fertilization with no competing vegetation control increased foliar N concentration but not the growth of the trees.

Applying fertilizer directly to the planting hole on sites with strong nutrient deficiencies might result in a positive seedling growth response (Walker 1999, 2002), but broadcast applications of fertilizer encourage a dense continuous shrub cover. Rose and Ketchum (2002) found that fertilizer applied in the planting hole increased tree growth for four years on two of five sites where moisture was plentiful. However, at four sites, tree growth increased with increasing intensity of weed control, regardless of whether or not fertilizer was applied. Cochran et al. (1991) observed significantly greater diameter and height growth in ponderosa pine seedlings after placing N, P, and Mg at the bottom of planting holes, a strategy frequently adopted to ensure maximal access of planted seedlings and minimal access of competing vegetation to the fertilizer.

Inconsistent responses of young trees to fertilization with N may be attributable to increased mineralization rates observed after clearcutting (Roberts et al. 2005).

Control of shrubs may also be important on moist sites, and results vary depending on the soils and the tree and shrub species involved. On podsols in British Columbia, a thick, mor humus layer and a dense cover of salal severely reduced or held in check the growth of Sitka spruce, western hemlock, and western redcedar seedlings (Weetman et al. 1989a, b). These authors found that fertilization with N and P increased Sitka spruce height growth, but control of salal did not improve seedling nutrition or increase height growth (Weetman et al. 1989a). Weevils are attracted to the most vigorous seedlings and therefore weevil damage was high on fertilized seedlings; however, it was not clear that the weevil damage offset the value of fertilization (Weetman et al. 1989b). On the same sites, both western hemlock and western redcedar seedlings responded to fertilization with N and P (Weetman et al. 1989b), and growth of both species increased when salal was removed. Cedar responded more to salal removal than did hemlock, and for both species response was greatest with both fertilization and salal removal. Mallik and Prescott (2001) demonstrated that competition for soil resources, and not allelopathy, was the primary means by which salal inhibited growth of western hemlock. The two species compete directly because their fine roots are concentrated in the upper forest floor (Bennett et al. 2002).

Arnott and Burdett (1988) report a similar increase of western hemlock growth after fertilization at time of planting on sites with little herb and shrub cover. Thinning and fertilization were effective in increasing the growth of the largest 100 trees per acre in a 15–20-yr-old western redcedar stand in western Washington (Harrington and Wierman 1990). The best treatment was thinning and fertilization with ammonium and dicalcium phosphate. This treatment increased height and diameter growth 65% and 106%, respectively, over no thinning and no fertilization.

On 22 high-elevation true fir sites, Chappell and Bennett (1993) found consistent increases in basal area growth, foliar N, and needle biomass of noble fir and silver fir natural regeneration (6–18 yr) fertilized with 225 and 450 lb/ac of N. Moisture may not be limiting on these sites, but N is probably low because litter on the forest floor decomposes slowly at the low temperatures and during the short growing seasons typical of high elevations.

The effects of fertilization on shrubs and herbs depend on the age and density of the fertilized stands. Fertilization of young stands (<10–15 yr old) appears to increase shrub density, especially if the fertilizer is broadcast rather than applied to each seedling (De Bell et al. 2002; Roth and Newton 1996b). Although Thomas et al. (1999) found that, compared to thinning of trees, fertilization reduced the cover and species richness of shrubs and herbs, their results were based on fertilization in dense, unthinned or lightly thinned young stands (80% canopy closure and 200–494 trees/ac) with only about a 4% cover of shrubs and a 25% cover of herbs. In those stands, thinning alone increased cover and species richness of herbs and shrubs, but fertilization in addition to thinning reduced the releasing effect of the thinning on the understory by accelerating the increase in overstory canopy density, in turn lowering light levels and reducing herb cover to 2–20% and shrub cover to 1–3% of what was found in unfertilized thinned stands.

In less dense or older stands, fertilization might have different effects on understory density and richness. VanderSchaff et al. (2000) found that the effect on species was quite variable from site to site and largely depended upon the species present before fertilization, with little generalized effect. In 15-22-yr-old plantations of Douglas-fir, Footen et al. (2009) reported 73% more understory biomass and 97% greater understory N content on plots that had been fertilized in the previous rotation.

WOOD QUALITY AND ECONOMICS

The economics of fertilization includes the cost of the fertilizer and applying it, the prevailing interest rate, period of investment, and the projected stumpage value at the time of harvest. The greatest biological demand for nutrients occurs in young stands (15–25+ yr) when the rates of volume growth and nutrient uptake are greatest (Miller 1981; figure 9-3). However, if the cost of fertilization (including the interest rate) is considered, fertilizing at 15–25 yr might be too early to be cost effective. If stands are to be harvested at age 55 or older, the cost of early fertilization plus interest compounded over 30–40 yr could amount to more than the value of the increased yield of wood. Fertilization later in the rotation improves the economics of fertilization because the interest charge is carried for a shorter period (Miller and Fight 1979). Fertilizing Douglas-fir later in the rotation may also increase growth of the stronger, more valuable mature wood on the outer part of the stem (Jozsa and Brix 1989), possibly making it more cost effective. In general, the most strategic use of fertilizers is probably in combination with mid- to late-rotation thinning, primarily due to its effect on accelerating site re-occupancy and minimizing the volume growth losses typically observed after thinning.

There is concern regarding the potential adverse effects of fertilization on wood quality. Various studies in Douglas-fir have shown that fertilization causes a slight decline in wood density immediately after fertilization, due to the combined effect of decreased early wood and latewood density and a decrease in the percentage of late wood (Jozsa and Brix 1989). However, Yang et al. (1988) found that, for the first 10 yr after fertilizing 70-yr-old lodgepole pine with N and S, no appreciable change in wood density and tracheid length could be detected. Fertilization is not likely to cause a substantial or lasting decrease in the quality of the outer core of mature wood.

Although exceptions have been documented to the general result that fertilization does not change stem form (e.g., Bi and Turner 1994, Younger et al. 2008), the degree of change is often subtle and probably of little economic significance. In contrast, thinning has been shown to promote an increase in taper of residual trees as a result of increased crown length. While overall diameter growth rates in thinned stands may be doubled by fertilization, proportional allocation of diameter growth along the stem remains unchanged, resulting in little perceptible change in stem form (Thomson and Barclay 1984, Haywood 2005). Thus, growth impacts of fertilization at the tree- and stand-level in thinned (low density) stands are a result of well-documented thinning responses that are accelerated by fertilization.

Past Disturbance and Response to Fertilization

Past disturbances to a site may affect its response to fertilizer. Brockley et al. (1992) found that fertilizing previously underburned 15-yr-old Engelmann spruce plantations in British Columbia with 100 lbs/ac of N increased their growth rate by 90 to 120%. As described above, fertilized lodgepole pine that regenerated after wildfire and was pre-commercially thinned also sustained superior growth 22 years after fertilization relative to unfertilized stands. DeBell et al. (2002) suggest that severe fire—both wildfire and prescribed fire for slash disposal—may have predisposed young Douglas-fir plantations to thinning shock (Harrington and Reukema 1983). Fertilizing those plantations with N seems to have helped them to recover more quickly, as evidenced by slower recovery of plantations that were thinned without being fertilized (DeBell et al. 2002). On similar sites that had not been burned, response to fertilization and thinning shock were both much less. These authors also found that thinning plus fertilization increased shrub and herb cover, in turn reducing tree height and diameter growth and greatly reducing the effectiveness of the fertilizer (DeBell et al. 2002).

PRESCRIBING FERTILIZER TREATMENTS

Ultimately the results from fertilization trials have to be translated into a number of decisions for operational fertilization. The first and most obvious decision involves identifying the limiting nutrient as the candidate for application. As stands grow, N tends to accumulate in the biomass and upper soil horizons. Although the total N pool in forest soils is typically quite large, relatively little is available for uptake by plants. When this N sequestration occurs, it may limit growth, especially in rapidly growing young stands where the canopy is closing and there is a high demand for N (Miller 1981, figure 9-3). On other sites, the total N pool may be small and even relatively rapid mineralization rates may not overcome N limits to productivity. In most coniferous forests N is so frequently limiting that its availability often is not formally assessed before fertilization.

After the limiting nutrients have been identified, the chemical form of the fertilizer and the rate of application must be prescribed. Although ammonium nitrate typically evokes a more rapid response in foliar N concentrations and direct growth effects (Barclay and Brix 1984), urea has become the preferred form of application due to its slower release rate and diminished opportunity for leaching loss if precipitation is heavy immediately after application. This underscores the need to consider season of application, frequency of application, and identification of responding sites.

PREDICTING RESPONSE TO FERTILIZATION

Currently, there is no certain way to predict the response of stands to fertilization other than through measurement or estimation of site index or site class. Miller and Fight (1979) and Peterson and Hazard (1990) concluded that thinned stands on moderately productive sites (III and IV, 50-yr site indices 75 to 155 ft; King 1966) are more likely to respond with greater volume growth than stands on highly productive sites, and particularly unthinned stands. Site class (based on the height of dominant and codominant trees) integrates several variables affecting productivity, including nutrient and water availability. Although site class appears to provide a good first approximation of response to N fertilization, it cannot account for nutrients directly, or for the potential response if combinations of fertilizers are used. Nor does site class account for site-to-site variation in maximum stand density index (e.g., Cochran et al. 1994). Curtis (2006) illustrated that even in coastal Douglas-fir, carrying capacity for basal area and volume at a given top height varies significantly among sites. It is conceivable that the carrying capacity for basal area/volume is a function of soil properties, probably water holding capacity to a large degree, but perhaps nutrient pool sizes as well.

Growth responses to N applications of 200 to 400 lbs/ac have varied by region in the Rocky Mountains. In central Idaho and Washington gross volume and basal area growth increased with up to 400 lbs N/ac (Moore et al. 1991). In other regions there was no response, suggesting that N was not limiting or that some other nutrient was more limiting than N. Site index or productivity (stand volume growth rate) were not predictors of response to fertilization in this region. The authors concluded that the growth of about 75% of the fertilized stands would increase by 10% for 6 yr following fertilization. Research on fertilizing ponderosa pine in this region suggested that growth responses and fertilizer blends will differ by parent material (Garrison-Johnston et al. 2005).

The effectiveness of N fertilization for increasing stand growth could depend on the availability of other nutrients or water limitations. The study mentioned above in young lodgepole pine stands showed that stimulating a volume growth response to applications of 100 and 150 lb/ac of N required a mixture of P, K, S, Ca, Mg, and micronutrients, with S appearing to be the nutrient in the mix required for tree growth responses to N (Kishchuk et al. 2002). Adequate sulfur has also been implicated in predicting response of coastal Douglas-fir to N fertilization (Turner et al. 1977, Turner and Gessel 1979). In more closely controlled laboratory experiments, balanced nutrition has been demonstrated to be effective in maximizing growth (Ingestad 1979, 1982, Ingestad and Ågren 1992); blended, or balanced, fertilization has been tested in the field, but results have not been promising for coastal Douglas-fir (Mainwaring et al. 2014).

Indices of potential response may help as diagnostic criteria for forest fertilization. For example, response has been related to C/N ratios in the forest floor. For thinned stands on high-quality sites, high C/N ratios (>30) indicate a more likely response than low ratios (Edmonds and Hsiang 1987). High ratios probably suggest that N and possibly other nutrients are tied up in organic matter. On low-quality sites and in unthinned stands, the amount of total N in the forest floor was the best predictor of site response to fertilization (Edmonds and Hsiang 1987). Optimal fertilization prescriptions will vary to some extent by soil type and parent material. Knowledge of past land use and the effects of severe fires may also help determine sites where fertilization would be effective (DeBell et al. 2002). Because not all stands respond to fertilizer and because fertilization with a single nutrient may not be effective, managers would be wise to gather site-specific information and conduct tests of fertilizer effects before beginning large-scale projects. Knowing the proper nutrient balances to increase tree growth for different species and sites may improve the effectiveness of fertilization.

Although some successes have been reported in predicting fertilization responses by foliar analysis (Hopmans and Chappell 1994, Brockley 2000), the dilution effect of the growth, both with and without fertilization, suggests that the efficacy of foliar analysis alone may be limited. Mainwaring et al. (2014) found that a significant amount of the variability in volume growth response of coastal Douglas-fir to fertilization by calcium and phosphorus could be explained by soil pH and foliar nutrient concentrations. Because the long-term growth responses appear attributable to elevated stand-level foliage mass (Brix 1981a, Albaugh et al. 2004) or, equivalently, total canopy nutrient content (for example, total canopy N; Kimmins et al. 1999), diagnostics that also consider current foliage mass may prove useful. However, accurate measurement of stand-level LAI presents significant logistical challenges.

Fertilization Effects on Disease, Insects, and Animal Damage

Predisposition of trees to pathogenic fungi might be controlled by nutritional status. However, work on the relationships between soil fertility and these pathogens is just beginning. A current example is the possibility of using fertilizer applications to control Swiss needle cast on Douglas-fir. One hypothesis is that the disease is most severe on moist sites on deeply weathered sandstone and less severe on drier sites and on basalt soils. Fertilization trials have been explored to assess whether nutritional imbalances predispose Douglas-fir to the disease and might therefore offer potential for lessening its intensity (e.g., Mainwaring et al. 2014). Ca/N ratios are very strongly correlated with foliage retention, but Ca fertilization has not been shown to significantly ameliorate disease intensity. Applications of elemental S have improved growth in some trials, possibly due mostly to its fungicidal effect rather than as a nutritional amendment. Regardless, excess N may increase the susceptibility to or severity of the disease if it accumulates in the foliage as free amino acids that can serve as a substrate for the causal fungus, *Phaeocryptopus gaeumannii* (El-Hajj et al. 2004). Other work suggests that annual weather conditions control the infection intensity and survival patterns of individual needle cohorts (Manter et al. 2005, Zhao et al. 2012). Given our currently limited knowledge, a wide range of environmental conditions, including soil and plant-nutrient status, need to be considered before routinely fertilizing to control Swiss needle cast.

It has been hypothesized that susceptibility to root diseases is increased by fertilization with N and decreased by fertilization with K. Increases in soil N availability stimulate tree photosynthate production, providing food for the pathogen, whereas increased K availability increases the ratio of phenolic compounds to sugar compounds, inhibiting root diseases like *Armillaria* (Entry et al. 1991). However, N and K fertilization at planting (Thies et al. 2006) or in year 4 (Miller et al. 2006b) did not increase or decrease mortality of Douglas-fir on sites infested with *Phellinus weirii* and *Armillaria* root disease up to 8 yr after treatment.

Fertilization of thinned grand fir stands with N was shown to increase the size and total biomass of spruce budworm larvae; however, their population size and dynamics appeared to be controlled by variables not affected by N fertilization (Mason et al. 1992). Consequently, numbers of insects and their reproduction and mortality rates were about the same in treated and untreated stands. In stands where defoliation had not killed the tops of severely damaged trees, fertilization helped the stand recover from budworm defoliation (Wickman et al. 1992). Apparently the increase in conifer leaf area following fertilization and thinning was more than the insects could consume, and the trees were able to increase their growth rate.

Black bears damaged young Douglas-fir stands that had been fertilized with urea at 225–445 lb/ac (Nelson 1989). They peeled the bark on the lower part of the tree and apparently consumed the inner bark, phloem, cambium, and some xylem tissue. Damage to young stands (25 yr) was four times greater in fertilized than in unfertilized stands, and trees larger than the stand average were preferred. Fertilization has been shown to attract bear damage by increasing the sugar content in the phloem, while having little effect on the terpenes that would dissuade bear consumption (Kimball et al. 1998).

REVIEW QUESTIONS

1. How are nutrients distributed throughout a forest stand, where do they occur? How do they vary by stand component? Describe and explain.
2. How does the demand for nutrients vary with stand age in even-aged stands? Explain. Compare with uneven-aged stands.
3. Explain the direct and indirect effects of forest fertilization.
4. Why is an understanding of the direct and indirect effects of forest fertilization important for implementing efficient forest fertilization prescriptions?
5. Explain how forest fertilization effects may vary by site quality.
6. How can the effects of fertilization be predicted? Explain the methods tried/used to predict response.
7. Why may fertilization effects last for longer time periods on low quality sites?
8. Explain how thinning and stand density are related to fertilization?
9. What effect may fertilization have on factors other than stand growth? Explain and discuss.
10. What are the pros and cons of fertilizing young and old stands?

CHAPTER 10
Multi-Aged Management and Complex Stands

Multi-aged management in western U. S. forests has traditionally focused on wood production (i.e. selective harvesting, including salvage of poor quality trees), similarly to even-aged management philosophies. In addition to ongoing societal interest in managing forests for commercial wood production, however, there is increasing interest in managing many forests for a broad range of ecosystem values, including improved aesthetics associated with reserved canopy cover; more heterogeneous habitat for native plants and animals; watershed integrity; and the restoration and maintenance of resistance to wildland fire, diseases, insects, and the potential effects of climate change (Puettmann et al. 2009; Franklin and Johnson 2012; O'Hara 2014). Wood production may be the main goal, or one of several goals, or a by-product of these other ecosystem management objectives that can be more readily accomplished using multi-aged management.

Management of multi-aged stands involves many of the same treatments as the management of even-aged stands, but often applied with less intensity and on a smaller temporal and spatial scale. Two-story systems (see chapter 2) with variable levels of green tree retention are the simplest structures in a spectrum of multi-aged management approaches, while multiple-species many-aged stands are the most structurally complex. All multi-aged approaches require treatments in order to maintain a continuous cover of trees of appropriate species and sizes This usually involves regular partial harvest entries (Smith et al. 1997) that are often supplemented with treatments like prescribed fire, planting, vegetation management, and precommercial thinning.

In multi-aged stands, we simultaneously manipulate the composition and growth of small, medium, and large trees together. Reduction of stand density is not only designed to provide growing space for the establishment of new cohorts of small trees, but also to stimulate the growth of medium and larger trees into progressively larger size classes. In essence, we combine regeneration methods (e.g., an irregular shelterwood with gap creation) with thinning methods common to even-aged management. As in even-aged management systems discussed in earlier chapters, there are predictable effects of altering stand density on residual tree growth and susceptibility to insects, diseases, wind throw and fire, while it is equally important to maintain healthy tree crowns and vigorous trees overall. The future quality of any multi-aged stand rests in: 1) the "preferred" growing stock, the real crop trees; and 2) the "reserved" growing stock around them, which may become preferred stock in the future (Smith et al. 1997). Multi-aged stands are like mid-rotation even-aged stands that have reduced canopy cover following thinning, but tree sizes are more highly variable, and on completely different management paths. See chapter 2 for more discussion of these contrasting silvicultural systems.

Uneven-Aged Management—Background and History

Complex multi-aged structure is best understood by first considering balanced uneven-aged stand structure and its management. The silviculture of uneven-aged management is built around two concepts: 1) maintenance of continuous canopy cover within some narrow range of stand density; and 2) manipulation of a broad distribution of tree size classes and (often) species. Total stand density, maximum tree size, and number of trees per size class define the particular structural goals for balanced uneven-aged stands. And as with even-aged stands, these values will depend on species composition, site productivity, and management objectives.

Because there is relatively little accumulated experience of long-term uneven-aged management in western forests, estimates of appropriate stand density are often based on growth and yield information from even-aged stands (see chapters 5 and 8). Stand Density Index (SDI) and basal-area values from even-aged stands can be useful for initial estimates of the densities that can maintain adequate tree and stand growth rates in uneven-aged stands, as well as reduce the probability of unacceptably severe insect attack or disease. This information will provide objective guidelines until more structure- and species-specific guidelines can be developed from experience with specific forest types and locations, although sufficient information has already accrued for some forest types to apply this system successfully (e.g., Bailey and Covington 2002). To make guidelines more site-specific, one could also measure nearby multi-aged stands that show desired properties (e.g. established regeneration, adequate growth and yield, and low pest populations) and set guidelines using their density and structure.

The range of tree sizes, from saplings to the largest diameter, depends primarily on the management objectives and capabilities of the species and site. Higher tree densities will produce more wood volume, deeper shade, and better hiding cover for some wildlife; in contrast, lower densities will promote faster growth of larger trees, improve resistance to fire and pests, offer more pleasing aesthetics to the majority of the public, and provide larger snags for nesting habitat. From a fiber production point of view alone, stand density and size class structure can be determined in part by the relative growth and/or growth efficiency (i.e.. volume growth relative to size or occupied growing space) of different sizes of trees under different stand densities. In even-aged stands, smaller trees often have greater relative growth or growth efficiency than larger trees (O'Hara 1988, Larocque and Marshall 1993), and both relative growth and eventually absolute growth decline with increasing age (Ryan et al. 1996). However, absolute growth per unit area and/or mean annual increment (see chapter 6) may still be increasing while relative growth is declining. Note that these trends may also occur for individual trees and for groups of trees of similar age within otherwise uneven-aged stands, but the relationship between tree size and relative growth or growth efficiency becomes more complicated across an entire stand.

In a stand with a balanced uneven-aged structure, or in a strongly stratified even-aged structure (Roberts et al. 1993), growth efficiencies are expected initially to increase with tree size, primarily due to associated increases in light availability, but then wane with further increases in size due to normal age-related decline; see O'Hara (1996) and Maguire et al. (1998) for details and exceptions. In uneven-aged stands managed for wood production, therefore, once trees have grown to a size that provides the desired wood product, and the economic return measured by percent volume growth falls below a the desired threshold, there is no further economic incentive to retain them in the stand. Replacing them with smaller and younger trees with faster relative growth rates is considered economically more efficient. Large trees, grown for aesthetics and wildlife values or resistance to fire, may be done so at a sacrifice of stand volume growth and efficiency.

SETTING STAND STRUCTURAL GOALS

We will offer two examples of setting and calculating stand structural goals to create and maintain a balanced uneven-aged ponderosa pine stand for which a residual SDI and basal area have been prescribed. Other methods also are available for determining appropriate density and structure goals for uneven-aged stands (e.g. O'Hara and Gersonde 2004). Cochran (1992), Long and Daniel (1990), Long (1998) and chapter 6 discuss assumptions behind application of SDI to uneven-aged stands, as well as the mechanics of calculating and applying it to density regulation.

ALLOCATING SDI

DeMars and Barrett (1987) allocated growing space across diameter classes using SDI. Assuming a maximum SDI for ponderosa pine of 365 in their example (table 10-1), they targeted a desired SDI of 120, or 33% of maximum, after each harvest entry. This low target density was appropriate for establishment and subsequent growth of regeneration (seedlings and saplings) of ponderosa pine on dry sites (Bailey and Covington 2002). Target residual SDI may vary by site index and/or stocking by size class distribution as a result of assessing that stand, or past growth rates and stocking (including natural regeneration) of nearby even-aged or other uneven-aged stands (Cochran et al. 1992).

Like calculating SDI for an even-aged stand (chapter 6), the equation for calculating SDI for a multi-aged stand is:

$$SDI = \Sigma\, TPA_i(D_{qi}/10)^b$$

where TPA_i is the number of trees per acre in each diameter class i; D_{qi} is the quadratic mean diameter of diameter class *i* (assumed to be the midpoint of each diameter class i), and b is the exponent on the TPA-D_q relationship (hence, the slope of a given SDI line on the logarithmically transformed scale of stand density management diagrams; see chapter 6). Note that b is 1.765 for ponderosa pine in eastern Oregon (Cochran et al. 1992), although a slope of 1.605 from Reineke's (1933) original work is traditionally used for defining SDI in absence of other information about this value for a given species (Daniels et al. 1979). Allocating SDI requires three basic steps (table 10-1):

1. Make a first approximation by allocating equal amounts of the SDI to the selected diameter classes (typically four as in the example, implying the number of trees/ac in each class. This yields 74 trees/ac, for example, in the six-inch diameter class (more than half of the trees).
2. Evaluate this first approximation. In this example, the six-inch diameter class has 15 ft^2/ac of basal area, considered high for trees of this size.
3. Reallocate SDI as needed, in this case 20 units of SDI from the six-inch size class to the largest class. This reallocation was based on experience and several important management considerations: the need to maintain vigor of regeneration by reducing competition among small trees, and to minimize fire hazard and ladder fuels that can lead to severe fire behavior. Other reallocations of SDI might have been tried (e.g. to the middle diameter classes); this reallocation did not have to go only to the largest class. Total basal area is only 6 ft^2/ac higher after reallocation, but there are 45 fewer trees/ac. Total SDI remains unchanged, of course.

Trees with less than four-inch diameter were not considered in this example, but note that saplings must be left at sufficient density and spacing/location that they can grow into this smallest size class. Growing space therefore must be provided to achieve successful establishment and vigorous growth of regeneration so as to ultimately provide a sufficient number of saplings. The number of trees in each successively smaller/younger size class must be equal to or greater than the target number of trees for the next larger size/age class.

Finally, trees larger than 32 inches diameter were present but infrequent on these sites—one or two every 5 acres; a low density of such individuals is unlikely to affect

Table 10-1. Allocation of SDI to provide goals for uneven-aged management of ponderosa pine. The SDI after a cutting cycle is allocated across four diameter classes in this example:

	First Approximation				Adjusted			
Diameter class	*SDI*	*Trees /ac*	*Basal area /diameter class (ft²)*	*Basal area /ac*	*SDI(a)*	*Trees /ac(a)*	*Basal area/ac(a)*	*Tree spacing (ft)*
4.1–8.0	30	74	0.196	15	10	25	5	42
8.1–16.0	30	22	0.785	17	30	22	17	45
16.1–24.0	30	9	2.181	20	40	12	26	60
24.1–32.0	30	5	4.276	21	40	7	30	66
Total	**120**	**108**		**72**	**120**	**63**	**78**	

the outcome of uneven-aged management, or only have very localized effects on stand dynamics, so they can be ignored where management objectives include the conservation of large trees. Where objectives include economic return from timber, such large trees are typically removed; where objectives include creation of habitat for cavity nesters, such large trees are preferred snags given their size and longevity in the landscape.

The last column in table 10-1 shows the equivalent average spacing of trees in the different size classes. This average spacing may be useful in selecting and marking residual trees to achieve the adjusted allocation of SDI, particularly in the smaller diameter classes. This should not imply the objective of perfect spacing among trees, however. Residual growing stock has to be left where it will survive and grow well into the future, and that rule should primarily guide the field marking.

Growth models can be used to: (a) project diameter growth of trees over time into larger-diameter classes, (b) determine the resultant rate of change in stand-level SDI, and thus (c) calculate an appropriate cutting cycle and associated yield. A reasonable upper limit at which to re-enter a stand is typically 45-60% of maximum SDI. Growing stands to a higher SDI will result in loss of established regeneration, as well as occasional mature trees, particularly those of shade-intolerant species, and may increase stand susceptibility to insects and diseases or wildland fire.

DETERMINING TARGET DIAMETER DISTRIBUTIONS WITH BDQ

Calculating a target diameter distribution using the "Basal area – maximum Diameter – diminution Quotient (BDQ)" method is a similar way to set structural goals for balanced uneven-aged stands. Like SDI allocation, silviculturists must: 1) specify diameter class size (e.g., 2- or 4-inch diameter classes); 2) set a smallest diameter class to manage, and evaluate how trees get to the size class, (natural or artificial regeneration, precommercial thinning); 3) prescribe the largest tree size the stand may have, even though there may currently be some trees beyond that size; and 4) estimate growth and stand dynamics over the time between harvest entries. The BDQ method specifically is described by:

1. Total basal area (B) to be left after partial harvesting. Basal area should be set low enough that trees have sufficient growing space throughout the period between harvest entries (e.g., 15- 25 years), and that regeneration can establish and grow into the smallest-diameter class. When studies of basal area-tree growth relationships in uneven-aged stands are lacking (as with SDI), a basal area target can be determined from even-aged stands, growth models, or measurements of nearby "reference" stands.
2. The largest diameter class (D), which thereby establishes the maximum residual tree size depending on the width of diameter classes. For example, 32 inches is the largest tree in the 30-inch diameter class when using 4-inch classes.
3. A diminution quotient (Q)—the ratio of the number of trees in a given diameter class to that in the next-larger class. For example, a Q of 1.1 for a stand with 10 trees/ac in the largest-diameter class would mean 11 and 12.1 trees/ac in the next two smaller diameter classes, respectively. For many western conifers, Q values of 1.1 to 1.4 appear to be satisfactory to account for gradual slowing of growth with size/age and low rates of mortality in the stand.

Developing a balanced uneven-aged diameter distribution using the BDQ approach is simple; see also Alexander and Edminster (1977), Guldin (1991) and Bailey and Covington (2002). As an example, a silviculturist can set a target basal area (B) and upper diameter limit (D) at 80 ft^2/acre and 24 inches, respectively, based on existing stand management objectives, with five 4-inch diameter classes (table 10-2). To construct the target diameter distribution:

1. Arbitrarily choose a number of trees/ac for the largest-diameter class in order to make a first approximation; in this example, 2.0 trees/ac in the 22-in class (20–23.9 inches).
2. Compute the number of trees in each progressively smaller diameter class by multiplying the number of trees in the adjoining higher diameter class by Q (1.2 in this example).
3. Calculate the basal area in each diameter class and sum for the stand. In this case our first approximation resulted in only 15.33 ft^2/ac, a factor of 5.22 less than the desired 80 ft^2/ac of basal area.
4. Multiply the arbitrary number of trees/ac in each diameter class by this factor to provide the actual number of trees/ac in each diameter class that sums to 80 ft^2/acre.

The same procedure is used to calculate distributions using Q values of 1.3 and 1.5 for comparison (table 10-2). Larger Q values result in larger numbers of trees and basal area in smaller diameter classes, which provide greater biomass and cover

Table 10-2. An example of calculating a diameter distribution for an uneven-aged stand with a basal area of 80 ft^2/acre, a q of 1.2 and an upper diameter of 24 in. Distributions using q values of 1.3 and 1.5 for the same basal area and diameter limits are included for comparison.

Calculating trees and basal area/acre for a q of 1.2.

Diameter class	*Estimate trees/acre (q=1.2)*	*Basal area/ class (ft^2)*	*Estimated basal area/ acre (q=1.2)*	*Adjusted trees/acre (q=1.2)*	*Adjusted basal area (ft^2/acre) (q=1.2)*
4.1-8.0	4.15	0.2	0.83	21.7	4.3
8.1-12.0	3.46	0.55	1.9	18.1	9.9
12.1-16.0	2.88	1.07	3.08	15.0	16.1
16.1 -20.0	2.40	1.77	4.24	12.5	22.1
20.1-24.0	2.00	2.64	5.28	10.4	27.6
Total			**15.33**	**77.7**	**80.0**

Correction factor: 80/15.33 = 5.22

Trees and basal area/acre for q's of 1.3 and 1.5. for comparison.

Diameter class	*Trees /acre (q=1.3)*	*Basal area (ft^2/acre) (q=1.3)*	*Trees /acre (q=1.5)*	*Basal area (ft^2/acre) (q=1.5)*
4.1-8.0	28	5.3	38.1	7.7
8.1-12.0	21	11.3	25.7	14.1
12.1-16.0	17	16.9	17.0	18.1
16.1 -20.0	12	21.5	11.3	20.1
20.1-24.0	9	24.9	7.6	19.9
Total	**87**	**79.9**	**99.7**	**79.9**

in lower canopy positions but can result in slow understory tree growth, a need to thin small unmerchantable trees, increased risk of fire from slash or dense seedling/sapling fuel ladders, and unnecessarily low overstory density that reduces total stand production. Lower Q values shift growth into larger trees and, when combined with lower basal area and higher upper diameter limits, lead to an open stand of large trees that is considered sustainable for most western forests. Too many trees in the smaller size classes make drier mixed-conifer stands particularly prone to crown fire and were historically isolated in fire-prone landscapes. Silviculturists therefore choose among B, D and Q options that best meet management objectives. It is important to realize that the stand structure implied by a Q-value will vary depending on the width of the diameter classes; for example, the Q-value needed to specify a similar target structure is lower for 2-inch diameter classes than for 4-inch diameter classes.

These two methods for developing structural goals for uneven-aged stands, SDI allocation and BDQ, are both flexible, because different amounts of SDI or basal area can be assigned to different diameter classes. The common problem of having too many trees in the smaller diameter classes and too little basal area in larger diameter classes can be addressed by increasing the SDI in the larger classes or by lowering Q.

In the past, BDQ targets were often based on the observed structure of natural stands rather than on the productive potential of alternative structures. As experience has accumulated in various forest types, however, some stand structures have been identified as more productive than others (Nyland 1996). Simulation of alternative structures with regional growth models allows comparison of growth efficiency as measured or predicted volume growth per unit SDI across different diameter classes. However, neither the SDI Allocation nor BDQ approach directly account for tree vigor, resistance to bark beetles, and stand growth.

OTHER APPROACHES

The two methods above focus on developing stand management guidelines by assigning stand density, numbers of trees and SDI/basal area, to diameter classes. O'Hara's (1996) approach is based mainly on leaf area and tree vigor rather tree number by diameter class. He developed a method of assigning leaf area to cohorts (roughly the same as age or size classes) in uneven-aged ponderosa pine stands. The method estimates that maximum two-sided leaf area index (foliage area per unit ground area) for ponderosa pine is approximately 10 m^2/m^2, and assigns about 60% of this maximum stocking for distribution among age or size classes, species, or canopy strata to achieve a desired residual structure (O'Hara 1996; O'Hara and Gersonde 2004). This approach may enable managers to design stand structures that are directly related to productivity, assuming that growth efficiencies or volume growth rates per unit leaf area are known and stable. It usually allocates a higher proportion of the leaf area to the larger trees, whose numbers decline with age.

O'Hara's (1996) approach does not guarantee a "balanced, sustainable structure" unless some portion of the 6 m^2/m^2 of residual leaf area is reserved for regeneration. However, if the assumed growth efficiencies hold, then it should be easier to design a vigorous, productive stand structure. As with SDI allocation, the leaf area allocation approach does not force the stand to conform to a smooth curve as the BDQ approach does; hence, in combination with information about growth efficiencies, it can take advantage of current stand structure to provide an efficient use of tree vigor and potential growth of different cohorts. This may avoid problems like heavily cutting vigorous trees in middle-size diameter classes, and thereby sacrificing stand productivity in order to achieve a perfect smooth diameter distribution.

O'Hara and others (2003) provide a model to implement their method of allocating growing space by leaf area. Like an individual-tree growth model, it estimates cohort and stand volume growth under alternative initial structures, as well as basal area, SDI, and other stand values at the end of a cutting cycle. Although individual-tree growth models offer greater precision, including characterization of stand structure, the relative accuracy of these approaches has not been tested. Leaf area allocation requires an intensive stand evaluation before application, as well as equations for estimating leaf area on standing trees. This latter relationship and that between leaf area and volume growth will vary somewhat by species, size class, and site, as do many of the relationships in other growth models.

Multi-Aged Management

The systematic approaches to balanced uneven-aged management presented above should be viewed as a working model subject to revision in most western forests because of the high variability in topography, soils, species/genetics, climate and periodicity of disturbance and regeneration processes (figure 10-1). The intent is not to establish a perfect balanced all-aged distribution of trees, but rather to provide goals for maintaining a multi-sized structure. Fortunately, the ecology of the forest and achievement of management objectives are likely insensitive to deviations from a smooth curve. In fact, those same forces that make a perfect distribution so difficult to achieve provide a myriad of opportunities for creating complex multi-aged stands that can persist indefinitely in a landscape.

Multi-aged management draws strongly on the inherent condition and potential of the existing stand. Concepts from the "Dauerwald" or "continuous forest" system were long ago proposed for European forests (Troup 1926; Smith et al. 1997) and may be applicable to western forests in some stand types and ownerships. This system is "opportunistic or eclectic," and the condition of the stand determines the treatments employed (i.e. a free mixture of thinning, release of advanced regeneration, harvest of diseased trees, and other treatments). There is no attempt to systematically regulate stand structure, growth, or yield.

Multi-aged management creates and maintains a broad range of stand conditions over time, including some habitat for early seral plant and animal species despite a relatively continuous cover of trees. Green tree retention, either dispersed or in groups, is used to 1) carry forward legacy structure, composition, and function from the past into the future forest; 2) provide for broad wildlife habitat needs; 3) disseminate lichen,

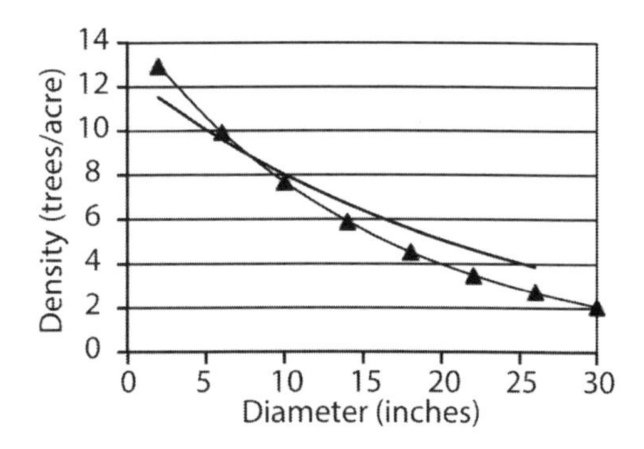

Figure 10-1. Two diameter distribution for unevenaged-stands: a) basal area 50ft^2/acre and upper diameter of 26 in. and a Q value of 1.2 (plain line); b) basal area of 50ft^2/acre, upper diameter of 30 in. and Q of 1.3 (triangles). In these examples the basal area remains constant. Setting Q at 1.2 and the diameter limit at 26 in. increases the basal area in 10 to 26 in. diameter classes. Basal area, Q and upper diameter can be adjusted to grow stands that produce different structures.

bryophyte, and liverwort spores into the next stand; and 4) improve aesthetics following harvest (Kohm and Franklin 1997). These roles extend well beyond a shelterwood: providing seed and shade for natural seedling establishment (Smith et al. 1997). Shelterwoods can contribute to the formation of old-growth structure over time, since large, quite old trees have the capacity to increase their growth rates and expand their crowns in response to removal of trees around them (Latham and Tappeiner 2002, Kolb et al. 2007). However, there is currently not enough experience with these stands to fully understand their development and potentials.

Managers should therefore proceed with caution. Treatments have many of the same considerations as thinning in even-aged stands (chapter 8), as used to regulate under- and mid-story development. Depending upon their densities, overstory trees reduce the growth of understory trees (Zenner et al. 1998; Acker et al. 1998; Huff 2009). Leaving residual large trees in aggregates typically concentrates both their value as habitat and the effect on the rest of the stand, may lessen the possibility of loss on windthrow-prone sites, and makes silvicultural treatments in the understory safer and more efficient. Leaving trees with low H:D ratios may also help lessen windthrow. If overstory trees are infected with diseases (e.g. dwarf mistletoe in ponderosa pine), then understory saplings and young trees are at risk of serious impact (systematic infections, slow growth and early death). Making snags of the large residual trees with infection is an option for reducing disease impacts while retaining some of their habitat value. Finally, since regeneration occurs regularly and throughout the understory, these stands may be more susceptible to wildland fire and, when left unmanaged over time, lose all early seral conditions as they grow in the absence of disturbance.

In multi-aged management using either regulated or unbalanced stand structures at any given point in time and space there may be an overabundance or lack of a given species and/or size class in a stand, depending on the effects of the most recent silvicultural treatments or disturbance(s). Silviculturists must understand the site (current condition and dynamics), evaluate the importance of any deviation from a desired diameter distribution and/or disturbance processes at work on the site, and plan the next stand entry appropriately—timing, activity, and intensity of the treatment(s). These approaches require the same intensity of management as even-aged approaches, but not the same degree of precision and control.

MARKING MULTI-AGED PRESCRIPTIONS

The above calculations and consideration focus on distinguishing between a) preferred and reserved growing stock to be left on-site versus b) the trees removed in order to provide for future growing space for that stock. Implementing prescriptions in the stand follows one of two fundamental approaches: **individual tree selection** or **group selection** (Smith et al. 1997), depending upon worker preference. Individual tree selection focuses on retaining the best single trees, at least within sight of the workers, which meet the selection criteria for preferred or reserved growing stock. This typically results in more uniform spacing among trees—a desired characteristic for some management objectives (e.g. equal distribution of growth among trees or more uniform canopy connectivity, and regeneration of shade tolerant species). Group selection focuses on the retention or removal of entire clusters of trees, which typically creates a clumpy residual stand or small gaps/openings, respectively, more appropriate for other objectives (e.g., altering fire behavior, promoting early-seral microsites and regenerating shade-intolerant species). Removing 25% of a stand's basal area via group selection likely would be a cut-tree mark dominated by scattered small openings with some partial harvests between gaps; removing 75% of the basal area would likely be a leave-tree mark with scattered individuals and clusters of trees left in a random pattern.

Beyond tree age/size, preferred and reserved growing stock is selected based on a tree's perceived potential to grow into the future to meet the management objectives. Crown classes (e.g. "dominant" and "co-dominant") are a less useful description of growing stock quality in multi-aged stands because trees of different sizes and ages are inherently mixed together. This mixing may occur when trees of species with rapid height growth, such as western larch or lodgepole pine, overtop Douglas-fir or grand fir (Cobb et al. 1993). It also occurs as regeneration becomes established in the understory or in openings. Thus, a tree with a full crown might be in a subordinate position to a larger tree, but classifying it as suppressed might be misleading if it has a high potential for increasing its growth if the larger trees around it are removed. A similar problem with crown classification occurs with variable horizontal structure that includes large gaps. Under these conditions, small trees may occupy the dominant canopy position of an opening, but they are much smaller than the overstory trees growing along the edge. Therefore, in multi-layered and complex stands, it is often best to use crown classes by canopy strata or layers in order to rate the growth potential of trees (Richards 1964).

Tree classification schemes have been developed that emphasize relative tree vigor, insect and disease resistance, and apparent predisposition to mortality to address the heterogeneous structure common in complex multi-aged/multi-sized stands and to better guide silvicultural choices and tree marking. These systems take into account the shape and condition of the crown and other tree characteristics along with crown class (table 10-3). Dunning's (1928) classification was developed for ponderosa pine in California, which often grows in mixed-aged/mixed-size stands with a mixture of shade-tolerant and -intolerant species. Conditions in such mixed-conifer stands require consideration of tree ages as well as their relative positions in the canopy. The major factors that enter into Dunning's classification system are: (1) age class; (2) degree of dominance within an age group; (3) crown shape and thrift or vigor, as judged from density and color of foliage, form at the top of the crown, and crown size (relative to

Table 10-3. Examples of tree classification systems to rate vigor, susceptibility to insects, and probability of mortality.

Author	*Species/forest type*	*Tree classes*	*Criteria*
Dunning (1928)	Ponderosa pine/ mixed conifer	7:1–7	Age, crown and bark characteristics, overtopping
Keen (1936, 1943)	Ponderosa pine/ insect risk	28: 7 vigor classes within 4 age classes	Age crown characteristics (branch angle, shape), bark texture
Salman & Bongberg (1942)	Ponderosa pine/ insect control	4: I–IV	Needle color, length, density, dead/dying branches
Ferrell (1980, 1983)	Red and white fir, mortality and growth	4: relative ranking	Crown raggedness, bark fissures/cortex, live crown ratio
Taylor, R. F. (1939)	Lodgepole pine	4:A–D	Crown profile, area, ratio, color, density, shape

age); and (4) color, thickness, and texture of the bark (table 10-3). Van Pelt (2008) has developed a similar guide for identifying old Douglas-fir trees from visual cues; Ferrell (1980 and 1983) developed a similar rating system for evaluating growth potential and likelihood of mortality of red and white fir in California.

Dunning's and Ferrell's systems both used many of the same crown and bark characteristics found in table 10-3, and were accompanied by tree-profile pictures or sketches to help determine tree classes. Ferrell's system assigned points to rate each characteristic, and the points were summed for an overall tree ranking. In another system, Keen (1943, figure 10-2) used tree ages and crown characteristics to rate ponderosa pine susceptibility to bark beetles. These tree classification systems can help select trees to cut or leave, because they are related to tree growth potential and the likelihood of mortality. All such classification systems and pictorial guides can be used to define cohorts in growing stock allocation methods (Ferrell 1983, O'Hara 1996). Selection of appropriate species for microsites, as well as individual tree vigor and potential effects of insects and diseases, are important criteria for successful use of multi-aged systems—more important than the establishment of an ideal diameter distribution (Graham and Smith 1983, Guldin 1991).

REGENERATION DYNAMICS

See chapter 7 for further discussion of regeneration in complex of multi-age stands.

Seedling establishment need not occur annually (or even every decade) for multi-aged management to be effective and sustainable over time. The system only requires that regeneration and growth is sufficient to replace trees removed from larger size classes. However, if regeneration is lacking in all or part of a stand, then planting may be a wise

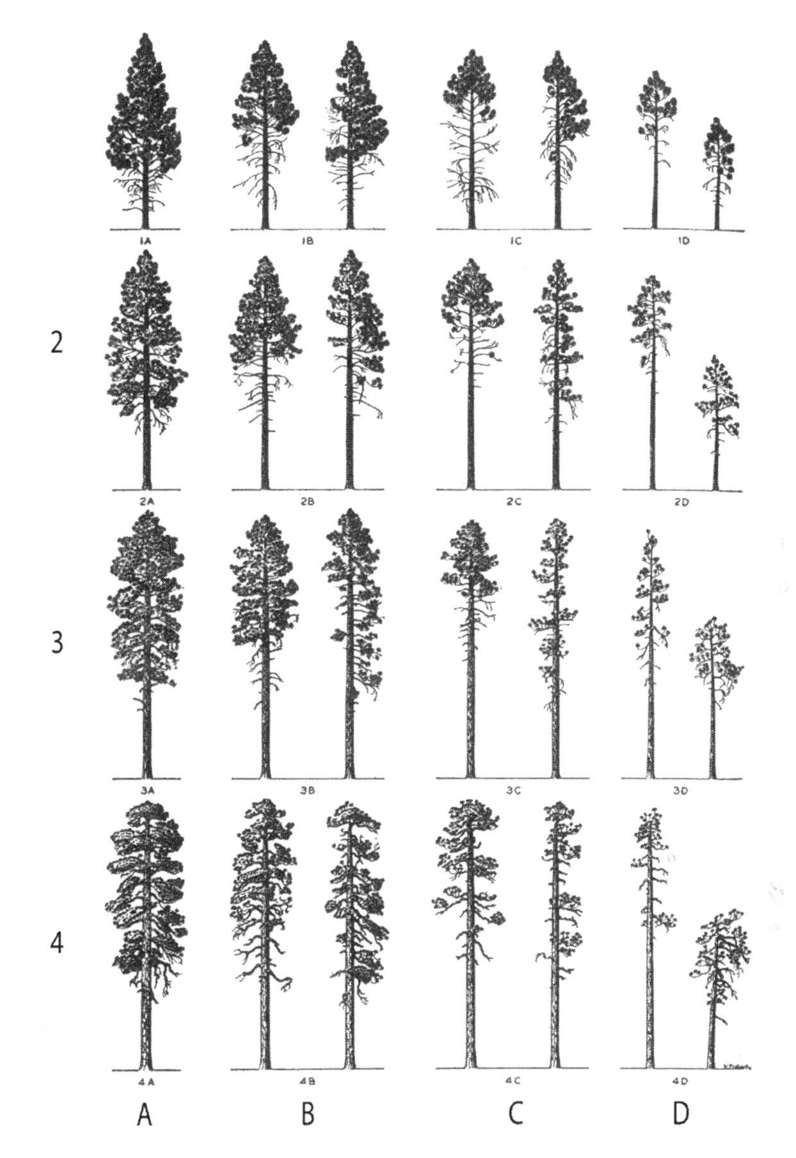

Figure 10-2. Keen's (1943) tree-classification system for evaluating ponderosa pine susceptibility to bark beetle attack. Tree profile in the upper left section of the figure are the most vigorous and least susceptible. Susceptibility increases proceeding down and to the right in the figure. This is a helpful relative ranking; susceptibility also varies with wet or dry periods, presence of root diseases, stand density, and other variables.

investment (especially if potential natural regeneration is not of the desired species). For example in mixed conifer stands natural regeneration of true fir and incense cedar may be abundant, but harvesting or prescribed fire and planting may be needed to maintain ponderosa pine in the stand. Regular removal of trees from the mid- and upper-range diameter classes is necessary to release seedlings to grow into the smaller

Figure 10-3. Western hemlock established in the understory of Douglas-fir stands thinned to about 50 (A) and 80 (B) trees/ac 20 yr previously. Hemlock was planted in A because there was no seed source in the stand. It was established by natural regeneration following thinning in B. Overstory density is high in both stands and needs to be reduced for substantial hemlock height growth.

diameter classes, as well as for crop tree development in successively larger size classes (like thinning in even-aged stands). Without maintaining appropriate tree density in all diameter classes, mid-size and larger trees will dominate a stand and likely create a two-storied stand with a suppressed understory.

Advanced regeneration (figure 10-3) and natural regeneration following harvest are fundamental in producing multi-aged and multi-story stands, but it must be of the desired species. In mixed-conifer forests of the West, shade-tolerant species such as true firs, incense-cedar, and black oak frequently occur in the understory as advanced regeneration, as do ponderosa and lodgepole pine seedlings on drier sites. Ideally, this seedling bank is of desirable composition and sufficient vigor to respond to release. These seedlings should have the potential to grow into saplings and poles, and eventually into overstory trees. Similarly, natural regeneration is often dominated by shade-tolerant species following partial harvest, particularly in the absence of regular surface fire. It may become necessary over time to thin smaller shade-tolerant saplings (similar to pre-commercial thinning in even-aged stands) as well as to plant shade-intolerant species to maintain desired stand composition. Prescribed burning can be used efficiently in drier forest types to remove small, poorly formed trees of fire-sensitive species, as well as to stimulate natural regeneration of desirable species.

Successful natural regeneration depends not only on tree and shrub species regeneration characteristics, but also on degree of disturbances, seed crops, etc. (see tables 7-5, 7-6. For example, natural regeneration of ponderosa pine, Douglas-fir, white fir, Engelmann spruce, and western white pine from seed, as well as aspen from root suckers, occurred in 1- to 2-ac openings in mixed-conifer forests in Arizona (Ffolliott and Gottfried 1991). Regeneration ranged from 1-11 years of age and density was 900 conifers and 100 aspen per acre. Density reduction treatments are likely needed to reduce inter-tree competition and favor shade-intolerant species on such sites. Regeneration surveys on two Sierra Nevada forests (McDonald 1976a, 1994; Olson and Helms 1996) indicated that natural seedlings of desired species occurred following both single tree and group selection approaches, although seedling growth in the single tree selection stands was considerably less than in the group selection stands (McDonald 1976a), and saplings were generally lacking (Olson and Helms 1996). There may not be sufficient growth of natural seedlings and saplings to ensure that these species remain major components of the overstory if stand density is not periodically reduced (Lilieholm et al. 1990).

Partial harvesting is also likely to result in abundant regeneration of shrub species, particularly on productive sites. Competition from manzanita, ceanothus, and salal is of particular concern in multi-aged management of Douglas-fir and pine species on productive sites, as are stump sprouts of tanoak and bigleaf maple. These woody species often preclude natural regeneration of conifer seedlings for decades unless they are controlled and/or desired tree species are planted (Brandeis 1999; Brandeis et al. 2001; Coates 2000).

Small openings associated more with group selection may favor natural seedling establishment, but not subsequent growth. Gray and Spies (1997), in a study in Douglas-fir stands where gaps ranged from 0.01 to 0.5 acre, found that establishment of western hemlock seedlings was greatest in the smaller gaps and was aided by shade in larger gaps. However, seedling size was greatest in large gaps, and western hemlock seedlings on the forest floor (mineral soil) grew more than those on dead wood. Seedlings are not expected to grow as rapidly in multi-aged systems with continuous canopy cover as they are in even-aged systems, because they are growing among larger trees. They must be able to respond to subsequent release, however. In mixed-conifer forests, trees increased their growth rates up to 40-50 ft from the gap edge (York and Heald 2008).

Promoting Complex Stands with Old-Growth Characteristics

There is increased interest in converting thousands of acres of young stands, primarily in Douglas-fir forests on federal land, to more structurally complex, old-growth-like

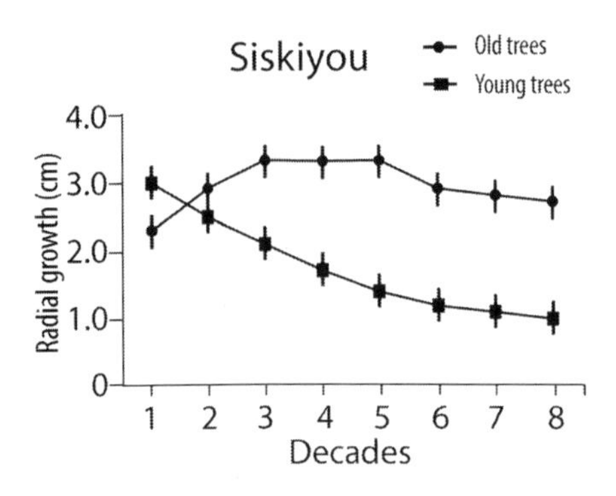

Figure 10-4. Radial growth rates of old-growth trees (circles) and dominant and codominant young trees (squares). After about 20 yr the old trees grew at a steady or slightly declining rate, while the young trees declined during the same period. From Sensenig et al. (2013).

stands (Franklin and Johnson 2012, Muir et al. 2002, Bailey and Tappeiner 1998, Spies and Franklin 1991). We use this example from the Douglas-fir region in the Pacific Northwest to illustrate the potential of adapting traditional silviculture to solve contemporary forest management issues. This new, adaptive silvicultural system is a "work in progress" and will likely be modified in the light of experience, emerging science and future policy changes. The details of growing complex stands with old-growth characteristics will vary widely by forest type, especially for fire-prone forests, since old-growth structure is not compatible with large-scale high-severity fire (Franklin et al. 2013). The primary old-growth characteristics in our example include:

- A main canopy supporting abundant trees with large diameters, wide/long crowns, large branches, and deeply furrowed bark;
- Mid- and understory tree canopy layers of conifers and hardwoods of varying densities, sizes and species composition (including scattered regeneration);
- Cover of native shrubs and herbs, particularly in canopy openings;
- Large, dead standing snags and fallen trees;
- High horizontal and vertical diversity in all of the above.

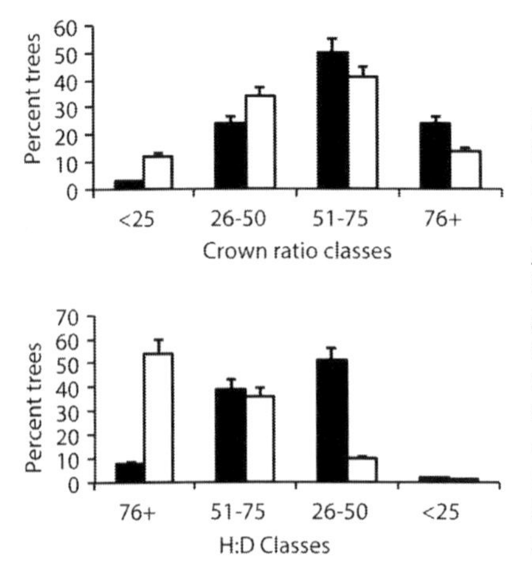

Figure 10-5. Percent trees by crown and H:D ratio classes in old-growth and young stands in southwestern Oregon. Populations of old-growth trees (black bars) had larger crowns and lower H:D ratios than populations of young trees (open bars). Nevertheless there were trees in young stands that could in time following thinning, produce crowns and stems like the old trees. Large, deep crowns are important characteristics of old-growth trees that contribute to stand structure and habitat for a variety of taxa. Trees with low H:D ratios are less susceptible to damage by wind and the effects of heavy loads of ice and snow than trees with thin stems and high ratios-—approximately 70+ (From Sensenig et al. (2013; See chapters 5 and 8).

These stand characteristics support many native fauna and flora that are not found, or are less abundant, in young forests (Ruggiero et al 1991). The density, species composition, and other details of old-growth forest structure vary throughout the Douglas-fir region (Poage and Tappeiner 2005, Spies and Franklin 1991). Western hemlock, western redcedar and bigleaf maple are common understory species in moist forest types; on drier sites, Pacific madrone, grand fir, tanoak and Douglas-fir are common understory trees. Local examples of old-growth forests, however, can be used to provide realistic goals for species composition and stand structure (Poage and Tappeiner 2005). The spatial heterogeneity of these older stands contrasts with that of younger stands (20 to 50 years old), which are typically well-stocked (>300 trees/ac), uniform, productive plantations regenerated following clearcutting and currently undergoing self-thinning (Sensenig et al. 2013, Tappeiner et al. 1997).

Old-growth stands often develop quite differently from contemporary young stands. The range of ages of the large old-growth trees suggest that they often established over periods of 100 years or more, especially on sites where fire was common (Sensenig et al. 2013). The large main canopy trees in these old growth stands grew at much lower densities; therefore, diameter growth at comparable ages was much greater in old-growth trees than in even the larger dominant and codominant trees of today's young stands (figure 10-4, Sensenig et al. 2013, Poage and Tappeiner 2002 and

Figure 10-6. Example of old-growth Douglas-fir trees with long crowns, large branches, and large stems with furrowed bark. Tree spacing is wide and irregular with space for understory trees, shrubs and grasses among them.

Tappeiner et al. 1997). Furthermore, height: diameter ratios were much lower and live crowns longer in old-growth trees, compared to the largest trees in the young stands (figure 10-5). Density reduction in young stands clearly promotes the development of trees with the characteristics of the large main canopy trees in old-growth forests. Many young stands have over 300 trees per acre, while there were often less than 40 to 50 trees per acre in the old stands. Self-thinning is not likely to produce trees with the large stems and crowns, and low H:D ratios typical of old-growth.

In addition to facilitating the development of individual large trees with the desired characteristics, lower stand density promotes other old-growth characteristics such as the establishment of understory shade-tolerant conifers, hardwoods, shrubs and other vascular and non-vascular plants (Bailey and Tappeiner 1998, Bailey et al. 1998). One or two heavy thinnings could lead to this type of old-growth structure, as could frequent, lighter entries. A variable density of residual trees will likely promote complex structure, since vertical and horizontal heterogeneity are important features of old-growth stands (figure 10-6). The establishment and growth of appropriate species (including hardwoods for mast production) are favored by reduced density and the existence of canopy openings (Bailey and Tappeiner 1998, Shatford et al. 2009); planting within low density stands is another way to achieve desired understory composition (e.g. Cole and Newton 2009).

Creating and maintaining complex old-growth stands in drier landscapes with the risk of stand-replacing fire requires careful management of stand structure, as structure determines fire behavior and risk (see chapter 11). Without substantial density reduction carried out regularly in these stands, it is not likely that many dense young stands will develop old-growth structure that includes large trees (figure 10-5, Sensenig et al. 2013). Deviation from strict spacing guidelines promotes variability in the fuel bed while promoting large main canopy trees, releasing understory trees for rapid growth through fire-susceptible size classes, and creating early seral plant communities in larger openings (Churchill et al. 2013, Franklin et al. 2013). Retaining larger living trees, particularly of unusual size, species and character (e.g. dead and/or broken tops, wind or ice damaged trees, or diseased trees), as well as large standing dead trees and downed wood, further promotes old-growth character in stands. Multi-aged management thus provides the basis for long-term maintenance of desired old-growth structural and compositional characteristics through regular disturbance processes.

REVIEW QUESTIONS

1. What are the fundamental requirements of any multi-aged forest management approach?
2. Compare and contrast multi-aged management with even-aged management in terms of its goals and characteristics. What are the advantages of multi-aged management ecologically and socially?
3. What is the main difference(s) between thinning in an even-aged system and harvesting trees in a multi-aged system?
4. Explain why diameter distributions are important in multi- or uneven-aged management.
5. Describe two methods for calculating specific structure goals for a balanced uneven-aged stands. Be able to use these two methods. Compare their advantages and disadvantages.
6. In addition to stand density and diameter distribution, what other variables are important in the decision to use multi-aged systems?
7. What are the three parameters that are used to set structural goals for balanced uneven-aged management? How are they determined?
8. Compare and contrast the use of tree and crown classes in multi-aged vs even-aged systems.
9. How does one take into account the need for regeneration in multi-aged stands?
10. Show how SDI is calculated and can be used to set structure guidelines for multi-aged stands.
11. Compare and contrast balanced uneven-aged stands with multi-aged stands.
12. What are the special challenges to multi-aged management in drier, fire-prone ecosystems? In areas with insects or diseases?

CHAPTER 11

Fire and Silviculture

Fire has been fundamental in shaping western U.S. forests for millennia, directly and indirectly affecting many aspects, including species composition, horizontal/vertical structure, habitat value in most forest types, and carbon, water and nutrient cycling over time (Agee 1993, DeBano et al. 1998, Arno and Fiedler 2005, Falk et al. 2011; table 3-4). Fire was common in drier forest types (containing ponderosa pine) throughout the West, where it effectively maintained relatively simple, open stand structures with

Figure 11-1. Dead trees and slash following a mountain pine beetle infestation in lodgepole pine make this stand and those around it susceptible to severe fire. The amount and arrangement of fuels predispose the stand to long flame lengths and a crown fire.

low tree densities and fuel accumulations (e.g., Covington and Moore 1994; Hagmann et al. 2013). Grasses and low shrubs provided fuel continuity for regular (decadal scale) surface fire, the severity and extent of which was primarily limited by fuel recovery since the previous fire. In comparison, more moist forest types and topographic positions, protected from such frequent fire, developed more complex structures, higher densities and greater fuel accumulations, leading to infrequent fires (at the century scale) that burned large percentages of those forests at high severity during particular climatic cycles. Native peoples likely added to the extent of this disturbance, its ecological role, and its frequency in the landscape; fire was and is not only a constant force of nature, but also a management practice (Ryan et al. 2013). Silviculturists are managers of this process and use it in many forest types.

Wildland Fire as a Forest Management Issue

Fire-management policy in the 20th century was generally directed at excluding and suppressing fire, and this remains a major consideration for most landowners today—to protect their investments. Fire exclusion began early in the West, via grazing and human settlement; while after 1950 active fire suppression with an arsenal of modern fire-fighting equipment (and Smoky Bear), combined with a relatively stable climatic period, led to a decrease in the annual acreage of burned forests (Donovan and Brown 2007). The result has been a widespread increase in tree density, including the density of fire-sensitive species, and fuel accumulation across multiple scales.

Gruell (1983) compared historic and recent photographs from the northern Rocky Mountains and thereby documented large increases in forest cover and density over the past century. Parsons and DeBenedetti (1979) similarly reported the accumulation of a large number (400+/ac) of shade-tolerant conifers in mixed-conifer stands in the Sierra Nevada. Sensenig et al. (2013) showed similar patterns in dry mixed-conifer in

Figure 11-2. After a severe fire in southwestern Oregon. Because of lack of conifer seed and competition from shrubs and hardwood sprouts, a sparse stand of conifers and a dense cover of shrubs and sprouting hardwoods are likely to develop. Successful planting would ensure a more rapid stand establishment.

southwestern Oregon. Such stand changes lead to accelerated tree mortality from self-thinning, insects, and drought (McNeil and Zobel 1980) and often the death of larger old trees (Biondi 1996), as well as a lack of regeneration of shade-intolerant species that reproduce on mineral soil in open microsites. Across multiple stands, entire hillsides and landscapes are now carpeted by trees of smaller diameters and shade-tolerant, fire-sensitive species.

The increase in leaf area/biomass, shifts in species composition, the development of high density stands, bark beetle mortality (figure 11-1), and the passage of time have combined to increase total available fuels (the accumulation of a "fire debt" from lack of fire) and made these forests more susceptible to large, severe fires (Stephens et al. 2014; figure 11-2). Today, we manage many western forests as high-risk assets, especially drier ponderosa pine and mixed-conifer forests that in earlier periods were normally characterized by low-severity, high-frequency fires (Agee 1990; Covington and Moore 1994; Hudec and Peterson 2012). Indeed, by the end of the 20th century there were major increases in large-fire activity, and this expanded in the early 21st century (see National Interagency Fire Center).

Society's views of forests, and the resource values associated with them, have changed greatly since the fire-exclusion policy was originally implemented. There are now conflicting ideas with respect to fire suppression (Donovan and Brown 2007; Stephens et al. 2013). Fire-susceptible stands often have habitat features (e.g., multiple stories, dead wood and snags) that are used by legally protected wildlife such as the northern spotted owl, and the presence of such protected species sharply limits the silvicultural choices available to managers for reducing fire hazard. Demand for housing sites, recreation, timber production, and watershed protection have all increased over the past century, and some of these activities inevitably take place in or near fire-prone forests. The public is more aware than it used to be of forest management activities in general, and people are increasingly concerned about potential wildfire damage to homes and other property, as well as the inconvenience of forest closures, evacuations and smoke. Finally, it is unlikely that agencies can afford to suppress wildland fire in landscapes, including large fires that can cost $1 million per day, while at the same time maintaining aggressive fuel reduction programs that cost money rather than yield it. Reducing fuels and fire risk is usually only one of the management objectives for a forest, and in some cases it may conflict with other objectives (Hurteau et al. 2013; Lindenmayer and Cunningham 2013; Wimberly and Liu 2013).

Silviculturists are therefore challenged to manage in a way that reduces the potential for very large, severe fires that are considered undesirable, while maintaining some presence of desired stand structures that may attract wildland fire. Stands or groups of small trees will always be more susceptible to fire than open stands of large fire-resistant trees. Their crowns are close to the ground and can be easily scorched, and their

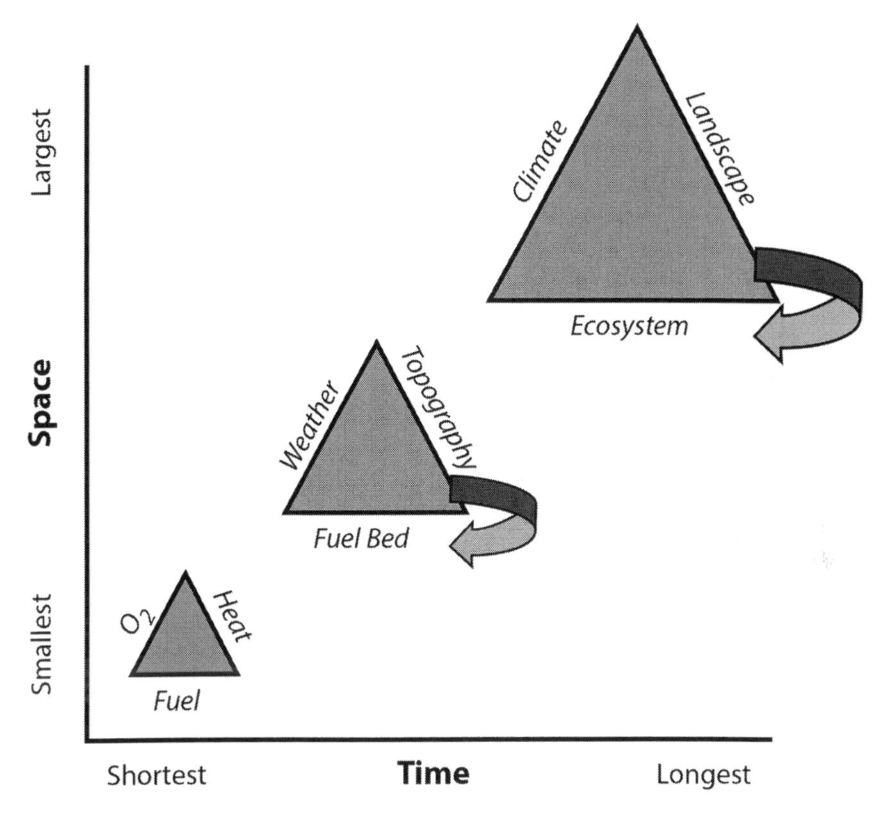

Figure 11-3. Factors regulating fire and burning intensity across multiple scales. The smallest/shortest scale might represent burning a small slash pile of even a campfire; the middle scale might represent a prescribed burn; the largest scales define fire regimes (i.e. a ranger district over centuries).

thin bark makes them susceptible to girdling by intense heat (Johnson and Miyanishi 2001). However, saplings and small trees are inevitably part of overall forest structure in time and space. Forestland under a single ownership very often deliberately includes a variety of stand structures and ages, representing varied interest in ecosystem services and varied susceptibility to wildland fire; such management across ownerships requires additional planning and collaboration.

Fire Intensity and Fuels

Fire ***intensity*** is the amount of energy released during combustion; intense fires have tall flames (above the ground surface) and high rates of spread, and therefore release large amounts of heat energy. This energy release is important in terms of drying and pre-heating adjacent fuel particles as well as actual ignition and spread of a flaming front—a line of fire advancing through a stand. Fire intensity is regulated at multiple scales by the combination of fuel and surrounding environmental conditions (figure

Table 11-1. Fuel categories are linked to stand-level fire behavior.

Fuel	*Definition*	*Examples*	*Importance*
Ground	Organic soil or biomass in immediate contact with mineral soil	• Peat • Roots • Buried logs • Humus	• Lingering combustion for many days that can be difficult to detect • Loss of deep organic accumulations
Surface	Living and dead biomass on the surface of the soil or in close proximity that burns as a defined layer, including fine and coarse materials	• Fermented and fresh litter • Dead branches • Grasses and forbs • Low shrubs • Needle drape	• Dead fine fuels (1-hour fuels) largely control ignition, flame heights, and rate of spread • High ecological impact on understory vegetation communities and nutrient cycling
Ladder	Material above the surface fuels, distributed sporadically or regularly, that assist in the transfer of flames to the canopy	• Larger shrubs • Lower branches • Saplings and small trees • Slash piles and jackpot accumulations	• Combined with sufficient surface fuels, these are fundamental to initiation of crown fires
Aerial	Burnable materials suspended well above the ground surface	• Tree crowns/ canopies	Rare but spectacular fire behavior; connected to the surface fire

11-3). At the smallest scale, the "fire triangle" is simply comprised of a fuel particle, oxygen and a heat/ignition source. At the stand scale, fire behavior is regulated by an entire fuel bed arranged on some topographic surface and interacting with current weather conditions—this is called the "Fire Behavior Triangle." Integrated over time and space, fire regimes are determined and defined by ecosystems (living and dead fuels) on complex landscapes that are maintained by climatic patterns that dictate drought severity, potential ignitions, and fire season length.

Silviculturists can manipulate fire intensity by affecting the amount, arrangement and character of fuels: ground fuels in the soil, surface and ladder fuels in the understory, plus overstory canopy fuels (table 11-1). Ground fuels contribute little to fire intensity and rate of spread in most western forest types, but are sometimes important because they support ignition when dry and are characterized by lingering, smoldering combustion. Ladder fuels are important when burning conditions allow for the transfer of fire from a surface to a forest canopy, and low crown heights and dense canopies facilitate fire movement into and among tree crowns. The most important fuel component for fire intensity and spread is predominantly surface fuel and, specifically, fine surface fuels (Agee 1993; Johnson and Miyanishi 2001). Large living and dead wood, which constitutes the vast majority of the biomass/carbon in a forest ecosystem, is not readily flammable nor consumed during a typical fire event.

Table 11-2. Examples of fuel models used to predict fire behavior (rate of spread and flame length) and resistance to control (tons/ac and depth). In this example, each fuel model assumes 8% fine-fuel moisture content and mid-flame wind speeds of 5 miles/hr. Adapted from Graham et al. (1999); see also Albini (1976) and Anderson (1982).

Fuel model	*Type of fuel*	*Dead <1.0 in (tons/ac)*	*Dead >3.0 in (tons/ac)*	*Fuel depth (ft)*	*Rate of spread (ft/min)*	*Flame length (ft)*
2	Timber/grass	3.0	0.5	2.5	39.0	6.0
8	Timber litter	2.5	5.5	0.2	1.8	1.0
9	Hardwood litter	3.3	0.15	0.2	8.0	2.6
10	Timber/understory	5.0	5.0	1.0	9.0	4.8
11	Light slash	6.0	5.5	1.0	7.0	3.5
12	Medium slash	18.0	16.5	2.3	14.3	8.0
13	Heavy slash	30.0	28.0	3.0	15.0	10.5

Fuels are typically classified according to the time required for them to respond to changes in their environment, principally relative humidity (RH). This lag time is generally related to piece size (Albini 1976; Agee 1993) but accounts for important factors like surface area and porosity, which make the fuel particles reactive to the weather, microclimate and fire. Fine materials ≤ 0.25 inches in diameter (e.g. needles, blades of grass, and small twigs) are classified as 1-hour fuels; slightly larger materials 0.25-1.0 inches are classified as 10-hour fuels; those 1-3 inches are classified as 100-hour fuels; and those ≥3.0 inches (e.g., logs) are 1000-hour fuels and beyond. This lag-time designation is defined by how long it takes a fuel particle to reach 63% of its new equilibrium moisture content (maximum percent moisture content for a given temperature and RH) when temperature is increased and relative humidity is decreased; i.e. 1-hour fuels can move 63% of the way within one hour. Assuming temperature and relative humidity are constant, moisture content will then move another 63% towards that new equilibrium, approximately 90% of the total way, in the next hour.

Moisture content of fine fuel responds rapidly to hour-by-hour (1-hour) and day-to-day (10-hour) weather conditions, including diurnal patterns (e.g., solar angle) and the passage of frontal systems. Fine fuels are also the most readily consumed during burning given the speed with which fire itself dries and heats them relative to larger fuels; more intense flaming fronts lead to more rapid drying, preheating and ignition of fuels and more rapid fire spread. In contrast, moisture content of large fuel changes slowly, following weekly (100-hour) and seasonal (1000-hour) trends. Large fuels, such as decaying snags and logs on the forest floor, slow the rate of spread at the flaming front but can burn for long times and emit large quantities of heat and embers.

Projecting Fire Behavior

Fire behavior models are used to project fire intensity under specified fuel, topographic, and weather conditions (figure 11-3). They are based predominantly on total amount by fuel type (lag time) and its moisture content; slope and aspect; and wind speed, temperature and relative humidity (Rothermel 1983; Agee 1993; Scott and Burgan 2005). Fire behavior projections help to determine the relative susceptibility of stands to fire and evaluate the effects of proposed pre-fire silvicultural treatments (Van Wagner 1977; Alexander 1988; Graham et al. 1999; Arno and Fiedler 2005; Knapp et al. 2011). They help answer such questions as: How susceptible is the current and/or future stand condition to wildland fire? What is the likelihood that a prescribed fire may become a crown fire? Will residual slash cause excessive flame lengths, and should it be removed mechanically?

"**Fuel models**" are used to enable easy categorization (table 11-2) of the types of fuelbeds commonly found in western U.S. forests according to their physical attributes and flammability, as well as to project potential flame length and rate of spread under a range of weather conditions (Albini 1976; Anderson 1982; Scott and Burgan 2005). When field measurements are not available, silviculturists can visually estimate a fuel model that best fits their stand(s) using photographic series maintained in a national database available on the web (Fuel Characteristic Classification System). These pictorial guides can help evaluate current fuel loads and the effectiveness of potential fuel-reduction treatments (Fischer 1981; Maxwell and Ward 1976).

For example, Fuel Model #8 (table 11-2) describes a stand with minimal amount and depth of fuel; it is predominantly in the form of surface litter accumulation. At 8% fine-fuel moisture content and a mid-flame windspeed of 5 miles/hour, fire will spread at 1.6 chains (105 feet) per hour on average, with flame heights averaging only 1 foot. By contrast, Fuel Model #10 represents a stand with an abundant understory, and fire under the same conditions will spread about five times faster and produce flame heights of averaging nearly 5 feet. Note that the potential for more intense fire increases as one moves further down the rows of this table.

Silvicultural treatments directly affect fuel models (table 11-2), including amounts and moisture content of surface fuel by category, canopy base height, and canopy bulk density. A light thinning from below and/or pruning will change canopy base height but have little effect on the canopy bulk density; a heavy thinning into larger diameters will further increase canopy base height and also decrease canopy bulk density. Both can have a range of impacts on surface fuels, depending on the amount of thinning slash produced, equipment used, and rates of fuel decay. Dense, multi-storied stands may be quite prone to severe fire because crown height is low (even zero), and canopy bulk density and surface fuels can be quite high (Graham et al. 1999). In uneven-aged and young even-aged stands, reducing future fire severity can be accomplished by controlling the amount of slash (which reduces surface fire flame length), reducing the density of smaller trees and ladder fuels, and encouraging the growth of larger trees with high canopy bases.

In very dense stands where there is potential for a crown fire, or where fire would be difficult to control, it might be prudent to use mechanical treatments or to pile and burn slash, small trees, and shrubs. However, on steep slopes, fire itself may be the most cost-effective alternative for reducing fuels. However, fire alone will often increase the density of sprouting shrubs and hardwoods, thus contributing to future fire hazard. Combining fire and herbicide treatments may control sprouts. Herbicides are most effective at reducing fuels when the shrubs are in the seedling/small sprout stage, when they are more easily controlled and produce little fuel when they die.

FIRE EFFECTS ON ECOSYSTEMS

Fire **severity** is the effect that fire has on an ecosystem, including anything of value: vegetation, soils, streams, timber, wildlife habitat, and human communities. It is commonly correlated with intensity; however, not every intense stand-replacing crown fire (also a stand-*initiating* fire) is severe, nor is every low-intensity surface fire harmless to an ecosystem. Severity is rooted in the intersection of plant and ecosystem adaptations to fire: the intensity of a particular fire and the resistance and resiliency of plants, soils and other parts of the ecosystem to that fire.

PLANT ADAPTATIONS

Fire kills woody vegetation by girdling the cambium near the base of the stem or by killing the foliage (and buds) in the crowns or the fine roots in the ground (Kauffman 1990; Peterson 1985; Ryan et al. 1988; Van Wagner 1973). Fires that are particularly intense, or burn for a long time on the surface, often girdle stems and kill fine roots in the upper soil layers, and associated flame heights (with ladder fuels) may reach the crowns of overstory trees. Forest vascular plants have various characteristics that enable them to withstand, recover from, and even take advantage of the effects of fire. These are genetic traits that have evolved in response to fire and other disturbances such as browsing, and are related to reproduction, size, and vigor (see chapter 4).

Plant size. Large plants are generally less prone to damage or mortality than small ones, given their height and diameter (see discussion of bark thickness below). Western conifers such as Douglas-fir or ponderosa pine may be more than 150 feet tall and, depending on stand density, their crowns may be 50-100+ feet from the ground. Surface fires are unlikely to damage these crowns. Tree response to crown scorching varies among tree species. In Wyant et al. (1986), Douglas-fir that died following underburning had an average of 85% crown scorch, whereas ponderosa pine that died had an average of 94%. In many hardwoods, large trees sprout more vigorously than small ones when their tops are killed by fire (Harrington et al. 1992; Tappeiner et al. 1984). Large trees are also more likely to produce seed following fire and thus provide new seedlings to replace trees killed by fire. Kauffman and Martin (1990) report lower rates of survival of shrubs and understory hardwoods as plant size decreased. Tappeiner (1979) found that rates of mortality for hazel seedlings and small clones were greater than for larger clones. The number of sprouts and the growth rate of survivors increased with the pre-burn size of seedlings or clones.

Returning fire to stands of shrubs from which fire has been excluded for many years is unlikely to produce the same effect as it will on those that have burned frequently. Regular burning keeps shrub density and their sprouting potential low. For example, in a stand where fire has been excluded for 50+ years, it is likely that hardwoods have developed large burls with the potential for vigorous sprouting after their tops are killed by fire. In such stands, an intense fire or possibly mechanical or herbicide treatments may be needed to reduce understory density to pre–fire suppression levels. Light underburning, similar to what is assumed to occur under natural fire regimes, may be sufficient to keep understory density low by controlling populations of small, low-vigor shrubs and hardwoods with low potential for sprouting after fire.

Bark thickness. Temperatures of about 150°F under the bark will kill living cambium and phloem cells. Photosynthate transport from crowns to roots is lost if phloem is killed around a tree's circumference; if the cambium is killed, then a tree can no longer produce new phloem or xylem cells, so roots will starve and a tree will ultimately die (it may take several years to deplete stored reserves). The bark of ponderosa pine, Douglas-fir, sugar pine, and redwood can be quite thick (Agee 1993), especially on larger trees, and provides insulation to protect cambial tissue from fire. Wyant et al. (1986) found that Douglas-fir and ponderosa pine that survived a prescribed burning averaged 9.0 to 11.0 in dbh, whereas those killed averaged 6.0 to 7.0 in, respectively. The bark of lodgepole pine, true firs, young conifers, and hardwoods such as aspen and alder is thin and a poor insulator even in larger trees. Shrubs and most hardwoods have thin bark, and therefore their tops are also likely to be killed by fire; exceptions include black oak and tanoak, which have relatively thick bark.

Bud scales. Like bark, bud scales protect living meristematic cells—growing points—at the ends of branches. Severe crown scorch generally kills the foliage, even though the buds may survive and grow in the following spring, producing new foliage. Because conifers generally maintain three or more cohorts of needles, the vigor of scorched trees may be reduced for several years after a fire while a full complement of foliage is being produced. Ponderosa pine has fire-resistant bud scales. The buds and cambium of hardwoods and shrubs are not as well protected from fire, but these species sprout from below-ground buds.

Sprouting. Forest plants may re-establish after fire by sprouting from their base, root system, or specialized structures (see chapter 4, table 4-1). Nearly all shrubs and hardwoods sprout after fire or other disturbances. Sprout vigor depends upon species, initial tree and shrub size, pre-existing vigor, and residual overstory density. Sprouting following fire varies with the time of the year. Burning (or cutting) shrub and hardwood tops in the late spring and early summer, just after new leaves are fully expanded, results in less sprouting than burning or cutting early in the spring, before the leaves have expanded, or in summer or fall, when growth has stopped (Zasada et al. 1994). Sprouting response probably depends more on the degree of fire damage to rhizomes, roots, or burls and stem bases than it does on plant phenology. Kauffman and Martin (1990) and Tappeiner (1979) found higher mortality of understory shrubs after late-fall burns, and attribute higher mortality to the greater fire intensity from burning when the moisture of the soil and litter layer was low. Feller (1998) points out that fire severity, as measured by the amount of forest floor consumed, is closely related to decrease in shrub biomass and the potential for shrubs to recover from a fire. This is probably because fires that consume the forest floor produce high (lethal) temperatures at the location of bud banks on burls and rhizomes and at the base of stems. Thus, fire that produces a high temperature just below the soil surface might be needed initially to reduce shrub populations to original, pre–fire suppression levels.

Residual Seed. On-site seed comes from surviving trees and shrubs, unburned seed stored in the forest floor, and/or serotinous cones - seeds retained in cones that are released when the cone scales open after exposure to high temperatures from fire or radiant heat (e.g. lodgepole or knobcone pines). Some shrubs and trees produce seed that remains viable in the forest floor, apparently for hundreds of years (Quick 1956, 1959), though there may be no evidence of its presence except for an occasional clump of dead stems. The forest floor in stands of western conifers may contain hundreds of thousands of seeds per acre of ceanothus or manzanita, for example, which were produced by former shrub communities. These communities persist for decades before being overtopped and shaded out by conifers. Fire and smoke sometimes facilitates overcoming seed dormancy by breaking the seed coat and making it permeable to water for germination with the result that within several years of the fire thousands of shrub seedlings, including non-native species like Scotch broom, may germinate from the stored seed. Very intense fires, however, consume much of the forest floor and therefore much of the seed stored there.

New Seeds. Seed may also be transported into burned sites, primarily by wind, but also by animals. Wind dispersal is largely dependent on seed weight. A number of light-seeded forb species such as thistle, foxglove, *Senecio,* and fireweed, as well as various grasses, are commonly found on forest sites after fire. Light-seeded tree species such as red alder, western hemlock, western larch, and lodgepole pine are more readily dispersed over large areas than are heavy-seeded species such as sugar pine, black oak, and tanoak. Heavy-seeded species are dependent on animals for dispersal. Animals are also the main vector for species whose seed is borne in berries, such as Pacific madrone, and small-seeded shrub species, such as bitterbrush and bearclover, which produce their seed close to the ground.

The frequency and size of seed crops are other important characteristics that determine how effectively a species can spread by seed after a fire. For example, ponderosa pine and Douglas-fir seed are relatively light, winged, and readily dispersed by wind. However, they produce large seed crops only infrequently: every 4-8 years. During periods of low seed production, other plants are more likely to occupy burned areas and may inhibit establishment of pine and Douglas-fir later. If the fire is especially intense, killing or damaging potential seed trees over large areas, re-colonization of the burned area will likely be slower, and therefore the ages and sizes in the resulting stand will be more variable. An exception, of course, is Rocky Mountain lodgepole pine, which often reproduces from seed stored in serotinous cones.

Grass seed or native seed mixes are sometimes sown on severely burned sites in an attempt to reduce soil erosion, especially on steep slopes. From the perspective of establishing natural and planted seedlings, it is important to restrict this practice to sites where it is needed. These grasses provide cover and food for mice, ground squirrels, and gophers that also eat tree seed and seedlings, as well as attractive forage for ungulates that also browse seedlings. Furthermore, grass develops a dense root system that severely competes with small conifer seedlings on moisture-limited sites, and grass competition is difficult to control on dry sites.

FIRE AND SOILS

The effects of fire on forest soils have generally focused on fire's effects on soil N, because it is relatively volatile compared to most nutrients and has been shown to limit productivity in these forests (Powers 1980). The initial effect of a fire is to reduce the organic matter on the forest floor and release some of the nutrients it contains (Covington and Sackett 1984). For the first several years after a fire, concentrations of NO_3 and NH_4 were frequently higher in burned soil than on unburned sites (Covington and Sackett 1986; Monleon et al. 1997; Ryan and Covington 1986; Stark 1977; White 1986; White et al. 1973). This release of nutrients from the consumption of forest floor and vegetation might be considered as fertilization. Harris and Covington (1982) found that nutrient concentration in understory vegetation increased one growing season

Figure 11-4. Treatments to reduce fire potential in ponderosa pine stands. The old-growth stand (upper) was thinned 10 yr previously to remove advance pine regeneration and was then broadcast burned. The young stand (lower) was thinned. Because the slash was so dense, it was piled for burning to reduce scorching the pine crowns.

following fire, suggesting that an understory captures and helps maintain nutrients on site that could be lost by leaching or soil movement. Harvey et al. (1989) and DeLuca and Zouhar (2000) showed that the upper soil layers in ponderosa pine forests are often low in organic matter, limiting nutrient storage, acquisition, and retention as well as water-holding capacity. Thus, high-intensity fire may not improve the nutrient status or productivity of these soils in the long term.

Fire effects on soils not only vary with fire intensity but also with soil properties, especially their nutrient content before burning (Boyle 1973; Mroz et al. 1980). An increase in soil nutrients after burning is likely dependent on organic matter content, texture, and total depth. Long-term N availability may decrease 12 years after a fire even though inorganic N concentrations in the soil increases in the near term. Organic matter depletion due to fire reduces the ability of the forest floor to supply N to the soil solution (Monleon et al. 1997). Additionally, the nutrient content of both burned and unburned fractions of the soil must be evaluated when considering fire's effects on soils. N-fixing plants such as ceanothus often establish after a fire and may help replace the N loss during the burning Additional work is needed to understand the effects of fire on cations and on the below-ground components of forest stands (Neary et al. 1999).

The accretion of organic matter and N from lack of fire may be part of the reason for the increased productivity of ponderosa pine stands that were unburned for several decades, as reported by Cochran and Hopkins (1991). Alternatively, though, White (1986) suggests that continued lack of fire might inhibit decomposition of the forest floor and mineralization of N. Most studies indicate that prescribed burnings on infertile sites should not be intense, because they might lead to a serious loss of nutrients (Harvey et al. 1989; Monleon et al. 1997; Stark 1977).

Managing Fire Effects

Managing fire intensity and severity appropriately are significant challenges to silviculturists (Walstad et al. 1990, Arno and Fiedler 2005), who often face two fundamental considerations:

1. The need to fit wildland fire hazard/risk into silvicultural prescriptions for stands and how collections of such stands affect management of the larger forest landscape; and
2. The opportunity to use prescribed fire within a stand prescription or forest management plan as a tool for reaching land management objectives.

Wildland fires are fires that burn without control, often with undesirable consequences, whether ignited by humans or by lightning. Though they are often perceived as uniformly severe, a considerable percentage of the area within a typical wildland fire perimeter burns with low or moderate severity (Hudec and Peterson 2012; Stephens et al. 2014). **Prescribed fires** are ignited intentionally (or sometimes started naturally and are then allowed or encouraged to spread); they are staged and actively manipulated to achieve forest management objectives. Such prescribed fires are generally low-severity and are designed, for example, to reduce fuels that might otherwise support severe wildland fires in the future, or to stimulate shrub sprouts for browse or cover for wildlife.

Silvicultural practices may either increase or reduce the fuel hazard that can lead to severe wildland fire. Silvicultural treatments such as thinning and multi-age management can increase/restore the resistance of stands by creating and maintaining structure and composition more conducive to wildland fire (figure 11-4). After treatment, particularly when it includes prescribed burning, surface and crown fuels are fundamentally altered (Arno and Fiedler 2005) and basal-area growth rates of residual large trees may improve (Latham and Tappeiner 2002; McDowell et al. 2003). The potential for a severe wildland fire is thereby reduced and, presumably, the resistance to insects and drought is enhanced, even in stands of very old trees. However, there is a potential for increased fire hazard if treatments increase surface fine fuels via slash and shrub growth, and/or lead to the development of abundant fuel ladders (Raymond and

Peterson 2005). Appropriate silvicultural treatments will address these factors in the near term and over time as stands develop.

Silvicultural treatments are also used to restore forests after severe wildland fire, especially those associated with reforestation and early vegetation control. Often there is a lack of seed for the major tree species, and a dense cover of shrubs and grasses may develop, interfering with the establishment of natural tree seedlings. In these cases, silviculturists may prepare the site for natural regeneration or for planting (See chapter 7). Disposal of slash from dead trees may be conducted as part of the reforestation effort, in order to improve access for planting, minimize mechanical damage from downslope migration of coarse wood on steep slopes, and reduce heavy fuel loadings related to a potential for severe future fire in the new forest (McIver and Starr 2001).

Fire Regimes

The need to develop stands that are resilient to landscape fire varies among forest types and their respective **fire regimes**, as does the appropriate application of prescribed burning. Fire regimes are typically described by a combination of forest types and the known or hypothesized effects of fires in them. Agee (1990, 1993) classified several forest types with respect to fire frequency, fire intensity and perceived severity. In addition, we often think of fire regimes as having common seasonality, fire sizes (stand to landscape), and complexity within the ultimate fire perimeter. Following are examples of Agee's description for common forest types by fire frequency and severity.

Infrequent high-severity fire. High-severity fires create major shifts in the structure, composition and function. Stands with high potential for stand-replacing fire disturbance generally occur on productive sites (high accumulation of fuels) but where seasonally dry weather regularly promotes sufficiently low fuel moisture; these fires spread rapidly during the hottest and driest parts of the season and, therefore, burn large areas with patch sizes ranging from 10 to 10,000 acres. Historically, major fires occurred every 100 years or more in these forest types, which include:

Douglas-fir/western hemlock and coastal redwood forests in the Coast Range of Oregon and California, western slopes of the Cascades, and Olympic Mountains of Oregon and Washington (and probably northwestern California). These forests are very productive, on average, given good soils and a climate that includes regular and plentiful precipitation. However, large and severe wildland fires occur during periods of extended drought and dry winds. Between stand-replacing events, surface fire occurred sporadically associated with drier microsites and indigenous burning practices. (Peter and Harrington. 2014).

Lodgepole pine forests in the interior West, Cascades, and Sierra Nevada. Lodgepole pine readily regenerates after stand-replacing fire and forms dense stands that are susceptible to severe fires after 60 years. Lodgepole pine stands become susceptible to fire because of the quantity of dead wood generated by self-thinning, and often because of insect mortality, as well as resultant understory growth (trees and shrubs) providing surface fuels.

True fir forests occurring at subalpine elevations with a persistent snowpack and a short growing season. These forests regenerate very slowly after a fire, and fires tend to occur at extremely long intervals (>400 yr). Fire may occur during extended years of drought, possibly associated with subtle changes in climate.

Mixed frequency and severity. Mixed-severity fire regimes generally cause ongoing and intricate temporal and spatial shifts in the character of a forest, creating more variability than is found in either high- or low-severity regimes (see below). Mixed fire regimes have regular fire within parts of a landscape, with mean fire return intervals <20 years (Sensenig et al. 2013), but fires burn less frequently into other parts of the forest (perhaps every 25-100 years) depending on season and weather conditions. A mixed-severity fire reduces stand density on average, creating a range of opening sizes, but leaves patches of trees unburned as well. Although the intervals between fires are short over most of the forest, they still accumulate biomass that fuels future fires. Topographic positions and plant communities with higher probabilities of burning tend to burn most frequently and the landscape thus develops an ecological memory. Examples of mixed-severity types include:

Mixed-evergreen forests in southwestern Oregon and northern California. These forests have fire-return intervals ranging from 60 years on wetter sites near the coast and on northern aspects, to 20 years on drier, interior sites and southern aspects. Tanoak in the understory carries intense, patchy fire to overstory Douglas-fir and pine. Tanoak sprouts after fire and is overtopped by Douglas-fir after 20-30 years.

Dry Douglas-fir forests in the eastern Cascades, west-central Cascades and the Puget Sound Trough. The fire return interval is 70-100 years. The forest structure is often patchy, because regeneration occurs in the openings caused by fire, and often with a hardwood component.

Red fir forests of the Sierra Nevada and southwestern Oregon. These forests have fire-return intervals of 40-80 years. Fire severity is mixed, with one fire producing areas of high, moderate, and low severity and unburned conditions. These forests often have a dense canopy with sparse to patchy understory development and deep litter layers.

Frequent low-severity fire. Low-severity fires historically occurred on drier sites, and their fire-return interval was short (5-25 years). Low-severity fire regimes might be called understory regimes, because their major effect is to remove understory trees, shrubs and grasses, and thereby maintain an open, park-like forest with few fuel ladders. There is little time for major fuel accumulation between fire events, and such drier sites have lower productivity on average anyway; however, fire seasons are predictably long and ignitions are abundant enough to produce widespread burning as soon as fuels re-accumulate. In these low-severity forest types, as well as within mixed-severity forests, bark beetles and other insects kill trees and add to periodic fuel accumulation between fires, especially during extended periods of drought. Forest types where low-severity fires are the norm include:

Ponderosa pine forests throughout the West. Sites with pronounced annual fire seasons that are dominated by ponderosa pine, with grass or shrub understories, normally have fire return intervals of 5-15 years. Seedlings and saplings occur in patches in the understory of large, thick-barked pine stands (figure 11-4). The normally frequent fires in this forest type keep smaller trees from becoming fuel ladders that would carry fire into the crowns of larger trees.

Mixed-conifer forests of the Sierra Nevada, southwestern Oregon at middle elevations, and on the east side of the Cascades. These forests are composed of ponderosa pine, sugar pine, Douglas-fir, incense-cedar, and white fir. Larger, thicker-barked pine and Douglas-fir are resistant to fire, while the thin-barked white fir is not. Exclusion of fire and the subsequent development of an understory of white fir, incense-cedar, and small Douglas-fir, as well as the high productivity of this forest type on many sites, create susceptibility to large, severe fires.

Oak woodland forests along the fringes of valleys in western Oregon, Washington, and California. These forests and woodlands had low-intensity fires that burned rapidly through an understory dominated by grasses. For millennia, these fires were typically initiated by Native Americans burning these and adjacent sites to create food and wildlife habitat, which would account for the short fire-return interval (<25 years). With the end of Native American burning practices and subsequent fire exclusion policies, Douglas-fir trees have often invaded these sites, ultimately overtopping and killing the oak trees.

Frequent high-severity fire. Shrub-dominated ecosystems have the potential for frequent severe fire. These include chaparral communities from southern Oregon to Mexico, and dense stands of ceanothus and other shrub species that often occur after severe fire in western forests. Subsequent fire in such stands typically results in dense communities of nearly the same species, regenerated from buried seed and sprouts. Shrub density and dead foliage/branches make these communities highly flammable within 20 years of the previous fire.

Fire regimes develop and express themselves over long time periods (many centuries) and often with high variability, but they are usually described by their average frequency and severity. Practically speaking, the broad classifications above are necessary; however, they are too often applied to very large forest types and areas that encompass

Figure 11-5 Lighting a prescribed under-burn in a Douglas-fir stand. The burn plan prescription called for 1-foot flame heights achieved under these fuel, slope and weather conditions with strip head fire and spot firing around tree bases.

tremendous variability in topography and microclimate, species composition and stand structure, total amount and arrangement of fuels, and probabilities of ignition/ spread. Thus, the actual fire occurrence and severity could be quite variable throughout forests that are assigned the same severity classification, requiring that silviculturists know their sites. For example, Arno (1976) studied fire history in a 10-mi^2 watershed and found that fire regimes varied considerably with plant association and elevation (3000-8600 feet). The mean fire return interval (MFRI) was 6-11 years (but ranged from 2-20 years) in lower-elevation ponderosa pine forests, 7-28 years (range 2-67) in Douglas-fir forests at mid elevation, and 30-40 years (range 2-78) in true fir forests at higher elevations. The extent of the 23 fires that occurred from 1734 to 1900 was quite variable, burning from <10% to > 90% of the watershed; there were no fires after 1900 (Arno 1976). Sensenig et al. (2013) reported MFRIs of 7-19 years (range 5-50) for 18 sites ranging from the Cascades to the coastal mountains in southwestern Oregon for 1700 to 1900; there was no evidence of fire from 1900 to 1995. The three longest fire-free periods before 1900 ranged from 9-50 years (figure 3-12).

There are important aspects of a fire regime that cannot be documented by MFRI alone (Agee 1993, Baker and Ehle 2001). Frequency alone does not offer direct evidence of how severe or widespread the fires might have been, or how often they re-burned the same areas. Sensenig et al. (2013) suggest that many southwestern Oregon fires were patchy and of low intensity because fire occurrence and conifer establishment occurred during the same decades, and there was no consistent pattern of how long it took for trees to become established after the fires. Likewise, no inference could be made regarding tree mortality related to fire: how many trees of what sizes were killed and at what frequency mortality occurred. It becomes very challenging to understand and implement a historic fire regime today.

Prescribed Burning as a Silvicultural Practice

Historically, prescribed burning has been used primarily for site preparation: to reduce the accumulation of logging debris and make planting or natural regeneration easier; to top-kill shrubs and hardwoods that compete with planted tree seedlings; to enhance availability of nutrients; and to reduce fuel load, thereby reducing potential flammability of the future plantation (Martin 1990; Martin and Dell 1978; Walstad and Seidel 1990; Walstad et al. 1990). Excessive quantities of slash can be a hindrance to safe, effective planting, and deep organic material on the forest floor can inhibit natural regeneration of many species. In some forest types (e.g. lodgepole pine), prescribed surface fire after harvest is essential for abundant natural regeneration. In landscapes with high fire risk, reducing slash during site preparation reduces future crown-scorch damage and tree mortality from wildland fires. For example, northwestern California pine plantations (10-15 years old) where slash had been piled and burned experienced less damage in subsequent wildland fire, and sites that had been broadcast-burned experienced the least (Weatherspoon and Skinner 1995). Sites that had received no preparation experienced the most severe damage, up to 100% tree mortality.

Today, prescribed burning is increasingly used in the understory of older stands (typically called "**underburning**") in order to reduce surface fuel loads during the stem exclusion stage of stand development, and thereby decrease the potential for future severe fire behavior. Underburning can also help control unwanted understory vegetation (when it is significantly competing with trees for limited moisture) and invasive species (figure 11-5), and promote regeneration of shade-intolerant overstory species, at least in more open stands. Prescribed burning is often used as a companion treatment after commercial thinning or any partial harvest, especially to reduce heavy concentrations of surface fuels (Weatherspoon and Skinner 1995; Keyes and O'Hara 2002). Prescribed burning several years before partial harvest is sometimes used to reduce the amount of non-merchantable understory biomass among the desired stems, but care must be exercised to minimize crop tree damage. And younger plantations with smaller trees are more likely to be affected by the intensity of prescribed underburning.

Surface fuel removal through prescribed underburning reduces the severity of subsequent wildland fires, even large ones (Helms 1979; Wagle and Eakle 1979; Buckley 1992; Strom and Fulé 2007). Cummings (1964) found that the number of damaged oak and ponderosa pine trees in stands that were underburned before a wildfire was

100-200% lower than in stands that were not underburned. The fuel reduction effect may last only for several years, however, before surface fuels re-accumulate (Finney et al. 2005). Davis and Cooper (1963) reported that five years after underburning, the incidence of wildland fire increased as understory shrubs and hardwoods recolonized stands.

PRESCRIBED BURNING ACROSS FIRE REGIMES

The application of prescribed burning varies across forest types and fire regimes, according to plant species adaptations and the inherent ecological resilience of the plant community. Prescribed underburning typically has the most applications in forests with frequent low-severity fire. Low-intensity prescribed burning is common after early fall precipitation or at the beginning of spring/summer drought, when fire intensity can be regulated by fuel moisture and weather conditions. Such fire might be used fairly regularly within stands to reduce fuels, renew browse production from sprouting shrubs, favor the regeneration of shade-intolerant conifers, reduce seedling density (and a few larger stems), and improve the vigor of large trees (Bailey and Covington 2002; Arno and Fiedler 2005). These objectives can be accomplished at the stand scale; by treating many stands, landscape-scale fire patterns can be altered.

Prescribed burning in forests with lower frequency and higher severity fire regimes is likely to be more difficult to mimic, because fuel accumulations typically will be greater and/or more variable, dry periods may be less common, and these regimes can be prone to intense fire behavior that is inherently difficult to contain (Ryan et al. 2013). Since natural fires killed part or most of the overstory in historic mixed- and high-severity fire regimes, underburning can vary in intensity, but it may be difficult to determine what the outcome of prescribed burning in these regimes ought to be. Regardless, forests in these regimes will benefit from low-intensity underburning used periodically to reduce concentrations of fuel (including dead wood from windthrow or logging), enhance reforestation, improve browse, stimulate understory species, and/or manage wildland fire risk and protect property.

Many acres of forest that developed with low- and mixed-severity regimes have, through fire exclusion, built up enough fuels to move them into the high-severity category (e.g., Gruell 1983). In these at-risk forests, surface fuels have accumulated to reach unprecedented levels and fuel ladders have established themselves in the absence of fire, making it difficult to simply reintroduce low-intensity fire (Ager et al. 2013). Other factors, too, may hinder the re-establishment of historical conditions through prescribed burning: competing objectives for wildlife habitat, wood production, and recreation; local air-quality standards; the concerns of neighbors; and the cost of burning operations. Furthermore, existing trees often have become larger and more resistant to fire over the decades of exclusion. It therefore may not be possible to achieve management objectives solely by re-introducing fire; mechanical methods of fuel removal are commonly necessary prior to underburning (Stephens et al. 2013).

Information about historic fire regimes provides general guidance to silvicultural prescriptions, although the challenge of basing any forest-management regime on a naturalistic/historic model is in how to apply it. Should fire be used at the stand or landscape level? Over how large an area should it be applied? Where should fire be used, at what intensity, and at what frequency? Should there be a focus on long fire-free periods or on frequent fire? What is the appropriate stand density and how often should density management occur? These are questions that will have to be resolved with local expertise and with an eye toward budgets, priorities for fuel reduction to protect private property and other resources and values, and concerns for maintaining wildlife habitat, among other factors. It may be a worthwhile goal to use historic fire regimes as a model for stand management, but natural fire dynamics are as yet poorly understood and do not lend themselves to standard formulas or easy-to-interpret guidelines (Baker and Ehle 2001), especially in moderate- or high-severity regimes.

Air-quality standards frequently regulate the timing and amount of prescribed burning due to smoke production, and these standards are likely to limit the use of fire regardless of ecological benefits and risk reduction. Smoke dispersion (i.e., lift and transport) is particularly important and requires a light wind and atmospheric instability. Smoke in populated areas, and particularly in the wildland/urban interface, ranges from an annoyance to a human health and safety hazard, especially along highways. Smoke management requires accurate weather forecasts, fuels estimates, and projections of fire behavior.

Ironically, one of the best times to use prescribed burning is during a wildland fire incident. "Backburning" (also "backfiring" and "black-lining"), which is used to expand fire lines and thereby manage the spread of a wildland fire, creates many acres of underburned stands within a fire perimeter; it can have major effects on the mix of fire severity within a landscape (or "fireshed"). Similarly, many acres within a wildland fire perimeter burn at low- and mixed intensity (given variability in topography, weather and fuel) with excellent ecological/silvicultural results, and those areas should be maintained over time. Indeed, the greatest irony in our current fire management policy is the focus on suppression during moderate fire weather conditions, when the percentage of area burned favorably would be at its maximum, only to delay fire until weather conditions are severe, when we are unable to suppress fires and that favorable percentage will be minimal. Continued use of prescribed underburning within these post-fire landscapes, as well as thinning/burning in adjacent stands, can promote the maintenance of a low-severity fire regime over time.

PRESCRIBED BURNING IN PRACTICE

The fire intensity associated with prescribed burning is dictated by the fire behavior triangle (figure 11-3) and can be regulated in the field through variables such as:

1. **Time of year**. Spring and late-fall burns are less intense because fine fuel moisture is typically high. Burning in late summer or early fall, when fuel moisture is low, tends to produce more intense fire and higher chances of escape. The amount of litter and dead wood consumed depends on pre-burn fuel moisture, resulting in variable exposure of mineral soil. At 200% moisture content by weight, there was very little fuel consumption; at 50% moisture there was >60% duff reduction and about 50% mineral soil exposure (Reinhardt et al. 1991). Coarse wood is preserved by spring burning since it contains moisture after fine fuels have dried sufficiently to burn.
2. **Time of day**. There are important diurnal variations in air temperature, relative humidity, and wind speed and direction—upslope with morning sunshine, and downslope in the cool evenings (Agee 1993; Martin 1990). Such timing issues are associated with important changes in fire behavior that can be used to manage an effective and safe prescribed fire.
3. **Timing of Precipitation**. It is safer to burn just before predicted precipitation, to reduce the possibility of the fire escaping and make it easier to extinguish residual fire. Likewise, light precipitation several days before a burn (i.e., "antecedent" precipitation) can reduce fire intensity by increasing fine fuel moisture.
4. **Wind**. Wind combined with available fine fuel regulates most fire behavior. Indeed, some wind is needed to help move fire through a stand, dissipate heat energy that accumulates under and scorches tree crowns, and disperse smoke. Wind velocities of 1-3 mph (measured at eye level) are desirable for prescribed burning, but higher speeds can be tolerated depending on burn plan objectives, or when burning piles of woody debris. Note that wind velocity within a stand is only about half that in the open. Wind direction should be stable and not vary more than about 45° during a burning operation. Burning when winds are strong and highly variable should, of course, be avoided.
5. **Slope and topography**. These variables are known and constant for a site and, along with surface winds, influence fire spread predictably (direction and rate). Fire planning and ignition pattern can take advantage of the slope, aspect and terrain (Martin 1990). "Backing" a fire down a slope produces lower flame lengths and slower rate of spread than allowing fire to burn uphill. However, fires burning down a slope have a longer residence time and more thorough combustion of fuels, and may generate higher temperatures/severity on a particular spot than upslope fires. Eastern and southern exposures warm earlier and burn differently than northern exposures.

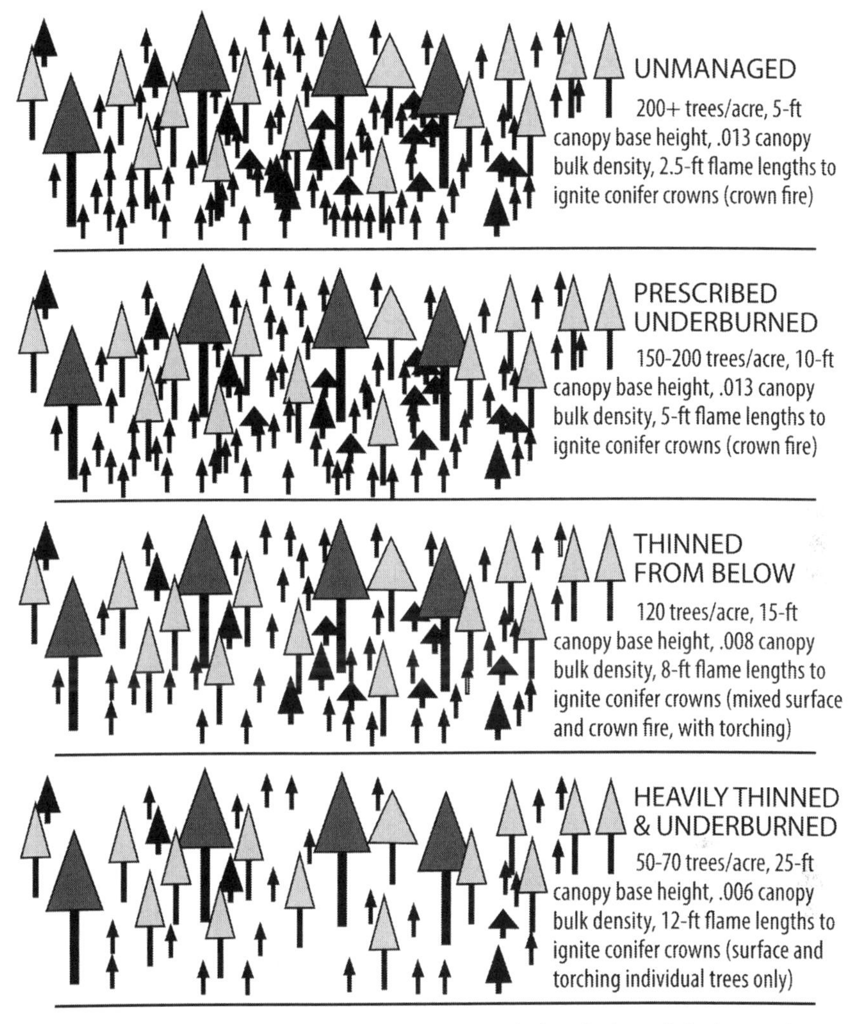

Figure 11-6. Effects of removing understory and mid-story trees/fuels on fire hazard of a dry mixed-conifer forest stand (fuel model 10, see table 11-1 adapted from Graham et al. 1999). Flame length estimated at a foliage moisture content of 75%; slash removal assumed. Tree number and distribution shown with and without prescribed burning and thinning treatments. Canopy base heights, canopy bulk density, and flame lengths are approximations; predicted fire behavior is summarized (crown, mixed crown/surface, torching and surface only) based on surface fuel loads, ladder fuels, and crown continuity.

Clearly, proper evaluation of all these variables is integral to preparing a burn plan and important for the successful use of prescribed burning. Beyond these broad variables, it is crucial to understand the spatial variability in micro-topography and amount/condition of fuels, and how that interacts with temporal changes in weather. Extensive training and experience in the use of fire is needed in order to fully develop prescribed burning as a land management tool—to manipulate vegetation and fuels for a silvicultural prescription in the present, as well as to regulate future fire in these forested landscapes.

Fire is efficient in thinning densities of smaller flammable materials, such as layers of shrubs and seedlings (Bailey and Covington 2002); however, using fire to thin stands of trees requires considerable skill, and it is not always even possible. We have seen stands in which prescribed burning was not intense enough to thin even dense groups of small trees, although it effectively reduced the density of shrubs and reduced 1- and 10-hour fuels. The line between effective thinning and whole-scale stand replacement can be very fine and extremely sensitive to weather conditions. Therefore, combining prescribed burning with prior mechanical treatments may be necessary under some conditions. For example, mechanical removal of trees for wood and felling of smaller trees can be followed by fire to reduce slash and density of shrubs. Depending on the amount and distribution of fuels that result from thinning, it may be prudent to allow 1-2 years for decomposition of finer fuels, or to chip the fuels, before prescribed burning is applied.

PRESCRIBED BURNING AND MECHANICAL TREATMENTS

Prescribed burning and mechanical treatments can be used separately or together to reduce fuels; both have advantages and disadvantages. Mechanical treatments are selective, giving excellent control over which trees are left after treatment based on size, species, quality, and stand-density goals (figure 11-6). Thinning and pruning predictably reduce ladder fuels and canopy bulk density, as well as increase canopy base height. Understory layers of shrubs and small trees also can be mowed or cut individually, then left onsite or removed depending on the fuel hazard. All mechanical treatments have inherently high costs and associated impacts on the site (e.g., compaction), however. Cable logging systems might be used to thin trees and remove slash on steep slopes, but mechanical treatment of slash will likely be most cost effective on gentler slopes.

Mechanical treatments can also affect fine surface fuel loading, depending on the treatment. Chipping and grinding woody fuels into small pieces reduces fire hazard without affecting the amount of organic matter and its eventual incorporation into the soil (Graham et al. 2004). Such masticated fuel beds have limited oxygen supply during combustion and burn differently (Knapp et al 2011). Frequently, layers of chips are not very deep (<1.0 in) or continuous. However, deep uniform layers of chips can insulate surface layers and reduce air circulation and temperature of the soil, resulting in slower rates of decomposition and restricted availability of N to plants on some sites (Graham et al. 2004). Chipping that makes large pieces can be used to disrupt the continuity of fuels without creating a uniform layer that changes soil properties.

Slash from thinning and pruning must be treated or it may leave a stand with higher potential for a severe fire than prior to the initial treatment. After evaluating the effects

Table 11-3. Effects of prescribed burning, and mixed-severity wildland fire, on stand dynamics over time: SI = stand initiation; SE = stem exclusion; UR = understory reinitiation; OG = old growth. Note that mixed-intensity effects are in addition to low-intensity effects.

Time	*Fire intensity*	*Key Fire Effects*	*Stand Dynamics*
Immediate	Low	Surface and ladder fuel reduction; mineral seed bed exposure; shrub community reduction; forage loss; limited overstory mortality	Creation of open understory conditions and fire resistance in SE and UR/OG stands
	Mixed	Additional impacts on overstory tree mortality (stand density), often in small openings; extensive snag creation for habitat value	Stimulation of UR/OG dynamics; monitor snag size distribution and dynamics
	High	Large openings, low density and early seral conditions; high numbers of snags; rich herbaceous communities	Creation of SI conditions; potential salvage and planting; monitor live tree size distribution and dynamics
Near term – first decade	Low	Stimulation of shrub and hardwood sprouting and early growth; nutrient cycling; high forage quality for wildlife	Maintenance of open under-understory conditions and fire resistance in SE and UR/OG
	Mixed	Additional creation of horizontal and vertical stand structure; habitat and landscape heterogeneity	Shifting mosaic pattern for UR dynamics and OG structure and composition
	High	High rates of snag fall and coarse wood accumulation; rich shrub communities; rapidly developing fuel beds	Implications for the speed of transition from SI to SE; monitor legacy structures
Long term – multiple decades	Low	*NA – unless burned again*	
	Mixed	Continued patch development and stratification; increased opportunities for shade-intolerant species	Loss of SI conditions; UR and OG development, enhanced by subsequent burning
	High	Loss of snags; high coarse wood accumulations and shrub cover; potential fire hazard until canopy lift	Loss of SI conditions; SE development

of a wildland fire in stands that had been thinned 7-10 years previously, Raymond and Peterson (2005) found that the mortality of overstory trees in thinned stands was 80 to 100% because of the surface slash accumulation, compared to only 53-54% mortality in untreated stands. In contrast, overstory mortality was only 5% in stands that had been thinned and underburned. Reduction of 1- to 100-hour fuels was needed to reduce fire intensity and prevent crown scorch.

Both mechanical and burning treatments can scar the boles of trees and may introduce decay fungi. Thus, treatments to reduce fuels and stand density in older, previously untreated stands may not always immediately achieve management objectives. Fire will often result in scorched tree crowns, black tree stems, standing dead trees, logs on the forest floor, and lingering fire-related mortality. In younger stands, with less well-developed understories and smaller accumulations of fuel, the effects of prescribed burning will be less obvious. Mechanical treatments can also kill or damage trees and cause unsightly accumulations of slash. With both methods, it is important to distinguish between temporary unsightly appearances and damage, such severe burning of soil organic layers, and damage to and mortality of overstory trees. As noted above, mortality of a small percentage (<5%) of overstory trees can be expected. Because they reduce stand density, thinning and prescribed burning both tend to result in a relatively open, park-like stand, with increased wind speed and decreased relative humidity and fuel moisture.

PRESCRIBED FIRE AND STAND DYNAMICS

The possible effects of prescribed burning include not only reduction of surface and ladder fuels, but also some mortality of small and large trees and other impacts on stand dynamics (table 11-3). Consumption of surface fuel around tree bases, crown scorch, and subsequent bark beetle attacks and weather events all relate to tree survival after prescribed fire (and wildland fire). In a study of a Sierra Nevada mixed-conifer forest, where a prescribed fire consumed 85% of all surface fuels, Mutch and Parsons (1998) reported reductions of 96%, 77%, and 60% in 1-, 10- and 100-hour fuels, respectively, as well as 97% reduction of the litter/duff layer, 82% of the rotten wood and 76% of the sound wood >3.0 in in diameter (1000-hour fuels). Tree mortality before the burn was 0.8%/yr; four years later, it was 1.4% in unburned areas but 17.2% in burned areas; 7.5% of all small (≤20 in dbh) sub-canopy trees were fire killed.

OVERSTORY MORTALITY

Thomas and Agee (1986) evaluated prescribed burning in southwestern Oregon and showed high rates of mortality in trees of lower crown class and those weakened by bark beetles. Fire reduced surface fuels by 60% or more on most plots, but they are expected to eventually exceed pre-fire amounts by about 20% as dead trees fall. The work of these researchers reinforces the idea that, in stands with a build-up of living and dead fuels, it may take several prescribed fires combined with mechanical removal of some trees to return fuels to historic and sustainable levels. Relying on prescribed burning alone in these ecosystems would be hazardous, because the intensity of fire required to kill mid-story fir, for example, could also kill desirable large pine.

Swezy and Agee (1991) found 19.5% mortality of ponderosa pine after prescribed burning, compared to 6.6% mortality on unburned areas; however, the highest mortality rates (38-50%) in their study were in the largest size classes (>40 inches). Mortality was associated with low tree vigor, drought, mistletoe, and bark beetles (associated with

Figure 11-7. Thinning and burning in ponderosa pine stands affects understory development on sites with manzanita and ceanothus. Stand A was thinned and not burned, stand B was thinned and burned, and stand C was heavily thinned and burned. The shrubs regenerated from seed in the forest floor. Stands A and B have low fire potential, because high stand density and/or lack of burning prevented establishment of a dense shrub understory. The shrubs in C could carry intense fire into the crowns of the remaining trees. Additional burning, mechanical, or herbicide treatment would be needed to reduce flammability in C. The large shrubs in C could sprout vigorously after burning or mechanical treatment, which might reduce the effectiveness of these treatments.

tree size and age), as well as the direct effects of fire scorch on the crowns. Trees in Keen's (1943) crown classes 3 and 4, which are considered low and poor vigor, were more susceptible than those in classes 1 and 2. Large tree mortality can also be associated with lingering combustion of localized fuel accumulations at a tree's base from litter fall and bark sloughing ("duff mounds"), which damage basal cambium (Swezy and Agee 1991). In general, most basal damage at the base of trees occurs on the uphill side, because fuels accumulate there and flames eddy around tree trunks during upslope/downwind fire spread. Crowns may be scorched or even torched if trees have long crowns and resinous boles, especially if surface fuel has accumulated beneath them.

The probability of mortality also varies by season of burn. Harrington (1987) found that the mortality of ponderosa pine in Arizona was 12%, 26%, and 29% after fall, spring, and summer burns, respectively, suggesting that trees may be more susceptible to fire during the growing season. Harrington found that 90% crown scorch in the fall caused only about 15% mortality, whereas the same amount of scorch in the spring (when buds were active) caused >40% mortality. Many of these studies cited above, however, report results from the initial re-introduction of fire into stands that had not burned for many decades. Tree mortality would likely have been lower, had fire regimes been maintained within the historic range of variability, with subsequent limits of surface fuel loading. Similarly, subsequent wildland fires and prescribed underburning will be lower-intensity and likely create less overstory mortality.

Increased stand density from exclusion of fire has probably decreased the vigor of the stand, especially the vigor of larger trees (Biondi 1996; Latham and Tappeiner 2002; McDowell et al. 2003). Although prescribed burning can kill overstory trees, it also reduces stand density (table 11-3), which will not only make remaining live trees less susceptible to fire but will also probably improve their resistance to insects and drought (Feeney et al. 1998). Leaf N in ponderosa pines increased after fire and after thinning, as did leaf toughness, basal-area growth, and resin flow. Leaf toughness is an indicator of resistance to sawfly, and basal-area growth and resin flow both indicate resistance to bark beetle attack (Feeney et al. 1998; see chapter 8 for a discussion of stand density and resistance to bark beetles).

UNDERSTORY DYNAMICS

The vigor of sprouts and seedlings is affected by fire severity and resultant overstory density. For example, the growth of sprouts of tanoak, Pacific madrone, California black oak (MacDonald 1978a), and bigleaf maple (Tappeiner et al. 1996) in an understory was only about 50% of that in the open. Ceanothus and manzanita seedlings that germinated after prescribed burning under a dense canopy grew slowly, whereas those germinating in a stand that had been thinned grew rapidly and formed a dense understory (figure 11-7). A dense understory might be desirable for improving browsing for wildlife or adding N to the soil, but the future shrub cover that develops may increase the flammability of a stand above pre-burning levels. Maintaining higher overstory density can help control the growth and density of understory seedlings and sprouts after prescribed underburning.

Repeated prescribed burning (at ~5-year intervals) might be used under sparse overstory tree canopies where fire stimulates a dense shrub understory from seeding and/or sprouting. The initial burn would stimulate seedling germination from below-ground seed banks and sprouting of understory shrubs, but subsequent burns would kill or reduce the vigor of those seedlings and sprouts. Herbicides would also be effective in controlling the shrub understory at the seedling stage (Tappeiner 1979). Understory shrubs may be more susceptible to herbicides than those growing in open conditions, such as following a clearcut harvest (Fredricksen 2005, Tappeiner 1979). Subsequent fires or other treatments also might be needed in a vigorous shrub understory in order to reduce its capacity to produce seed.

Shrub communities must be considered in the application of prescribed burning. Underburning in a ponderosa pine/bitterbrush forest reduces surface and ladder fuels, but a major part of that is often a reduction in bitterbrush, which is important winter forage for deer. Bitterbrush produces low-vigor sprouts, and much of its regeneration after a fire is from new seedlings that, unlike manzanita and ceanothus, grow slowly. Furthermore, its seed is not stored in the forest floor, but is dispersed by animals (Nord 1965; Vander Wall 1994). An intense fire over a large area could severely reduce sprouting and cause a poor distribution of seed-producing plants, requiring several decades to re-establish bitterbrush and good winter forage for deer. On the other hand, a low-intensity, patchy burn would leave a bitterbrush seed source scattered among burned areas and would not severely affect sprouting. But by leaving patches of highly flammable understory shrubs and trees, this type of burn would result in less overall fuel reduction.

Developing Prescription Details

As with all silvicultural prescriptions, it is important to determine the following before beginning treatments that use prescribed fire: (1) current stand condition and where it fits in the physical and social landscape; (2) stand-management objectives and desired future stand structure, species composition, and fire dynamics; and (3) potential treatment(s), including but not limited to prescribed burning, that can be used to achieve the desired conditions. Often a series of treatments lasting for a decade or more may be needed. Following is a summary of the most important silvicultural considerations that should be evaluated before proposing fuel-reduction treatments.

OVERSTORY CONSIDERATIONS

1. Thick-barked species such as ponderosa pine, western larch, Douglas-fir, and giant sequoia are more resistant to fire than thin-barked species such as lodgepole pine, true fir, and western hemlock. These latter species are more susceptible to pathogens when their stems are damaged by fire or harvesting, as well.
2. Larger, older trees have thicker bark and are more resistant to fire unless they have very low vigor and/or thick basal fuel accumulations, or where they are located near concentrations of fuel where crown scorch or high soil temperatures can occur.
3. Stands with uneven-size distributions of overstory trees are less susceptible to fire if trees in the smaller-size classes are spatially separated from larger, older trees. On sites with regular fire, it may be difficult to maintain dense, uneven-aged stands with many small and medium-sized trees as ladder fuels under larger trees.

4. Open canopies with many gaps are less likely to support fires than are dense canopies with few gaps. However, stands with open canopies will have higher temperatures and lower relative humidity and therefore drier fuels, greater mid-flame wind speeds, and typically more ladder fuels (e.g., from denser shrub and hardwood understories) than closed-canopy stands.

FUEL CONSIDERATIONS

1. Fine surface fuels regulate fire spread and intensity because they dry rapidly and are loosely arranged, often in a fairly deep and continuous layer (within 20 inches of the forest floor). Common examples include herbaceous vegetation and small shrubs draped with needles from overstory conifers, and accumulations of dry branches with dead foliage.
2. Concentrations of logs on the forest floor can burn with high intensity and long duration; they ignite easily when dry and intermixed with fine fuels to initiate combustion. High-intensity fire can alter soil horizons, girdle trees, and scorch or spread to the overstory. Burning snags spread fire to adjoining stands by wind-born embers.
3. In stands with high concentrations of coarse fuels, several burns and/or mechanical treatments may be needed to reduce fuels to a desirable level.

UNDERSTORY CONSIDERATIONS

1. Pre-burn plant communities strongly influence post-fire communities. Larger and more vigorous shrubs or hardwoods have greater potential for sprouting, and dense cover can develop from seedlings germinating from buried seed banks on treated areas or from seed dispersal into the area treated from nearby. Repeated burning will progressively reduce sprouting opportunities within stands.
2. The density of residual overstory canopy will affect the regeneration and growth of seedlings and sprouts after fire (and other treatments) through the modification of microsite characteristics associated with plant productivity.
3. Burning generally results in the sprouting or seedling establishment of shrubs and herbaceous plants to produce high-quality browse for wildlife, at least for a while. But this vegetation can develop into fire ladders.

TREATMENT CONSIDERATIONS

1. Consider what treatments or combination of treatments will best accomplish total forest management objectives over time. A single prescribed-fire or mechanical treatment may not be sufficient or applicable to all sites.

2. Commercial thinning can be used to reduce canopy bulk density, but additional treatments such as prescribed burning or chipping of slash may be needed to reduce surface and ladder fuels.
3. Topography is important. Mowing shrubs or crushing dead wood and other mechanical treatments may be effective on flat ground but impractical on steep slopes. Fire behavior and methods of prescribed burning are also affected by slope.
4. An initial fire may kill understory trees and shrubs and cause shrubs and hardwoods to sprout or seed themselves, both of which will increase fuels onsite over time. Several burns may be needed to reduce fire hazard.
5. Herbicide treatments can be used effectively to reduce the density and size of sprouting shrubs and hardwoods before (as a "brown and burn" treatment) or after prescribed burning.
6. Mechanical treatments and fire need to be used with care to avoid, respectively, soil excessive disturbance and removal of nutrient-rich organic layers in the soil.

Fire Across Landscapes

Reducing fuel accumulations and reintroducing fire as an ecological process is an important goal for individual stands but must be considered within a landscape context. Across any forested landscape, there is great variability in the potential for a wildland fire in terms of extent and severity (Ager et al. 2013). Chances of ignition, rates of spread, and fluctuations in fire intensity over space lead to **occurrence** on a particular site, for which the probability of burning relates as much to its landscape position (its neighbors) as its own fuel hazard. Fire risk is this probability of occurrence combined with the resource values potentially lost. Two other important factors in considering fire risk are the availability of early detection and equipment access for fire suppression, both of which can vary markedly throughout forests. Thus, it can be more important to prioritize stands for fuel reduction than to select the most appropriate treatment strategies within a particular stand.

In general, reducing fuels over multiple stands and cumulatively larger areas make landscapes less susceptible to fire (Weatherspoon and Skinner 1996; Agee et al. 2000; Finney 2001). Modeling exercises, as well as recent experience with large wildland fires (e.g. the Wallowa Fire in Arizona), document this potential. As a rule of thumb, treating more than two-thirds of a given landscape ensures that future fire flow through that area will be modified, regardless of exact placement and silvicultural detail; treating less than one-third is typically ineffective at modifying fire flow through the fireshed, though it may be effective at altering fire behavior in those particular stands. In the middle third, where land managers often operate, stand treatments (and the details of those silvicultural prescriptions), where they are, and how they fit together in space and time can make a major difference in landscape-level fire behavior and extent (Finney 2001). And these stand treatments can be augmented and connected with strategically placed linear fuel breaks.

Shaded fuel breaks are strips of land >100 feet in width where potential fire intensity has been decreased (Agee et al. 2000) Their attributes include: reduced fine surface fuels and ladder fuels; reduced canopy bulk density and stem density; an overstory of large, fire-resistant trees; and elevated canopy base height. Open and simple stand structures are ideal for fuel breaks by 1) limiting flame heights to low-intensity surface fire (instead of crown fire) that can then be actively suppressed, or 2) creating an area within which managers can easily backfire from established fire lines. Fuel breaks cannot usually stop fire completely, and once fire spreads through the break it regains intensity (Agee et al. 2000).

Maintaining fire-resistant treated stands and shaded fuel breaks over time will always be a challenge. When the density of overstory trees is reduced, an understory of shrubs and trees can become quite vigorous, given that the regeneration of shrubs is often stimulated by mechanical and prescribed burning treatments (figure 11-7), as is sprouting of burned or cut hardwood trees. Before treatment, overstory and understory density and species composition varies, and therefore treatment options will need to vary. Non-native plant species may invade these disturbed sites and greatly alter species composition, abundance, and fuel loading in the future understory (Chmura et al. 2011). Post-burn evaluation and periodic reduction of fuels will be needed to ensure that stand density and structure compatible with reduced fire potential are achieved and maintained.

Individual stands and shaded fuel breaks alone are unlikely to be sufficient to reduce fire severity throughout a forest; however, they can be places from which to "grow" fuel reduction over larger areas within the forest and thus reduce fire severity overall. They are likely to be most effective in forest types within low- and moderate-severity fire regimes. In reducing the potential for a severe fire across a landscape, fuel reduction will be one objective in many silvicultural prescriptions, including:

1. Site preparation, aimed at reducing slash loads for ease of planting, can also be designed to reduce the likelihood of future severe wildland fire;
2. Thinning prescriptions can be modified to enrich the proportion of fire-resistant tree species and encourage their growth into larger size classes, raise canopy base heights, and reduce canopy bulk density (as well as yield wood);
3. Slash can be reduced or rearranged after any harvest to minimize flame heights and resultant basal/crown scorch during subsequent prescribed burning(s); and
4. Group selection openings can be intermixed with individual tree selection to create discontinuities in surface, ladder and aerial fuels appropriate to topographic position to alter the severity pattern of wildland fire.

Because high fuel volumes are currently common throughout the West, especially in pine and mixed-conifer forests, and constitute a massive "fire debt," concerted efforts at fuel reduction spanning several decades will likely be needed to re-establish reasonable levels of fuels and fire-resilient landscapes (Stephens et al. 2013). The large variability in stand structures and densities, condition of fuels in various stands, regeneration potential of shrubs and hardwoods, as well as major issues associated with ignition potential, topography, property boundaries, and values, all call for different treatment priorities and techniques. In addition, a continuous evaluation of treatment cost vs effectiveness will be necessary, and the area treated will likely be limited by available funds. Careful use of a variety of treatments, including fuel chipping, cutting of small trees and shrubs, logging of merchantable trees, and herbicide applications to control shrubs, will ensure the greatest area treated for the funds available.

REVIEW QUESTIONS

1. Discuss the effects of wildland fire before the beginning of contemporary forest management. How has fire in the forest changed in the last 100 years? 50 years? 15 years?
2. Why was the 20th-century fire suppression policy enacted, what were its component parts, and in what ways was it effective or not?
3. What makes up the fire triangle? The fire behavior triangle?
4. What is the difference between fire intensity and fire severity?
5. What are fire regimes? Be able to give examples and discuss. How do they apply to the use of prescribed fire in today's forests?
6. What is meant by fire frequency? What are other ways than mean fire return interval to characterize fire regimes, and how are they important?
7. What are the limitations of using fire history studies to develop prescribed burning/silviculture prescriptions?
8. Can natural/pre-management fire regimes be models for use of fire in today's forests? Discuss these questions.
9. How are fuels classified in terms of location? Size? Characteristics? What are 1-hour, 10-hour, and other such fuels?
10. How is fuel classification helpful in silviculture/prescribed burning prescriptions?
11. What are fuel models?
12. What stand parameters (structure) can be manipulated to reduce fire severity? How does that work?
13. Explain how the first prescribed burning may make a stand more susceptible to fire, and why a long-term multi-treatment approach may be needed in order to successfully control fuels?
14. Discuss the possible effects of an initial prescribed fire on large/old trees.
15. Discuss the potential effects of thinning and prescribed burning in stands that have a seed bank of ceanothus or manzanita.

CHAPTER 12
Case Histories

Introduction

Silviculture is an applied science practiced in an ecosystem context (i.e., "applied forest ecology"). The environmental variables of each ecosystem are constantly changing, often in unpredictable ways. For this and other reasons, they are not entirely understood. In addition, forests are subject to the effects of plant competition, wind, frost, insects, and pathogens, as well as treatments such as prescribed fire, thinning, and fertilization. Together these treatments affect stand growth and development, but their effects may not be recognizable until years after they occur. We use the examples below to show how a combination of long-term, first-hand experience in managing a forest and knowledge derived from well-designed studies provides an understanding of the effects of both the environment and silvicultural treatments on the development of forest stands. These examples also illustrate how research and experience together improve and update silvicultural practices.

Case History 1: Intensive Management of Conifer Plantations for Wood Products

In many private forest properties throughout the West, management objectives call for producing high yields of commercial wood under an intensive management regime. The details of the silvicultural systems employed vary somewhat, depending upon forest type and species, and policies. For example, companies that grow wood to support their own mills may use different systems and rotation lengths than those that just sell wood from their lands. Family-owned companies might have different policies than publicly traded companies. We use an example from Douglas-fir forests to illustrate an even-aged silviculture system. Intensive management in other forest types may use very similar systems, or possibly uneven-aged systems (see case 4), and also include plantations of hybrid poplar and red alder with rotations of ±10-30yr.

REFORESTATION

Reforestation closely follows clearcutting and is focused on establishment of fast growing Douglas-fir seedlings. The goal is to have seedlings "free to grow" within 2-3 yr after planting. Because of rapidly invading forbs and shrubs that compete with planted seedlings for water and light in this forest type, site preparation is intensive. Ideally, logging is completed by early spring to allow time for competing vegetation to develop prior to an herbicide treatment. If the density of hardwoods is high, they may be controlled by stem injection or cut stump treatments prior to or just following logging. Logging slash may be piled and burned to facilitate hand planting, although excessive soil exposure and disturbance can favor certain invasive species, such as Scotch broom (Harrington and Schoenholtz 2010). Herbicides are applied the first growing season after logging to prevent overtopping by shrubs and hardwoods, as well as to reduce abundance of forbs and grasses below 20% cover (Dinger and Rose 2009)—a competition threshold that ensures plentiful availability of soil water to the newly planted seedlings. Thus, seedlings are planted on relatively "weed free" sites.

Seed is from seed orchards often established from vigorous trees selected from the local native forest and then from the best trees of the second generation of their progeny (second-generation progeny tested parents). Nursery production of container and bare-root stock ensures that seedlings have high root-growth potential and can quickly grow roots after planting, as soon as soil temperature and moisture are adequate. Seedlings planted in the late winter or spring begin root growth when soil temperatures reach 40°F and are rising. Some landowners practice fall planting, but its success depends on adequate soil moisture following summer drought. Fall planting enables seedlings to have well-developed root systems the following spring and increases early seedling growth compared to spring-planted seedlings. It also requires nursery practices that control seedling dormancy to enable fall root growth.

Additional vegetation control may not be used, unless sprouting shrubs or hardwoods proliferate and threaten to overtop the planted seedlings (figure 12-1).

Fertilization with nitrogen, and sometimes phosphorus and micronutrients, is typically applied soon after crown closure (i.e., 15-20 years after planting)—and after precommercial thinning if early density management is deemed necessary.

COMMERCIAL THINNING AND DENSITY MANAGEMENT

Density control of conifers depends on management objectives. Since reforestation techniques usually promote high survival, planting density may be reduced to eliminate the costs of pre-commercial thinning and to grow large trees in a shorter period of time. However, new logging techniques and markets for chips/biomass are enabling commercial use of formerly non-commercial wood. Use of small wood is limited by distance to manufacturing facilities and hauling/logging costs, particularly on steep

Figure 12-1. View of the Molalla, Oregon Long-Term Soil Productivity Study in the fifth year after planting Douglas-fir seedlings. Primary competitor species include trailing blackberry, salal, cascara, and velvet grass. On this site of average productivity typical of the Cascade Mountain foothills, about 94% of planted trees have survived. Douglas-fir height ranged from 7.0 to 7.8 ft and diameter at breast height (DBH at 4.5 ft) ranged from 0.7 to 0.9 at five years. At age 10 years, tree height ranged from 21.3 to 22.9 ft and DBH ranged from 3.3 to 3.8 in.

slopes. Silvicultural systems including density control and early thinning are adapted to these variables.

Commercial thinning also varies with company policies, and with markets and demand for wood. The demand for large logs (>20 in dbh) has decreased in general. Some companies grow stands at high densities for short rotations (30-40 yr) to supply highly automated, efficient mills that use small logs. Companies that sell logs grow stands on somewhat longer rotations (>50-60 yr), with several light- to moderate-intensity thinnings. These thinnings prevent mortality from competition, and generally keep stands well stocked and volume growth rates high. They may yield high value logs for poles, pilings, and other products.

Case 2: Black Rock—Heavy Thinning and Underplanting to Grow Large Trees and Complex Stands

The Black Rock study is an unreplicated trial established by Alan Berg of Oregon State University in 1957. The initial purpose of this trial was to determine whether the commercial timber yield of a young Douglas-fir stand on site class II land could be improved by a heavy commercial thinning and underplanting with western hemlock, a shade-tolerant species. The yield from the stand would include the thinning and underplanting at 48 yr, plus the harvest of the overstory Douglas-fir and the understory hemlock 40–50 yr after thinning and underplanting. The yield proved high, but only for the Douglas-fir. Though this had not been anticipated, the combination of heavy thinning and the planting of hemlock produced a stand structure with some of the characteristics of old forests (See figure 8-8).

THE PRESCRIPTION

A 48-yr-old naturally regenerated Douglas-fir stand was thinned from below from 294 trees/ac to 40–50 trees/ac in 1957. The average diameter was 11.9 in before thinning and 16.1 in after thinning. The volume of timber removed was 6,918 ft^3/ac, or 24,169 bd ft/ac. Western hemlock seedlings were planted immediately after thinning.

At the last remeasurement in 1990, the average diameter of the overstory trees at age 81 was 29 in. The net volume in the overstory trees was 12,174 ft^3/ac, or 66,343 bd ft/ac. The total production in the overstory trees from the thinnings plus the standing volume at age 81 was 19,902 ft^3/ac, or 90,512 bd ft/ac. The periodic annual increment in volume was high from 1985 to 1990 (326 ft^3/ac/yr), well above the mean annual increment (Curtis 1995; Curtis and Marshall 1993). Some of the large Douglas-fir (30+ in in diameter) died from root disease. The western hemlock planted in the understory were about 50 ft tall and only 3–6 in in diameter.

HABITAT/OLD FOREST CHARACTERISTICS IN 2005

In 2005, the large overstory Douglas-fir were well over 32 in in diameter and about 200 ft tall. They had deep crowns and were developing large branches. The death of some of the large Douglas-fir from root disease had caused gaps in the canopy, which accommodated the growth of the hemlock, and the fallen trees and snags provided dead wood on the forest floor and habitat for cavity-nesting animals. The heavy thinning also stimulated the growth of some bigleaf maple into the mid-canopy and the development of salal, vine maple, and huckleberry in the understory.

INTERPRETATION

Some foresters thought that thinning to 50 trees/ac "ruined the stand," because there were so few trees and their crowns were not well developed (probably <35% live crown ratio). However, heavy thinning stimulated the diameter growth of the overstory Douglas-fir; its average diameter at age 81 was 29 in, compared to 17–18 in in nearby stands in which the Douglas-fir were growing at 125 trees/ac. For the period between 48 and 81 yr, volume production was quite high. The periodic volume growth plus the yield from thinning was about 1,500+ bd ft/ac/yr. The net periodic annual increment remained well above the mean annual increment during this time (Curtis 1995); it was 2065 bd ft/ac/yr during the last 5 yr of the period. The relatively high stand density before thinning and the planted hemlock understory may have improved wood quality, because only a few small branches persisted on the lower stems of the overstory Douglas-fir.

In this stand, a hemlock understory would likely not have become established without planting, because there was no seed source and no evidence of hemlock regeneration in surrounding thinned stands. On nearby sites with seed-producing hemlock present, the initial thinning established a hemlock understory, and there was no need to plant.

This case history shows an example of how forest stands might be grown for multiple forest management objectives. The stand has produced a large volume of wood and provided a yield of timber at a relatively early age. Earlier heavy thinnings might have produced somewhat less yield, but they might have improved the economics of the treatments by providing income earlier in the life of the stand. Similarly, another thinning with irregular spacing, perhaps between 1995 and 2005, would have produced additional yields and stimulated an irregular stand structure more typical of old-growth Douglas-fir forests by facilitating variable development of the hemlock, bigleaf maple, and shrubs in the understory.

The stand should become increasingly valuable as habitat for species associated with old forests. It currently has large snags and pieces of wood on the ground, large overstory trees, and the beginnings of a diverse understory, which are important characteristics of an old-growth Douglas-fir stand. Thinning for wood products has resulted in forests with old-forest characteristics on similar sites (Newton and Cole 1987). Because of the long rotation and the value in the large volume of standing trees, this type of management might not be suitable for industrial forestlands except in special cases in which the objective is to produce large logs. On the other hand, this management strategy may be quite suitable for public forests.

On these sites root disease (*Phellinus weirii*) kills trees of all sizes, whether the stands have been thinned or not. Because thinning reduces the number of smaller trees but does not affect the disease, one effect of thinning on such sites is that large trees die from root disease rather than small ones. This, however, provides the forest manager with options. The dead trees could provide large pieces of dead wood or snags for habitat, or they might be salvaged for timber, whereas small dead trees are probably not suitable either for salvage or as long-lasting snags or logs on the forest floor. It is important to note that nearby unthinned stands of the same age do not have trees, dead wood, or snags of large diameter, or a diverse understory.

SUMMARY

This case history illustrates how forest stands can be managed over time for a variety of objectives. When the stand was young, the goal was growing timber volume; after thinning it began to produce old-forest habitat while continuing to produce large volumes of commercial wood. Thinning to less than 50 trees/acre or using irregular spacing would likely have provided a better developed understory. The stand will likely continue to develop habitat and yield more wood. After an additional thinning, this stand could be left indefinitely for habitat, or it could be harvested for wood production at some time in the future.

Case 3: Regenerating Forest Stands on the Dead Indian Plateau in Southwestern Oregon

This summary is provided by Dave Russell and Tom Sensenig, who, with other Bureau of Land Management and Forest Service silviculturists, worked in these forests for 20+ yr.

Productive mixed-conifer forests are quite common in the southern Cascade Mountains in Oregon and California at elevations ranging from about 4,000 to 6,000 ft. They occur to the north near Diamond Lake; to the south near Mount Shasta; and on flat sites east of the Cascades and Sierra Nevada.

Species composition on these sites varies with elevation and includes ponderosa pine, Douglas-fir, white and grand fir, sugar pine, western white pine, and lodgepole pine. These forests merge with ponderosa pine forests at lower elevations and true fir forests at higher elevations.

Much of the terrain in the southern Oregon Cascades is in relatively flat, saucer-like plateaus surrounded by gently sloping mountains. Temperatures during the winter and the early growing season are cold, especially on the plateaus, where cold air from the surrounding mountains settles (Stein 1984). The growing season is short at the higher elevations, where shoot growth of conifers may not start before July. Summers are dry, with little precipitation from May through September, and drought tends to limit growth by mid-August.

Early efforts to regenerate these forests following fire or clearcutting included site preparation to control shrubs and planting of ponderosa pine and Douglas-fir. This worked well at lower elevations, but at higher elevations Douglas-fir regeneration failed altogether, and sufficient ponderosa pine stocking was lacking on many sites.

ANALYSIS OF THE SITES AND REGENERATION EFFORTS

Several formal analyses were made by PNW researchers (Minore 1978; Stein 1984; Williamson and Minore 1978) to determine the environmental variables affecting seedling survival on the Dead Indian Plateau near Medford. These analyses showed that the following variables affected reforestation efforts on this site:

1. Cold air temperatures early in the growing season routinely kill foliage of planted trees, especially Douglas-fir. Trees typically remain susceptible to frost until they are more than 10–15 ft tall. Cold soil temperatures during the spring shorten the planting window because soil temperatures may remain below 40°F and suppress root growth and probably water uptake until June or later.
2. Summer drought is a major variable and requires control of shrubs and grasses to increase available soil water.
3. Browsing by gophers, cattle, rabbits, deer, and elk is common. Gophers severely damaged seedling crowns under the snow pack and roots below ground, which prevented plantation establishment on some sites (Williamson and Minore 1978). Deer and elk mainly browsed the shoots of smaller seedlings, and their damage was not thought to be of major importance. Rabbits girdled seedlings where there was a cover of shrubs (Gratkowski and Anderson 1968). In areas where cattle had access to plantations, they killed many small seedlings up to about 5 yr old. In addition, there was a reported one-time outbreak of grasshoppers that severely damaged some plantations.
4. The competitive effects of shrubs and grasses were very important. Grasses can grow in cold soils and use the soil water before the soil is warm enough for root growth and water uptake by conifer seedlings. Root growth of planted seedlings was slow, which probably limited water uptake in cold soils. Grasses harbor gophers and probably helped sustain their population at damaging levels; they also provide habitat for grasshoppers.
5. Early establishment of shrub seedlings created a cover that competed with conifer seedlings for water and, on some sites, provided habitat for rabbits (Gratkowski and Anderson 1968). White fir gradually became established under the shrubs, but not Douglas-fir, ponderosa pine, or lodgepole pine. Pure stands of true fir began to grow on sites that formerly had mixed-conifer stands.
6. Strong winds on these sites blew down shelterwood trees that had been left to mitigate low spring and high summer temperature extremes.
7. Natural regeneration occurred in partially cut stands where trees of low vigor or trees killed by insects had been removed. Regenerated trees consisted mainly of advance regeneration, but some new seedlings were also established after logging. These seedlings could be used as part of the regeneration of the next stand. In a survey of about 100 sites on the plateau, Stein (1984) found that, after clearcutting, about 33% of the regeneration was established prior to harvest. Also, uneven-aged management was a viable way to harvest timber and obtain regeneration, but regeneration is often patchy, and some planting was needed to ensure that the stand was fully stocked and pine species were maintained. Precommercial thinning of small trees may be needed in dense patches of advance regeneration.

8. Pathogens affect the choice of species and regeneration method. Partial cutting has the potential to introduce stem pathogens into true firs from logging damage. When it is present, dwarf mistletoe may increase in overstory trees and spread to regeneration in the understory. Dwarf mistletoe is species-specific, however, so its effects may be reduced in mixed-species stands.
9. White pine blister rust is present, which discourages the planting of western white pine, a cold-tolerant species, and also sugar pine, which occurs at low density in these forests. Rust-resistant sugar pine seedlings have been planted on some sites.

CURRENT REFORESTATION PRACTICES

Current reforestation practices are aimed at establishing mixed-conifer stands of pine species and white fir and/or Douglas-fir depending on the effects of elevation, aspect, and cold temperatures. Based on evaluation of the environment, the following reforestation practices are used.

1. For areas with cold temperatures: A mix of Douglas-fir and ponderosa pine can be planted below about 4,000 ft in elevation as well as up to about 4,500 ft if the site has about a 15% slope, enabling the drainage of cold air. On flat sites above about 4,500 ft, mixtures of ponderosa and lodgepole pine are planted, but no Douglas-fir. At very cold sites at 5,500 ft, the mixture includes Jeffrey pine, a species that is especially cold tolerant, from nearby seed sources.

2. For establishing Douglas-fir and white fir: On sites where Douglas-fir was not in the original mix of species established in the plantations, it can be underplanted in 15-yr-old pine stands after thinning. On some sites, seed from adjoining stands will allow Douglas-fir to become established naturally. In both cases, the larger pine will lessen extreme cold temperatures, and the low level of shrubs and grasses beneath the pine should result in reduced gopher populations. On most sites, there will be sufficient white fir if there is a seed source. The overstory pine will be too small (10 in dbh) at this time for commercial thinning, so thinning to underplant or release naturally established white fir, grand-fir, and Douglas-fir was deferred. Precommercial thinning is expensive and would create large volumes of slash that would predispose the stands to bark beetles and fire.

3. For managing competition for water and animal damage: At planting and at least once afterward, grasses and shrubs are scalped from a 2-ft radius around the seedlings to reduce competition for soil water. Competing species are successfully controlled by herbicides on industrial lands. Planting mixtures of ponderosa, lodgepole, and Jeffrey pines helps reduce browsing from deer and elk. Ponderosa pine appears to be the preferred browse species, but Jeffrey pine grows rapidly and is therefore less susceptible to browsing than the other pines. Gophers are trapped out of areas where they are found to cause severe damage to plantations.

4. For managing regeneration density: Planting high densities of seedlings (500+/ac) helps compensate for expected mortality from cold, drought, and animal damage, especially from gophers. The objective is to achieve a stand of 50–100 trees/ac at 20–30 yr of age. It should be noted that, on industrial land on the plateau, pine plantations are successfully established on similar sites using herbicide vegetation control, which reduces the effects of grass and shrub competition for water and also reduces damage by animals and consequently the need to plant so many seedlings.

SUMMARY

This case history illustrates the range of variables that can affect regeneration and the various silvicultural practices that can be used to ensure it is successfully established. On some sites, mixed-conifer regeneration can be established in one step; on others, establishment requires time and several steps, possibly including planting cold-tolerant species for the overstory and establishing cold-intolerant conifers in the understory. It appears that both even-aged and uneven-aged regeneration methods will work, but each requires its own site-specific adaptations to the effects of browsing, summer drought, competition from grass, cold temperature extremes, windthrow, pathogens, and overstory density.

Case 4: Management of a Sierra Nevada Mixed Conifer Forest—Initial Logging and Conversion of Even-Aged Stands to Uneven-Aged Management

This case history comes from the University of California's Blodgett Forest in the Sierra Nevada mixed-conifer forest. The following scenarios are based on the experience and observations of Bob Heald, from 1975 to 2005. During this period, the forest was divided into stands that were targeted for management under even-aged and uneven-aged systems. Heald was responsible for silvicultural prescriptions, logging, reforestation, and ensuring compliance with California forest practice rules.

SETTING

Blodgett Forest is located on the Georgetown divide, at about 4,500 ft on the west slope of the Sierra Nevada. The forests are quite productive (ponderosa pine site class II+, Dunning 1942). Stands are a mixture of ponderosa pine, sugar pine, Douglas-fir, white fir, incense-cedar, and black oak. Other woody species include shrubs (bearclover, deerbrush, gooseberry, and manzanita) and hardwood trees (tanoak and chinkapin).

HISTORY

The last fire occurred in about 1870. The forest was logged for high-value logs from about 1910 to 1920. The logged trees were generally 24–60 in in diameter. Logging, which included ground skidding of logs to a central point, caused widespread soil disturbance and prepared an ideal seedbed for natural regeneration of all the conifer species. Cut or damaged black oak sprouted.

Forest inventories from 1930 indicate that approximately one tree of each species per acre with diameter of about 24 in was left after logging. There is no evidence of fire after logging. The six seed trees per acre of varying species and the soil disturbance led to plentiful natural regeneration of all species, and in 1960 the forest was generally composed of well-stocked, even-aged stands with an age ranged of 40–60 yrs. Canopy closure was more than 60%, and there were very few trees or shrubs in the understory.

At this point the forest was divided into stands of 40–60 ac. Stands were management units delineated by topography and aspect. Each stand was assigned to be managed under either an even-aged or an uneven-aged system. For the stands designated for uneven-aged management, the goal was to use single-tree or group-selection regeneration methods to economically convert them to sustainable uneven-aged stands. There was little prior research upon which to base these uneven-aged silvicultural systems. Data from inventories and growth and yield studies in even-aged stands, general knowledge of regeneration principles, and site-specific information were used in combination to develop the silvicultural system (figure 12-2).

SINGLE-TREE SELECTION: INITIAL TRIALS

1. Regeneration: A main concern was how to achieve regeneration in a dense even-aged stand, especially of shade-intolerant species such as ponderosa pine, sugar pine,

Figure 12-2. Examples of two uneven-aged stands in the Sierra Nevada mixed conifer type. Species include ponderosa and sugar pine, Douglas-fir, white fir, incense-cedar, and black oak. In photo A the ponderosa pine in the foreground were planted as in a group selection opening. In photo B all species are regenerating in a small gap from single tree selection, which will have to be widened for the seedlings to grow. White fir readily regenerates in the small gaps made in single-tree selection.

and relatively shade-intolerant Douglas-fir. The initial approach was to make small gaps (≤0.25 ac) in the canopy by removing one to three trees. These gaps would release advance regeneration, if it were present, or enable conifer seedlings to become established. Small gaps were used for two important reasons: (a) they minimize establishment of shrubs, and (b) conifer seedlings are relatively shade-tolerant when they are small, so that little light is needed for them to become established. Additional gaps were made later to encourage growth.

2. Stand density and allocation of growing space: The cutting cycle was about 10 yr. At each cutting, the overall stand density was reduced to a basal area of about 150 ft^2/ac for forests of this type. This residual basal area is about 50% of normal (the basal area of well-stocked stands; Dunning and Reineke 1933). Growth and yield principles suggested that thinning to 50–55% of normal basal area would keep PAI high (200 ft^3/ac/yr on these sites). However, the basal area left after cutting varied throughout the stand. It was lower (120 ft^2) in those parts of the stand composed mainly of pine and those with regeneration to release. Higher basal area (170 ft^2) was left on sites where the species composition was mainly true fir and those on which the stand had grown at a high density. No more than 30% of the basal area was removed at each cutting, to avoid reducing growing stock too severely or exposing a stand to severe damage from wind or snow. Thus, if stand density was 250 ft^2/ac, 75 ft^2/ac would be removed, leaving the average basal area after thinning at about 170 ft^2/ac.

The basal area was allocated among tree size classes by leaving trees of a given size class at a prescribed spacing. For example, 150 ft^2/ac for trees 6+ in dbh might be allocated as shown in table 12-1. Other methods for allocating stand density to tree size classes are now available (see chapter 7).

More than half the basal area was allocated to the larger size classes in order to favor the larger trees, which are more valuable economically and are accumulating in volume and value at the greatest rate, and also to enhance the forest's aesthetic quality, provide some resistance to fire, and improve wildlife habitat. Thinning to 150 ft^2/ac reduced the canopy density so that trees less than 6 in dbh grew into the 6- to 12-in

Table 12-1. An example of the distribution of trees per acre and basal area for single-tree selection.

Diameter (in)	*Basal area (sq ft/ac)*	*Trees/ acre*	*Spacing (ft)*
24+	40	10	66
18–14	50	21	46
12–16	45	47	30
6–12	15	34	36

class and new seedlings became established. Where there were patches of shrubs, site preparation and planting were done to establish shade-intolerant conifer species.

3. Marking guidelines: Selecting trees to cut or leave after logging focused on leaving vigorous trees of the proper species and size. Marking was based on the principle that what was left was more important than what was taken. Vigorous trees were cut only to provide growing space for other, more vigorous trees or to release advanced regeneration of shade-intolerant species. Leave trees were chosen with the following priorities: (a) quality, (b) spacing, and (c) species. It was determined that thinning from below and leaving the best trees in each size class would meet the target diameter or size class distribution. Thinning from above, in contrast, might have achieved the correct size class distribution, but the trees left could have been of poor vigor.

A concept called size class area occupancy was used to evaluate the size class distribution or stand structure before marking. This method enables silviculturists to estimate the size classes that are under- or over-represented in a stand. On transects through a stand, a grid of temporary toe-point plots is established. At each point, the size classes present are estimated visually according to the classes in table 10-1. The goal is to have about 16% of the stand in each class. This semi-formal stand exam also helps silviculturists decide which size classes to favor when marking trees, and it is an excellent way for them to familiarize themselves with other important aspects of a stand, such as the variation in density and species composition, the presence of pathogens, etc., before marking.

The upper diameter limit was set at 24–28 in because at this size the volume growth rate of trees drops below 5–6%, considered a reasonable return on investment in growing stock. However, each tree >28 in dbh was kept unless there was a better tree to replace it. These retained large trees served as seed trees of high genetic quality. Currently there are trees >36 in in these stands. Sugar pines that appeared to be resistant to blister rust were also kept, with the hope of obtaining rust-resistant regeneration.

As management progressed, the following concerns became apparent and adaptations were made as needed:

a) There were many trees in the small size classes, and initially there was no market for them. However, a market for small trees (10–14 in dbh) developed and greatly facilitated implementation of this system. Had there been no market, it would have been necessary for the trees in this size class to be precommercially thinned (PCT).

b) PCT is difficult in this system, and yet it is often needed to meet density and species composition goals and to release trees <6 in to grow into larger size classes. Guidelines for PCT must account for tree quality, species, and density of trees in the larger size classes, as well as of those in the size classes being thinned.

c) Fuel reduction is needed and has to be planned as part of thinning. Piling and burning slash appears to be more practical than underburning the stand, but in retrospect, underburning to reduce fuels present from self-thinning and from the initial logging would have been an efficient way to begin fuel management before this system was implemented to start the regeneration process.

d) Shrubs, mainly *Ceanothus* spp. and *Ribes* spp., have invaded as canopy density has been reduced. Generally these are not dense enough to inhibit regeneration, and *Ceanothus* may provide forage for deer. However, *Ribes*, an alternate host of white pine blister rust, has probably caused an increase of this pathogen on sugar pine.

e) With the exception of blister rust, pathogens and insects, while present, have not been a problem in this system. Insects and root diseases occur in even-aged ponderosa pine stands. However, they have not caused significant mortality in these mixed size/species stands. For example, although *Heterobasidium* spp. is present in both white fir and ponderosa pine, there are few of the spreading root disease centers typical of single-species stands. Mistletoe was not present in the trees left after the initial logging, and so today's stands have few infected trees. Should root disease and mistletoe centers occur, they could be controlled by making canopy gaps (removing 10 or 12 trees) and regenerating them with the five conifer species. Severe drought could cause a greater impact of insects and pathogens.

f) There are few snags because stands have been kept at low densities and because the effects of insects and pathogens have been minimal. Therefore, where snags are needed for wildlife habitat, trees are killed.

g) Timber markers need training to implement this uneven-aged system. It is initially difficult to space trees of different sizes while keeping in mind tree quality and species, changes in density and species composition throughout the stand, and the need to provide gaps for regeneration. Timber markers need to evaluate trees by size class, starting with the larger trees. This means first choosing the larger trees to leave, and then choosing trees to leave in smaller size classes, deciding at the same time where to make gaps to release seedlings or saplings. Initially this is an interactive top-down process. Soon markers learn to evaluate size classes, tree quality, and species simultaneously as they move through the stand. Prisms (BAF 20) are helpful in estimating basal area from point to point. It is important that tree markers understand logging operations and the capabilities of logging systems. Otherwise they might specify tree or volume removal that would cause unacceptable damage to the remaining trees.

h) Logging is done by hand felling and ground skidding with grapple-rigged skidders. Permanent skid trails (10–12 ft wide) and landings are established. Loggers are paid a premium of 10–15% for skidding smaller volumes. Average volume removed is about 8,000 bd ft/ac. Less than 5,000 is not economical and more than 10,000 causes too much damage. The logging crew must be skilled to minimize tree damage and remove volume efficiently.

GROUP SELECTION: INITIAL TRIALS

The major criteria for group selection were as follows:

1. Gap location, size, and regeneration. Locations of gaps were chosen to (a) promote reforestation of unstocked areas, (b) reduce concentration of fuels, (c) release advance regeneration, (d) control dwarf mistletoe and other diseases, and (e) reforest extremely dense aggregations of small trees.

Group selection gaps were planted with shade-intolerant sugar pine, ponderosa pine, and Douglas-fir. The goal was to have a mixture of species in each group, and it was expected that white fir and incense-cedar would regenerate naturally, including release of advance regeneration. Gap size ranged from 0.5 ac to 2.5 ac, and both planting and natural regeneration were tried.

2. Tree growth and thinning. Within each gap, trees were projected to be 20–30 in in diameter when the gap was harvested a second time. The strategy chosen to achieve this size at harvest was to let trees grow for 80–100 yr, conduct precommercial and commercial thinning when needed, and harvest about 10% of the area by group selection every 10–12 yr. The matrix (remnants of the initial even-aged stand) was thinned about every 10 yr as described above to an average basal area of 150 ft^2/ac (120–170 ft^2/ac) depending upon species composition and density.

Results of 35 yr of experience and current practices have revealed the following:

1. Gap size of 0.25 ac was sufficient to establish regeneration of all species. However, gaps of less than 1 ac and ≤100 ft wide are very difficult to log without serious damage to adjoining trees from felling and skidding. Therefore, gaps are now about 1 ac and are laid out so that the narrow axis is no less than one-half of the longer axis. Narrow or irregular shapes are avoided.

2. About 90% of the gaps regenerated naturally. Those with dense shrub cover generally were restocked with white fir. Therefore, to ensure a good mix and growth of all species with minimal shrub control, gaps are now planted with vigorous nursery-grown seedlings of all five species at about 150 trees/ac.

3. There was very little edge effect (York et al. 2003 and 2008). Growth of trees within the gaps was primarily affected by tree and shrub density within the gaps rather than by older trees surrounding the gaps. Surprisingly, older trees over 100 ft from the edge of the gap increased their growth rates. There was no need to widen gaps to ensure the growth of trees at the edges.

4. In the matrix between the gaps, thinning from below was successful, and trees there are approaching 30+ in in diameter. Because the trees are relatively large, only a few trees/ac need to be removed every 10 yr to reduce the density to 150 ft^2/ac.

5. In 20–30 yr, the gaps tend to merge with the matrix and disappear. At this age, trees are 14–16 in in diameter and can be thinned along with the older trees in the matrix. At each 10-yr cutting cycle, about 2000-3000 ft^3/ac (10,000 board ft/ac) is produced from thinning the matrix and former gaps and from making new gaps.

6. If alternative structures are required for habitat or aesthetics, regeneration of parts of the matrix can be delayed to provide patches of shrubs; or areas of large trees can be maintained by deferring group selection cutting in them.

7. Precommercial thinning, shrub control, and planting cost less in this method than in the single-tree selection method, but they are somewhat more expensive than in even-aged stands using the clearcut method. Marking trees to cut or leave is easier than with the single-tree selection method.

8. Plant species diversity is higher in group selection than in single-tree selection, and somewhat lower than in clearcutting. The group selection areas have the lowest occurrence of exotic plant species, and the most late-seral wildlife species occur in the group selection matrix.

9. Fuels are controlled by piling and pruning slash in the groups and underburning in the matrix. Pruning is done to reduce fuel ladders in young, recently regenerated gaps. Care must be used when burning to prevent killing black oak stems.

Case 5: Silviculture of Young Ponderosa Pine Stands Established from Advanced Regeneration

This case history is derived from a study in ponderosa pine conducted on the Pringle Falls Experimental Forest near Bend, Oregon (Cochran and Barrett 1999). This valuable long-term study shows how environmental variables may change through time and how difficult it may be to explain their effects on stand growth and development.

Advanced ponderosa pine regeneration was released and thinned in about 1958 to uniform spacings of 6.6, 9.3, 13.2, 18.9, and 26.4 ft, with six replicate plots for each spacing. Understory vegetation, mainly greenleaf manzanita and snowbrush ceanothus, was removed by cutting from half of the plots at each spacing. Tree and stand growth were measured every 4–5 yr from 1960 to 1994, eight consecutive growth periods. As expected, tree characteristics and stand development were strongly affected by initial stand density, but also by the presence or absence of shrubs in the understory (figures 12-3, 12-4).

After 34 yr, the average diameter at breast height ranged from about 5.6 in in the 6.6-ft spacing to 14.2 in in the 26.4-ft spacing. Tree diameter in the 6.6-ft spacing with

Figure 12-3. Stand A grew at 1000 trees per ac (6.6 ft spacing). It will likely stagnate and trees will be too small for attack by bark beetles. Stand B grew at 250 trees per ac (approximately 13 ft spacing) with no understory. Stand C grew at 125 trees/ac (approximately 20 ft spacing) and its understory is being shaded out by the overstory. Stands B and C will likely be infested with bark beetles as their density increases.

shrubs was nearly the same as in plots without shrubs. However, at the 26.4-ft spacing, tree diameter was 12.4 in with shrubs present and 14.1 in without shrubs. The cubic volume growth and the volume after 34 yr varied with tree density, as expected, and with the presence or absence of shrubs. In the 6.6-ft spacing, the average cubic volume in plots with shrubs was only 4% less than in plots where shrubs were absent (2,341 ft^3/ac compared to 2,447 ft^3/ac). However, the influence of shrubs increased as spacing increased, and at the 26.4-ft spacing, the cubic volume in plots with shrubs present was 34% less than in plots with shrubs absent (910 ft^3/ac compared to 1,389 ft^3/ac). Thus, shrubs had a greater effect on stand growth at the wider tree spacings.

For the first 20 yr, average volume growth for all spacings was greater on the plots without shrubs. However, this relationship changed over time, and in the last 14 yr, volume growth at all spacings was nearly the same, or even greater, on the plots with

Figure 12-4. Stands thinned to 62 trees/ac (26 x 26 ft spacing) about 35 yr previously with understory removed (A) and no understory removal (B). Understory has added N and organic matter to the soil and has provided cover and browse for some wildlife species. It has reduced the overstory growth by about 35%, and it has the potential to fuel crown fires.

shrubs. One possible cause may be the effects of shrubs on soil nutrients. Plots with understory had 189 lb/ac more N in the 0-horizon and upper 9.5 in of mineral soil, and they had more microbial biomass than plots with no understory (Busse et al. 1996). This may help explain the increase in volume growth over the past two decades. However, more total N does not necessarily mean more N available for overstory tree growth because much of the N added by the shrubs is at least temporarily fixed in soil organic matter.

Busse et al. (1996) reported that overstory density had no effect on shrub cover except at the most dense 6.6-ft spacing. However, shrub cover may not be strongly correlated to leaf area, shrub vigor, or nitrogen fixation, and it may not accurately reflect the effects of the shrubs' competition with the trees for soil resources. Shrubs beneath a dense overstory have less leaf area and use less nutrients and water than those under a low-density overstory, and that difference in shrub vigor may explain the greater relative reduction in ponderosa pine volume growth in plots with shrubs present but with low overstory density.

The increase in PAI in the plots with an understory present may have been caused by the combination of a sudden decrease in shrub vigor and an increase in soil N. The understory vegetation cover was reduced by a severe winter kill in 1978–79, the period just before growth increased in the understory-present plots (Cochran and Barrett 1999). The frost and the overstory competition may have combined to reduce shrub competition. Also, the relationship between shrubs and tree growth was complicated by attacks of the pandora moth, which defoliated the pine twice during this period (Cochran 1998a).

The PAI curves for volume growth in the plots with and without shrubs suggests that the presence of shrubs delayed the growth of the ponderosa pine for about 20 yr, and now the trees in the shrubs-present plots are reaching the same high PAI rates that were achieved earlier on the shrub-free plots. The net effect, however, is that the trees in plots without shrubs currently have 5–35% more volume than those in the plots with shrubs. This effect seems to be borne out in the two lowest-density spacings. During the past 15 yr in the 18.7-ft and 26.4-ft spacings, plots without an understory have produced 25–38% more volume than those with an understory.

Mortality of ponderosa pine on these sites has been quite low. About 10% of trees died at the 6.6-ft spacing, 1.3–4.2% died at the 9.3-ft spacing, 1.3% died at the 13.2-ft spacing, and no trees died at the 18.9-ft and 26.4-ft spacings. As the trees grow, however, the stands will increase in density. Those at the mid-range densities, in particular, will become increasingly susceptible to mortality from bark beetles (Cochran and Barrett 1999) and will build up more fuels as trees become stressed and die.

This discussion of the presence and absence of shrubs has focused on their competitive effects on ponderosa pine stand growth and on their alteration of soil conditions and processes. However, the presence or absence of shrubs also has important implications for other forest resources. Shrubs are ladder fuels that increase the chance of a severe crown fire. They also provide forage and cover for wildlife, and they may lead to improvement in soil fertility in the long run. Prescribed fire would likely reduce the amount of N and soil organic matter fixed by the shrubs. On the other hand, a severe wildfire would probably destroy not only the shrubs and much of the soil fertility attributed to their influence, but also the overstory trees and snags.

Thus forest managers are faced with a complex array of ecosystem variables, including soil fertility, shrubs, stand growth, and potential for severe fire. Currently the effects of increased soil N and organic matter on stand growth are uncertain. As this study demonstrates, thinning can be a tool to reduce stand density and achieve some predictability about the future dynamics of the stand. Together with careful use of fire or mechanical treatments to control fuels, well-planned thinning can render these multiple variables a little more manageable.

References

Acker, S. A., E. K. Zenner, and W. E. Emmingham. 1998. Structure and yield of two-aged stands on the Willamette National Forest, Oregon: Implications for green tree retention. *Canadian Journal of Forest Research* 28:749-58.

———, C. B. Halpern, M. E. Harmon, and C. T. Dyrness. 2002. Trends in bole biomass accumulation, net primary production, and tree mortality in *Pseudotsuga menziesii* forests of contrasting age. *Tree Physiology* 22: 213–17.

Adams, P. W., and J. R. Boyle. 1982. The quantity and quality of nutrient cations in some Michigan spodosols. *Soil Science* 133:383–89.

———, A. L. Flint, and R. L. Fredriksen. 1991. Long-term patterns in soil moisture and re-vegetation after a clearcut of a Douglas-fir forest in Oregon. *Forest Ecology and Management* 41:249–63.

Adams, W. T. 1992. Gene dispersal within forest tree populations. *New Forests* 6:217–40.

———, R. Campbell, and J. Kitzmiller. 1992. Genetic considerations in reforestation. In: S. Hobbs et al., eds,. *Reforestation practices for southern Oregon and northern California.* Corvallis OR: Forest Research Laboratory.

———, J. Zuo, J. Y. Shimizu, and J. C. Tappeiner. 1998. Impact of alternative regeneration methods on genetic diversity in Douglas-fir. *Forest Science* 44:390–96.

Agee, J. K. 1990. The historical role of fire in Pacific Northwest forests. Pp. 25–38 in J. D. Walstad, S. R. Radosevich, and D. V. Sandberg, eds., *Natural and Prescribed Fire in Pacific Northwest Forests*. Oregon State University Press, Corvallis.

———. 1991. Fire history along an elevational gradient in the Siskiyou mountains, Oregon. *Northwest Science* 65: 188–99.

———. 1993. *Fire Ecology of Pacific Northwest Forests*. Island Press, Washington DC.

———, B. Bahro, M. A. Finney, P. N. Omi, D. B. Sapsis, C. N. Skinner, J. W. vanWagtendouk, and C. P. Weatherspoon. 2000. The use of fuel breaks in landscape fire management. *Forest Ecology and Management* 127:55–66.

Ager, A. A., N. M. Vaillant and A. McMahan. 2013. Restoration of fire in managed forests: A model to prioritize landscapes and analyze tradeoffs. *Ecosphere* 4(2): 29.

Aho P. E., G. Fiddler, and M. Srago. 1983. *Logging damage in thinned, young-growth true fir stands in California and recommendations for prevention.* Research Paper PNW–304, USDA Forest Service Pacific Northwest Forest and Range Experiment Station, Portland OR.

Alaback, P. B. 1982. Dynamics of understory biomass in Sitka spruce–western hemlock forests of southeast Alaska. *Ecology* 63:1932–48.

———, and F. R. Herman. 1988. Long-term response of understory vegetation to stand density in *Picea–tsuga* forests. *Canadian Journal of Forest Research* 18:1522–30.

———, and J. C. Tappeiner. 1991. Response of western hemlock and early huckleberry to forest windthrow. *Canadian Journal of Forest Research* 21:534–39.

Albaugh, T.J., H.L. Allen, P.M. Dougherty, and K.H. Johnsen. 2004. Long term growth responses of loblolly pine to optimal nutrient and water resource availability. *Forest Ecology and Management* 192:3–19.

Albaugh, T.J, Berghm, T. Lundmark, U. Nilsson, J.L. Stape, H.L. Allen, and S. Linder. 2009. Do biological expansion factors adequately estimate stand-scale aboveground component biomass for Norway spruce? *Forest Ecology and Management* 258:2628–2637.

Albaugh, T. J., T. R. Fox, C. E. Blinn, H. L. Allen, R. A. Rubilar, and J. L. Stape. 2013. Developing a new foliar nutrient-based method to predict response to competing vegetation control in *Pinus taeda. Southern Journal of Applied Forestry* 37:196–201.

Albini, F. 1976. *Estimating wildfire behavior and effects.* USDA Forest Service General Technical Report GTR-INT-156, Ogden, UT.

Albrektson, A. 1980. Relations between tree biomass fractions and conventional silvicultural measurements. Structure and function of northern coniferous forests—An ecosystems study. *Ecological Bulletin* (Stockholm) 32:315–27.

Alexander, M. E. 1988. Help with making crown fire assessments. In W. C. Fischer and S. F. Arno, compilers, *Protecting people and homes from wildfire in the interior West: Proceedings of the symposium and workshop*, 1987 October 6–8, Missoula, MT. General Technical Report INT-251. USDA Forest Service Intermountain Forest and Range Experiment Station, Ogden UT.

Alexander, R. R. 1964. Minimizing windfall around clear cuttings in spruce–fir forests. *Forest Science* 10:130–42.

———. 1986. *Engelmann spruce seed production and dispersal and seedling establishment in the central Rock Mountains.* General Technical Report RM-134, USDA Forest Service, Rocky Mountain Forest and Range Experiment Station, Fort Collins, CO.

———, and C. B. Edminster. 1977. *Regulation and control of cut under unevenage management.* Research Paper RM-182, USDA Forest Service, Rocky Mountain Forest and Range Experiment Station, Fort Collins, CO.

———, D. Tackle, and W.G. Dahms. 1967. *Site indexes for lodgepole pine with corrections for stand density.* Research Paper RM-29, USDA Forest Service Rocky Mountain Forest and Range Experiment Station, Fort Collins CO.

Allen, G. S. 1941 A basis for forecasting seed crops of some coniferous trees. *Journal of Forestry* 39:1014–16.

Amman, G. D., M. D. McGregor, R. F. Schmitz, and R. D. Oakes. 1988. Susceptibility of lodgepole pine to infestation by mountain pine beetle stands following partial cutting of stands. *Canadian Journal of Forest Research* 18:688–95.

Amaranthus, M. P., and D. A. Perry. 1987. Effect of soil transfer on ectomycorrhiza formation and the survival and growth of conifer seedlings on old, non-forested clear-cuts. *Canadian Journal of Forest Research* 19:944–50.

Anderson, H. E. 1982. *Aids to determining fuels models for estimating fire behavior*. USDA Forest Service General Technical Report, GTR-INT-122. 22p.

Anhold, J. A., M. J. Jenkins, and J. N. Long. 1996. Management of lodgepole pine stand density to reduce susceptibility to mountain pine beetle attack. *Western Journal of Applied Forestry* 11:50–53.

Antos, J. A., and J. R. Habeck. 1981. Successional development in *Abies grandis* (Dougl.) Forbes forests in the Swan Valley, western Montana. *Northwest Science* 55:26–39.

Arno, S. F. 1976. *The historical role of fire on the Bitterroot National Forest*. USDA Forest Service Research Publication INT-187, Intermountain Forest and Range Experiment Station, Ogden, UT.

———, and C. E. Fiedler. 2005. *Mimicking Nature's Fire: Restoring Fire-Prone Forests in the West*. Island Press, Washington DC.

Ares, A., T.A. Terry, R.E. Miller, H.W. Anderson, and B.L. Flaming. 2005. Ground–Based Forest Harvesting Effects on Soil Physical Properties and Douglas-Fir Growth. *Soil Science Society of America Journal* 69:1822–1832.

———, Terry, C. Harrington, W. Devine, D. Peter and J. Bailey. 2007. Biomass removal, soil compaction, and vegetation control effects on five–year growth of Douglas-fir in coastal Washington. *Forest Science* 53:600–610.

———, T.A. Terry, R.B. Harrison, K. Piatek, R.E. Miller, B.L. Flaming, C. Licata, B. Strahm, C.A. Harrington, R. Meade, H.W. Anderson, L.C. Brodie, and J.M. Kraft. 2007. The Fall River long term site productivity study in coastal Washington: Site characteristics, experimental design, and biomass, carbon, and nitrogen stories before and after harvest. USDA-FS Gen. Tech. Rep. PNW-691. 85 p.

Arnott, J. T, and A. N. Burdett. 1988. Early growth of planted western hemlock in relation to stock type and controlled release fertilizer application. *Canadian Journal of Forest Research* 18:710–17.

Assmann, E. 1970. *Principles of Forest Yield Study*. Pergamon Press, New York.

Atzet, T., and R. H. Waring. 1970. Selective filtering of light by coniferous forests and minimum light requirements for regeneration. *Canadian Journal of Botany* 48:2136–67.

Aubry, K. B., M. P. Amaranthus, C. B. Halpern, J. D. White, B. L. Woodard, C. E. Peterson, C. A. Lagoudakis, and A. J. Horton. 1999. Evaluating the effects of varying levels and patterns of green-tree retention: Experimental design of the DEMO study. *Northwest Science* 73 (Special Issue):12–26

Aussenac, G., and A. Granier. 1988. Effects of thinning and water stress on growth of Douglas-fir. *Canadian Journal of Forest Research* 18:10–105.

Avery, T. E, and H. E. Burkhart. 2002. *Forest Measurements*. McGraw-Hill, New York.

Bailey J. D., and L. Liegel. 1997 Response of Pacific yew (*Taxus brevifolia*) to removal of the overstory. *Western Journal of Applied Forestry* 12:41–43.

———, and J. C. Tappeiner. 1998. Effects of thinning on structural development in 40- to 100-year-old Douglas-fir stands in western Oregon. *Forest Ecology and Management* 108:99–113.

———, and W. W. Covington. 2002. Evaluating ponderosa pine regeneration rates following ecological restoration treatments in northern Arizona, USA. *Forest Ecology and Management* 155:271–78.

———, C. Marysohn, P. Doescher, E. St. Pierre, and J. C. Tappeiner. 1998. Understory species in old and young forests in western Oregon. *Forest Ecology and Management* 112:289–302.

Bailey, R. L., and T. R. Dell. 1973. Quantifying diameter distributions with the Weibull function. *Forest Science* 19:97–104.

Baird, M., D. Zabowski, and R. L. Evert. 1999. Wildfire effects on carbon and nitrogen in inland coniferous forests. *Plant and Soil* 209:233–43.

Baker, F. S. 1929. Effect of excessively high temperatures on coniferous reproduction. *Journal of Forestry.* 1929; 27:949–75.

———. 1944. Mountain climates of the western United States. *Ecological Monographs* 14:223–54.

———. 1950. *Principles of Silviculture*. McGraw-Hill, New York.

———. 1955. California's forest regeneration problems. *Report to Department of Natural Resources, Division of Forestry, State Board of Forestry, Regeneration Committee, Sacramento CA.*

Baker, W. L., and D. Ehle. 2001. Uncertainty in surface fire history: The case of ponderosa pine forests in the western United States. *Canadian Journal of Forest Research* 31:1205–26.

Baldwin, V. C., K. D. Peterson, H. E. Burkhart, R. L. Amateis, and P. M. Dougherty. 1997. Equations for estimating loblolly pine branch and foliage weight. *Canadian Journal of Forest Research* 27:918–27.

Balough, J. C., and D. F. Grigal. 1987. Age-density distributions of tall shrubs in Minnesota. *Forest Science* 33:846–57.

Balster, N.J. and J.D. Marshall. 2000. Decreased needle longevity of fertilized Douglas-fir and grand fir in the northern Rockies. *Tree Physiology* 20:1191–1197.

Barclay, H.J. and H. Brix. 1984. Effects of urea and ammonium nitrate fertilizer on growth of a young thinned and unthinned Douglas-fir stand. *Canadian Journal of Forest Research* 14:952–955.

———, and H. Brix. 1985. Fertilization and thinning effects on a Douglas-fir ecosystem at Shawnigan Lake: 12–year growth response. Pacific Forestry Centre, Victoria, BC. Information Report BC-X-271. 34 p.

———,P. C. Pang, and D. F. W. Pollard. 1986. Aboveground biomass distribution among trees and stands in thinned and fertilized Douglas-fir. *Canadian Journal of Forest Research* 16:438–42.

Barg, A. K., and R. L. Edmonds. 1999. Influence of partial cutting on site microclimate, soil nitrogen dynamics, and microbial biomass in Douglas-fir stands in western Washington. *Canadian Journal of Forest Research* 29:705–13.

Baron, F. J. 1962. *Effects of different grasses on ponderosa pine seedling establishment.* Research Note PSW-199, USDA Forest Service Pacific Southwest Research Station, Berkeley CA.

Barrett J. W. 1982. *Twenty-year growth of ponderosa pine thinned to five spacings in Central Oregon.* USDA Forest Service Research Paper PNW-301. USDA Forest Service Pacific Northwest Forest and Range Experiment Station, Portland OR.

———, and L. F. Roth. 1985. *Response of dwarf mistletoe–infested ponderosa pine to thinning.* Research Paper PNW-330, USDA Forest Service Pacific Northwest Forest and Range Experiment Station, Portland OR.

Bassman, J. H. 1989. Influence of two site preparation treatments on ecophysiology of planted *Picea engelmannii* x *glauca* seedlings. *Canadian Journal of Forest Research.* 19:1359–70.

———, J. C. Zwier, J. R. Olson, and J. D. Newberry. 1992. Growth of advanced regeneration in response to residual overstory treatment in northern Idaho. *Western Journal of Applied Forestry* 7:78–81.

Beaufait, W. R. 1966. *Prescribed fire planning in the interior west.* USDA Forest Service Research Paper INT–26, Intermountain Forest and Range Experiment Station, Ogden UT.

Beese, W.J. and A.A. Bryant. 1999. Effects of alternative silvicultural systems on vegetation and bird communities in coastal montane forests of British Columbia, Canada. *Forest Ecology and Management* 115:231–242.

Bega, R. V. 1978. *Diseases of Pacific Coast Conifers.* Agriculture Handbook 521, U.S. Department of Agriculture, Washington DC.

Belz, D. 2003. Severing red alder: Timing the cut to achieve the best mortality. *Western Journal of Applied Forestry.* 18:199–201.

Bengston, G. W. 1979. Forest fertilization in the United States: Progress and outlook. *Journal of Forestry* 78:222–29.

———. 1981. Nutrient conservation in forestry: A perspective. *Southern Journal of Applied Forestry* 5:50–58.

Bennett, J. N., B. Andrew, and C. E. Prescott. 2002. Vertical fine root distributions of western redcedar, western hemlock, and salal in old-growth cedar-hemlock forests on northern Vancouver Island. *Canadian Journal of Forest Research.* 32:1208–16.

Berryman, A. A. 1982. Mountain pine beetle outbreaks in Rocky Mountain lodgepole pine forests. *Journal of Forestry,* 80:410–13.

Bi, H. and J. Turner. 1994. Long-term effects of superphosphate fertilization on stem form, taper, and stem volume estimation of *Pinus radiata*. *Forest Ecology and Management* 70:285–297.

Binkley, D. 1986. *Forest Nutrition Management*. John Wiley & Sons, New York.

———, and L. Husted. 1983. Nitrogen accretion, soil fertility, and Douglas-fir nutrition in association with red stem ceanothus. *Canadian Journal of Forest Research* 13:122–25.

———, and P. Reid. 1984. Long-term responses of stem growth and leaf area to thinning and fertilization in a Douglas-fir plantation. *Canadian Journal of Forest Research* 14:656–60.

———, and P. Reid. 1985. Long–term increase of nitrogen from fertilization of Douglas-fir. *Canadian Journal of Forest Research* 15:723–24.

Biondi, F. 1996. Decadal-scale dynamics at the Gus Pearson Natural Area: Evidence for inverse (a)symetric competition. *Canadian Journal of Forest Research* 26:1397–1406.

Black, H. C. 1992. *Silvicultural approaches to animal damage management in Pacific Northwest forests*. General Technical Report PNW-GTR-287, USDA Forest Service Pacific Northwest Research Station, Portland OR.

Blanco, J.A., J.B. Imbert, and F.J. Castillo. 2009. Thinning affects nutrient resorption and nutrient-use efficiency in two Pinus sylvestris stand in the Pyrenees. *Ecological Applications* 19:682–698

Briggs, D. and J. Trobaugh. 2001. Management practices on Pacific Northwest West-side industrial forest lands, 1991–2000: with projections to 2005. University of Washington, Stand Management Cooperative, Seattle WA. SMC Working Paper No. 2.

Brockley, R.P. 1991. Response of thinned, immature lodgepole pine to nitrogen fertilization: Six-year growth response. British Columbia Ministry of Forests, Vernon, BC. FRDA Report 184.

———. 1996. Lodgepole pine nutrition and fertilization: A summary of B.C. Ministry of Forests research results. British Columbia Ministry of Forests, Vernon, BC. FRDA Report 266.

———. 2000. Using foliar variables to predict the response of lodgepole pine to nitrogen and sulphur fertilization. *Canadian Journal of Forest Research* 30:1389–1399.

———. 2005. Effects of post-thinning density and repeated fertilization on the growth and development of young lodgepole pine. *Canadian Journal of Forest Research* 35:1952–1964.

———2006. Effects of fertilization on the growth and foliar nutrition of immature Douglas-fir in the interior cedar–hemlock zone of British Columbia: Six-year results. British Columbia Ministry of Forests and Range, Victoria, BC. Research Report 27.

Brodie, L.C. and D.S. DeBell. 2013. Residual densities affect growth of overstory trees and planted Douglas-fir, western hemlock, and western redcedar: results from the first decade. *Western .Journal of Applied Forestry* 28:121–127.

Bork, J. 1985. *Fire History in Three Vegetation Types on the Eastside of the Oregon Cascades*. PhD dissertation. Department of Forest Science, Oregon State University, Corvallis.

Bormann, B. T., and J. C. Gordon. 1989. Can intensively managed forest ecosystems be self-sufficient in nitrogen? *Forest Ecology and Management* 29:95–103.

Boyd, R. J., and G. H. Deitschman. 1969. *Site preparation aids natural regeneration in western larch–Engelmann spruce strip clearcuts.* USDA Forest Service Research Paper Int–Rp 64 USDA Forest Service Intermountain Forest and Range Experiment Station, Ogden, UT

Boyle, J. R. 1973. Forest soil chemical changes following fire. *Communications in Soil Science and Plant Analysis.* 4:369–74.

———, J. J. Phillips, and A. R. Ek. 1973. "Whole tree" harvesting nutrient budget evaluation. *Journal of Forestry* 71:760–62.

Brandeis, T. J. 1999. *Underplanting and Competition in Thinned Douglas-fir.* Ph.D. dissertation, Department of Forest Science, Oregon State University, Corvallis.

———, M. Newton, and E. C. Cole. 2001. Underplanted conifer seedling survival and growth in thinned Douglas-fir stands. *Canadian Journal of Forest Research.*; 31:302–12.

Bravo, F., D. W. Hann, and D. A. Maguire. 2005. Impact of competitor species composition on predicting diameter growth and survival rates of Douglas-fir trees in southwestern Oregon. *Canadian Journal of Forest Research* 31:2237–47.

Briegleb, P. A. 1952. An approach to density management in Douglas-fir. *Journal of Forestry* 50:529–36.

Briggs, D. G., and W. R. Smith. 1986. Effects of silvicultural practices on wood properties of conifers. Pp. 108–17 in C. D. Oliver, D. P. Hanley, and J. A. Johnson, eds., *Douglas-fir: Stand Management for the Future.* Institute of Forest Resources, College of Forest Resources, University of Washington, Seattle.

———, and R. D. Fight. 1992. Effects of silvicultural practices on product quality and value of coastal Douglas-fir trees. *Forest Products Journal* 42:40–46.

Brix, H. 1971. Effects of nitrogen fertilization on photosynthesis and respiration in Douglas-fir. *Forest Science* 17:407–14.

———. 1981a. Effects of thinning and nitrogen fertilization on branch and foliage production in Douglas-fir. *Canadian Journal of Forest Research* 11:502–511.

———. 1981b. Effects of nitrogen fertilizer source and application rates on foliar nitrogen concentration, photosynthesis, and growth of Douglas-fir. *Canadian Journal of Forest Research* 11:775–80.

———. 1983. Effects of thinning and nitrogen fertilization on growth of Douglas-fir: Relative contribution of foliage quantity and efficiency. *Canadian Journal of Forest Research* 13:167–75.

———, and A. K. Mitchell. 1983. Thinning and nitrogen fertilizer effects on sapwood development and relationships of foliage quantity to sapwood area and basal area in Douglas-fir. *Canadian Journal of Forest Research* 13:384–89.

———. 1986. Thinning and nitrogen fertilization effects on soil and tree water stress in a Douglas-fir stand. *Canadian Journal of Forest Research* 16:1334–38.

———, and L. F. Ebell. 1969. Effects of nitrogen fertilizer on growth, leaf area, and photosynthesis in Douglas-fir. *Forest Science* 15:189–96.

———, and A. K. Mitchell. 1980. Effects of thinning and nitrogen fertilization on xylem development in Douglas-fir. *Canadian Journal of Forest Research* 10:121–28.

———,1991. Mechanisms of response to fertilization. II. Utilization by trees and stands. Pp. 76–93 in J.D. Lousier (ed). Proc. Forest Fertilization Workshop, March 2–3, 1988, Vancouver, BC. B.C Ministry of Forests, Victoria, BC.

———. 1993. Fertilization and thinning effects on a Douglas-fir ecosystem at Shawnigan Lake: A synthesis of project results. Forestry Canada, Pacific Forestry Centre, Victoria, BC. FRDA Report 196.

Brockley, R. P. 1991. Response of thinned immature lodgepole pine to nitrogen fertilization: six-year growth response. British Columbia Ministry of Forests, Vernon B. C. FRDA Report 184.

———.1992. Effects of fertilization on the nutrition and growth of a slow-growing Engelmann spruce plantation in south central British Columbia. *Canadian Journal of Forest Research* 22:1617–22.

———.2000. Using foliar variables to predict the response of lodgepole pine to nitrogen and sulphur fertilization. *Canadian Journal of Forest Research* 30: 1389–99.

———. 2005. Effects of post thinning density and repeated fertilization on the growth and development of young lodgepole pine. *Canadian Journal of Forest Research*. 35:

———, H. C. Black, E. J. Dimock II, J. Evans, C. Kao, and J. A. Rochelle. 1979. *Animal damage to coniferous plantations in Oregon and Washington.* Research Bulletin 26, Forest Research Laboratory, Oregon State University, Corvallis.

Brooks, M. A., R. W. Campbell, J. J. Colbert, R. G. Mitchell, and R. W. Stark, eds. 1987. *Western spruce budworm.* UDSA Forest Service Coop State Research. Tech. Bulletin 1695.

Brooks, R.J., F. C. Meinzer, R. Coulombe and J. Gregg. 2002. Hydraulic redistribution of soil water during summer drought in two contrasting Pacific Northwest coniferous forests. *Tree Physiology* 22:1107–1117.

Broun, A. F. 1912. *Sylviculture in the Tropics.* Macmillan, London, UK.

Brown, E. R., and J. H. Mandery. 1962. Planting and fertilization as a means of controlling big game animals. *Journal of Forestry* 60:33–35.

Buckley, A. J. 1992. Fire behavior and fuel reduction burning: Bemm River wildfire. *Australian Forestry* 55:135–47.

Buckman, R. E. 1962. *Growth and yield of red pine in Minnesota.* Technical Bulletin 1272, USDA Forest Service Lake States Forest Experiment Station, St. Paul MN.

———, Bishaw, B , Hanson, T.J., Benford, F. A. 2006. *Growth and yield of red pine in the Lake States.* Gen. Tech. Rep. NC-271. St. Paul, MN: U.S. Department of Agriculture, Forest Service, North Central Research Station. 114 p.

Burns, R. M., and B. H. Honkala. 1990. *Silvics of North America.* USDA Agriculture Handbook 654, US Department of Agriculture, Washington DC.

Büsgen, M., and E. Münch. 1929. *The Structure and Life of Forest Trees* (3rd ed.). Translated by T. Thompson. John Wiley & Sons, New York.

Busse, M. D., P. H. Cochran, and J. W. Barrett. 1996. Changes in ponderosa pine site productivity following removal of understory vegetation. *Soil Science Society of America Journal* 60:614–21.

Cafferata, S. L. 1986. Douglas-fir stand establishment overview: Western Oregon. Pp. 211–18 in W. Oliver and H. Johnson, eds. *Douglas-fir stand management for the future.* College of Forest Resources, University of Washington, Seattle.

Cahill, J. M., T. A. Snellgrove, and T. D. Fahey. 1986. The case for pruning young-growth stands of Douglas-fir. Pp. 123–31 in W. Oliver and H. Johnson, eds. *Douglas-fir stand management for the future.* College of Forest Resources, University of Washington, Seattle.

Cain, M. D. 1994. *Importance of seed year, seedbed, and overstory for establishment of natural loblolly and shortleaf pine in southern Arkansas.* Research Paper SO-268, USDA Forest Service Southern Forest Experiment Station, New Orleans LA.

Campbell, R. K. 1979. Genecology of Douglas-fir in a watershed in the Oregon Cascades. *Ecology.* 1979; 60:1036–50.

———. 1986. Mapped genetic variation of Douglas-fir to guide seed transfer in southwestern Oregon. *Silvae Genetica* 35:85–96.

———, and J. F. Franklin. 1981. A comparison of habitat type and elevation for seed zone classification in western Oregon. *Forest Science* 27:49–59.

Cannell, M. G. R. 1982. *World Forest Biomass and Primary Production Data.* Academic Press, London, UK.

———, and F. T. Last. 1976. *Tree Physiology and Yield Improvement.* Academic Press, London, UK.

Caprio, A. C. and T. W. Swetnam 1996. Historic fire regimes along along an elevation gradient on the west slope of the Sierra Nevada. Pp. 173–79 in J. K. Brown et al., tech. coords., *Proceedings of a symposium on fire in wilderness and park management.* General Technical Report INT-GTR-320, USDA Forest Service Intermountain Research Station, Ogden UT.

Carey, A. B., and R. O. Curtis. 1996. Conservation of biodiversity: A useful paradigm for forest ecosystem management. *Wildlife Society Bulletin* 24:610–20.

———, D. R. Thysell, and A. W. Brodie. 1999. *The forest ecosystem study: Background, rationale, implementation, baseline conditions, and silvicultural assessment.* USDA Forest Service General Technical Report PNW-GTR-457, Pacific Northwest Research Station, Portland OR.

Carlyle, J.C. 1995. Nutrient management in a Pinus radiate plantation after thinning: the effect of thinning and residues on nutrient distribution, mineral nitrogen fluxes, and extractable phosphorus. *Canadian Journal of Forest Research* 25:1278–1291.

Carmean, W. H. 1956. Suggested modification of the standard Douglas-fir site curves for certain soils in southwestern Washington. *Forest Science* 2:242–50.

Carter, G. A., J. H. Miller, D. E. Davis, and R. M. Patterson. 1984. Effects of vegetative competition on the moisture and nutrient status of loblolly pine. *Canadian Journal of Forest Research* 14:1–9.

Carter, R. E., and K. Klinka. 1992. Variation in the shade tolerance of Douglas-fir, western hemlock, and western red cedar in coastal British Columbia. *Forest Ecology and Management* 55:87–105.

Champion, H. G., and G. Trevor. 1938. *Manual of Indian Silviculture*. Oxford University Press, Calcutta, India.

Chan, S. S., and J. D. Walstad. 1987. Correlations between overtopping vegetation and development of Douglas-fir saplings in the Oregon Coast Range. *Western Journal of Applied Forestry* 2:177–79.

Chappell, C. B. 1991 *Fire Ecology and Seedling Establishment in Shasta Red Fir Forests of Crater Lake National Park, Oregon*. MS thesis, College of Forest Resources, University of Washington, Seattle.

Chappell, H. N., and W. S. Bennett. 1993. Young fir trees respond to nitrogen fertilization in western Washington and Oregon. *Soil Science Society of America Journal* 57:834–38.

Chappell, H.N., C.E. Prescott, and L. Vesterdal. 1999. Long-term effects of nitrogen fertilization on nitrogen availability in coastal Douglas-fir forest floors. *Soil Science Society of America Journal* 63:1448–1454.

Childs, S. W., and L. E. Flint. 1987. Effects of shade cards, shelterwoods, and clearcuts on temperature and moisture environments. *Forest Ecology and Management* 18:205–17.

Chisman, H. H., and F. X. Schumacher. 1940. On the tree-area ratio and certain of its applications. *Journal of Forestry* 38:311–17.

Christie, J. E., and R. N. Mack. 1984. Variation in the demography of juvenile *Tsuga heterophylla* across a substrate mosaic. *Journal of Ecology* 72:75–91.

Churchill, D. C., A. J. Larson, M. C. Dalhgreen, J. F. Franklin, P. F. Hessburg, and J. A. Lutz. 2013. Restoring forest resilience: From reference spatial patterns to silvicultural prescriptions and monitoring. *Forest Ecology and Management* 291:442–457.

Chung, H. H., and R. L. Barnes. 1977. Photosynthesis in *Pinus tadea*. I. Substrate requirements for synthesis of shoot biomass. *Canadian Journal of Forest Research* 7:106–11.

Chmura, D. J., P. D. Anderson, G. T. Howe, C. A. Harrington, J. E. Halofsky, D. L. Peterson, D. C. Shaw and J. B. St. Clair. 2011. Forest responses to climate change in the northwestern United States: Ecophysiological foundations for adaptive management. *Forest Ecology and Management* 261:1121–1142.

Cissel, J. H., and F. J. Swanson. 1999. Landscape management using historical fire regimes: Blue River, Oregon. *Ecological Applications* 9:1217–31.

Cleary, B. D., and J. B. Zaerr. 1980. Pressure chamber techniques for monitoring and evaluating seedling water status. *New Zealand Journal of Forest Science* 10:133–41.

———, R. D. Greaves and R. K. Hermann, eds. 1978. *Regenerating Oregon's forests.* Oregon State University Extension, Oregon State University.

Coates, K. D. 2000. Conifer seedling response to northern temperate forest gaps. *Forest Ecology and Management* 127:249–69.

Cobb, D. F., K. L O'Hara, and C. D. Oliver. 1993. Effects of variation in stand structure on development of mixed-species stands in eastern Washington. *Canadian Journal of Forest Research* 23:545–52.

Cochran, P. H. 1983. Stocking levels for east-side white or grand fir. In: Oliver, C.D. and R.M. Kenaday, eds. *Biology and management of true fir in the Pacific Northwest*: Proceedings of a symposium: 1981 Feb 24–26; Seattle WA. Pp. 186–189. U.S. Department of Agriculture Forest Service Experiment Station, Portland, OR.

———. 1985 *Site index, height growth, normal yields, and stocking levels for larch in Oregon and Washington.* Research Note PNW–424, USDA Forest Service Pacific Northwest Forest and Range Experiment Station Portland OR.

. ———.1989. Growth rates after fertilizing lodgepole pine. *Western Journal of Applied Forestry* 4:18–20.

———. 1992. *Stocking levels and underlying assumptions for uneven-aged ponderosa pine stands.* Research Note PNW-RN-509, USDA Forest Service Pacific Northwest Research Station, Portland OR.

———1998a. *Reduction in growth of pole-sized ponderosa pine related to a Pandora moth outbreak in central Oregon.* USDA Forest Service Research Note, PNW-RN-526, USDA Forest Service Pacific Northwest Research Station, Portland OR.

———.1998b. *Examples of mortality and reduced annual increments of white fir induced by drought, insects, and disease at different stand densities.* USDA Forest Service Research Note, PNW–RN–525, USDA Forest Service Pacific Northwest Research Station, Portland OR.

———,1991. *Response of thinned white fir stands to fertilization with nitrogen plus sulfur.* USDA–FS Pacific Northwest Research Station, Portland OR. Res. Note PNW-RN-501

Cochran, P.H., C.T. Youngberg, E.C. Steinbrenner, and S.R. Webster. 1979. Response of ponderosa pine and lodgepole pine to fertilization. Pp. 89–94 in S.P. Gessel, R.M. Kenady, and W. A. Atkinson (Eds.). Proceedings of the Forest Fertilization Conference. September 25–27, 1979, Union, WA. Institute of Forest Resources, College of Forest Resources, University of Washington, Seattle, WA.

———, and W. E. Hopkins. 1991. Does fire exclusion increase productivity of ponderosa pine? In A. E. Harvey, and L. F. Neuenschwander, compilers. *Management and productivity of western montane soils.* USDA Forest Service General Technical Report GTR-INT-280, USDA Forest Service Intermountain Research Station, Ogden, UT.

———, J.M. Geist, D.L. Clemens, R.R. Clausnitzer, and D.C. Powell. 1994. Suggested stocking levels for forest stands in northeastern Oregon and southeastern Washington. USDA-FS Pacific Northwest Research Station, Portland OR. Res. Note PNW-RN-513.

———, and J. W. Barrett. 1993. Long-term response of planted ponderosa pine to thinning in Oregon's Blue Mountains. *Western Journal of Applied Forestry* 8:126–32.

———, and J. W. Barrett. 1999. *Thirty-five year growth of ponderosa pine saplings in response to thinning and understory removal.* USDA Forest Service Pacific Northwest Research Station Research Paper PNW-RP-512, Portland OR.

———, J. M. Geist, D. L. Clemens, R. R. Clausnitzer, and D. C. Powell. 1994. *Suggested stocking levels for forest stands in northeastern Oregon and southeastern Washington.* Research Note PNW-RN-513, USDA Forest Service Pacific Northwest Research Station, Portland OR.

Cole, D.W., and S.P. Gessel (eds). 1988. *Forest site evaluation and long-term productivity.* University of Washington Press, Seattle WA. 196 p.

———, Cole, D. M., and W. C. Schmidt. 1986. *Site treatments influence development of a mixed species western larch stand.* USDA Forest Service Research Paper INT 364, Intermountain Forest and Range Experiment Station, Ogden UT.

———, Cole, D. W. and D. Johnson. 1979. The cycling of elements within forests. In: Heilman, P. E., H. W. Anderson and D. M. Baumgartner, eds *Forest Soils of the Douglas-fir Region.* pp 185–198. Washington State University Cooperative Extension Service. Pullman, WA.

Cole, E. C., and M. Newton. 1987. Fifth-year response of Douglas-fir to crowding and non–coniferous competition. *Canadian Journal of Forest Research* 17:181–86.

———, and M. Newton. 2009. Tenth-year survival and size of underplanted seedlings in the Oregon Coast Range. *Canadian Journal of Forest Research* 39:580–595.

Comeau, P. G., T. F. Braumandl and C. Y. Xie. 1993. Effects of overtopping vegetation on light availability and growth of Engelmann spruce seedlings. *Canadian Journal of Forest Research* 23:2044–48.

Conard, S. G., and S. R. Radosevich. 1982a. Post-fire succession in white fir *Abies concolor* vegetation of the northern Sierra Nevada. *Madroño* 29:42–56.

———, and S. R. Radosevich. 1982b. Growth responses of white fir to decreased shading and root competition of montane shrubs. *Forest Science* 28:309–20.

Conckle, T. 1973. Growth data for 29 years from the California elevational transect of ponderosa pine. *Forest Science* 19:31–39.

Coutts, M. P., and J. Grace. 1995. *Wind and Trees.* Cambridge University Press.

Covington, W. W., and S. S. Sackett. 1984. The effect of a prescribed burn in southwestern ponderosa pine on organic matter and nutrients in woody debris and forest floor. *Forest Science* 30:183–92.

———, and S. S. Sackett. 1986. Effect of periodic burning on soil nitrogen concentrations in ponderosa pine. *Soil Science Society of America Journal* 50:452–57.

———, and M. M. Moore. 1994. Southwestern ponderosa pine forest structure: Changes since Euro-American settlement. *Journal of Forestry* 92(1):39–47.

Cressie, N. 1991. *Statistics for Spatial Data.* John Wiley & Sons, New York.

Cronemiller, F. P. 1959. The life history of ceanothus, a fire type. *Journal of Range Management* 12:21–25.

Crouch, G. L. 1971. Susceptibility of ponderosa pine and lodgepole pine to pocket gophers. *Northwest Science* 45:252–56.

———. 1979. Atrazine improves survival and growth of ponderosa pine threatened by vegetative competition and pocket gophers. *Forest Science* 25:99–111.

———. 1986. *Effects of thinning pole-sized lodge pole pine on understory vegetation and large herbivore activities in central Colorado.* USDA Forest Service, Rocky Mountain Forest and Range Experiment Station Research Paper RM 294, Fort Collins, CO.

Cummings, J. A. 1964. Effectiveness of prescribed burning in reducing wildfire damage during periods of abnormally high fire danger. *Journal of Forestry* 62:535–37.

Curtis, R. O. 1970. Stand density measures: An interpretation. *Forest Science* 16:403–14.

———. 1982. A simple index for density of Douglas-fir. *Forest Science* 28:92–94.

———. 1983. *Procedures for establishing and maintaining permanent plots for silvicultural yield and research.* General Technical Report GTR PNW-155, USDA Forest Service, Pacific Northwest Forest and Range Experiment Station, Portland OR.

———. 1992. A new look at an old question—Douglas-fir culmination age. *Western Journal of Applied Forestry* 7:97–99.

———. 1995. *Extended rotations and culmination ages of coast Douglas-fir: Old studies speak to current issues.* Research Paper PNW-RP-485, USDA Forest Service Pacific Northwest Research Station, Portland OR.

———. 1998. "Selective cutting" in Douglas-fir: History revisited. *Journal of Forestry* 96(7):40–46.

Curtis, R.O. 2006. Volume growth trends in a Douglas-fir Levels-of-Growing-Stock study. *Western Journal of Applied Forestry* 21:79–86.

———, and D. L. Reukema. 1970. Crown development and site estimates in a Douglas-fir plantation spacing test. *Forest Science* 16:287–301.

———, and D. D. Marshall. 1993. Douglas-fir rotations … time for reappraisal. *Western Journal of Applied Forestry* 8:81–85.

———, F. R. Herman, and D. J. DeMars. 1974. Height growth and site index estimates for Douglas-fir in high-elevation forests of the Oregon and Washington Cascades. *Forest Science* 20:307–16.

———, D. DeBell, C. Harrington, D. Lavender, J. St. Clair, J. Tappeiner, and J. Walstad. 1998. *Silviculture for multiple objectives in the Douglas-fir region.* USDA Forest Service General Technical Report PNW-GTR-435, USDA Forest Service Pacific Northwest Forest and Range Experiment Station, Portland OR.

Daniel, T. W., and J. Schmidt. 1971. Lethal and non-lethal effects of organic horizons of forested soils on the germination of seeds from several associated conifer species of the Rocky Mountains. *Canadian Journal of Forest Research* 2:179–84.

———, J. A. Helms, and F. S. Baker. 1979. *Principles of Silviculture.* McGraw-Hill, New York.

Davies, G. 1980 *Site class at culmination of mean annual increment.* Adapted from USDA Forest Service FSM 2409.6, Region 5 supplement 232, San Francisco CA.

Davis, L. S., and R. W. Cooper. 1963. How prescribed burning affects wildfire occurrence. *Journal of Forestry.* 1963; 61:915–17.

———, K. N. Johnson, P. S. Bettinger, and T. E. Howard. 2001. *Forest Management.* McGraw-Hill, Boston MA.

Deal, R. L. 2001. The effects of partial cutting on forest plant communities of western hemlock-Sitka stands in southeast Alaska. *Canadian Journal of Forest Research* 31:2067–79.

———, and J. C. Tappeiner. 2002. The effects of partial cutting on stand structure and growth of western hemlock-Sitka spruce stands in southeast Alaska. *Canadian Journal of Forest Research* 159:173–86.

———, C. D. Oliver, and B. T. Borman. 1991. Reconstruction of mixed hemlock-spruce stands in coastal southeast Alaska. *Canadian Journal of Forest Research* 21:643–54.

———, J. C. Tappeiner and P. E. Hennon. 2002. Developing silvicultural systems based on partial cutting in western hemlock–Sitka spruce stands of southeast Alaska. *Forestry* 75:425–31.

Dean, T. J., S. D. Roberts, D. W. Gilmore, D. A. Maguire, J. N. Long, K. L. O'Hara, and R. S. Seymour. 2002. An evaluation of the uniform stress hypothesis based on stem geometry in selected North American conifers. *Trees* 16:559–68.

DeBano, L. F., D. G. Neary, and P. F. Folliott. 1998. *Fire's Effects on Ecosystems.* John Wiley & Sons, New York NY.

DeBell, D. S., and T. C. Turpin. 1989. *Control of red alder by cutting.* USDA Forest Service Research Paper PNW-RP-414, USDA Forest Service Pacific Northwest Research Station, Portland OR.

———, C. A. Harrington, and J. Shumway. 2002. *Thinning shock and response to fertilizer less than expected in young Douglas-fir stand at Wind River Experimental Forest.* Research Paper PNW-RP-547, USDA Forest Service Pacific Northwest Research Station, Portland OR.

DeBell, J. D., and B. L. Gartner. 1997. Stem characteristics on the lower log of 35-year old western red cedar. *Western Journal of Applied Forestry* 1997; 12:9–14.

———, J. C. Tappeiner, and R. L Krahmer. 1994a. Branch diameter of western hemlock: Effects of precommercial thinning and implications for log grades. *Western Journal of Applied Forestry* 9:88–90.

———, J. C. Tappeiner, and R. L. Krahmer. 1994b. Wood density of western hemlock: effect of ring width. *Canadian Journal of Forest Research* 24:638–41.

DeByle, N. V. 1964. Detection of functional intraclonal root connections by tracers and excavations. *Forest Science* 10:386–96.

Del Rio, E., and A. Berg. 1979. *Growth of Douglas-fir reproduction in the shade of a managed forest.* Research Paper 40, Forest Research Laboratory, Oregon State University, Corvallis OR.

DeLuca, T. H., and K. L. Zouhar. 2000. Effects of selection harvest and prescribed fire on the soil nitrogen status of ponderosa pine forests. *Forest Ecology and Management* 138:263–71.

DeMars, D. J., and J. W. Barrett. 1987. *Ponderosa pine managed–yield simulator: PPSIM users' guide.* General Technical Report PNW-GTR-203, USDA Forest Service Pacific Northwest Research Station, Portland OR.

Devine, W. D., and C. A. Harrington. 2004. Garry oak woodland restoration in the Puget Sound region: Releasing oaks from overtopping conifers and establishing oak seedlings. *Proceedings of the 16th International Conference of the Society for Ecological Restoration* (CD–ROM), Aug. 24–26, Victoria BC, Canada. Northwest Research Station, Portland OR.

———, and R. Rose. 2009. Integration of soil moisture, xylem water potential, and fall-spring herbicide treatments to achieve the maximum growth response in newly planted Douglas-fir seedlings. *Canadian Journal of Forest Research* 39:1401–1414

———, T.B. Harrington, T.A. Terry, R.B. Harrison, R.A. Slesak, D.H. Peter, C.A. Harrington, C.J. Shilling, and S.H. Schoenholtz. 2011. Five-year vegetation control effects on aboveg-round biomass and nitrogen content and allocation in Douglas-fir plantations on three contrasting sites. *Forest Ecology and Management* 262:2187–2198.

———, and C.A. Harrington. 2013. Restoration release of overtopped Oregon white oak increases 1 0-year growth and acorn production. *Forest Ecology and Management* 291:87–95

Dietrich, J. H. 1980. *Chimney Springs Forest fire history.* General Technical Report GTR-RM-220, USDA Forest Service Rocky Mountain Research Station, Ogden UT.

Dinger, E.J. and R. Rose. 2009. Integration of soil moisture, xylem water potential, and fall–spring herbicide treatments to achieve the maximum growth response in newly planted Douglas-fir seedlings. *Canadian Journal of Forest Research* 39:1401–1414.

Doer, J. G., and N. H. Sandberg. 1986. Effects of precommercial thinning on understory vegetation and deer habitat utilization in southeast Alaska. *Forest Science* 32:1092–95.

Doescher, P. S., S. D. Tesch, and M. Alejandro-Castro. 1987. Livestock grazing: A silvicultural tool for plantation establishment. *Journal of Forestry* 85:29–37.

———, S. D. Tesch, and W. E. Drewien. 1989. Water relations and growth of conifer seedlings during three years of cattle grazing on a southwest Oregon plantation. *Northwest Science* 63:232–40.

Dolph, L. D., S. R. Mori, and W. W. Oliver. 1995. Long-term response of old-growth stands to varying levels of partial cutting in the eastside pine type. *Western Journal of Applied Forestry* 10:101–8.

Donner, B. L., and S. W. Running. 1986. Water stress after thinning *Pinus contorta* stands in Montana. *Forest Science* 32:614–25.

Donovan, G. H., and T. C. Brown. 2007. Be careful what you wish for: the legacy of Smokey Bear. *Frontiers in Ecology* 5(2): 73–79.

Dowling, C. D. 2003. *Comparing Structure and Development of Douglas-fir Old-growth, Plantations, and Young Natural Forests in Western Oregon*. M.S. thesis, Department of Forest Resources, Oregon State University, Corvallis.

Drever, C. R., and K. P. Lertzman. 2001. Light-growth responses of coastal Douglas-fir and western redcedar saplings under different regimes of soil moisture and nutrients. *Canadian Journal of Forest Research* 31:2124–33.

———, and K. P. Lertzman. 2003. Effects of a wide gradient of retained tree structure on understory light in Coastal Douglas-fir. *Canadian Journal of Forest Research* 33:137–46.

Drew, T. J., and J. W. Flewelling. 1977. Some recent Japanese theories of yield–density relationships and their application to Monterey pine plantations. *Forest Science* 23:517–34.

——, and J. W. Flewelling. 1979. Stand density management: An alternative approach and its application to Douglas-fir. *Forest Science* 25:318–32.

Duane, T. P. 1996. Human settlement, 1850–2040. In *Status of the Sierra Nevada, volume II: Assessments and scientific basis for management options*. Report No. 38, Sierra Nevada Ecosystem Project—Final Report to Congress. Centers for Water and Wildland Resources, University of California, Davis.

Dubrasich, M. E., D. W. Hann, and J. C. Tappeiner. 1997. Methods for evaluating crown area profiles of forest stands. *Canadian Journal of Forest Research* 27:385–92.

Duff, G. H., and N. J. Nolan. 1953. Growth and morphogenesis in the Canadian forest species: 1. Controls of cambial and apical activity in *Pinus resinosa* Ait. *Canadian Journal of Botany* 31:471–513.

Dunlap J. M., and J. A. Helms. 1983. First year growth of planted Douglas-fir and white fir seedlings under different shelterwood regimes in California. *Forest Ecology and Management* 5:25–68.

Dunning, D. 1923. *Some results of cutting in the Sierra forests of California*. Bulletin 1176, US Department of Agriculture, Washington DC.

———. 1928. A tree classification for the selection forests of the Sierra Nevada. *USDA Journal of Agriculture Research* 36:755–71.

———. 1942. *A site classification for the mixed conifer selection forests of the Sierra Nevada.* Research Note 28, USDA Forest Service California Forest and Range Experiment Station, Berkeley CA.

———, and L. H. Reineke. 1933. *Preliminary yield tables for second-growth stands in the California pine region.* USDA Technical Bulletin 354, US Department of Agriculture, Washington DC.

Duryea, M. L., and T. D. Landis, eds. 1984. *Forest Nursery Manual: Production of Bareroot Seedlings.* Martinus Nijhoff/Dr W. Junk, The Hague/Boston/Lancaster.

Dyrness, C. T. 1973. Early stages of plant succession following logging in the western Cascades of Oregon. *Ecology* 54:57–69.

Eckberg, T. B., J. M. Schmid, S. A. Mata, and J. E. Lundquist. 1994. *Primary focus trees for the mountain pine beetle in the Black Hills.* Research Note RM 531, USDA Forest Service Rocky Mountain Forest and Range Experiment Station, Ogden UT.

Edmonds, R. L., and T. Hsiang. 1987. Forest floor soil influence on response of Douglas-fir to urea. *Soil Science Society of America Journal* 51:1332–37.

Eglitis, A., and P. E. Hennon. 1997. Porcupine damage in precommercially thinned conifer stands of central southeast Alaska. *Western Journal of Applied Forestry* 12:115–21.

El-Hajj, Z., K. Kavanagh, C. Rose, and Z. Kanaan-Atallah. 2004. Nitrogen and carbon dynamics of a foliar biotrophic fungal parasite in fertilized Douglas-fir. *New Phytologist* 163:139–147.

Emmingham, W. H. 1977. Comparison of selected Douglas-fir seed sources for cambium and leader growth patterns in four western Oregon environments. *Canadian Journal of Forest Research* 7:154–64.

———, and R. H. Waring. 1977. An index of photosynthesis for comparing forest sites in western Oregon. *Canadian Journal of Forest Research* 7:165–74.

———, M. Bondi, and D. E. Hibbs. 1989. Underplanting western hemlock in a red alder stand: Early survival, growth, and damage. *New Forests* 3:31–43.

Entry, J. A., K. Cromack, R. G. Kelsey, N. E. Martin. 1991. Response of Douglas-fir to infection by *Armillaria ostoyae* after thinning or thinning plus fertilization. *Phytopathology* 81:682–89.

Espinosa-Bancalari, M. A., D. A. Perry, and J. D. Marshall. 1987. Leaf area-sapwood area relationships in adjacent young Douglas-fir stands with different early growth rates. *Canadian Journal of Forest Research* 17:174–80.

Evert, F. 1964. Components of stand volume and its increment. *Journal of Forestry* 62:810–13.

Fahey, T. D., J. M. Cahill, T. A. Snellgrove and L. S. Heath. 1991. *Lumber and veneer recovery from intensively managed young-growth Douglas-fir.* Research Paper PNW-RP-437, USDA Forest Service Pacific Northwest Research Station, Portland OR.

Fahnestock, G. R., and J. K. Agee. 1983. Biomass consumption and smoke production by prehistoric and modern fires in western Washington. *Journal of Forestry* 81:653–57.

Falk, D. A., E. K. Heyerdahl, P. M. Brown, C. Farris, P. Z. Fulé, D. M. McKenzie, T. W. Swetnam, A. H. Taylor, and M. L. Van Horne. 2011. Multi-scale controls of historical forest-fire regimes: new insights from fire-scar networks. *Frontiers in Ecology* 9(8):446–454.

Feeney, S. R., T. E. Kolb, W. W. Covington and M. R. Wagner. 1998. Influence of thinning and burning restoration treatments on presettlement ponderosa pines at the Gus Pearson Natural Area. *Canadian Journal of Forest Research* 28:1295–1306.

Feller, M. C. 1998. The influence of fire severity, not fire intensity, on understory vegetation biomass in British Columbia. Pp. 335–48 in *Proceedings of the 13th fire and forest meteorology conference* (1996). IAFW, Lorne, Australia.

Ferguson, D. E. 1999. Effects of pocket gophers, bracken fern, and western cone flower on planted conifers in northern Idaho—an update on two more species. *New Forests* 18:199–217.

———, A. R. Stage, and R. J. Boyd. 1986. Predicting regeneration in the grand fir-cedar-hemlock ecosystem of the northern Rocky Mountains. *Forest Science* Monograph 26, Society of American Foresters, 41p.

Fernow, B. F. 1914. *A Brief History of Forestry*. University of Toronto Press and Forestry Quarterly, Cambridge MA.

Ferrell, G. T. 1980. *Risk-rating systems for mature red and white fir in northern California*. General Technical Report PSW-GTR-39, USDA Forest Service Pacific Southwest Forest and Range Experiment Station, Berkeley CA.

———. 1983. *Growth classification systems for red fir and white fir in northern California*. General Technical Report PSW-GTR-72, USDA Forest Service Pacific Southwest Forest and Range Experiment Station, Berkeley CA.

———, and R. C. Hall. 1975. *Weather and tree growth associated with white fir mortality cased by fir engraver and round headed fir borer*. USDA Forest Service Research Paper PSW-109, USDA Forest Service Pacific Southwest Forest and Range Experiment Station, Berkeley CA.

———, W. J. Otrosina, and C. J. DeMars Jr. 1994. Predicting susceptibility during a drought-associated outbreak of the fir engraver *Scolytus ventralis*, in California. *Canadian Journal of Forest Research* 24:302–5.

Ffolliott, P. F., and G. Gottfried. 1991. *Mixed conifer and aspen regeneration in small clearcuts within a partially harvested Arizona mixed conifer forest*. USDA Forest Service, Research Paper RM-294, Rocky Mountain Forest and Range Experiment Station, Fort Collins CO.

Fiddler, G. O., T. A. Fiddler, D. R. Hart, and P. M. McDonald. 1989. *Thinning decreases mortality and increases growth of ponderosa pine in northeastern California*. Research Paper PSW–194, USDA Forest Service Pacific Northwest Research Station, Portland OR.

Filip, G. M. 1986. Symptom expression of root diseased trees in mixed conifer stands in central Washington. *Western Journal of Applied Forestry* 1:46–48.

———, D. J. Goheen, and D. W. Johnson. 1989a. Precommercial thinning in a ponderosa pine stand affected by Armillaria root disease: 20 years of growth and mortality in central Oregon. *Western Journal of Applied Forestry* 4:58–59.

———, J. J. Colbert and C. Parks. 1989b. Effects of thinning on volume growth of western larch infected with dwarf mistletoe. *Western Journal of Applied Forestry* 4:143–45.

———, C. L. Schmitt, and K. P. Hosman. 1992. Effect of harvesting, season and stump size on incidence of annosus root disease of true fir. *Western Journal of Applied Forestry* 7:54–56.

———, S. A. Fitzgerald, and L. Ganio. 1999. Precommercial thinning in a ponderosa pine stand affected by Armillaria root disease: 30 years of growth and mortality. *Western Journal of Applied Forestry* 14:144–48.

———, A. Kanaskie, K. Kavanagh, G. Johnson, R. Johnson, and D. Maguire. 2000a. *Silviculture and Swiss needlecast research and recommendations.* Research Contribution 30. Forest Research Laboratory, Oregon State University, Corvallis.

———, C. L. Schmitt, and C. G. Parks. 2000b. Mortality of mixed conifer regeneration surrounding stumps infected with *Heterobasidion annosus. Western Journal of Applied Forestry* 15:189–94.

Finney, M. A. 2001. Design of regular landscape fuel treatment patterns for modifying fire growth and behavior. *Forest Science* 47:219–28.

Fischer, W. C. 1981. *Photo guide for appraising down-woody fuels in Montana forests: Grand fir–larch, western hemlock, western hemlock–redcedar, and redcedar cover types.* General Technical Report INT-GTR-98, USDA Forest Service Intermountain Forest and Range Experiment Station, Ogden UT.

Fisher, R. F., and D. Binkley. 2000. *Ecology and Management of Forest Soils.* John Wiley & Sons, New York.

Ford, E. D. 1982. High productivity in a pole stage Sitka spruce, Picea sitchensis stand and its relation to canopy structure. *Forestry* 55:1–18.

Footen, P.W., R.B. Harrison, and B.D. Strahm. 2009. Long-term effects of nitrogen fertilization on the productivity of subsequent stands of Douglas-fir in the Pacific Northwest. *Forest Ecology Management* 258:2194–2198.

Fowells, H. A., and G. Schubert. 1956. *Seed crops of forest trees in the pine region of California.* Technical Bulletin 1150, US Department of Agriculture, Washington DC.

Frank, E. C., and R. Lee. 1966. *Potential solar beam irradiation on slopes: Tables for 30 to 50 degree latitude.* Research Paper RM-18, USDA Forest Service Rocky Mountain Forest and Range Experiment Station, Fort Collins CO.

Franklin, J. F. 1963. *Natural regeneration of Douglas-fir and associated species using modified clear-cutting systems in the Oregon Cascades.* Research Paper PNW-3, USDA Forest Service Pacific Northwest Forest and Range Experiment Station, Portland OR.

———, and J. Hoffman. 1968. *Two tests of white pine, true fir and Douglas-fir seed spotting in the Cascade Range.* Research Note PNW 80, USDA Forest Service Pacific Northwest Forest and Range Experiment Station, Portland OR.

———, D. R. Berg, D. A. Thornburg, and J. C. Tappeiner. 1997. Alternative silviculture approaches to timber harvesting: Variable retention systems. Pp. 111–49 in K. A. Kohm and J. F. Franklin, eds., *Creating a Forestry for the 21st Century.* Island Press, Washington DC.

———, T. A. Spies, R. VanPelt, A. B. Carey, D. A. Thornberg, D. R. Berg, D. B. Lindenmayer, M. E. Harmon, W. S. Keeton, D. C. Shaw, K. Bible, and J. Chen. 2002. Disturbances and structural development of natural forest ecosystems with silvicultural implications, using Douglas-fir forests as an example. *Forest Ecology and Management* 155:399–423.

———, and N. Johnson. 2012. A restoration framework for federal forests in the Pacific Northwest. Journal of Forestry 110:429–439.

———,N. K. Johnson, D. J. Churchill, K. Hagmann, D. Johnson, and J. Johnston. 2013. Restoration of Dry Forests in Eastern Oregon: A Field Guide. The Nature Conservancy of Oregon, Portland OR.

Freeland, R.O. 1952. Effect of age of leaves upon the rate of photosynthesis in some conifers. Plant Physiology 27:685–690.

Fried, J., J. C. Tappeiner II, and D. E. Hibbs. 1988. Bigleaf maple seedling establishment and early growth in Douglas-fir forests. *Canadian Journal of Forest Research* 18:1226–33.

———, J. R. Boyle, J. C. Tappeiner, and K. Cromack, Jr. 1989. Effects of bigleaf maple on soils in Douglas-fir forests. *Canadian Journal of Forest Research* 20:259–66.

Furniss, R. L., and V. M. Carolin. 1977. *Western Forest Insects.* USDA Forest Service, Miscellaneous Publication 1339, Washington DC.

Garber, S. M., and D. A. Maguire. 2003. Modeling stem taper of three central Oregon species using non-linear mixed effects models and autoregressive error structures. *Forest Ecology and Management* 179:507–22.

———, and D. A. Maguire. 2004. Stand productivity and development in two mixed species spacing trials in the central Oregon Cascades. *Forest Science* 50:92–105.

________, and D. A. Maguire. 2011. Growth and mortality of residual Douglas-fir after regeneration harvests under group selection and two-story silvicultural systems. *Western Journal of Applied Forestry* 26:64–70. Gardner, E.R. 1990. Fertilization and thinning effects on a Douglas-fir ecosystem at Shawnigan Lake: 15-year growth response. Forestry Canada, Pacific Forestry Centre, Victoria, BC. Information Report BC–X–319. 42 p.

Gardner, E.R. 1990. Fertilization and thinning effects on a Douglas-fir ecosystem at Shawnigan Lake: 15-year growth response. Forestry Canada, Pacific Forestry Centre, Victoria, BC. Information Report BC–X–319. 42 p.

Garrison–Johnston, M.T., T.M. Shaw, P.G. Mika, and L.R. Johnson. 2005. Management of ponderosa pine nutrition through fertilization. Pp. 123–143 in Ponderosa Pine: Issues, Trends, and Management. USDA-FS Pacific Southwest Research Station, Redding CA. Gen. Tech. Rep. PSW-GTR-198.

Gashweiler, J. S. 1969. Seedfall of three conifers in west-central Oregon. *Forest Science* 15:290–95.

———. 1970. Further study of conifer seed survival in a western Oregon clearcut. *Ecology* 51:849–54.

George, L. O., and F. A. Bazzaz. 1999. The fern understory as an ecological filter: Emergence and establishment of canopy-tree seedlings. *Ecology* 80:833–45.

Gessel, S.P., D.W. Cole, E.C. Steinbrenner. 1973. Nitrogen balances in forest ecosystems of the Pacific Northwest. *Soil Biology and Biochemis*try 5:19–34.

Gholz, H. L. 1982. Environmental limits on aboveground net primary production, leaf area, and biomass in vegetation zones of the Pacific Northwest. *Ecology* 54:152–59.

Giardina, C. P., and C. C. Rhoades. 2001. Clearcutting and burning affect nitrogen supply, phosphorus fractions and seedling growth in soils from a Wyoming lodgepole pine forest. *Forest Ecology and Management* 140:19–28.

Gilmore, D. W., and R. S. Seymour. 1997. Crown architecture of *Abies balsamea* from four crown positions. *Tree Physiology* 17:71–80.

Gingrich, S. F. 1967. Measuring and evaluating stocking and stand density in upland hardwood forests in the Central States. *Forest Science* 13:38–53.

Ginn, S. E., J. R. Seiler, B. H. Cazell, and R. E. Kreh. 1991. Physiological and growth responses of eight-year-old loblolly pine. *Forest Science* 37:1030–40.

Giordano, P. A., and D. E. Hibbs. 1993. Morphological response to competition in red alder: The role of water. *Functional Ecology* 7:462–68.

Goldberg, D. E. 1990. Components of resource competition in plant communities. In J. B. Grace and D. Tilman, eds., *Perspectives on Plant Competition*. Academic Press, New York.

Golding, D. L., and R. H. Swanson. 1978. Snow accumulation and melt in small forest openings in Alberta. *Canadian Journal of Forest Research* 8:380–88.

Gomez, A., R. F. Powers, M. J. Singer, and W. R. Horwath. 2002. Soil compaction effects on growth of young ponderosa pine following litter removal in California's Sierra Nevada. *Soil Science* 66:1334–43.

Gordon, D. T. 1962. *Growth response of eastside pine poles to removal of low vegetation.* Research Note PSW 209, USDA Forest Service Pacific Southwest Forest and Range Experiment Station, Berkeley CA.

———. 1970. *Natural regeneration of white and red fir—influence of several factors.* Research Paper PSW-58, USDA Forest Service Pacific Southwest Forest and Range Experiment Station, Berkeley CA.

———. 1973. *Released advanced regeneration of white and red fir—growth, damage, mortality.* Research Paper PSW-95, USDA Forest Service Pacific Southwest Forest and Range Experiment Station, Berkeley CA.

———. 1979. *Successful natural regeneration cuttings in California true firs.* Research Paper PSW–140, USDA Forest Service Pacific Southwest Forest and Range Experiment Station, Berkeley CA.

Gould, P. J., C. A. Harrington, and J. B. St. Clair. 2012. Growth phenology of coast Douglas-fir seed sources planted in diverse environments *Tree Physiology* 32:1482–1496. Online at http://www.treephys.oxfordjournals.org

———,and D. D. Marshall. 2010. *Incorporation of genetic gain into projections of Douglas-fir using ORGANON and the Forest Vegetation Simulator. Western Journal of Applied Forestry* 25:55–61.

Gourley, M., M. Vomicil, and M. Newton. 1987. Mitigating deer browse damage—Protection or competition control? P. 113 in D. M. Baumgartner, R. L. Mahoney, J. Evans, eds., *Animal damage management in Pacific Northwest forests*. Washington State University Cooperative Extension, Pullman.

Gower, S.T., Vogt, K.A., and Grier, C.G. 1992. Carbon dynamics of Rocky Mountain Douglas-fir: influence of water and nutrient availability. *Ecological Monographs* 62:43–65.

———, B. E. Haynes, K. S. Faschnacht, S. W. Running, and E. R. Hunt, Jr. 1993. Influence of fertilization on the allometric relations of two pines in contrasting environments. *Canadian Journal of Forest Research* 23:1704–11.

Grah, R. F. 1961. Relationship between tree spacing, knot size, and log quality in young Douglas-fir stands. *Journal of Forestry* 59:270–72.

Graham, R. T. 1988. Influence of stand density on development of western white pine, redcedar, hemlock, and grand fir in the Rocky Mountains. In W.C. Schmidt, ed. *Proceeding—Future forests of the Mountain West: A stand culture symposium*. General Technical Report INT GTR 243, USDA Forest Service Intermountain Forest and Range Experiment Station, Ogden UT.

———, and R. A. Smith. 1983. *Techniques for implementing individual tree selection method in the grand fir-cedar-hemlock ecosystems of northern Idaho*. USDA Forest Service Research Note INT-322, Intermountain Forest and Range Experiment Station, Ogden UT.

———, A. E. Harvey, and M. F. Jurgensen. 1989. Effect of site preparation on survival on survival and growth of Douglas fir seedlings. *New Forests* 3:89–98.

———, A. E. Harvey, T. B. Jain, and J. R. Tonn. 1999. *The effects of thinning and similar stand treatments on fire behavior in western forests*. USDA Forest Service General Technical Report PNW-GTR-463 USDA Forest Service Pacific Northwest Research Station, Portland OR.

———, S. McCaffery, and T. B. Jain. 2004. *Science basis for changing forest structure to modify wildfire behavior and severity*. General Technical Report GTR-RMRS-120, USDA Forest Service Rocky Mountain Research Station, Fort Collins CO.

———, T. B. Jain, and P. Cannon. 2005. Stand establishment and tending in the Inland Northwest. In C. A. Harrington and S. H. Schoenholtz, eds. *Productivity of western forests: A forest products focus*. USDA Forest Service General Technical Report Pacific PNW-GTR-642, Northwest Research Station, Portland OR.

Gratkowski, H. J. 1956. Windthrow around staggered settings in old-growth Douglas-fir. *Forest Science* 2:60–74.

———. 1961. Brush seedlings after controlled burning of brushlands in southwestern Oregon. *Journal of Forestry* 59:885–88.

———, and L. Anderson 1968. *Reclamation of non-sprouting greenleaf manzanita brush-fields in the Cascade Range.* USDA Forest Service Pacific Northwest Forest and Range Experiment Station Research Paper PNW–72, Portland OR.

Graves, H. S. 1911. *Principles of Handling Woodlands.* John Wiley & Sons, New York.

Gray, A, N.2005, Eight nonnative plants in western Oregon forests: associations with environment and management. *Environmental Monitoring and Assessment.* 100: 109–127.

———, and T. A. Spies. 1997. Microsite control on tree seedling establishment in conifer forest canopy gaps. *Ecology* 78:2458–73.

Gray, H. R. 1956. *The Form and Taper of Forest Tree Stems.* Imperial Forestry Institute, Oxford University, Oxford, UK.

Green, S. R. J. Grace and N. J. Hutchings. 1995. Observations of turbulent air flow in three stands of widely spaced Sitka Spruce. *Agriculture and Forest Meteorology* 74:205–25.

Grier, C. C., and R. H. Waring. 1974. Conifer foliage mass related to sapwood area. *Forest Science* 20:205–6.

Grigal, D. F. 2000. Effects of extensive forest management on soil productivity. *Forest Ecology and Management* 138:167–85.

Grime, J. P. 1981. *Plant Strategies and Vegetation Processes.* John Wiley & Sons, New York.

Grubb P. J. 1977. The maintenance of species richness in plant communities: The importance of regeneration niches. *Biology Review* 52:107–45.

Gruell, G. E. 1983. *Fire and vegetative trends in the northern Rocky Mountains: Interpretations from 1871 to 1982 photographs.* USDA Forest Service General Technical Report INT–158,, Intermountain Forest and Range Experiment Station, Ogden UT.

Guldin, J. M. 1991. Uneven-aged *BDq* regulation of Sierra Nevada mixed conifers. *Western Journal of Applied Forestry* 6:27–32.

———, and J. B. Baker. 1988. Yield comparisons from even–aged and uneven–aged loblolly–shortleaf pine stands. *Southern Journal of Applied Forestry* 12:107–14.

Haase, D. L., J. H. Batdorff, and R. Rose. 1993. Effect of root form on 10–yr survival and growth of planted Douglas-fir trees. *Tree Planters' Notes* 44:53–57.

Haase, S. M. 1986. Effect of prescribed burning on soil moisture and germination of southwestern ponderosa pine seed on basaltic soils. USDA Forest Service Research Note RM–462, Rocky Mountain Forest and Range Experiment Station, Fort Collins CO.

Haeussler, S., and J. C. Tappeiner. 1993. Effects of light environment on seed germination of red alder *Alnus rubra. Canadian Journal of Forest Research* 23:1487–91.

———, J. C. Tappeiner, and B. Greber. 1995. Germination and first-year survival of red alder seedlings in the central Oregon Coast Range. *Canadian Journal of Forest Research* 25:1487–91.

Hafley, W. L., and H. T. Schreuder. 1977. Statistical distributions for fitting diameter and height data in evenage stands. *Canadian Journal of Forest Research* 7:481–87.

Hagar, J. C., W. C. McComb, and W. H. Emmingham. 1996. Bird communities in commercially thinned and unthinned Douglas-fir stands of western Oregon. *Wildlife Society Bulletin* 24:353–66.

Hagmann, R. K., J. F. Franklin, and K. N. Johnson. 2013. Historical structure and composition of ponderosa pine and mixed-conifer forest in south-central Oregon. *Forest Ecology and Management* 304:492–504.

Haig, I. T. 1936. *Factors controlling initial establishment of western white pine and associated species.* School of Forestry Bulletin 41, Yale University, New Haven CT.

———, K. R. Davis, and R. H. Weidman. 1941. *Natural regeneration in the western white pine type.* USDA Technical Bulletin 767, Washington DC.

Hall, F. C. 1973. *Plant communities of the Blue Mountains of eastern Oregon and southeastern Washington.* Area Guide 3-1, USDA Forest Service Pacific Northwest Forest and Range Experiment Station, Portland OR.

———. 1976. Fire and vegetation in the Blue Mountains: Implications for land managers. *Proceedings of the Tall Timbers fire ecology conference* 15:155–170. Tall Timbers Research Station, Tallahassee FL.

———. 1983. Growth basal area: A field guide for appraising forest site potential for stockability. *Canadian Journal of Forest Research* 13:70–77.

Halpern, C. B. 1988. Early successional pathways and the resistance and resilience of forest communities. *Ecology* 69:1703–15.

———, and T. A. Spies. 1995. Plant species diversity in natural and managed forests of the Pacific Northwest. *Ecological Applications* 5:913–34.

———,J.A. Antos, M.A. Geyer, and A.M. Olson. 1997. Species replacement during early secondary succession: the abrupt decline of a winter annual. *Ecology* 78:621–631.

———, S. A. Evans, and S. Nielson. 1999. Soil seed banks in young closed canopy forest of the Olympic Peninsula, Washington: Potential contributors to understory reinitiation. *Canadian Journal of Botany* 77:922–35.

Hanley, T. 1983. Black-tailed deer, elk and forest edge in a western Cascades watershed. *Journal of Wildlife Management* 47:237–42.

Hann, D. W. 1997. *Equations for predicting the largest crown width of stand-grown trees in western Oregon.* Research Contribution 17, Forest Research Laboratory, Oregon State University, Corvallis.

———, 2005. *ORGANON User's Manual, Edition 8.0.* Oregon State University, Department of Forest Resources, Corvallis, OR. 129p.

———, and J. A. Scrivani. 1987. *Dominant height growth and site index equations for Douglas-fir and ponderosa pine in southwest Oregon.* Research Bulletin 59, Forest Research Laboratory, Oregon State University, Corvallis.

———, and C. H. Wang. 1990. *Mortality equations for individual trees in the mixed-conifer zone of southwest Oregon.* Research Bulletin 67, Forest Research Laboratory, Oregon State University, Corvallis.

———, and M. L. Hanus. 2002. *Enhanced height growth rate equations for undamaged and damaged trees in southwest Oregon.* Forest Research Laboratory, Research Contribution 41, Oregon State University, Corvallis.

———, D. D. Marshall, and M. L. Hanus. 2003. *Equations for predicting height to crown base, 5-year diameter growth rate, 5-year height growth rate, 5-year mortality rate, and maximum size density trajectory for Douglas-fir and western hemlock in the coastal region of the Pacific Northwest.* Research Contribution 40, Forest Research Laboratory, Oregon State University, Corvallis.

Hansen, E. M., J. K. Stone, B. R. Capitano, P. Rosso, W. Sutton, L. Winton, A. Kanaskie, and M.G. McWilliams. 2000. Incidence and impact of Swiss needle cast in forest plantations of Douglas-fir in coastal Oregon. *Plant Disease* 84:773–778.

Hanus, M. L., D. W. Hann and D. D. Marshall. 1999. *Predicting height for undamaged and damaged trees in southwestern Oregon.* Forest Research Laboratory, Research Contribution 27, Oregon State University.

Happe, P. J., K. J. Jenkins, E. E. Starkey, and S. H. Sharrow. 1990. Nutritional quality and tannin astringency of browse in clearcuts and old-growth forests. *Journal of Wildlife Management* 54:557–66.

Harmon, M. E., and J. F. Franklin. 1989. Tree seedlings on logs in *Picea-tsuga* forests of Washington and Oregon. *Ecology* 70:48–59.

Harper, J. 1977. *Population Biology of Plants.* Academic Press, New York.

Harrington, C. A. 1984. Factors influencing the initial sprouting of red alder. *Canadian Journal of Forest Research* 14:357–61.

———,1991. Retrospective shoot growth analysis for three seed sources of loblolly pine. Canadian Journal of Forest Research 21:306–317

———, and D. L. Reukema. 1983. Initial shock and long-term thinning in a Douglas-fir plantation. *Forest Science* 29:33–46.

———, and R. O. Curtis. 1986. *Height growth and site index curves for red alder.* Research Paper PNW-358, USDA Forest Service Pacific Northwest Research Station, Portland OR.

———, and C. A. Wierman. 1990. Growth and foliar nutrient response to fertilization and precommercial thinning in a coastal western red cedar stand. *Canadian Journal of Forest Research* 20:764–73.

Harrington, M. G. 1987. Ponderosa pine mortality from spring, summer and fall crown scorching. *Western Journal of Applied Forestry* 2:14–16.

Harrington, T. B. and J. C. Tappeiner. 2009. Long-term effects of tanoak competition on Douglas-fir stand development. *Canadian Journal of Forest Research* 39:765–776.

———, 2006. Five-year growth responses of Douglas-fir, western hemlock, and western redcedar seedlings to manipulated levels of overstory and understory competition. *Canadian Journal of Forest Research* 36:2439–2453.

———, and J. C. Tappeiner. 1991. Competition affects shoot morphology, growth duration and relative growth rates of Douglas-fir saplings. *Canadian Journal of Forest Research* 21:474–81.

———, and L. Parendes, eds. 1993. *Forest vegetation management without herbicides; Proceedings of a workshop* (Feb. 18–19, 1992, Corvallis OR). Forest Research Laboratory, Oregon State University, Corvallis OR.

———, and J. C. Tappeiner. 1997. Growth responses of young Douglas-fir and tanoak 11 years after various levels of hardwood removal and understory suppression in southwestern Oregon, USA. *Forest Ecology and Management* 96:1–11.

———, J. C. Tappeiner, and J. D. Walstad. 1984. Predicting leaf area and biomass of 1- to 6-year-old tanoak and Pacific madrone sprout clumps in southwestern Oregon. *Canadian Journal of Forest Research* 14:209–13.

———, J. C. Tappeiner, and T. F. Hughes. 1991. Predicting average growth and size distributions of Douglas-fir saplings competing with sprouts of tanoak or Pacific madrone. *New Forests* 5:109–30.

———, J. C. Tappeiner, and R. Warbington. 1992. Predicting crown size and diameter distributions of tanoak, Pacific madrone, and giant chinkapin sprout clumps. *Western Journal of Applied Forestry* 7:103–8.

———, R. J. Pabst, and J. C. Tappeiner. 1994. Seasonal physiology of Douglas-fir saplings: response to microclimate in stands of tanoak or Pacific madrone. *Forest Science* 40:59–82.

———, R. G. Wagner, S. R. Radosevich, and J. D. Walstad. 1995. Interspecific competition and herbicide injury influence 10-year responses of coastal Douglas-fir and associated vegetation to release treatments. *Forest Ecology and Management* 76:55–67.

——— K.D. Howell. 1998. Planting cost, survival, and growth one to three years after establishing loblolly pine seedlings with straight, deformed, and pruned taproots. *New Forests* 15:193–204.

———. and A.A. Bluhm. 2001. Tree regeneration responses to microsite characteristics following a severe tornado in the Georgia Piedmont, USA. *Forest Ecology and Management* 140:265–275.

———. and S.H. Schoenholtz. 2010. Effects of logging debris treatments on five-year development of competing vegetation and planted Douglas-fir. *Canadian Journal of Forest Research* 40:500–510.

———. R.A. Slesak, and S.H. Schoenholtz. 2013. Variation in logging debris cover influences competitor abundance, resource availability, and early growth of planted Douglas-fir. *Forest Ecology and Management* 296:41–52.

———. 2014. Synthetic auxin herbicides control germinating Scotch broom (*Cytisus scoparius*). *Weed Technology* 28(2):435–442.

Harrington, T.C., Reinhart, C., Thornburgh, D.A., and Cobb, Jr., F.W. 1983. Association of black-stain root disease with precommercial thinning of Douglas-fir. *Forest Science* 29:12–14

Harris, A.S. and W.A. Farr. 1992. Response of western hemlock and sitka spruce to fertilization in southeastern Alaska. Pp. 78–81 in S.P. Gessel, R.M. Kenady, and W. A. Atkinson (Eds.). Proceedings of the Forest Fertilization Conference. September 25–27, 1979, Union, WA. Institute of Forest Resources, College of Forest Resources, University of Washington, Seattle, WA.

Harris, G. R., and W. W. Covington. 1982. The effect of a prescribed fire on nutrient concentration and standing crop of understory vegetation in ponderosa pine. *Canadian Journal of Forest Research* 13:501–7.

Harris,L. D.1984. *The Fragmented Forest.* University of Chicago Press, Chicago IL.

Harvey, A. E., M. F. Jurgensen, R. Martin, and R. T. Graham. 1989. Fire-soil interactions governing site productivity in the northern Rocky Mountains. Pp. 9–18 in D. M. Baumgartner, D. Breuer, W. Zamora, A. Benjamin, L. F. Neuenschwander, and R. H. Wakamoto, eds., *Proceedings: Prescribed fire in the intermountain region.* Washington State University, Pullman.

Hasse, W. B., and A. R. Ek. 1981. A simulated comparison of yields for even- versus uneven-aged northern hardwood stands. *Journal of Environmental Management* 12:235–40.

Hatiya, K., I. Takeuchi, and K. Tochiaki. 1989. Primary productivity of high density stands of *Pinus densiflora*. *Bulletin of the Forestry and Forest Products Research Institute* 345:39–97 (in Japanese).

Hawksworth, F. G, and D. Wiens. 1996. *Dwarf mistletoes: Biology, pathology, and systematics.* Agricultural Handbook 709, USDA Forest Service, Washington DC.

Hawley, R. C. 1921. *The Practice of Silviculture.* Wiley and Sons, New York.

Hayes, J. P., S. S. Chan, W. H. Emmingham, J. C. Tappeiner, L. D. Kellogg, and J. D. Bailey. 1997. Wildlife responses to thinning young forests in the Pacific Northwest. *Journal of Forestry* 95:28–33.

Haygreen, J. G., and J. L. Bowyer. 1982. *Forest Products and Wood Science.* Iowa State University Press, Ames.

Haywood, J.D. 2005. Influence of precommercial thinning and fertilization on total stem volume and lower stem form of loblolly pine. *Southern Journal of Applied Forestry* 29:215–220.

Heilman, P. E., and S. P. Gessel. 1963. Nitrogen requirements and biological cycling in Douglas-fir stands in relationship to the effects of nitrogen fertilization. *Plant and Soil* 18:386–402.

Helgerson, O. T. 1990a. Alternate types of artificial shade increase survival of planted Douglas-fir in clearcuts. *New Forests* 3:117–22.

———. 1990b. Heat damage in tree seedlings and its prevention. *New Forests* 3:17–42.

———, S. D. Tesch, S. D. Hobbs and D. McNabb. 1991. Survival and growth of Douglas-fir seedlings after prescribed burning of a brushfield in southwest Oregon. *Western Journal of Applied Forestry* 6:55–59.

———, M. Newton, D. DeCalesta, T. Schowalter, and E .Hansen. 1992. Protecting young regeneration. Pp. 384–420 in S. D. Hobbs, S. D. Tesch, P. W. Owston, R. E. Stewart, J. C. Tappeiner II, and G. E. Wells, eds., *Reforestation practices in southwestern Oregon and northern California.* Forest Research Laboratory, Oregon State University, Corvallis.

Helms, J. A. 1964. Apparent photosynthesis of Douglas-fir in relation to silvicultural treatment. *Forest Science* 10:432–42.

———. 1972. Environmental control of net photosynthesis in naturally growing ponderosa pine *Pinus ponderosa* Laws. *Ecology* 53:92–101.

———. 1979. Positive effects of prescribed burning on wildfire intensities. *Fire Management Notes* 403:10–13.

———. 1998. *The Dictionary of Forestry.* Society of American Foresters, Bethesda MD.

———, and R. B. Standiford. 1985. Predicting release of advance reproduction of mixed conifer species in California. *Forest Science* 31:3–15.

———, and C. Hipkin. 1986. Effects of soil compaction on tree volume in a California ponderosa pine plantation. *Western Journal of Applied Forestry* 1:121–24.

———, C. Hipkin and E. B. Alexander. 1986. Effects of soil compaction on height growth of a California ponderosa pine plantation. *Western Journal of Applied Forestry* 1:104–7.

Hemstrom, M. A., and J. F. Franklin. 1982. Fire and other disturbances of the forests in Mount Rainier National Park. *Quaternary Research* 18:32–51.

Heninger, R.L., W. Scott, A. Dobkowski, R. Miller, H. Anderson, and S. Duke. 2002. Soil disturbance and 10-year growth response of coast Douglas-fir on nontilled and tilled skid trails in the Oregon Cascades. *Canadian Journal of Forest Research.* 32:233–46.

Herman, F. R., C. D. Curtis, and D. J. DeMars. 1978. *Height growth and site index estimates for noble fir in high-elevation forests of the Oregon–Washington Cascades.* Research Paper PNW–243, USDA Forest Service Pacific Northwest Forest and Range Experiment Station, Portland OR.

Hermann R. K. 1967. Seasonal variation in sensitivity of Douglas-fir seedlings to the exposure to roots. *Forest Science* 13:140–49.

———, and W. W. Chilcote. 1965. *Effects of seedbeds on germination and survival of Douglas-fir.* Forest Research Lab. Oregon State University Research Paper 4.

———, and D. P. Lavender. 1968. Early growth of Douglas-fir from various altitudes and aspects in southwestern Oregon. *Silvae Genetica* 17:141–153.

———, and R. K. Petersen. 1969. Root development and height increment of ponderosa pine in pumice soils of eastern Oregon. *Forest Science* 15:226–37.

———, and D. P. Lavender. 1999. Douglas-fir planted forests. *New Forests* 17:53–70.

Hett, J. M., and O. Loucks. 1971. A dynamic analysis of age in sugar maple seedlings. *Ecology* 59:507–20.

Hibbs, D. E., B. Withrow-Robinson, D. Brown, and R. Fletcher. 2003. Hybrid Poplar in the Willamette Valley. *Western Journal of Applied Forestry* 18:281 285.

_______, P. A. Giordano. 1996. Vegetation characteristics of alder-dominated riparian buffer strips in the Oregon Coast Range. *Northwest Science* 70:213–22.

———, D. S. DeBell and R. F. Tarrant. 1984. *Biology and Management of Red Alder*. Oregon State University Press, Corvallis.

———, W. H. Emmingham, and M. C. Bondi. 1989. Thinning red alder: Effects of method and spacing. *Forest Science* 35:16–29.

———, S. S. Chan, M. Castellano, and C. Niu. 1995a. Response of red alder seedlings to CO_2 enrichment and water stress. *New Phytology* 129:569–577.

———, W. H. Emmingham, and M. C. Bondi. 1995b. Response of red alder to thinning. *Western Journal of Applied Forestry* 10:17–23.

Hinckley, T. M., J. P. Lassioe, and S. W. Running. 1978. Temporal and spatial variations in the water status of forest trees. *Forest Science Monograph* 20.

Hobbs S. D., and K. A. Wearstler, Jr. 1985. Effects of cutting sclerophyll brush on sprout development and Douglas-fir growth. *Forest Ecology and Management* 13:69–81.

———, S. D. Tesch, P. W. Owston, R. E. Stewart, J. C. Tappeiner II, and G. E. Wells. 1992. *Reforestation practices in southwestern Oregon and northern California*. Forest Research Laboratory, Oregon State University, Corvallis.

Hodgson, G. E. 1990. Prescription for Wayrot Stand, 358. Prairie City Ranger District, Malheur National Forest. Master of Forestry Paper, College of Forestry, Oregon State University, Corvallis OR.

Hoff, R. J., G. I. McDonald, and R. T. Bingham. 1976. *Mass selection for blister resistance: a method for natural regeneration of western white pine*. USDA Forest Service Research Note INT 202, Intermountain Forest and Range Experiment Station, Ogden UT.

Hoffman, J. V. 1924. *Natural regeneration of Douglas-fir in the Pacific Northwest.* USDA Bulletin 1200, US Department of Agriculture, Washington DC.

Holbo, H. R., and S. W. Childs. 1987. Summertime radiation balances of clearcut and shelterwood slopes in southwest Oregon. *Forest Science* 33:504–16.

Holmes, J. R. B., and D. Tackle. 1962. *Height growth of lodgepole pine in Montana related to soil and stand factors*. Bulletin 21, Montana Forest and Conservation Experiment Station, School of Forestry, Montana State University, Missoula.

Holub, S.M., T.A. Terry, C.A. Harrington, R.B. Harrison, and R. Meade. 2013. Tree growth ten years after residual biomass removal, soil compaction, tillage, and competing vegetation control in a highly productive Douglas-fir plantation. *Forest Ecology and Management* 305:60–66.

Homma, K., N. Akasi, T. Abe, M. Hasegawa, K. Harada, Y. Hirabuki, K. Irie, M. Kaji, H.

Miguchi, N. Mizoguchi, H. Mizunaga, T. Nakashizuka, S. Natume, K. Niiyama, T. Ohkubo, S. Sawada, H. Sugita, S. Takatuski, and N. Yamanaka. 1999. Geographical variation in the early regeneration process of Siebold's beech (*Fagus crenata*) in Japan. *Plant Ecology* 140:129–38.

Hooven, E. F., and H. C. Black. 1976. Effects of some clearcutting practices on small mammal populations in western Oregon. *Northwest Science* 50:189–208.

Hopmans, P. and H.N. Chappell. 1994. Growth of young, thinned Douglas-fir stands to nitrogen fertilizer in relation to soil properties and tree nutrition. *Canadian Journal of Forest Research* 24:1684–1688.

Horton, T., T. D. Burns, and V. T. Parker 1999. Ectomycorrhizal fungi associated with *Arctostaphylos* contribute to *Pseudotsuga menziesii* establishment. *Canadian Journal of Botany* 77:93–102.

Houle, G. 1995. Seed dispersal and seedling recruitment: the missing links. *Ecoscience* 2:238–44.

Howard, K. M., and M. Newton. 1984. Overtopping by successional Coast-Range vegetation slows Douglas-fir seedlings, *Journal of Forestry* 82:178–80.

Howard, W. E., and H. E. Childs. 1959. Ecology of pocket gophers with special reference to *Thomomys bottae mewa. Hilgardia* 29:277–358.

Hudec, J. L., and D. L. Peterson. 2012. Fuel variability following wildfire in forests with mixed severity fire regimes, Cascades Range, USA. *Forest Ecology and Management* 277:11–24.

Hummer, K. E. 2000. History of the origin and dispersal of white pine blister rust. *Horticulture Technology* 10:1–3.

Huff, T. 2009. *Conifer regeneration, understory vegetation and artificially topped conifer responses to alternative silvicultural treatments.* M.S. thesis, Department of Forest Resources, Oregon State University, Corvallis.

Huffman, D. W., and J. C. Tappeiner. 1997. Clonal expansion and seedling recruitment of Oregon grape (*Berberis nervosa)* in Douglas-fir (*Pseudotsuga menziesii*) forests: Comparison with salal (*Gaultheria shallon). Canadian Journal of Forest Research* 27:1788–93.

———, J. C. Tappeiner, and J. C. Zasada. 1994. Regeneration of salal in the central Coast Range forests of Oregon. *Canadian Journal of Botany* 72:39–51.

Hughes, T. F., C. R. Latt, J. C. Tappeiner, and M. Newton. 1987. Biomass and leaf area estimates for varnishleaf ceanothus, deerbrush, and whiteleaf manzanita. *Western Journal of Applied Forestry* 2:124–28.

———, J. C. Tappeiner, and M. Newton. 1990. Relationship of Pacific madrone sprout growth to productivity of Douglas-fir seedlings and understory vegetation. *Western Journal of Applied Forestry* 5:20–24.

Hunter, M. L., ed. 1999. *Biological Diversity*. Cambridge University Press, Cambridge, UK.

Hurteau, M. D., J. B. Bradford, P. Z. Fulé, A. H. Taylor and K. L. Martin. 2013. Climate change, fire management, and ecological services in the southwestern US. *Forest Ecology and Management*. 327:280–289

Husch, B., T. W. Beers, and J. A. Keershaw, Jr. 2003. *Forest Mensuration* (4th ed.). John Wiley & Sons, Hoboken NJ.

Hyink, D. M., W. Scott, and R. M. Leon. 1987. Some important aspects in the development of a managed stand growth model for western hemlock. Pp. 9–21 in A. R. Ek, S. R. Shifley, and T. E. Burk, eds. *Forest growth modeling and prediction*. USDA Forest Service General Technical Report. NC-120, St Paul MN.

Hytönen, J. and P. Jylhä. 2011. Long-term response of weed control intensity on Scots pine survival, and growth and nutrition on former arable land. *European Journal of Forest Research* 130: 91–98.

Ingestad, T. 1979. Mineral nutrient requirements of Pinus sylvestris and Picea abies seedling. *Physiologia Plantarum* 45:373–380.19

———. 1982. Relative addition rate and external concentration: Driving variable used in plant nutrition research. *Plant, Cell & Environment* 5:443–453.

———. and G.I. Ågren. 1992. Theories and methods on plant nutrition and growth. *Physiologia Plantarum* 84:177–184.

Isaac, L. A. 1930. Seed flight in the Douglas-fir region. *Journal of Forestry* 28:492–99.

———. 1938. *Factors affecting the establishment of Douglas-fir seedlings*. USDA Circular 486, US Department of Agriculture, Washington DC.

———. 1940. Vegetative succession following logging in the Douglas-fir region, with special reference to fire. *Journal of Forestry* 38:716–21.

———. 1956. *Place of partial cutting of old-growth stands of the Douglas-fir region*. Research Paper 16, USDA Forest Service Pacific Northwest Forest and Range Experiment Station, Portland OR.

Jacobs, M. R. 1954. The effect of wind sway on the form and development of *Pinus radiata* D. Don. *Australian Journal of Botany* 2:35–51.

Jahnke, L. S., and D. B. Lawrence. 1965. Influence of photosynthetic crown structure on potential productivity of vegetation, based primarily on mathematical models. *Ecology* 46:319–26.

Jain, T.B., R. T. Graham, and P. Morgan. 2004. Western white pine growth relative to forest openings. *Canadian Journal of Forest Research* 34:2187–97.

James, R. L., F. W. Cobb, P. R. Miller, and J. R. Parmeter. 1980. Effects of oxidant air pollution on susceptibility of pine roots to *Fomes annosus*. *Phytopathology* 70:560–63.

Jaramillo, A. E. 1988. *Growth of Douglas-fir in southwestern Oregon following removal of competing vegetation*. Research Note PNW-470, USDA Forest Service Pacific Northwest Forest and Range Experiment Station, Portland OR.

Jarmer, C. B., J. W. Mann, and W. A. Atkinson. 1992. Harvesting timber to achieve reforestation objectives. Pp. 202–30 in S. D. Hobbs, S. D. Tesch, P. W. Owston, R. E. Stewart, J. C. Tappeiner II, and G. E. Wells, eds., *Reforestation practices in southwestern Oregon and northern California.* Forest Research Laboratory, Oregon State University, Corvallis.

Jenkinson, J. L. 1980. *Improving plantation establishment by optimizing root growth capacity and planting time of western yellow pine.* Research Paper PSW-154, USDA Forest Service Pacific Southwest Forest and Range Experiment Station, Berkeley CA.

———, J. A. Nelson, and M. E. Huddleson. 1993. *Improving planting stock quality—the Humboldt experience.* General Technical Report PSW-GTR-143, USDA Forest Service Pacific Southwest Research Station, Albany CA.

Jenny, H. 1941. *Factors of soil formation.* McGraw Hill, New York.

Jensen, E. C., and J. N. Long. 1983. Crown structure of a codominant Douglas-fir. *Canadian Journal of Forest Research* 13:264–69.

Johnson, D. W., and P. S. Curtis. 2001. Effects of forest management on soil C and N storage: Meta analysis. *Forest Ecology and Management* 140:227–3

Johnson, E. A., and K. Miyanishi. 2001. *Forest Fires: Behavior and Ecological Effects.* Academic Press, New York NY.

Jonard, M., L. Misson, and Q. Ponette. 2006. Long-term thinning effects on the forest floor and the foliar nutrient status of Norway spruce stands in the Belgian Ardennes. *Canadian Journal of Forest Research* 36:2684–2695.

Jozsa, L. A, and H. Brix. 1989. Effects of fertilization and thinning on wood quality of a 24-year-old Douglas-fir stand. *Canadian Journal of Forest Research* 19:1137–45.

Julin, K. R., G. Segura, M. Hinkley, C. G. Shaw III, and W. A. Farr. 1998. The fluted western hemlock of southeast Alaska III: Six growing seasons after treatment. *Forest Ecology and Management* 103:272–85.

Kahn, S., R. Rose, D. L. Haase, and T. E. Sabin. 2000. Effects of shade on morphology, chlorophyll concentration and chlorophyll florescence of four Pacific Northwest conifers. *New Forests* 19:171–86.

Kärki, L. 1985. Genetically narrow-crowned trees combine high timber quality and stem wood production at low cost. Pp. 245–56 in P. M. Tigerstedt, A. P. Puttonen, and V. Koski, eds., *Crop Physiology of Forest Trees.* University of Helsinki, Finland.

Kauffman, J. B. 1990. Ecological relationships of vegetation and fire in Pacific Northwest forests. In J. D. Walstad, S. R. Radosevich, and D. V. Sandberg, eds. *Natural and Prescribed Fire in the Pacific Northwest.* Oregon State University Press, Corvallis.

———, and R. E. Martin. 1990. Sprouting shrub response to different seasons and fuel consumption levels of prescribed fire in Sierra Nevada mixed conifer ecosystems. *Forest Science* 36:748–64.

Kaufmann, M. R., and J. D. Weatherred. 1982. *Determination of potential direct beam solar irradiance*. Research Paper RM-242, USDA Forest Service Rocky Mountain Forest and Range Experiment Station, Fort Collins CO.

———, and M. G. Ryan. 1986. Physiographic stand and environment effects on individual tree growth and growth efficiency in subalpine forests. *Tree Physiology* 2:47–59.

Keen, F. P. 1936. Relative susceptibility of ponderosa pine to bark-beetle attack. *Journal of Forestry* 34:919–27.

———. 1943. Ponderosa pine tree classes redefined. *Journal of Forestry* 41:854–48.

———. 1958. *Cone and seed insects of western forests*. USDA Forest Service Technical Bulletin 1169.

Kellog, L. D. 1980. *Thinning young timber stands in mountainous terrain*. Forest Research Lab Research Bulletin 34, Oregon State University College of Forestry, Corvallis.

———, P. Bettinger, and R. M. Edwards. 1996. A comparison of logging planning, felling, and skyline yarding costs between clearcutting and five group selection harvest methods. *Western Journal of Applied Forestry* 11:90–96.

Kellomäki, P. 1986. A model for the distribution of branch number and biomass in *Pinus sylvestris* crowns and the effect of crown shape and stand density on branch and stem biomass. *Scandinavian Journal of Forest Research* 1:545–472.

Kemperman, J. A., and B. V. Barnes. 1976. Clone size in American aspens. *Canadian Journal of Botany* 54:2603–07.

Kelpsas, B., M. Newton, and C. Landgren. 2014. Forestry. In: Peachey, E., editor. *Pacific Northwest Weed Management Handbook* [online]. Corvallis, OR: Oregon State University. http://pnwhandbooks.org/weed/other-areas/forestry-and-hybrid-cottonwoods/forestry (accessed 02 December 2014).

Kershaw, J. A., and D. A. Maguire. 2000. Influence of vertical foliage structure on the distribution of stem cross-sectional area increment in western hemlock and balsam fir. *Forest Science* 46:86–94.

Ketchum, J. S., and R. Rose. 2003. Preventing establishment of exotic shrubs (*Cytisus scoparius* (L. Link) and *Cytisus striatus* (Hill)) with soil active herbicides (hexazinone, sulfometeron, and meta sulfuron). *New Forests* 25:83–92.

———, R. Rose, and B. Kelpsas. 1999. Weed control in spring and summer after fall application of sulfometron. *Western Journal of Applied Forestry* 14:80–85.

Keyes, C. R., and K. L. O'Hara. 2002. Quantifying stand targets for silvicultural prevention of crown fires. *Western Journal of Applied Forestry* 17:101–9.

Kilgore, B. M., and D. Taylor. 1979. Fire history of a Sequoia-mixed conifer forest. *Ecology* 60:129–41.

Kimball, B.A., E.C. Turnblom, D.L. Nolte, D.L. Griffin, and R.M. Engeman. 1998. Effects of thinning and nitrogen fertilization on sugars and terpenes in Douglas-fir vascular tissues: Implications for black bear foraging. *Forest Science* 44:599–602.

Kimmins, J.P. 1990. Modelling the sustainability of forest production and yield for a changing and uncertain future. *Forestry Chronicle* 66, 271–280.

———1992. *Balancing Act: Environmental Issues in Forestry*. University of British Columbia Press, Vancouver BC, Canada.

———. D. Mailly, and B. Seely. 1999. Modelling forest ecosystem net primary production: the hybrid simulation approach used in FORECAST. *Ecological Modelling* 122:195–22.

———. 2004. *Forest Ecology: A Foundation for Sustainable Forest Management and Environmental Ethics in Forestry*. Prentice-Hall, Upper Saddle River NJ.

King, J. E. 1966. *Site index curves for Douglas-fir in the Pacific Northwest*. Weyerhaeuser Forestry Paper 8, Centralia WA.

Kingery, J. L., and R.T. Graham. 1991. The effects of cattle grazing on ponderosa pine regeneration. *Forestry Chronicle* 67:245–48.

Kinloch, B., B. M. Marosy, and M. E. Huddleston. 1992. *Sugar pine: Status, values, and roles in ecosystems*. Publication 3362, Division of Agriculture and Natural Resources, University of California, Berkeley CA.

Kira, T. 1975. Primary productivity of forests. In J. P. Cooper, ed., *Photosynthesis and Productivity in Different Environments*. Cambridge University Press, Cambridge, UK.

———, and T. Shidei. 1967. Primary production and turnover of organic matter in different forest ecosystems of the western Pacific. *Japanese Journal of Ecology* 17:70–87.

Kirby, K. J., and C. Watkins, eds. 1998. *The Ecological History of European Forests*. CAB International, New York.

Kirkland, B. P., and A. J. F. Brandstrom. 1936. *Selective timber management in the Douglas-fir region*. USDA Forest Service Document, Washington DC.

Kishchuk, B. E., G. W. Weetman, R. P. Brockley, and C. E. Prescott. 2002. Fourteen-year growth response of young lodgepole pine to repeated fertilization. *Canadian Journal of Forest Research* 32:153–60.

Kitzmiller, J. H. 1990. Managing genetic diversity in a tree improvement program. *Forest Ecology and Management* 35:111–49.

———. 2005. Provenance trials of ponderosa pine in northern California. *Forest Science* 51:595–607.

Knapp, E. E., J. M. Varner, M. D. Busse, C. N. Skinner and C. J. Shestak. 2011. Behaviour and effects of prescribed fire in masticated fuelbeds. *International Journal of Wildland Fire* 20:932–945.

Knowe, S. A., and D. E. Hibbs. 1996. Stand structure and dynamics of young red alder as affected by planting density. *Forest Ecology and Management* 82:69–85.

Knutson, D., and R. Tinnin. 1999. Effects of dwarf mistletoe on the response of young Douglas-fir to thinning. *Canadian Journal of Forest Research* 16:30–35.

Kohm, K. A., and J. F. Franklin. 1997. *Creating a Forestry for the 21st Century*. Island Press, Washington DC.

Kolb, T. E., J. K. Agee, P. Z. Fulé, N. G. McDowell, K. Pearson, A. Sala and R. H. Waring. 2007. Perpetuating old ponderosa pine. *Forest Ecology and Management* 249:141–157.

Koroleff, A.1954. Leaf litter as killer. *Journal of Forestry* 52:178–82.

Korpela, E. J., and S. D.Tesch. 1992. Plantations vs. advanced regeneration; height growth comparisons for southwestern Oregon. *Western Journal of Applied Forestry* 7:44–47.

Kostler, J. 1956. *Silviculture.* Oliver & Boyd, Edinburgh, UK.

Kozlowski, T. T., P. J. Kramer, and S. G. Pallardy. 1991. *The Physiological Ecology of Woody Plants*. Academic Press, New York.

Kraft, G. 1884. *Beiträge zur Lehre von den Durchforstungen, Schlagstellungen und Lichtungshieben*. Klingworth, Hannover, Germany.

Krajicek, J. E., K. A. Brinkman, and S. F. Gingrich. 1961. Crown competition—a measure of density. *Forest Science* 7:36–42.

Krinard, R. M. 1985. *Cottonwood development through 19 years in a Nelder's design.* Research Note SO-322, USDA-Forest Service.

Kurmis, V., and E. Sucoff. 1989. Population height and density distribution of *Corylus cornuta* in undisturbed forests of Minnesota. *Canadian Journal of Botany* 67:2409–13.

Laacke, R. J., and G. O. Fiddler. 1986. *Overstory removal: Stand factors related to success and failure.* Research Paper PSW-RP-183, USDA Forest Service Pacific Southwest Research Station, Berkeley CA.

———, and J. H. Tomascheski. 1986. *Shelterwood regeneration of true fir after 8 years.* Research Paper PSW-184, USDA Forest Service Pacific Southwest Forest and Range Experiment Station, Berkeley CA.

Laclau, J. P., J.C.R. Almeida, J. L.M. Gonçalves, L. Saint-André, M. Ventura, J. Ranger, R.M. Moreira, and Y Nouvellon. 2009. Influence of nitrogen and potassium fertilization on leaf lifespan and allocation of above-ground growth in *Eucalyptus* plantations. *Tree Physiology* 29:111–124.

Lambert, M. 1986. Sulphur and nitrogen nutrition and their interactive effects on Dothistroma infection in *Pinus radiata. Canadian Journal of Forest Research* 16:1055–1062.

———. and J. Turner. 1977. Dieback in high site quality Pinus radiata stands—the role of sulphur and boron deficiencies. *New Zealand Journal of Forestry Science* 7:333–348.

Landsberg, J. J., and S. T. Gower. 1997. *Applications of Physiological Ecology to Forest Management*. Academic Press, San Diego CA.

Lange, K. E., 1998. Nutrient and tannin concentrations in managed and unmanaged forests of the Oregon Coast Range. MS thesis, Department of Forest Science, Oregon State University, Corvallis.

Lanini, W. T., and S. R. Radosevich. 1986. Response of three conifer species to site preparation and shrub control. *Forest Science* 32:61–77.

Lanner, R. M. 1985. On the insensitivity of height growth to spacing. *Forest Ecology and Management* 13:143–48.

———. 1996. *Made for Each Other: A Symbiosis of Birds and Pines*. Oxford University Press, New York.

Larocque, G. R., and P. L. Marshall. 1993. Crown development in red pine stands. I. Absolute and relative growth measures. *Canadian Journal of Forest Research* 24:762–74.

Larson, P. J. 1963. Stem form development of forest trees. *Forest Science Monographs* 5:42.

Latham, P., and J. C. Tappeiner. 2002. Response of old-growth conifers to reduction in stand density in western Oregon forests. *Tree Physiology* 22:137–46.

Lavender, D. P. 1958. *Effects of ground cover on seedling germination and survival.* Research Note 34, Oregon Department of Forestry, Salem OR.

Lawrence, W. H, and J. H Rediske. 1962. The fate of sown Douglas-fir seed. *Forest Science* 8:210–18.

Ledig, F. T., and J. H. Kitzmiller. 1992. Climate change and genetic strategies for reforestation. *Forest Ecology and Management* 60:153–69.

Leininger, W. C., and S. H. Sharrow. 1987. Seasonal diets of herded sheep grazing Douglas-fir plantations. *Journal of Range Management* 40:551–55.

———. and S. H. Sharrow. 1989. Sheep production in coastal Oregon Douglas-fir plantations. *Northwest Science* 63:195–200.

Levy , L. R. Deal, and J. Tappeiner 2010. *The density and distribution of Sitka spruce and western hemlock seedling banks in partially harvested stands in southeast Alaska.* Research Paper PNW-RP-585 US Forest Service Pacific Northwest Research Station, Portland, OR.

Lieffers, V. J., C. Meissier, K. Stadt, F. Gendron, and P. Comeau. 1999. Predicting and managing light in the understory of boreal forests. *Canadian Journal of Forest Research* 29:796–811.

Lieth, H. 1975. Some prospects beyond production management. Pp. 285–304 in H. Lieth and R. H. Whittaker, eds., *Primary Productivity of the Biosphere.* Springer-Verlag, New York.

Lilieholm, R. J., L. S. Davis, R. C. Heald, and S. P. Holmen. 1990. Effects of single-tree selection harvests on stand structure, species composition, and understory tree growth in a Sierra Nevada mixed-conifer forest. *Western Journal of Applied Forestry* 4:43–47.

Lindenmayer, D. B., and S. A. Cunningham. 2013. Six principles for managing forests as ecologically sustainable ecosystems. *Landscape Ecology* 28:1099–1110.

Linder, S. and B. Axelsson. 1982. Changes in carbon uptake and allocation patterns as a result of irrigation and fertilization in a young *Pinus sylvestris* stand. Pp. 38–44 in R.H. Waring (ed). *Carbon Uptake and Allocation in Sub-alpine Ecosystems as a Key to Management.* Oregon State University, Corvallis.

———, and D.A. Rook. 1984. Effects of mineral nutrition on the carbon dioxide exchange of trees. Pp. 211–238 in G.D. Bowen and E.K.S. Nambiar (eds). Nutrition of Forest Trees in Plantations. Academic Press, London, UK.

Lindquist, J. L., and M. N. Palley. 1963. *Empirical yield tables for young-growth redwood.* Agriculture Experiment Station Bulletin 831, University of California College of Agriculture, Berkeley CA.

Lindzey, F. G, and C. E. Meslow. 1977. Population characteristics of black bears on an island in Washington. *Journal of Wildlife Management* 41:408–12.

Liu, X., U. Silins, V. Lieffers, and R. Man. 2003. Stem hydraulic properties and growth in lodgepole pine stands following thinning and sway treatments. *Canadian Journal of Forest Research* 33:1295–1303.

Lotan, J. E. 1967. Cone serotiny in lodgepole pine near West Yellowstone Montana. *Forest Science* 13:55–59.

———, and C. E. Jensen. 1970. *Estimating seed stored in serotinous cones of lodgepole pine*. Research Paper INT-83, USDA Forest Service Intermountain Forest and Range Experiment Station, Ogden UT.

Long, J. N. 1985. A practical approach to density management. *Forestry Chronicle* 61:23–27.

———. 1996. A technique for the control of stocking in two-storied stands. *Western Journal of Applied Forestry* 11:59–61.

———, and J. Turner. 1975. Above-ground biomass of understory and overstory in an age sequence of four Douglas-fir stands. *Journal of Applied Ecology* 12:178–88.

———, and F. W. Smith. 1984. Relation between size and density in developing young stands: A description and possible mechanisms. *Forest Ecology and Management* 7:191–206.

———, and T. W. Daniel. 1990. Assessment of growing stock in uneven-aged stands. *Western Journal of Applied Forestry* 5:93–96.

———, and J. D. Shaw. 2005. A density management diagram for even-aged ponderosa pine stands. *Western Journal of Applied Forestry* 20:205–15.

———, J. B. McCarter, and S. B. Jack. 1988. A modified density management diagram for coastal Douglas-fir. *Western Journal of Applied Forestry* 3:88–89.

Ludovici, K. H, and L. A Morris. 1996. Responses of loblolly pine, sweetgum, and crab-grass roots to localized increases in nitrogen in two watering regimes. *Tree Physiology* 16:933–39.

Lusk, C. H. 1995 Seed size, establishment sites and species coexistence in a Chilean rainforest. *Vegetation Science* 6:249–56.

Maguire, D. A. 1983. Suppressed crown expansion and increased bud density after pre-commercial thinning in California Douglas-fir. *Canadian Journal of Forest Research* 13:1246–48.

———. 1985. The effect of sampling scale on the detection of interspecific patterns in a hemlock-hardwood forest herb stratum. *The American Midland Naturalist* 113:138–45.

———, and R. T. Forman. 1983. Herb cover effects on tree seedling patterns in a mature hemlock-hardwood forest. *Ecology* 64:1367–80.

———, and D. W. Hann. 1989. The relationship between gross crown dimensions and sapwood area at crown base in Douglas-fir. *Canadian Journal of Forest Research* 19:557–65.

———, and J. L. F. Batista. 1996. Sapwood taper models and implied sapwood volume and foliage profiles for coastal Douglas-fir. *Canadian Journal of Forest Research* 26:849–63.

———, and W. S. Bennett. 1996. Patterns in vertical distribution of foliage in young coastal Douglas-fir. *Canadian Journal of Forest Research* 26:1991–2005.

———, and A. Kanaskie. 2002. The ratio of live crown length to sapwood area as a measure of crown sparseness. *Forest Science* 48:93–100.

———, J. A. Kershaw Jr., and D. W. Hann. 1991. Predicting the effects of a silvicultural regime on branch size and crown wood core in Douglas-fir. *Forest Science* 37:1409–28.

———, J. Brissette, and L. Gu. 1998. Canopy structure and growth efficiency of red spruce in uneven-aged, mixed species stands in Maine. *Canadian Journal of Forest Research* 28:1233–40.

———, D. B. Mainwaring, and C. B. Halpern. 2006. Stand dynamics after variable-retention harvesting in mature Douglas-fir forests of western North America. *Allgemeine Forst- und Jagd-Zeitung* 177:20–31.

Maguire, W. P. 1955. Radiation, surface temperatures, and seedling survival. *Forest Science* 1:277–85.

Mailly, D., and J. P. Kimmins. 1997. Growth of *Pseudotsuga menziessi* and *Tsuga heterophylla* seedlings along a light gradient: Resource allocation and morphological acclimatization. *Canadian Journal of Botany* 75:1424–35.

Mainwaring, D.B., D.A. Maguire, and S.S. Perakis. 2014. Three-year growth response of young Douglas-fir to nitrogen, calcium, phosphorus, and blended fertilizers in Oregon and Washington. *Forest Ecology and Management* 327:178–188.

Mallik, A. U., and C. E. Pescott. 2001. Growth inhibitory effects of salal on western hemlock and western red cedar. *Agronomy Journal* 93:85–92.

Manter, D.K., P.D. Reeser, and J.K. Stone. 2005. A climate-based model for predicting geographic variation in Swiss needle cast severity in the Oregon coast range. *Phytopathology* 95:1256–1265.

Margolis, H. A., and D. G. Brand. 1990. An ecophysiological basis for understanding plantation establishment. *Canadian Journal of Forest Research* 20:375–90.

Marshall, D. D. 1990. *The Effects of Thinning on Stand and Tree Growth in a Young, High-site Douglas-fir Stand in Western Oregon.* Ph.D. dissertation, Department of Forest Science, Oregon State University, Corvallis.

———, and R. O. Curtis. 2002. *Levels-of-growing-stock Cooperative Study in Douglas-fir: Report 15—Hoskins 1963–1998.* Research Paper PNW-RP-537, USDA Forest Service Pacific Northwest Research Station, Portland OR.

———, J. F. Bell, and J. C. Tappeiner. 1992. *Levels-of-growing-stock Cooperative Study in Douglas-fir: Report 10—Hoskins 1963–1983.* Research Paper PNW-RP-448, USDA Forest Service Pacific Northwest Research Station, Portland OR.

Marschner, H. 1995. Mineral Nutrition of Higher Plants. Academic Press, London, UK. 889 p.

Martin, P. 1971. Movement and activities of the mountain beaver (*Aplodontia rufa*). *Journal of Mammalogy* 52:717–23.

Martin, R. E. 1990. Goals, methods and elements of prescribed burning. Pp. 55–66, in J. D. Walstad, S. R. Radosevich, and D. V. Sandberg, eds. *Natural and Prescribed Fire in Pacific Northwest Forests.* Oregon State University Press, Corvallis.

———, and J. D. Dell. 1978. *Planning for prescribed burning in the Pacific Northwest.* General Technical Report GTR-PNW-76, USDA Forest Service Pacific Northwest Forest and Range Experiment Station, Portland OR.

Mason, A. C., and D. L. Adams. 1989. Black bear damage to thinned timber stands in northwest Montana. *Western Journal of Applied Forestry* 4:10–13.

Mason, R., B. E. Wickman, R. C. Beckwith, and H. G. Paul. 1992. Thinning and nitrogen fertilization in a grand fir stand infested with western spruce budworm. I. Insect response. *Forest Science* 38:235–51.

Mathiasen, R. L. 1998. Infection of young western larch by larch dwarf mistletoe in northern Idaho and western Montana. *Western Journal of Applied Forestry* 13:41–46.

Matthews, J. D. 1991. *Silvicultural Systems.* Oxford University Press, Oxford, UK.

Maxwell, W. G, and F. R. Ward. 1976. *Photo series for quantifying forest residues in the ponderosa pine type, ponderosa and associated species type, and lodgepole pine type.* General Technical Report PNW-GTR-52, USDA Forest Service Pacific Northwest Forest and Range Experiment Station, Portland OR.

McArdle, R. E., W. H. Meyer, and D. Bruce. 1961. *The yield of Douglas-fir in the Pacific Northwest.* USDA Technical Bulletin 201, US Department of Agriculture, Washington DC.

McCarter, J. B., and J. N. Long. 1986. A lodgepole pine density management diagram. *Western Journal of Applied Forestry* 1:6–11.S

McCaughey, W., C. E. Fielder, and W. C. Schmidt. 1991. *Twenty year natural regeneration following five silvicultural prescriptions in spruce-fir forests of the inter mountain west.* Research Paper INT-439, USDA Forest Service Intermountain Research Station, Ogden UT.

McCay, N. 1985. *A stockability equation for forest land in Siskiyou County, California.* Research Note PNW-435, Pacific Northwest Forest and Range Experiment Station, Portland OR.

MacClean, J. B., and C. L. Bolsinger. 1973. *Estimating productivity on sites with a low stocking capacity.* Research Paper PNW-152, USDA Forest Service Pacific Northwest Forest and Range Experiment Station, Portland OR.

McCreary, D. D., and D. A. Perry. 1983. Strip thinning and selective thinning in Douglas-fir. *Journal of Forestry* 81:375–77.

McDonald, P. M. 1976a. *Forest regeneration and seedling growth from five major cutting methods in north-central California.* Research Paper PSW-RP-115, USDA Forest Service Pacific Southwest Forest and Range Experiment Station, Berkeley CA.

———. 1976b. *Shelterwood cutting in a young-growth mixed conifer stand in north-central California.* Research Paper PSW-RP-117, USDA Forest Service Pacific Southwest Forest and Range Experiment Station, Berkeley CA.

———. 1978a. *Silviculture-ecology of Three Native California Hardwoods on High Sites in Northern Central California.* Ph.D. dissertation, Department of Forest Science, Oregon State University, Corvallis.

———. 1978b. Inhibiting effect of ponderosa pine seed trees on seedling growth. *Journal of Forestry* 74:220–24.

———. 1983. *Clearcutting and natural regeneration ... management implications for the northern Sierra Nevada.* USDA Forest Service General Technical Report PSW-GTR-70, Albany CA.

———. 1992. Estimating seed crops of conifer and hardwood species. *Canadian Journal of Forest Research* 22:832–838.

———, and O. T. Helgerson. 1991. *Mulches aid in regenerating Oregon and California forests: Past, present and future.* USDA Forest Service General Technical Report PSW-12, USDA Forest Service Pacific Southwest Research Station, Albany CA.

———, and G. O. Fiddler. 1993a. Feasibility of alternatives to herbicides in young conifer plantations in California. *Canadian Journal of Forest Research* 23:2015–22.

——— and G. O. Fiddler. 1993b. *Vegetative trends in a young conifer plantation after 10 years of grazing by sheep.* Research Paper PSW-RP-215, USDA Forest Service, Pacific Southwest Research Station, Albany CA.

———, and C. S. Abbot. 1994. *Seed fall regeneration and seedling development in group selection openings.* Research Paper PSW-220, USDA Forest Service Pacific Southwest Research Station, Albany CA.

_______, and J. H. Kitzmiller. 1994. Genetically improved ponderosa pine seedlings outgrow nursery-run seedlings with and without competition—early findings. *Western Journal of Applied Forestry* 9:57–61.

——— and G. O. Fiddler. 2001. *Timing and duration of release affect vegetation development in a young ponderosa pine plantation.* Research Paper PSW-RP-245, USDA Forest Service Pacific Southwest Research Station, Albany CA.

———, and J. C. Tappeiner. 2002. *California's hardwood resource: Seeds, seedlings, and sprouts of three important forest-zone species.* General Technical Report PSW-GTR-185. USDA Forest Service Pacific Southwest Research Station, Albany CA.

———, C. Skinner, and G. O. Fiddler. 1992. Ponderosa pine needle length: An early indicator of release treatment effectiveness. *Canadian Journal of Forest Research* 22:761–64.

McDonald, S., R. J. Boyd, and D. D. Sears. 1983. *Lifting storage and planting practices influence growth of conifer seedlings in the northern Rockies.* Research Paper INT-300. USDA Forest Service Intermountain Forest and Range Experiment Station, Ogden UT.

McDowell, N., J. R. Brooks, S. A. Fitzgerald, and B. J. Bond. 2003. Carbon isotope discrimination and growth response of old *Pinus ponderosa* trees to stand density reduction. *Plant Cell and Environment* 26:631–44.

McIver, J., and L. Starr. 2001. A literature review on the environmental effects of post-fire logging. *Western Journal of Applied Forestry* 16(4): 159–168.

McNay, R. S., L. D. Peterson, and J. B. Nyberg. 1988. The influence of forest stand characteristics on snow interception in the coastal forest of British Columbia. *Canadian Journal of Forest Research* 18:556–73.

McNeil, R. C., and D. B. Zobel. 1980. Vegetation and fire history of ponderosa pine-white fir forest in Crater Lake National Park. *Northwest Science* 54:30–46.

Mead, D.J., S.X. Chang, and C. M. Preston. 2008. Recovery of ^{15}N-urea 10 years after application to a Douglas-fir pole stand in coastal British Columbia. *Forest Ecology and Management* 256:694–701

Means, J. E., P. C. MacMillan, and K. Cromak Jr. 1992. Biomass and nutrient content of Douglas-fir logs, and other detrital pools in an old-growth forest, Oregon, USA. *Canadian Journal of Forest Research* 22:1536–46.

Megahan, W. F., and R. Steele. 1987. An approach for predicting snow damage to ponderosa pine plantations. *Forest Science* 33:485–503.

Megraw, R. A. 1986. Douglas-fir wood properties. Pp. 81–96 in C. D. Oliver, D. P. Hanley, and J. A. Johnson, eds. *Douglas-fir: Stand management for the future.* Institute of Forest Resources, College of Forest Resources, University of Washington, Seattle.

Meiman, J. R. 1987. *Influence of forests on snow pack accumulation.* General Technical Report RM-GTR-149. USDA Forest Service, Rocky Mountain Forest and Range Experiment Station, Fort Collins CO.

Messier, C. R., and J. P. Kimmins. 1991. Above- and below-ground vegetation recovery in recently cut and burned sites dominated by *Gaultheria shallon* in coastal British Columbia. *Forest Ecology and Management* 46:275–94.

———, T. W. Honer, and J. P. Kimmins. 1989. Photosynthetic photon flux density, red:far-red ratio and minimum light requirements for survival of *Galtheria shallon* in western red cedar-western hemlock stands in coastal British Columbia. *Canadian Journal of Forest Research* 19:1470–77.

———, J. C. Doucet, Y. Ruel, C. Claveau, C. Kelly, and M. Lechowicz. 1999. Functional ecology of advanced regeneration in relation to light in boreal forests. *Canadian Journal of Forest Research* 29:812–23.

Meyer, C. A. 1963. Vertical distribution of annual increment in thinned ponderosa pine. *Forest Science* 9:39–43.

Meyer, H. A., and D. D. Stevenson. 1943. The structure and growth of virgin beech, birch, maple, hemlock forests in northern Pennsylvania. *Journal of Agricultural Research* 67:465–84.

Meyer, W. H. 1938. *Yield of even-age stands of ponderosa pine.* Technical Bulletin 630, US Department of Agriculture, Washington DC.

Miesel, J. R. 2012. Differential responses of *Pinus ponderosa* and *Abies concolor* foliar characteristics and diameter growth to thinning and prescribed fire treatments. *Forest Ecology and Management* 284:163–173.

Mika, P.G., and J. VanderPloeg. 1991. Six year fertilizer response of managed second-growth

Douglas-fir stands in the Intermountain Northwest. Pp. 293–301 in D.M. Baumgartner (ed). Interior Douglas-fir: The species and its management. Washington State University Cooperative Extension, Pullman.

Miller, H. G. 1981. Forest fertilization: Some guiding concepts. *Forestry* 54:157–67.

Miller, J. H., H. L. Allen, B.R. Zutter, S.M. Zedaker, R.A. Newbold. 2006a. Soil and pine foliage nutrient responses 15 years after competing-vegetation control and their correlation with growth for 13 loblolly pine plantations in the southern United States. *Canadian Journal of Forest Research* 36:2412–2425.

Miller, M., and W. H. Emmingham. 2001. Can selection thinning convert even-age Douglas-fir stands to uneven-age structures? *Western Journal of Applied Forestry* 16:35–43.

Miller P. R., J. R. Parmeter Jr., O. C. Taylor and E. A. Cardiff. 1963. Ozone injury to foliage of ponderosa pine. *Phytopathology* 53:1072.

———, and A. A. Millecan. 1971. Extent of oxidant air pollution damage to some pines and other conifers in California. *Plant Disease Reporter* 55:555–59.

Miller, R. E. and R. D. Fight. 1979. *Fertilizing Douglas-fir forests.* General Technical Report PNW-83, USDA Forest Service Pacific Northwest Forest and Range Experiment Station, Portland OR.

———, and R. Tarrant. 1983. Long-term growth response of Douglas-fir to ammonium nitrate fertilizer. *Forest Science* 29:127–37.

———, P. R. Barker, C. E. Peterson, and S. R. Webster. 1986. Using nitrogen fertilizers in management of coast Douglas-fir: Regional trends of response. In: *Douglas-fir: Stand management for the future.* University of Washington, College of Forest Resources, Seattle.

———, D. L. Reukema, and H. W. Anderson. 2004. *Tree growth and soil relations at the 1925 Wind River spacing test in coast Douglas-fir.* USDA Forest Service Research Paper PNW-RP 558, Pacific Northwest Research Station, Portland OR.

———, T. B. Harrington, J. Madsen, and W. Thies. 2006b. *Laminated root rot in a western Washington plantation: Eight-year mortality and growth of Douglas-fir as related to infected stumps, tree density, and fertilization.* USDA Forest Service, Research Paper, Pacific Northwest Research Station, Portland OR. 37p.

Minore, D. 1978. *The Dead Indian Plateau: A historical summary of forestry observations and research in a severe southwestern Oregon environment.* General Technical Report PNW-GTR-72, USDA Forest Service Pacific Northwest Forest and Range Experiment Station, Portland OR.

———.1986. *Germination, survival and early growth of conifer seedlings in two habitat types.* Research Paper PNW-348, USDA Forest Service Pacific Northwest Forest and Range Experiment Station, Portland OR.

———, and R. J. Laacke. 1992. Natural regeneration. In S. Hobbs et al., eds. *Reforestation practices for southwestern Oregon and northern California.* Forest Research Laboratory,

Oregon State University, Corvallis.

———, A. W. Smart, and M. E. Dubrasich. 1979. *Huckleberry ecology and management in the Pacific Northwest.* General Technical Report GTR-PNW-93, USDA Forest Service Pacific Northwest Forest and Range Experiment Station, Portland OR.

Mitchell, A.K. and T.M. Hinckley. 1993. Effects of foliar nitrogen concentration on photosynthesis and water use efficiency in Douglas-fir. *Tree Physiology* 12:403–410.

Mitchell, R. G., and H. K. Preisler. 1991. Analysis of spatial pattern of lodgepole pine attacked by the mountain pine beetle. *Forest Science* 37:1390–1408.

———, R. Waring, and G. Pitman. 1983. Thinning lodgepole pine increases tree vigor and resistance to mountain pine beetle. *Forest Science* 29:204–11.

Mitchell, S. J. 2000. Stem growth responses in Douglas-fir and Sitka spruce following thinning: Implications for assessing wind-firmness. *Forest Ecology and Management* 135:105–14.

———, and W. J. Beese. 2002. The retention system: Reconciling variable retention with the principles of silvicultural systems. *Forestry Chronicle* 78:397–403.

Moeur, M. 1981. *Crown width and foliage weight of northern Rocky Mountain conifers multiple linear regression models,* Pseudotsuga, Picea, Abies, Pinus, Larix, Thuja, Idaho, Montana. USDA Forest Service, Research Paper INT-283, Intermountain Forest and Range Experiment Station, Fort Collins CO.

———, 1993. Characterizing spatial patterns of trees using stem-mapped data. *Forest Science* 39:756–775

Monleon, V. J., K. Cromak, and J. D. Landsberg. 1997. Short-and long-term effects of prescribed burning on nitrogen availability in ponderosa pine stands in central Oregon. *Canadian Journal of Forest Research* 27:369–78.

———, D. Azuma, and D. Gedney. 2004. Equations for predicting uncompacted crown ratio based on compacted crown ratio and tree attributes. *Western Journal of Applied Forestry* 19:260–267.

Monserud, R. A. 2002. Large-scale management experiments in the moist maritime forests of the Pacific Northwest. *Landscape and Urban Planning* 59:159–80.

———, and J. D. Marshall. 1999. Allometric crown relations in three northern Idaho conifer species. *Canadian Journal of Forest Research* 29:521–35.

Monsi, M., Z. Uchijuma, and T. Oikawa. 1973. Structure of foliage canopies and photosynthesis. *Annual Review of Ecology and Systematics* 4:301–27.

Moore, J. A., P. G. Mike, J. L. Vander Ploeg. 1991. Nitrogen fertilizer response of Rocky Mountain Douglas-fir by geographic area across the Inland Northwest. *Western Journal of Applied Forestry* 6:94–98.

Morgan, P., and L. F. Neuenschwander. 1988. Seed-bank contributions to regeneration of shrub species after cutting and burning. *Canadian Journal of Botany* 66:169–72.

Morisita, M. 1959. Measuring dispersion and analysis of distribution patterns. Pp. 215–35 in *Memoirs of the Faculty of Science: Kyushu University, Series E: Biology*. Japan.

Morrison, P., and F. J. Swanson. 1990. *Fire history and pattern in a Cascade Range landscape*. General Technical Report PNW-GTR-254, USDA Forest Service Pacific Northwest Research Station, Portland OR.

Mroz, G. D., M. F. Jurgensen, A. E. Harvey, and M. J. Larsen. 1980. Effects of fire on nitrogen in forest floor humus. *Soil Science Society of America Journal* 44:395–400.

Muelder, D. W., D. O. Hall, and R. G. Skolman. 1963. *Root growth and first-year survival of* Pinus ponderosa *seedlings in second-growth stands in the Sierra Nevada*. Publication 32, University of California Forestry and Forest Products, Berkeley CA.

Muir, P. S., R. L. Mattingly, J. C. Tappeiner II, J. D. Bailey, W. E. Elliot, J. C. Hagar, J. C. Miller, E. B. Peterson, and E. E. Starkey. 2002. *Managing for biodiversity in young Douglas-fir forests in western Oregon*. Biological Science Report USGS/BRD-2002-0006, US Geological Survey, Forest and Range Ecosystem Science Center, Corvallis OR.

Munger, G. T., R. E. Will, and B. E. Borders. 2003. Effects of competition control and annual nitrogen fertilization and irrigation on gas exchange of different-aged *Pinus tadea*. *Canadian Journal of Forest Research* 33:1076–83.

Munger T. T. 1911. *Growth and management of Douglas-fir in the Pacific Northwest*. USDA Forest Service Circular 195, U.S. Department of Agriculture, Washington DC.

———. 1950. A look at selective cutting in Douglas-fir. *Journal of Forestry* 4:97–99.

Murthy, R. P., M. Dougherty, S. J. Zarnoch, and H. L. Allen. 1996. Effects of carbon dioxide, fertilization, and irrigation on photosynthetic capacity of loblolly pine trees. *Tree Physiology* 16:537–46.

Mutch, L. S., and D. Parsons. 1998. Mixed conifer forest mortality and establishment before and after prescribed fire in Sequoia National Park, California. *Forest Science* 44:341–55.

Neal, F. D., and J. E. Borreco. 1981. Distribution and relationship of mountain beaver to openings in sapling stands. *Northwest Science* 55:79–86.

Nason, G.E. and D.D. Myrold. 1992. Nitrogen fertilizers: Fates and environmental effects in forests. Pp. 67–81 in H.N. Chappell, G.F. Weetman, and R.E. Miller (eds). Forest Fertilization: Sustaining and improving nutrition and growth of western forests. Institute of Forest Resources, University of Washington, Seattle WA. Contribution Number 73.

———, D. J. Pluth, and W.B. McGill. 1988. Volatization and foliar recapture of ammonia following spring and fall application of nitrogen-15 urea to a Douglas-fir ecosystem. *Soil Science Society of America Journal* 52:821–828.

Neary, D. G., C. C. Klopatek, L. F. Debano, and P. F. Ffolliott. 1999. Fire effects on belowground sustainability: a review and synthesis. *Forest Ecology and Management* 122:51–71.

Nelson, E. E. 1989. Black bears prefer urea-fertilized trees. *Western Journal of Applied Forestry* 4:13–15.

Nelson, L. R., R. C. Pedersen, L. L. Autry, S. Dudley, and J. D. Walstad. 1981. Impacts of herbaceous weeds in young loblolly pine plantations. *Southern Journal of Applied Forestry* 5:153–158.

Newton, M. B. Kelpsas, and C. Lundgren. 2012. Forestry. In: Peachey, E., editor. *Pacific Northwest Weed Management Handbook* [online]. Corvallis: Oregon State University. http://pnwhandbooks.org/weed/other-areas/forestry-and-hybrid-poplars/forestry (accessed 16 December 2013).

———, 2006. Forest–land brush control. In: R. William, A. Dailey, D. Ball, J. Colquhoun, R. Parker, J. Yenish, T. Miller, D. Morishita, and J. Hutchinson, compilers, *Pacific Northwest weed management handbook*, Extension Service, Oregon State University, Corvallis.

———, and D. S. Preest. 1988. Growth and water relations of Douglas-fir seedlings under different weed control regimes. *Weed Science* 36:653–62.

———, and E. C. Cole. 1987. A sustained yield scheme for old-growth Douglas-fir. *Western Journal of Applied Forestry* 2:22–25.

———, and F.B. Knight. 1981. Handbook of weed and insect control chemicals for forest resource managers. Timber Press, Beaverton OR.

———, B. Kelpsas, and C. Lundgren. 2012. Forestry. In: Peachey, E., editor. *Pacific Northwest Weed Management Handbook* [online]. Corvallis: Oregon State University. http://pnwhandbooks.org/weed/other-areas/forestry-and-hybrid-poplars/forestry (accessed 16 December 2013).

———, and E. C. Cole. 1991. Root development in planted Douglas-fir under varying competitive stress. *Canadian Journal of Forest Research* 21:25–31.

Nigh, G. D. 1996. Growth intercept models for species without distinct annual whorls: Western hemlock. *Canadian Journal of Forest Research* 26:1407–15.

Nord, E. C. 1965. Autecology of bitterbrush in California. *Ecological Monographs* 35:307–34.

Nyland, R. D. 2002. *Silviculture Concepts and Applications*. McGraw-Hill, New York.

O'Dea, M., J. Zasada, and J. Tappeiner. 1995. Vine maple clonal growth and reproduction in coastal Douglas-fir forests. *Ecological Applications* 5:63–73.

O'Hara, K. L. 1988. Stand structure and growing space efficiency following thinning in even-aged Douglas-fir stands. *Canadian Journal of Forest Research* 18:859–66.

———. 1996. Dynamics and stocking-level relationships of multi-aged ponderosa pine stands. *Forest Science Monograph* 33.

———, and R. F. Gersonde. 2004. Stocking control concepts in uneven-aged silviculture. *Forestry Chronicle* 77:131–43.

———, and L. M. Nagel. 2006. A functional comparison of productivity in even-aged and multi-aged ponderosa pine stands. *Forest Science* 52:290–203.

———, R. S. Seymour, S. D. Tesch, and J. M. Guldin. 1994. Silviculture and our changing profession: Leadership for shifting paradigms. *Journal of Forestry* 92:8–13.

———, N. I. Valappil, and N. M. Nagel. 2003. Stocking control procedures for multi-aged ponderosa pine in the inland northwest. *Western Journal of Applied Forestry* 18:5–14.

———. 2014. *Multiaged Silviculture*. Oxford University Press, United Kingdom.

Oliver, C. D. 1980. Even-age development of mixed-species stands. *Journal of Forestry* 78:201–3.

———. 1981. Forest development in north America following major disturbances. *Forest Ecology and Management* 3:153–68.

———, and B. C. Larson. 1996. *Forest Stand Dynamics*. McGraw-Hill, New York.

Oliver, W. W. 1972. *Height intercept for estimating site index in young ponderosa pine plantations and natural stands*. Research Note PSW-276, USDA Forest Service Pacific Southwest Forest and Range Experiment Station, Berkeley CA.

———.1984. *Brush reduces growth of thinned ponderosa pine in northern California*. Research Paper PSW-RP-172, USDA Forest Service Pacific Southwest Research Station, Berkeley CA.

———. 1986. *Growth of California red fir advanced regeneration after overstory removal and thinning*. Research Paper PSW-RP-180, USDA Forest Service Pacific Southwest Forest and Range Experiment Station, Berkeley CA.

———. 1988. Ten-year growth response of a California red and white fir sawtimber stand to several thinning intensities. *Western Journal of Applied Forestry* 3:41–43.

———. 1990. Spacing and shrub competition influence 20-year development of planted ponderosa pine. *Western Journal of Applied Forestry* 5:79–82.

———.1995. Is self-thinning in ponderosa pine ruled by *Dendroctonous* bark beetles? Pp. 213–18 in L. G. Eskew, compiler, *Forest health through silviculture*. General Technical Report GTR-RM-267, USDA Forest Service Rocky Mountain Research Station, Fort Collins CO.

———.1997. Twenty-five-year growth and mortality of planted ponderosa pine repeatedly thinned to different stand densities in northern California. *Western Journal of Applied Forestry* 12:122–30.

———, and F. C. C. Uzoh. 2002. Little response of true fir seedlings to understory removal. *Western Journal of Applied Forestry* 15:5–8.

———, J. L. Lindquist, and R. O. Strothmann. 1994. Young-growth redwood stands respond well to various thinning intensities. *Western Journal of Applied Forestry* 9:106–12.

Olsen, W. K., J. M. Schmid, and S. A. Mata. 1996. Stand characteristics associated with mountain pine beetle infestations in ponderosa pine. *Forest Science* 42:310–27.

Olson, C. M., and J. A. Helms. 1996. Forest growth and stand structure at Blodgett Research Station. Pp. 681–732 in *Status of the Sierra Nevada, volume III: Assessments and scientific basis for management options*. Report No. 38, Sierra Nevada Ecosystem Project—Final Report to Congress. Centers for Water and Wildland Resources, University of California, Davis CA.

Olszyk, D. M., M. G. Johnson, D. T. Tingey, P. T. Rygiewicz, C. Wise, E. VanEss, A. Benson, M. J. Storm, and R. King. 2003. Whole seedling biomass allocation, leaf area, and tissue chemistry for Douglas-fir exposed to elevated CO_2 and temperature for 4 years. *Canadian Journal of Forest Research* 33:269–78.

Oren, R., E. D. Schulze, R. Matyssek, and R. Zimmerman. 1986. Estimating photosynthetic rate and annual carbon gain in conifers from specific leaf weight and leaf biomass. *Oecologia* 70:187–93.

———, R. H. Waring, S. G. Stafford, and W. Barrett. 1987. Twenty-four years of ponderosa pine growth in relation to canopy leaf area and understory competition. *Forest Science* 33:538–47.

Osawa, A., and R. B. Allen. 1993. Allometric theory explains self-thinning relationships of mountain beech and red pine. *Ecology* 74:1020–32.

Otrosina, W. J., and F. W. Cobb Jr. 1989. Biology ecology and control of *Heterobasidium annosum.* In: W. J. Otrosina and R. F. Scharpf. *Proceedings of the symposium on research and management of annosus root disease in western North America:* USDA Forest Service General Technical Report PSW-116, Albany CA.

Ottmar, R. D., C. C. Hardy, and R. E. Vihnanek. 1990. *Stereo photo series for quantifying forest residues in the Douglas-fir hemlock type of the Willamette National Forest.* General Technical Report PNW-GTR-258, USDA Forest Service Pacific Northwest Research Station, Portland OR.

Ovington, J. D. 1962. Quantitative-ecology and the woodland concept. Pp. 103–92 in J. B. Cragg, ed., *Advances in Ecological Research.* Academic Press, London, UK.

Owens, J. N., and S. J. Morris. 1998. Factors affecting cone and seed development in Pacific silver fir. *Canadian Journal of Forest Research* 28:1146–63.

Owston, P. W., G. A. Walters, and R. Molina. 1992. Selection of planting stock, inoculation with mycorrhizal fungi, and use of direct seeding. Pp. 310–27 in S. D. Hobbs, S. D. Tesch, P. W. Owston, R. E. Stewart, J. C. Tappeiner II, and G. E. Wells, eds., *Reforestation practices in southwestern Oregon and northern California.* Forest Research Laboratory, Oregon State University, Corvallis.

Pabst, R. J., J. C. Tappeiner, and M. Newton. 1990. Varying densities of Pacific madrone in a young stand alter soil water potential, plant moisture stress and growth of Douglas-fir. *Forest Ecology and Management* 37:267–83.

Paine, D. P., and D. W. Hann. 1982. *Maximum crown width equations for southwestern Oregon Tree Species.* Research Paper 46, Forest Research Laboratory, Oregon State University, Corvallis.

Pang, P.C. and K. McCullough. 1982. Nutrient distribution in forest soil leachates after thinning and fertilizing Douglas-fir forest. *Canadian Journal of Soil Science* 62:197–208.

———,H. J. Barclay, and K. McCullough. 1987. Aboveground nutrient distribution within trees and stands in thinned and fertilized Douglas-fir. *Canadian Journal of Forest Research* 17:1379–84.

Panshin, A. J., and C. deZeeuw. 1964. *Textbook of Wood Technology*, 3rd ed. McGraw Hill, New York.

Parkash, R., and L. S. Khana. 1991. *Theory and Practice of Silviculture Systems*. International Book Distributors, Dehra Dun, India.

Parks, C. G., E. L. Bull, R. O. Tennin, J. F. Shepard, and A. K. Blumton. 1999. Wildlife use of dwarf mistletoe brooms in Douglas-fir. *Western Journal of Applied Forestry* 14:100–105.

Parmeter, J. R., Jr. 1978. Forest stand dynamics and ecological factors in relation to dwarf mistletoe spread, impact, and control. Pp. 16–30 in R. F. Scharpf and J. R. Parmeter, Jr., eds., *Proceedings of the symposium on dwarf mistletoe control through forest management* (April 11–13). General Technical Report PSW-131, USDA Forest Service Pacific Southwest Research Station, Berkeley CA.

Parsons, D. J., and S. H. DeBenedetti. 1979. Impact of fire suppression on a mixed-conifer forest. *Forest Ecology and Management* 2:23–33.

Pearson, G. A. 1923. *Natural reproduction of western yellow pine in the southwest.* USDA Technical Bulletin 1105, U.S. Department of Agriculture, Washington DC.

Pearson, J. A., D. H. Knight, and T. J. Fahey. 1987. Biomass and nutrient accumulation during stand development in Wyoming lodgepole pine forests. *Ecology* 68:1966–73.

Peet, R. K., and N. L. Christensen. 1987. Competition and tree death. *Bioscience* 37:586–95.

Peltola, H., S. Kellomäki, H. Väisänen, and V. Ikonen. 1999. A mechanistic model for assessing the risk of wind and snow damage to single trees and stands of Scots pine, Norway spruce, and birch. *Canadian Journal of Forest Research* 29:647–61.

Perry, D. A. 1984. A model of physiological and allometric factors in the self-thinning curve. *Journal of Theoretical Biology* 106:383–401.

———, Meurisse, R.; Thomas, B.; Miller, R.; Boyle, J.; Means, J.; Perry, C. R.; Powers, R. F. (eds). 1989. *Maintaining long-term productivity of Pacific Northwest ecosystems*. Timber Press, Portland OR.

———. 1994. *Forest Ecosystems*. Johns Hopkins University Press, Baltimore MD.

———, and J. E. Loftan. 1977. *Regeneration and early growth on strip clearcuts in lodgepole pine bitterbrush habitat type*. Research Note INT 238, USDA Forest Service Intermountain Forest and Range Experiment Station, Ogden UT.

Peter, D.H. and T.B. Harrington. 2012. Herbicide and logging debris effects on development of plant communities after forest harvesting in the Pacific Northwest. Res. Pap. PNW-RP-589. Portland OR: U.S. Department of Agriculture, Forest Service, Pacific Northwest Research Station. 37 p.

Petersen, T. D. 1988. Effects of interference from *Clamagrostis pubesence* on size distributions in stands of Ponderosa pine. *Journal of Applied Ecology* 25:265–72.

———, M. Newton, and S. M. Zedaker. 1988. Influence of *Ceanothus velutinus* and associated species on the water stress and stem wood production of Douglas-fir. *Forest Science* 34:333–43.

Peterson, C. E., and J. W. Hazard. 1990. Regional variation in growth response of coastal Douglas-fir to nitrogen fertilizer in the Pacific Northwest. *Forest Science* 36:625–40.

Peterson, D. L. 1985. Crown scorch volume and scorch height: Estimates of post fire tree condition. *Canadian Journal of Forest Research* 15:596–98.

———, and M. J. Arbaugh. 1986. Post fire survival of Douglas-fir and lodgepole pine comparing the effects of crown and bole damage. *Canadian Journal of Forest Research* 16:1175–79.

Peterson, E. B., N. M. Peterson, G. F. Weetman, and P. J. Martin. 1997. *Ecology and Management of Sitka Spruce*. University of British Columbia Press, Vancouver, BC.

Peterson, J. A., J. R. Seiler, J. Nowak, S. E. Ginn, and R. E. Kreh. 1997. Growth and physiological responses of young loblolly pine stands to thinning. *Forest Science* 43:529–34.

Peterson, W. C., and D. E. Hibbs. 1989. Adjusting stand density management guides for stands with low stocking potential. *Western Journal of Applied Forestry* 4:62–65.

Piekielek, W. 1975 A black bear population study in northern California. *California Fish and Game* 61:4–25.

Pienaar, L. V., and B. D. Shiver. 1993. Early results from an old-field loblolly pine spacing study in the Georgia Piedmont with competition control. *Southern Journal of Applied Forestry* 17:193–96.

Piene, H. 1978. Effects of increased spacing on carbon mineralization rates and temperature in a stand of young balsam fir. *Canadian Journal of Forest Research* 8:398–406.

Pinno, B.D., V.J. Lieffers, and S.M. Landhäusser. 2012. Inconsistent growth response to Fertilization and Thinning of Lodgepole Pine in the Rocky Mountain Foothills Is Linked to Site Index. International Journal of Forestry Research doi:10.1155/2012/193975.

Pitcher, C. C. 1987. Fire history and age structure in red fir forests in Sequoia National Park, California. *Canadian Journal of Forest Research* 17:582–87.

Poage, N.J. and P.D. Anderson. 2007. Large-scale silviculture experiments of western Oregon and Washington. USDA Forest. Service., PNW Research Station, Portland OR. PNW-GTR-713

———J. C. Tappeiner. 2002. Long-term patterns of diameter and basal area growth of old-growth Douglas-fir trees in western Oregon. *Canadian Journal of Forest Research* 32:1232–43.

———, 2005. Tree species and size structure of old-growth Douglas-fir forests in central western Oregon. *Forest Ecology and Management 204*:329–343.

Powers, R.F. 1979. Response of California true fir to fertilization. Pp. 95–101 in S.P. Gessel, R.M. Kenady, and W. A. Atkinson (Eds.). Proceedings of the Forest Fertilization Conference. September 25–27, 1979, Union, WA. Institute of Forest Resources, College of Forest Resources, University of Washington, Seattle, WA.

———,, 1980. Mineralizable soil nitrogen as an index on nitrogen availability to forest trees. *Soil Science Society of America Journal* 44:1314–20.

———, and G. Jackson. 1978 *Ponderosa pine response to fertilizer: Influence of brush removal and soil type*. USDA Forest Service Research Paper PSW-RP-132, Berkeley CA.

———. and G. T. Ferrell. 1996. Moisture, nutrient, and insect constraints on plantation growth: The "Garden of Eden" study. *New Zealand Journal of Forestry Science* 26:126–44.

———, and P. E. Reynolds. 1999. Ten-year responses of ponderosa pine plantations to repeated vegetation and nutrient control along an environmental gradient. *Canadian Journal of Forest Research* 29:1027–38.

———, D.A. Scott, F. G. Sanchez. R. A. Voldseth, D. Page-Dumroese, J. D. Elliot, and D. M. Stone. 2005. The North American long-term soil productivity experiment: Findings from the first decade of research. *Forest Ecology and Management* 220:31–50.

———,Powers, R.F. 2006. Long-term Soil Productivity: Genesis of the concept and principles behind the program. *Canadian Journal of Forest Research* 36:519–528.

Price, D. T,., T. A. Black, and F. M. Kelliher. 1986. Effects of salal understory removal on photosynthetic rate and stomatal conductance of young Douglas-fir trees. *Canadian Journal of Forest Research* 16:90–97.

Protz, C.G., U. Silins, and V.J. Lieffers. 2000. Reduction in branch sapwood hydraulic permeability as a factor limiting survival of lower branches of lodgepole pine. Canadian. Journalof Forest. Research. 30:1088–1095

Puettmann, K. J., D. S. DeBell, and D. E. Hibbs. 1993a. *Density management guide for red alder*. Res. Contrib. 2. Forest Research Lab,, Oregon State University, Corvallis.

———, D. W. Hann, and D. E. Hibbs. 1993. Evaluation of the size density relationships for pure red alder and Douglas-fir stands. *Forest Science* 39:7–27.

———, and Berger, C.A. 2006 Development of tree and understory vegetation in young Douglas-fir plantations in western Oregon. *Western Journal of Applied Forestry* 21, 94–101.

———, K. D. Coates, and C. Messier. 2009. A Critique of Silviculture, Island Press, Washington,

———, and J. C. Tappeiner. 2013. Multi-scale assessments highlight silvicultural opportunities to increase species diversity and spatial variability in forests. *Forestry* 87:1–10.

Quick, C. R. 1956. Viable seeds from duff and soil of sugar pine forests. *Forest Science* 2:26–42.

———. 1959. Ceanothus seeds and seedlings on burns. *Madroño* 15:75–81.

Radwan, M.A., J.S. Shumway, D.S. DeBell, and J.M. Kraft. 1991. Variance in response of pole-size trees and seedlings of Douglas-fir and western hemlock to nitrogen and phosphorus fertilizers. *Canadian Journal of Forest Research* 21:1431–1438.

Raupach, M. R. 1989. Turbulent transfer in plant canopies. Pp. 41–61 in G. Russell, G. B. Marshall, and P. G. Jarvis, eds., *Plant Canopies: Their Growth, Form, and Function*. Cambridge University Press, Cambridge, UK.

Raymond, C. L., and D. L. Peterson. 2005. Fuel treatments alter the effects of wildfire in a mixed evergreen forest, Oregon, USA. *Canadian Journal of Forest Research* 35:2981–95.

Reader, T. G., and E. A. Kurmes. 1996. The influence of thinning to different stocking levels on compression wood development in ponderosa pine. *Forest Products Journal* 46:92–100.

Reeves, G. H., L. E. Benda, K. M. Burnett, P. A. Bisson, and J. R. Sedell. 1995. A disturbance-based ecosystem approach to maintaining and restoring freshwater habitats of evolutionarily significant anadrodmous salmonids in the Pacific Northwest. *American Fisheries Society Symposium* 17:334–49.

Rehfeldt, G. E. 1974. Local differentiation of populations of Rocky Mountain Douglas-fir. *Canadian Journal of Forest Research* 4:399–406.

———.1989. Ecological adaptations in Douglas-fir (*Pseudotsuga menziesii* var *glauca)*: A synthesis. *Forest Ecology and Management* 28:203–15.

Reineke, L. H. 1933. Perfecting a stand-density index for even-age forests. *Journal of Agriculture Research* 46:627–38.

Reinhardt, P. R., J. K. Brown. W. C. Fischer, and R. T. Graham. 1991. *Woody fuel and duff consumption by prescribed fire in northern Idaho mixed conifer logging slash.* USDA Forest Service Research Paper INT-443, Intermountain Forest and Range Experiment Station, Ogden UT.

Renée, B., F. C. Meinzer, R. Coulombe, and J. Gregg. 2004. Hydraulic redistribution of soil water during summer drought in two contrasting Pacific Northwest coniferous forests. *Tree Physiology* 22:1107–17.

Reukema, D. L. 1979. *Fifty-year development of Douglas-fir stands planted at different densities.* Research Paper PNW-253, USDA Forest Service Pacific Northwest Research Station, Portland OR.

———. 1982. Seedfall in a young Douglas-fir stand: 1950–1978. *Canadian Journal of Forest Research* 12:249–54.

Richards, P. W. 1964. *The Tropical Rain Forest.* Cambridge University Press, London, UK.

Richter, D. D., and D. Markewitz. 2001. *Understanding Soil Change.* Cambridge University Press, Cambridge, UK.

Riegel, G. M., R. F. Miller, and W. C. Krueger. 1992. Competition for resources between understory vegetation and overstory *Pinus ponderosa* in northeastern Oregon. *Ecological Applications* 2:71–85.

———, R. F.. Miller, and W. C. Krueger. 1995. The effects of aboveground and belowground competition on understory species composition in a *Pinus ponderosa* forest. *Forest Science* 41:864–89.

Ritchie, G. A. 1997. Evidence of red:far red signaling and photomorphogenic response in Douglas-fir (*Pseudotsuga menziesii*) seedlings. *Tree Physiology* 17:161–68.

Roberts, S. D., C. A. Harrington, and T. A. Terry. 2005. Harvest residue and competing vegetation affect soil moisture, soil temperature, N availability, and Douglas-fir seedling growth. *Forest Ecology and Management* 205:333–50.

———, J. N Long, and F. W. Smith. 1993. Canopy stratification and leaf area efficiency: A conceptualization. *Forest Ecology & Management* 60:143–156.

Roorbach, A. H. 1999. *The Ecology of Devil's Club* (Oplopanax horridum) *in Western Oregon*. MS thesis, Department of *Forest Science*, Oregon State University, Corvallis.

Rose, R. 1992. Seedling handling and planting. Pp. 328–45 in S. Hobbs et al., eds. *Reforestation practices for southern Oregon and northern California*. Forest Research Lab, Oregon State University, Corvallis.

———Gleason J, Atkinson M, Sabin T. 1991. Grading ponderosa pine seedlings for outplanting according to their root volume. Western Journal of Applied Forestry 6:11–15.

———, and D. L. Haase. 1993. Soil moisture stress induces transplant shock in stored and unstored 2+0 Douglas-fir seedlings of varying root volumes. *Forest Science* 39:275–94.

———, and J. Coate. 2000. Reforestation rules in Oregon. *Journal of Forestry* 98:24–28.

———, and J. S. Ketchum. 2002. Interaction of vegetation control and fertilization on conifer species across the Pacific Northwest. *Canadian Journal of Forest Research* 32:136–52.

———, S. Campbell, and T. D. Landis, eds. 1990. *Target seedling symposium proceedings, combined meeting of the Western Forest Nursery Associations*. General Technical Report GTR-RM-200, USDA Forest Service Rocky Mountain Forest and Range Experiment Station, Fort Collins CO.

———, D. L. Haase, F. Kroiher, and T. Sabin. 1997. Root volume and growth of ponderosa pine and Douglas-fir seedlings: a summary of eight growing seasons. *Western. Journal of Applied Forestry* 12:69–73.

———, J. S. Ketchum, and D. E. Hanson. 1999. Three-year survival and growth of Douglas-fir seedlings under various vegetation-free regimes. *Forest Science* 45:117–26.

Rosner, L. S., and R. Rose. 2006. Synergistic volume response to combinations of vegetation control and seedling size in conifer plantations in Oregon. *Canadian Journal of Forest Research* 36:1–16.

Ross, D. W., W. Scott, R. Henninger, and J. D. Walstad. 1986. Effects of site preparation on ponderosa pine associated vegetation, and soil properties in southwestern Oregon. *Canadian Journal of Forest Research* 16:612–18.

Roth, B. E., and M. Newton. 1996a. Role of lammas growth in recovery of Douglas-fir seedlings from deer browsing, as influenced by weed control, fertilization, and seed source. *Canadian Journal of Forest Research* 26:936–44.

———, and M. Newton. 1996b. Survival and growth of Douglas-fir relating to weeding, fertilization, and seed source. *Western Journal of Applied Forestry* 11:62–69.

Roth, L. F. 2001. Dwarf mistletoe-induced mortality in Northwest ponderosa pine growing stock. *Western Journal of Applied Forestry* 16:136–41.

———, C. G. Shaw III, and L. Rolph. 2000. Inoculum reduction measures to control Armillaria root disease in a severely infested stand of ponderosa pine in south-central Washington. *Western Journal of Applied Forestry* 15:92–100.

Rothermal, R. C. 1983. *How to predict the spread and intensity of wildfires*. General Technical Report GTR-INT-143, USDA Forest Service Intermountain Forest and Range Experiment Station, Ogden UT.

Roy D. F. 1960. *Douglas-fir seed dispersal in northwestern California.* Technical Paper PSW-RP-49, USDA Forest Service, California Forest and Range Experiment Station, Berkeley CA.

Rudnicki, M., V. Lieffers, and U. Silins. 2003. Stand structure governs the crown collisions of lodgepole pine. *Canadian Journal of Forest Research* 33:1238–44.

———, V. Lieffers, and U. Silins. 2004. Crown cover is correlated with relative density, tree slenderness, and tree height in lodgepole pine. *Forest Science* 50:356–63.

Runkle, J. R. 1985. Disturbance regimes in temperate forests. Pp. 17–33 in S. T. A. Pickett and P. S. White, eds., *The Ecology of Natural Disturbance and Patch Dynamics.* Academic Press, New York.

Running, S. W., and C. P. Reid. 1980. Soil temperature influences on root resistance of *Pinus contorta* seedlings. *Plant Physiology* 65:635–40.

Russell, G., P. G. Jarvis, and J. L. Monteith. 1989. Absorption of radiation by canopies and stand growth. In G. Russell, G. B. Marshall, and P. G. Jarvis, eds., *Plant Canopies: Their Growth, Form, and Function.* Cambridge University Press, Cambridge, UK.

Ruth, R. H. 1956. *Plantation survival and growth in two brush-threat areas in coastal Oregon.* Research Paper 17, USDA Forest Service, Pacific Northwest Forest and Range Experiment Station, Portland OR.

———, and A. S. Harris. 1979. *Management of western hemlock-Sitka spruce forests for timber production.* General Technical Report PNW-88, USDA Forest Service, Pacific Northwest Research Station, Portland OR.

Ryan, K. C., and E. D. Reinhart. 1988. Predicting post fire mortality of seven western conifers. *Canadian Journal of Forest Research* 18:1291–97.

———, D. L. Petersen, and E. D. Reinhardt. 1988. Modeling long-term fire-caused mortality of Douglas-fir. *Forest Science* 18:190–99.

———, E. E. Knapp and J. M. Varner.2013. Prescribed fire in North American forests and woodlands: History, current practice and challenges. *Frontiers in Ecology* 11:e15–e24.

Ryan, M.G., D. Binkley, and J.H. Fownes. 1996. Age-related decline in forest productivity: pattern and process. *Advances in Ecological Research* 27:213–262.

———, and B. J. Yoder. 1997. Hydraulic limits to tree height and tree growth. *BioScience* 47:235–42.

———, and W. W. Covington. 1986. *Effect of a prescribed burn in ponderosa pine on inorganic nitrogen concentrations of mineral soil.* Research Note RM-464, USDA Forest Service Rocky Mountain Forest and Range Experiment Station, Fort Collins CO.

Safranyik, L., R. Nevill, and D. Morrison. 1998. *Effects of stand management on forest insects and diseases.* Technology Transfer Note 12, Pacific Forestry Center, Canadian Forest Service, Vancouver BC, Canada.

Salman, K., and J. W. Bongberg. 1942. Logging high-risk trees to control insects in pine stands of northeastern California. *Journal of Forestry* 40:533–39.

Samuelson, L. J. 1998. Influence of intensive culture on leaf net photosynthesis and growth of sweetgum and loblolly pine seedlings. *Forest Science* 44:308–16.

Sands, R., and R. L. Correll. 1976. Water potential and leaf elongation in radiata pine and wheat. *Plant Physiology* 37:293–97.

Sassaman, R.W., J. W. Barrett, and A. D. Twombly 1977. *Financial precommercial thinning guides for Northwest Ponderosa pine stands.* Paper PNW-226, USDA Forest Service Pacific Northwest Forest and Range Experiment Station Research, Portland OR.

Sartwell, C., and R. E. Stevens. 1975. Mountain pine beetle—prospects for silvicultural control in second-growth stands. *Journal of Forestry* 73:136–40.

Satterlund, D. R., and P. W. Adams. 1992. *Wildland Watershed Management.* John Wiley & Sons, New York.

Savage, M., and T. W. Swetnam. 1990. Early nineteenth-century fire decline following sheep pasturing in a Navajo ponderosa pine forest. *Ecology* 71:2374–78.

Schaap, W., and D. DeYoe. 1986. *Seedling protectors for preventing deer browse.* Research Bulletin 54, Forest Research Laboratory, Oregon State University, Corvallis.

Scharpf, R. F. 1993. *Diseases of Pacific coast conifers.* USDA Agriculture Handbook Washington DC.

———, and J. R. Parmeter, Jr. 1976. *Population buildup and vertical spread of dwarf mistletoe on young red and white firs in California.* Research Paper PSW-122, USDA Forest Service Pacific Southwest Forest and Range Experiment Station, Berkeley CA.

———, and R. V. Bega. 1981. *Elytroderma disease reduces growth and vigor and increases mortality of Jeffrey pines at Lake Tahoe Basin.* Research Paper PSW-155, USDA Forest Service Pacific Southwest Forest and Range Experiment Station, Berkeley CA.

Schearer, R. C., and W. C. Schmidt. 1970. *Natural regeneration of ponderosa pine in western Montana.* Research Paper INT-86, USDA Forest Service Intermountain Forest and Range Experiment Station, Ogden UT.

Schlesinger, W.H. 1991. *Biogeochemistry: An analysis of global change.* Academic Press, San Diego CA. 443 p.

Schoettle, A.W. 1990. The interaction between leaf longevity and shoot growth and foliar biomass per shoot in *Pinus contorta* at two elevations. Tree Physiology 7:209–214.

———, and Smith, W.K. 1991. Interrelation between shoot characteristics and solar irradiance in the crown of Pinus contorta ssp. latifolia. *Tree Physiology* 9:245–254.

Scott, J. H. and R. E. Burgan. 2005. *Standard fire behavior fuel models: a comprehensive set for use with Rothermel's surface fire spread model.* General Technical Report RMRS-GTR-153, USDA Forest Service Rocky Mountain Research Station, Fort Collins, CO.

Schopmeyer, C. S. (tech. coord.). 1974. *Seeds of woody plants in the United States.* Agriculture Handbook 450, U.S. Department of Agriculture, Washington DC.

Schmid, J. M., S. A. Mata, and R. A. Obedzinski. 1994. *Hazard rating ponderosa pine stands for mountain pine beetles in the Black Hills.* Research Note RM-529, USDA Forest Service Rocky Mountain Forest and Range Experiment Station, Fort Collins CO.

Schmidt, and K. Seidel 1988. Western larch and spruce thinning to optimize growth. In Schmidt, W., ed. *Future forests of the mountain west: a stand culture symposium.*, General Technical Report INT-243, USDA Forest Service, Intermountain Research Station Ogden UT.

Schoenholz, S. H., H. Van Miegroet, and J. A. Burger. 2000. A review of chemical and physical properties as indicators of forest soil quality: Challenges and opportunities. *Forest Ecology and Management* 138:335–56.

Schoonmaker, P., and A. McKee. 1988. Species composition and diversity during secondary succession of coniferous forests in western Cascade mountains of Oregon. *Forest Science* 34:960–79.

Schreuder, H. T., and W. T. Swank. 1974. Coniferous stands characterized with the Weibull distribution. *Canadian Journal of Forest Research* 4:518–23.

Schwilk, D. W., J. E. Keeley, E. E. Knapp, J. McIver, J. D. Bailey, C. J. Fettig, C.E. Fiedler, R. J. Harrod, J. J. Moghaddas, K. W. Outcalt, C. N. Skinner, S. L. Stephens, T. A. Waldrop, D. A. Yaussy and A. Youngblood. 2009. The national Fire and Fire Surrogate study: effects of fuel reduction methods on forest vegetation structure and fuels. *Ecological Applications* 19:285–304.

Schumacher, F. X. 1928. *Yield stand and volume tables for red fir in California.* Station Bulletin 456, Agricultural Experiment Station, University of California, Berkeley.

Scott, W., R. Meade, R. Leon, D. Hyink, and R. Miller. 1998. Planting density and tree size relations in coastal Douglas-fir. *Canadian Journal of Forest Research* 28:74–78.

Sean, C. T., C. Halpern, D. A. Falk, D. A. Liguori, and K. A. Austin. 1999. Plant diversity in managed forests: responses to thinning and fertilization. *Ecological Application* 9:864–79.

Seidel, K.W. 1979. *Regeneration in mixed conifer clearcuts in the Cascade Range and Blue Mountains of eastern Oregon.* Research Paper PNW 248, USDA Forest Service Pacific Northwest Forest and Range Experiment Station, Portland OR.

———. 1983. *Regeneration in mixed conifer and Douglas-fir shelterwood cuttings in the Cascade Range of Washington.* Research Paper PNW-RP-314. USDA Forest Service Pacific Northwest Forest and Range Experiment Station, Portland OR.

———, and Cochran. 1981. *Silviculture of mixed-conifer forests in eastern Oregon and Washington.* General Technical Report PNW-121, USDA Forest Service, Pacific Northwest Forest and Range Experiment Station Portland OR.

———, and S. C. Head. 1983. *Regeneration in mixed conifer partial cuttings in the Blue Mountains of Oregon and Washington.* Research Paper PNW-310, USDA Forest Service Pacific Northwest Forest and Range Experiment Station, Portland OR.

Seiwa, K., and K. Kikuzawa. 1996. Importance of seed size for the establishment of seedlings of five deciduous broad-leaved tree species. *Vegetatio* 123:51–64.

Selter, C. M., W. D. Pitts, and M. G. Barbour. 1986. Site microenvironment and seedling survival of Shasta red fir. *American Midland Naturalist* 115:288–300.

Sensenig, T., J. D. Bailey, J. C. Tappeiner 2013. Stand development, fire and growth of old-growth and young forests in southwestern Oregon, USA. Forest Ecology and Management 291 (2013) 96–109.

———, 2002. *Development, Fire History, and Current and Past Growth Rates of Old-growth and Young-growth Forest Stands in the Cascade, Siskiyou, and Mid-coast Mountains of Southwestern Oregon.* Ph.D. dissertation, Department of Forest Science, Oregon State University, Corvallis.

Senyk, J. P., and R. B. Smith. 1989. Forestry practices and soil degradation in B.C. *Canadian Forest Industries* 109(3):36–43.

Seymour, R. S., and M. L. Hunter. 1999. Pp. 22–61 in M. L. Hunter, ed., *Principles of Forest Ecology*. Cambridge University Press, Cambridge, UK.

Shafii, B., J.A. Moore, and J.R. Olson. 1989. Effects of nitrogen fertilization on growth of grand fir and Douglas-fir stands in northern Idaho. *Western Journal of Applied Forestry* 4:54–57.

Shainsky, L. J., and S. R. Radosevich. 1986. Growth and water relationships of *Pinus ponderosa* seedlings in competitive regimes with *Arctostaphylos patula* seedlings. *Journal of Applied Ecology* 23:957–66.

Sharrow, S. H., W. C. Leininger, and B. D. Rhodes. 1989. Sheep grazing as a silvicultural tool to suppress brush. *Journal of Range Management* 42:2–4.

———, W. C. Leininger, and K. A. Osman. 1992. Sheep grazing effects on a coastal Douglas-fir forest: A ten-year perspective. *Forest Ecology and Management* 50:75–84.

Shatford, J. P. A., J. D. Bailey, and J. C. Tappeiner. 2009. Understory tree development with repeated with repeated stand density treatments in coastal Douglas-fir forests. *Western Journal of Applied Forestry* 24:11–16.

Shaw, J. D. 2000. Application of stand density index to irregularly structured stands. *Western Journal of Applied Forestry* 15:40–42.

Shearer, R. C., and W. C. Schmidt. 1971. Ponderosa pine cone and seed losses. *Journal of Forestry* 69:370–72.

———, and ———. *Natural regeneration in ponderosa pine in forests of western Montana.* Research Paper INT-86, USDA Forest Service Forest and Range Experiment Station, Ogden, UT.

Shindler, B., A. M. Brunson, and G. H. Stankey. 2002. *Social acceptability of forest conditions and management practices: a problem analysis.* General Technical Report PNW-GTR-537, USDA Forest Service Pacific Northwest Research Station, Portland OR.

Shinozaki, K., K. Yoda, K. Hozumi, and T. J. Kira. 1964a. A quantitative analysis of plant form—the pipe model theory: I. Basic analyses. *Japanese Journal of Ecology* 14:97–105.

Shinozaki, K., K. Yoda, K. Hozumi, and T. Kira.. 1964. A quantitative analysis of plant form—the pipe model theory: II. Further evidence of the theory and its application in forest ecology. *Japanese Journal of Ecology* 14:133–139.

Silen, R. R., and J. G. Wheat. 1979. Progressive tree improvement program in coastal Douglas-fir. *Journal of Forestry* 77:78–83.

Simard, S. W., D. A. Perry, M. D. Jones, D. D. Myrold, D. M. Durall, and R. Molina. 1997. Net transfer of carbon between ectomychorrizal tree species in the field. *Nature* (London), 338:579–82.

Slaughter, G. C., J. R. Parmeter, Jr., and J. T. Kliejunas. 1991. Survival of saplings and pole-sized conifers near true fir stumps with annosus root disease in northern California. *Western Journal of Applied Forestry* 6:102–5.

———, and J. R. Parmeter Jr. 1989. Annosus root disease in true firs in northern and central California forests. In J. Otrosina and R. F. Scharpf, eds. *Proceedings of the symposium on research and management of annosus root disease in western North America.* USDA Forest Service General Technical Report PSW-116, Albany CA.

Slesak, R.A., T.B. Harrington, and S.H. Schoenholtz. 2010. Soil and Douglas-fir (Pseudotsuga menziesii) foliar nitrogen responses to variable logging-debris retention and competing vegetation control in the Pacific Northwest. *Canadian Journal of Forest Research* 40:254–264.

Sloan, J. P., L. H. Jump, and R. A. Ryker. 1987. Container-grown ponderosa pine seedlings outperform bare root seedlings on harsh sites in southern Utah. Research Paper Int 384, USDA Forest Service, Intermountain Forest and Range Experiment Station Ogden UT.

Smith, D. M. 1970. *Applied ecology and the new forest: Proceeding of the Western Reforestation Coordinating Committee.* Western Forestry and Conservation Association, Portland OR.

———, B. C. Larson, M. J. Kelty, and P. M. S. Ashton. 1997. *The Practice of Silviculture.* John Wiley & Sons, New York.

Smith, F. W., and J. N. Long. 1987. Elk hiding and thermal cover guidelines in the context of lodgepole pine stand density. *Western Journal of Applied Forestry* 2:6–10.

Smith, G. W. 1982. Habitat use by porcupines in a ponderosa pine/Douglas-fir forest in northeastern Oregon. *Northwest Science* 56:236–40.

Smith, J. H. G. 1980. Influences of spacing on radial growth and percentage latewood of Douglas-fir, western hemlock, and western red cedar. *Canadian Journal of Forest Research* 10:169–75.

Smith, N. J. 1989. A stand-density control diagram for western redcedar, *Thuja plicata*. *Forest Ecology and Management* 27:235–44.

Smithwick, E. A., M. E. Harmon, S. M. Remillard, S. A. Acker, and J. F. Franklin. 2002. Potential upper bounds of carbon stores of forests in the Pacific Northwest. *Ecological Applications* 12:1303–17.

Sollins, P. C., C. Grier, F. M. McCorrson, K. Cromack Jr, R. Fogel, and R. L. Fredriksen. 1980. The internal element cycles of an old-growth Douglas-fir ecosystem in western Oregon. *Ecological Monographs* 50:261–85.

Son, Y, W. Lee, S.E. Lee, and S. R. Ryu. 1999. Effects of thinning on soil nitrogen mineralization in a Japanese larch plantation. *Communications in Soil Science and Plant Analysis* 30:2539–2550.

Spies, T. A., and J. F. Franklin. 1991. The structure of natural young and old-growth forests in Oregon and Washington. Pp. 91–109 in L. F. Ruggiero, K. B. Aubrey, A. B. Carey, and M. H. Huff, tech. coords., *Wildlife and vegetation of unmanaged Douglas-fir forests.* General Technical Report PNW-GTR-285, USDA Forest Service Pacific Northwest Research Station, Portland OR.

Spittlehouse, D. L., R. S. Adams, and R. D. Winkler. 2004. *Forest edge and opening microclimate at Sicamous Creek, B.C.* Research Paper 24:43, Ministry of Forestry Research Branch, Victoria BC, Canada.

Spurr, S. 1956. German silvicultural systems. *Forest Science* 2:75–80.

St. Clair, J. B., N. L. Mandel, K. J. S. Jayawickrama. 2004. Early realized genetic gains for Coastal Douglas-fir in the northern Oregon Cascades. *Western Journal of Applied Forestry* 19:195–201.

Staebler, G. R. 1960. Theoretical derivation of numerical thinning schedules for Douglas-fir. *Forest Science* 2:98–109.

———. 1963. Growth along the stem of full-crowned Douglas-fir trees after pruning to different heights. *Journal of Forestry* 61:124–27.

Stark, N. M. 1963. *Thirty-year summary of climatological measurements from the central Sierra Nevada.* Research Note PSW-RN-36, USDA Forest Service, Pacific Northwest Forest and Range Experiment Station, Portland OR.

———. 1977. Fire and nutrient cycle in a Douglas-fir/larch forest. *Ecology* 58:16–30.

Stein, W. I. 1955. Pruning to different heights in young Douglas-fir. *Journal of Forestry* 52:352–55.

———. 1984. *Regeneration outlook on BLM land in the southern Oregon Cascades.* Research Paper PNW-RP-284, USDA Forest Service Pacific Northwest Forest and Range Experiment Station, Portland OR.

———. 1992. Regeneration surveys and evaluation. Pp. 346–82 in S. D. Hobbs, S. D. Tesch, P. W. Owston, R. E. Stewart, J. C. Tappeiner II, and G. E. Wells, eds., *Reforestation practices in southwestern Oregon and northern California.* Forest Research Laboratory, Oregon State University, Corvallis.

———. 1995. *Ten-year development of Douglas-fir and associated vegetation after different site preparation on coast range clearcuts.* Research Paper Res. Pap. PNW-RP-473. USDA Forest Service Pacific Northwest Research Station, Portland OR.

———. 1997. Ten-year survival and growth of planted Douglas-fir and western redcedar after seven site preparation treatments. *Western Journal of Applied Forestry* 12:74–80.

Stegemoeller, K.A. and H.N. Chappell. 1990. Growth response of unthinned and thinned Douglas-fir stand to single and multiple applications of nitrogen. *Canadian Journal of Forest Research* 20:343–349.

———. and H.N. Chappell. 1991. Effects of fertilization and thinning on 8-year growth responses of second-growth Douglas-fir stands. *Canadian Journal of Forest Research* 21:516–521.

Stephens, S. L. 1998. Evaluation of the effects of silvicultural and fuels treatments on potential fire behavior in the Sierra Nevada mixed-conifer forests. *Forest Ecology and Management* 105:21–35.

———. and M. A. Finney. 2002. Prescribed fire mortality of mixed conifer tree species: effects of crown damage and forest floor consumption. *Forest Ecology and Management* 162:261–71.

———, J. K. Agee, P. Z. Fulé, M. P. North, W. H. Romme, T. W. Swetnam and M. G. Turner. 2013. Managing forests and fire in changing climates. *Science* 342: 41–42.

———, N. Burrows, A. Buyantuyev, R. W. Gray, R. E. Keane, R. Kubian, S. Liu, F. Seijo, L. Shu, K. G. Tolhurst and J. W. van Wagtendonk. 2014. Temperate and boreal forest mega-fires: Characteristics and challenges. *Frontiers in Ecology* 12(2):115–122.

Sterba, H., and R. A. Monserud. 1993. The maximum density concept applied to uneven-aged, mixed species stands. *Forest Science* 39:432–52.

Stickney, P. F. and R. B. Campbell, Jr. 2000. *Data base for early post-fire succession in Northern Rocky Mountain forests.* General Technical Report GTR-61CD, Ogden, U.S. Department of Agriculture, Forest Service, Rocky Mountain Research Station. UT.

Stiell, W. M. 1982. Growth of clumped versus equally spaced trees. *Forestry Chronicle* 58:23–25.

Stoszek, K. J. 1973. Damage to ponderosa pine plantations by the western pine-shoot borer. *Journal of Forestry* 71:701–5.

Strader, R. H., and D. Binkley. 1989. Mineralization and immobilization of soil nitrogen in two Douglas-fir stands 15 and 22 years after nitrogen fertilization. *Canadian Journal of Forest Research* 19:798–801.

Strand, R.F. 1970. The effect of thinning on soil temperature, soil moisture, and root distribution of Douglas-fir. P. 295–304 in Youngberg, C.T. and Davey, C.B. (eds.), Tree growth and forest foils. *Proceedings of the Third North American Forest Soils Conference,* North Carolina State University, Raleigh. August 1968, Oregon State University Press, Corvallis.

Strom, B. A., and P. Z. Fulé. 2007. Pre-wildfire fuel treatments affect long-term ponderosa pine forest dynamics. *International Journal of Wildland Fire* 16:128–138.

Strothman, R. O., and D. F. Roy. 1984. *Regeneration of Douglas-fir in the Klamath Mountain region, California, Oregon.* General Technical Report PSW-GTR81, USDA Forest Service Pacific Southwest Forest and Range Experiment Station, Berkeley CA.

Stuart, J. D. 1984. Hazard rating of lodgepole pine stands to mountain pine beetle outbreaks in south central Oregon. *Canadian Journal of Forest Research* 14:666–671.

———, J. K. Agee, and R. I. Gara. 1989. Lodgepole pine regeneration in an old-growth self-perpetuating forest in south central Oregon. *Canadian Journal of Forest Research* 19:1096–1104.

Sucoff, E. C., and S. G. Hong. 1974. Effects of thinning on needle water potential in red pine. *Forest Science* 20:25–29.

Sullivan, K. F., G. M. Filip, J. V. Arean, S. A. Fitzgerald, and S. D. Tesch. 2001. Incidence of infection caused by *Heterobasidium annosum* in managed noble fir on the Warm Springs Indian Reservation, Oregon. *Western Journal of Applied Forestry* 16:106–13.

Sullivan, T. P. 1979. The use of alternative foods to reduce conifer seed predation by the deer mouse. *Journal of Applied Ecology* 16:475–95.

———, and D. S. Sullivan. 1982. Population dynamics and regulation of the Douglas squirrel with supplemental food. *Oecologia* 53:264–70.

———, and A Vyse. 1987. Impact of red squirrel feeding damage on juvenile spacing of lodgepole pine in the Cariboo region of British Columbia. *Canadian Journal of Forest Research* 17:166–674.

———, W.T. Jackson, J. Pojar, and A. Banner. 1986. Impact of feeding damage by porcupine on western hemlock-Sitka spruce forests of north coastal British Columbia. *Canadian Journal of Forest Research* 16:642–47.

Sundahl, W. E. 1971. Seedfall in young-growth ponderosa pine. *Journal of Forestry* 69:790–92.

Swanson, F. J., and J. F. Franklin. 1992. New forestry principles from ecosystem analysis of Pacific Northwest forests. *Ecological Applications* 2:262–74.

Swezy, D. M., and J. K. Agee. 1991. Prescribed fire effects on fine root and tree mortality in old-growth ponderosa pine. *Canadian Journal of Forest Research* 21:626–34.

Sword, M.A., A.E. Tiarks, J.D. Haywood. 1998. Establishment treatments affect the relationships among nutrition, productivity and competing vegetation of loblolly pine saplings on a Gulf Coastal Plain site. *Forest Ecology and Management* 105_175–188.

Tadaki, Y. 1966. Some discussion of the leaf biomass of forest stands. *Bulletin of the Government Forest Experiment Station* (Tokyo) 184:135–61.

Tan, X., S.X. Chang, P. G. Comeau, and Y. Wang. 2008. Thinning effects on microbial biomass, N mineralization, and tree growth in a mid-rotation fire-origin lodgepole pine stand in the lower foothills of Alberta, Canada. *Forest Science* 54:465–474.

Tappeiner, J. C., II 1967. *Natural regeneration of Douglas-fir and white fir on Blodgett Forest in the Sierra Nevada of California*. School of Forestry, University of California, Berkeley CA.

———. 1971. Invasion and development of beaked hazel in red pine stands in northern Minnesota. *Ecology* 52:514–19.

———.1979. Effect of fire and 2,4–D on the early stages of beaked hazel understories. *Weed Science* 27:162–66.

———, and J. A. Helms. 1971. Natural regeneration of Douglas-fir and white fir on exposed sites in the Sierra Nevada of California. *American Midland Naturalist* 86:358–70.

———, and H. H. John. 1973. Biomass and nutrient content of hazel undergrowth. *Ecology* 54:1342–48.

———, and A. A. Alm. 1975. Undergrowth vegetation effects on the nutrient content of litterfall and soils in red pine and birch stands in northern Minnesota. *Ecology* 56:1193–1200.

———, and S. R. Radosevich. 1982. Effect of bearmat (*Chamaebatia foliosa*) on soil moisture and ponderosa pine (*Pinus ponderosa*) growth. *Weed Science* 30:98–101.

———, and P. M. McDonald. 1984. Development of tanoak understories in conifer stands. *Canadian Journal of Forest Research* 14:271–77.

———, and P. A. Alaback. 1989. Early establishment and vegetative growth of understory species in the western hemlock-Sitka spruce forests of southeast Alaska. *Canadian Journal of Botany* 67:318–26.

———, and J. C. Zasada. 1993. Establishment of salmonberry, salal, vine maple and bigleaf maple seedlings. *Canadian Journal of Forest Research* 23:1175–80.

———, J. Bell, and D. Brodie. 1982. *Response of young Douglas-fir to 16 years of intensive thinning.* Research Bulletin 38, Forest Research Laboratory, Oregon State University, Corvallis.

———, T. B. Harrington, and J. D. Walstad. 1984. Predicting recovery of tanoak and Pacific madrone after cutting or burning. *Weed Science* 32:413–17.

———, P. M. McDonald, and T. F. Hughes. 1986. Survival of tanoak and Pacific madrone seedlings in forests of southwestern Oregon. *New Forests* 1:43–55.

———, R. Pabst, and M. Cloughsey. 1987a. Stem treatments to prevent tanoak sprouting. *Western Journal of Applied Forestry* 2:41–45.

———, T. F. Hughes, and S. D. Tesch. 1987b. Bud production of Douglas-fir seedlings: Response to shrub and hardwood competition. *Canadian Journal of Forest Research* 17:1300–1304.

———, J. C. Zasada, P. Ryan, and M. Newton. 1991. Salmonberry clonal and population structure: The basis for persistent cover. *Ecology* 72:609–18.

———, M. Newton, P. M. McDonald, and T. B. Harrington. 1992. Ecology of hardwoods, shrubs, and herbaceous vegetation: Effects on conifer regeneration. Pp. 137–64 in S. Hobbs et al., eds., *Reforestation practices in southwestern Oregon and northern California.* Forest Research Lab, Oregon State University, Corvallis.

———, J. Zasada, D. W. Huffman, and B. Maxwell. 1996. Effects of cutting time, stump height, parent tree characteristics and harvest variables on bigleaf maple sprout clumps. *Western Journal of Applied Forestry* 11:120–24.

———, D. Huffman, D. Marshall, T. A. Spies, and J. D. Bailey. 1997. Density, ages and growth rates in old-growth and young-growth forests in coastal Oregon. *Canadian Journal of Forest Research* 27:638–48.

———, J. C. Zasada, D. W. Huffman, and L. M. Ganio. 2001. Salmonberry and salal annual aerial stem production: The maintenance of shrub cover in forest stands. *Canadian Journal of Forest Research* 31:1629–38.

Taylor, R. F. 1939. The application of a tree classification in marking lodgepole pine for selection cutting. *Journal of Forestry* 37:77–82.

Teensma, P. D. A. 1987. *Fire History and Fire Regimes of the Central Western Cascades of Oregon*. Ph.D. dissertation, University of Oregon, Eugene.

Teklehamainot, Z., P. G. Jarvis, and D. C. Ledger. 1991. Rainfall interception and boundary layer conductance in relation to tree spacing. *Journal of Hydrology* 123:261–78.

Tesch S. D., and S. D. Hobbs. 1989. Impact of shrub sprout development on Douglas-fir seedling development. *Western Journal of Applied Forestry* 4:89–92.

———, and J. W. Mann. 1991. *Clearcut and shelterwood reproduction methods for regenerating southwest Oregon forests*. Bulletin 72, Forest Research Laboratory, Oregon State University, Corvallis.

———, and J. Helms. 1992. Regeneration methods. Pp. 166–201 in S. Hobbs et al., eds., *Reforestation practices in southwestern Oregon and northern California*. Forest Research Lab, Oregon State University, Corvallis.

———, and E. J. Korpela. 1993. Douglas-fir and white fir advanced regeneration for renewal of mixed conifer forests. *Canadian Journal of Forest Research* 23:1427–37.

———, K. B. Katz, and E. J. Korpela. 1993a. Recovery of Douglas-fir seedlings and saplings wounded during overstory removal. *Canadian Journal of Forest Research* 23:1684–94.

———, E. J. Korpela, S. D. Hobbs. 1993b. Effects of sclerophyllous shrub competition on root and shoot development and biomass partitioning of Douglas-fir seedlings. *Canadian Journal of Forest Research* 23:1415–26.

Teskey, R. O., T. M. Hinckley, and C. C. Grier. 1984. Temperature-induced change in the water relations of *Abies amabilis* (Dougl.) Forbes. *Plant Physiology* 74:77–80.

Tevis, L. P., Jr. 1956. Responses of small mammal populations to logging of Douglas-fir. *Journal of Mammalogy* 37:189–96.

Thies, W. G., and R. N. Sturrock. 1995. *Laminated root rot in western North America*. Resource Bulletin PNW-GTR-349, USDA Forest Service Pacific Northwest Research Station, Portland OR.

———, and E. E. Nelson. 1997. Laminated root rot: New considerations. *Western Journal of Applied Forestry* 12:49–51.

———, R.G. Kelsey, D. J. Westlind, and J. Madsen. 2006. Potassium fertilizer applied immediately after planting has no impact on Douglas-fir seedling mortality caused by laminated root rot on a forested site in Washington State. *Forest Ecology and Management* 229:195–201.

Thomas, S. C., C. B. Halpern, D. A. Falk, D. A. Liguori, and K. A. Austin. 1999. Plant diversity in managed forests: Understory responses to thinning and fertilization. *Ecological Applications* 9:864–79.

Thomas, T. L., and J. K. Agee. 1986. Prescribed fire effects on a mixed conifer forest at Crater Lake, Oregon. *Canadian Journal of Forest Research*. 16:1082–87.

Thompson, W. A., and A. M. Wheeler. 1992. Photosynthesis by mature needles of field-grown *Pinus radiata. Forest Ecology and Management* 52:225–42.

Thomson, A.J. and H.J. Barclay. 1984. Effects of thinning and urea fertilization on the distribution of area increment along the boles of Douglas-fir at Shawnigan Lake, British Columbia. *Canadian Journal of Forest Research* 14:879–884.

Timmer, V. R. 1997. Exponential nutrient loading: A new fertilization technique to improve seedling performance on competitive sites. *New Forests* 13:279–99.

Tinnin, R. O, C. G. Parks, and D. M. Knutson. 1999. Effects of Douglas-fir dwarf mistletoe on trees in thinned stands in the Pacific Northwest. *Forest Science* 45:359–65.

Toumey, J. W., and C. F. Korstian. 1937. *Foundations of Silviculture upon an Ecological Basis* (2nd ed.). John Wiley & Sons, New York.

Troendle, C. A., and R. M. King. 1987. The effect of partial cutting and clearcutting on stream flow at Deadhorse Creek, Colorado. *Journal of Hydrology* 90:145–57.

Trofymow, J. A., H.J. Barclay, K.M. McCullough 1991. Annual rates and elemental concentrations of litter fall in thinned and fertilized Douglas-fir. *Canadian Journal of Forest Research 21:1601–1615.*

Troup, R. S. 1928. *Silvicultural Systems.* Oxford University Press, London, UK.

Tu, M., C. Hurd, and J.M. Randal. 2001. *Weed control methods handbook: tools and techniques for use in natural areas.* The Nature Conservancy. http://www.invasive.org/gist/products/handbook/methods-handbook.pdf (accessed 16 December 2013).

Turner, J., M. J. Lambert, and S. P. Gessel. 1977. Use of foliage sulphur concentrations to predict response to urea fertilization by Douglas-fir. *Canadian Journal of Forest Research* 7:476–80.

———, and S.P. Gessel. 1979. Sulfur requirements of nitrogen fertilized Douglas-fir. *Forest Science* 25:461–467.

———, M.J. Lambert. 1986. Nutrition and nutritional relationships of *Pinus radiata.* Annual Review of Ecology and Systematics 17:325–350.

Valinger, E., and N. Pettersson. 1996. Wind and snow damage in a thinning and fertilization experiment in *Picea abies* in southern Sweden. *Forestry* 69:25–33.

Van Keulen, H., J. Goudrian, and N. G. Seligman. 1989. Modeling the effects of nitrogen on canopy development and crop growth. In G. Russell, B. Marshall, and P. G. Jarvis, eds., *Plant Canopies: Their Growth, Form, and Function.* Cambridge University Press, Cambridge, UK.

Van Wagner, C. E. 1973. Height of crown scorch in forest fires. *Canadian Journal of Forest Research* 3:373–78.

———. 1977. Conditions for the start and spread of crown fire. *Canadian Journal of Forest Research* 7:23–24.

Vander Wall, S. B. 1992. Establishment of Jeffrey pine seedlings from animal caches. *Western Journal of Applied Forestry* 7:14–20.

———.1994. Seed fate pathways of antelope bitterbrush: Dispersal by seed-caching yellow pine chipmunks. *Ecology* 1911–26.

———. 1995. Dynamics of yellow pine chipmunk (*Tamias amoenus*) seed caches: Underground traffic in bitterbrush seeds. *Ecoscience* 2:261–66.

VanderSchaaf, C. L., J. A. Moore and J. L. Kingery. 2000. The effect of multi-nutrient fertilization on understory plant diversity. *Northwest Science* 74:316–314.

Van Pelt, R. 2008. Identifying old trees and forests in eastern Washington. Washington State Department of Natural Resources, Olympia WA.

Verdyla, D., and R. F. Fischer. 1989. Ponderosa pine habitat types as indicators of site quality in the Dixie National Forest, Utah. *Western Journal of Applied Forestry* 4:52–54.

Vincent, A. B. 1961. Is height/age a reliable index of site? *Forestry Chronicle* 37:144–49.

Vose, J.M. and H.L. Allen. 1988. Leaf area, stemwood growth, and nutrition relationships in loblolly pine. *Forest Science* 34:547–563

Wagg, J. W. B., and R. K. Hermann. 1962. *Artificial seeding of pine in central Oregon.* Research Note 47, Oregon State University Forest Research Lab, Corvallis.

Wagle, R. F., and T. W. Eakle. 1979. A controlled burn reduces the impact of a subsequent wildfire in a ponderosa pine vegetation type. *Forest Science* 25:123–29.

Wagner, R. G., and S. R. Radosevich. 1991a. Interspecific competition and other factors influencing the performance of Douglas-fir saplings. *Canadian Journal of Forest Research* 21:829–35.

———.1991b. Neighborhood predictors of interspecific competition in young Douglas-fir plantations. *Canadian Journal of Forest Research* 21:821–28.

———, and J. C. Zasada. 1991. Integrating plant autecology and silvicultural activities to prevent forest vegetation management problems. *Forestry Chronicle* 67:506–13.

———, and M. W. Rogozynski. 1994. Controlling sprout clumps of bigleaf maple with herbicides and manual cutting. *Western Journal of Applied Forestry* 9:118–124.

Walker, R. B. and S. B. Gessel. 1991. Mineral deficiencies of coastal Northwest conifer. College of Forest Resources, University of Washington, Seattle WA, USA. Institute of Forest Resources, Contribution No. 70,63p.

Walker, R. F. 1999. Artificial regeneration of Jeffrey pine in the Sierra Nevada: Growth, nutrition, and water relations as influenced by controlled release fertilization and solar protection. *Journal of Sustainable Forestry* 9:23–38.

———. 2002. Response of Jeffrey pine on a surface mine site to fertilizer and lime. *Restoration Ecology* 10:204–12.

Walstad, J. D., and F. Dost. 1984. *The health risk of herbicides in forestry: A review of the scientific record.* Special Publication No. 10, Forest Research Lab, Oregon State University, Corvallis OR.

———, and P. J. Kuch. 1987. *Forest Vegetation Management for Conifer Production.* John Wiley & Sons, New York.

Weidman, P. H. 1939.. Evidences of racial influence in a 25-year test of ponderosa pine. *Journal of Agriculture Research* 59: 885–87.

———, and K. W. Seidel. 1990. Use and benefits of prescribed fire in reforestation. Pp. 67–80 in J. D. Walstad, S. R. Radosevich, and D. V. Sandberg, eds., *Natural and Prescribed Fire in Pacific Northwest Forests.* Oregon State University Press, Corvallis.

———, S. R. Radosevich, and D. V. Sandberg, eds. 1990. *Natural and Prescribed Fire in Pacific Northwest Forests.* Oregon State University Press, Corvallis.

Wang, G., and Kemball, K. J., 2005. Effects of fire severity on early development of understory vegetation. *Canadian Journal of Forest Research* 35:254–62.

———, G. H. Qian, and K. Klinka. 1994. Growth of *Thuja plicata* seedlings along a light gradient. *Canadian Journal of Botany* 72:1749–57.

Wang, Z. Q., M. Newton, and J. C. Tappeiner. 1995. Competitive relations between Douglas-fir and Pacific madrone on shallow soils in a Mediterranean climate. *Forest Science* 41:744–57.

Waring, R. H. 1983. Estimating forest growth and efficiency in relation to canopy leaf area. Pp. 327–54 in A. McFadyen and E. D. Ford, eds, *Advances in Ecological Research.* Academic Press, New York.

———, and J. F. Franklin. 1979a. Distinctive features of the northwestern coniferous forest: Development, structure, and function. In R. H. Waring, ed., *Forests: Fresh Perspectives from Ecosystem Analysis.* Proceedings of the 40th Annual Biology Colloquium. Oregon State University Press, Corvallis.

———. 1979b. Evergreen coniferous forests of the Pacific Northwest. *Science* 204:1380–86.

———, and G. B. Pitman. 1980. *A simple model of host resistance to bark beetles.* Research Note 65, Forest Research Laboratory, Oregon State University, Corvallis.

———, and J. D. Schlesinger. 1985. *Forest Ecosystems—Concepts and Management.* Academic Press, New York.

———, W. G. Thies, and D. Muscato. 1980. Stem growth per unit of leaf area: A measure of tree vigor. *Forest Science* 26:112–17.

———, K. Newman, and J. Bell. 1981. Efficiency of tree crowns and stemwood production at different canopy leaf densities. *Forestry* 54:129–37.

———, J. J. Landsberg, and M. Williams. 1998. Net primary production of forests. *Tree Physiology* 18:129–34.

Watkins, C., ed. 1998. *European Woods and Forests: Studies in Cultural History.* CAB International, New York.

Weatherspoon, C. P., and C. N. Skinner. 1995. An assessment of factors associated with damage to tree crowns from 1987 wildfires in northern California. *Forest Science* 41:430–51.

———, and C. N. Skinner.1996. Landscape-level strategies for forest fuel management. *Sierra Nevada Ecosystem Project: Final report to Congress, Vol. II Assessments and scientific basis for management.* Center for Water and Wildland Resources, Davis CA.

Weaver, H. 1959. Ecological changes in the ponderosa pine forest of the Warm Springs

Indian Reservation. *Journal of Forestry* 57:15–20.

———. 1961. Ecological changes in the ponderosa pine forest of Cedar Valley in southern Washington. *Ecology* 42:416–20.

Weetman, G. F. and R. M. Fournier. 1982. Graphical diagnoses of lodgepole pine response to fertilization. *Soil Science Society of America Journal* 46:1280–1289.

———, R. Fournier, J. Barker, and E. Schnorbus-Panozzo. 1989a. Foliar analysis and response of chlorotic western hemlock and western red cedar reproduction on salal-dominated cedar-hemlock cutovers on Vancouver Island. *Canadian Journal of Forest Research* 19:1512–20.

———, R. Fournier, J. Barker, E. Schnorbus-Panozzo, and A. Germain. 1989b. Foliar analysis and response of fertilized chlorotic Sitka spruce plantations on salal-dominated cedar-hemlock cutovers on Vancouver Island. *Canadian Journal of Forest Research* 19:1501–11.

Wehtje, G., R. Dickens, J.W. Wilcut, and B.F. Hajek. 1987. Sorption and mobility of sulfometuron and imazapyr in five Alabama soils. *Weed Science* 35:858–864.

Weidman, P.H. 1939. Evidences of racial influence in a 25-year test of ponderosa pine. *Journal of Agricultural Research* 59:855–887.

Wender, B. W., C. A. Harrington, and J. C. Tappeiner. 2004. Flower and fruit production of understory shrubs in western Washington and Oregon. *Northwest Science* 78:124–40.

White, C. S. 1986. Effects of prescribed fire on rates of decomposition and nitrogen materializations in a ponderosa pine ecosystem. *Biology and Fertility of Soils* 2:87–95.

White, D. E., and M. Newton. 1989. Competitive interactions of whiteleaf manzanita, herbs, Douglas-fir and ponderosa pine in southwestern Oregon. *Canadian Journal of Forest Research* 19:232–38.

White, E. M., W. W. Thompson, and F. R. Gartner. 1973. Heat effect on nutrient release from soils under ponderosa pine. *Journal of Range Management* 26:22–24.

Whitehead, D. 1986. Dry matter production and transpiration by *Pinus radiata* stands in relation to canopy architecture. Pp. 243–63 in T. Fujimora and D. Whitehead, eds., *Crown and Canopy Structure in Relation to Productivity*. Forestry and Forest Products Research Institute, Tsukuba, Japan.

Wickman, B. E. 1988. *Tree growth in thinned and unthinned white fir stands 20 years after a tussock moth outbreak*. Research Note PNW-RN-477, USDA Forest Service Pacific Northwest Research Station, Portland OR.

———, and C. B. Eaton. 1962. *The effects of sanitation-salvage cutting on insect-caused mortality at Blacks Mountain Experimental Forest 1938–1959*. Technical Paper 66, USDA Forest Service California Forest and Range Experiment Station Berkeley CA.

———, R. R. Mason, and H. G. Paul. 1992. Thinning and fertilization in a grand fir stand infested with western spruce budworm. II. Tree growth response. *Forest Science* 38:252–64.

Wierman, C. A., and C. D. Oliver. 1979. Crown stratification by species in even-aged mixed stands of Douglas-fir and western hemlock. *Canadian Journal of Forest Research* 9:1–9.

Williamson, D. 1973. *Results of shelterwood harvesting of Douglas-fir in the Cascades of western Oregon.* Research Paper PNW 616. USDA Forest Service Pacific Northwest Forest and Range Experiment Station, Portland OR.

———, and D. Minore. 1978. *Survival and growth of planted conifers on the Dead Indian Plateau east of Ashland, Oregon.* Research Paper PNW-RP-242, USDA Forest Service Forest and Range Experiment Station, Portland OR.

Williamson R. L. 1982. *Response to commercial thinning in a 110-year-old Douglas-fir stand.* Research Paper PNW-296, USDA Forest Service Pacific Northwest Forest and Range Experiment Station, Portland OR.

———, and F. E. Price. 1971. *Initial thinning effects in 70- to 150-year-old Douglas-fir—Western Oregon and Washington.* Research Paper PNW-117, USDA Forest Service Pacific Northwest Forest and Range Experiment Station, Portland OR.

Wilson, F. G. 1979. Thinning as an orderly discipline: A graphic spacing schedule for red pine. *Journal of Forestry* 77:483–86.

Wilson, J. S., and C. D. Oliver. 2000. Stability and density management in Douglas-fir plantations. *Canadian Journal of Forest Research* 30:910–20.

Wimberly, M. C., and Z. Liu. 2013. Interactions of climate, fire and management in future forests of the Pacific Northwest. *Forest Ecology and Management.* 327:270–279

Winter, L. E., L. B. Brubaker, J. F. Franklin, E. A. Miller, and D. Q. DeWitt. 2002. Initiation of an old-growth Douglas-fir stand in the Pacific Northwest: A reconstruction from tree ring records. *Canadian Journal of Forest Research* 32:1039–56.

Witcosky, J. J., T. D. Schowalter, and E. M. Hansen. 1986. The influence of precommercial thinning on the colonization of three species of root-colonizing insects. *Canadian Journal of Forest Research* 16:745–49.

Witmer, G. W. 1981. *Roosevelt Elk Habitat Use in the Oregon Coast Range.* Ph.D. dissertation, Department of Fisheries and Wildlife, Oregon State University, Corvallis.

Wohletz., L. R. 1968. *Guides for calculating available water-holding capacity.* USDA, Soil Conservation Service, TN-Soils-15, Berkeley CA.

Wollum, A. G., and G. H. Schubert. 1975. Effect of thinning on the foliage and forest floor properties of ponderosa pine stands. *Soil Science Society of America Proceedings* 39:968–72.

Wonn, H. T., and K. L. O'Hara. 2001. Height:diameter ratios and stability relationships for four northern Rocky Mountain tree species. *Western Journal of Applied Forestry* 16:87–94.

Woodman, J. N. 1971. Variation of net photosynthesis within the crown of a large forest-grown conifer. *Photosynthetica* 5:50–54.

Woodruff, D. R., B. J. Bond, G. A. Ritchie, and W. Scott. 2002. Effects of stand density on young Douglas-fir trees. *Canadian Journal of Forest Research* 32:420–27.

Wright, E. F., K. D. Coates, and P. Bartemucci. 1998. Regeneration from seed of six tree species in the interior cedar-hemlock forests of British Columbia as affected by substrate and canopy gap position. *Canadian Journal of Forest Research* 26:1352–68.

Wyant, J. G., P. N. Omi, and R. O. Laven. 1986. Fire induced mortality in a Colorado ponderosa pine/Douglas-fir stand. *Forest Science* 32:49–59.

Yang, R. C. 1998. Foliage and stand growth responses of semi-mature lodgepole pine to thinning and fertilization. *Canadian Journal of Forest Research* 28:1794–1804.

———, E. I. C. Wang and M. M. Micko. 1988. Effects of fertilization on wood density and tracheid length on 70-year-old lodgepole pine in western Alberta. *Canadian Journal of Forest Research* 18:945–56.

Yoda, K., T. Kira, H. Ogawa, and K. Hozumi. 1963. Interspecific competition among higher plants. XI. Self-thinning in overcrowded pure stands under cultivated and natural conditions. *Osaka University Journal of Biology* 14:107–29.

York, R.A., J. J. Battles, and R. C. Heald. 2003. Edge effects in mixed conifer group selection openings: Tree height response to resource gradients. *Forest Ecology and Management* 179:107–21.

_________, and R. C. Heald 2008. Growth response of mature trees versus seedlings to gaps associated with group selection management in the Sierra Nevada, California. *Western Journal of Applied Forestry* 23:94–98.

Youngberg, C. T., and A. G. Wollum, II. 1976 Nitrogen accretion in developing *Ceanothus velutinus* stands. *Soil Science Society of America Journal* 40:109–12.

Youngblood, A., and D. E. Ferguson. 2003. Changes in needle morphology of shade-tolerant seedlings after partial overstory canopy removal. *Canadian Journal of Forest Research* 33:1315–22.

Younger, N.L., H. Temesgen, and S.M. Garber. 2008. Taper and volume responses of Douglas-fir to sulfur treatments for control of Swiss needle cast in the Coast Range of Oregon. *Western Journal of Applied Forestry* 23:142–148.

Zasada, J. C. 1986. Natural regeneration of trees and tall shrubs on forest sites of interior Alaska. Pp. 44–73 in F. S. Van Cleve, P. Chapin, W. Flannigan, L. A. Viereck, and C. T. Dyrness, eds., *Forest Ecosystems in the Alaska Taiga*. Springer-Verlag, New York.

———, T. L. Sharik, and M. Nygren. 1992. The reproductive process in boreal forest trees. In H. Shugart, R. Leemans, and G. Bonan, eds. *Systems Analysis of the Global Boreal Forest.* Cambridge University Press, UK.

———, J. C. Tappeiner, B. D. Maxwell, and M. A. Radwan. 1994. Seasonal changes in shoot and root production and in carbohydrate content of salmonberry (*Rubus spectabilis*) rhizome segments from the central Oregon Coast Ranges. *Canadian Journal of Forest Research* 24:272–77.

Zavitkowski, J., and M. Newton. 1968. Ecological importance of snowbrush in the Oregon Cascades. *Ecology* 49:1134–45.

Zenner, E. K., S. A. Acker, and W. H. Emmingham. 1998. Growth reduction in harvest-age coniferous forests with residual trees in the western central Cascade Range of Oregon. *Forest Ecology and Management* 102:75–88.

Zhang, S. and H.L. Allen. 1996. Foliar nutrient dynamics of 11-year-old loblolly pine (*Pinus taeda*) following nitrogen fertilization. *Canadian Journal of Forest Research* 26:1426–1439.

Zhao, J., D. A. Maguire, D. B. Mainwaring and A. Kanaskie. 2012. Climate influences on needle cohort survival mediated by Swiss needle cast in coastal Douglas-fir. *Trees* 26:1361–71

Ziemer, R. R. 1968. *Soil moisture depletion patterns around scattered trees.* Research Note RN-PSW-166, USDA Forest Service Pacific Southwest Research Station, Berkeley CA.

Zimmerman, M. H., and C. L. Brown. 1971. *Trees: Structure and Function.* Springer Verlag, New York.

Zlatnik, E. H. DeLuca, K. S. Milner and D. F. Potts. 1999. Site productivity and soil conditions on terraced ponderosa pine sites in western Montana. *Western Journal of Applied Forestry* 14:35–40.

Zobel, D. B. 1980. Effect of forest floor disturbance on seedling establishment of *Chamaecyparis lawsoniana. Canadian Journal of Forest Research* 10:441–46.

———, L. F. Roth, and G. M. Hawk. 1985. *Ecology, pathology, and management of Port-Orford-cedar.* General Technical Report GTR-PNW-184, USDA Forest Service Pacific Northwest Forest and Range Experiment Station, Portland OR.

Zutter, B.R., J.H. Miller, H.L. Allen, S.M. Zedaker, M.B. Edwards, and R.A. Newbold. 1999. Fascicle nutrient and biomass responses of young loblolly pine to control of woody and herbaceous competitors. *Canadian Journal of Forest Research* 29:917–925.

Zwieniecki, M. A., and M. Newton. 1994. Root distribution of 12-year-old forests at rocky sites in southwestern Oregon. *Canadian Journal of Forest Research* 24:1791–96.

———, and M. Newton. 1995. Roots growing in rock fissures: Their morphological adaptation. *Plant and Soil* 172:181–87.

———, and M. Newton. 1996a. Seasonal patterns of water depletion from soil/rock by selected ecosystems in a Mediterranean climate. *Canadian Journal of Forest Research* 26:1346–52.

———, and M. Newton. 1996b. Water-holding characteristics of meta-sedimentary rock in selected forest ecosystems in southwestern Oregon. *Soil Science Society of America Journal* 60:1578–82.

Index

Page numbers in italics refer to figures and tables.